# COMPENDIUM OF ZEOLITE FRAMEWORK TYPES

## BUILDING SCHEMES AND TYPE CHARACTERISTICS

# COMPENDIUM OF ZEOLITE FRAMEWORK TYPES

## BUILDING SCHEMES AND TYPE CHARACTERISTICS

Henk van Koningsveld
DCT-Ceramic Membrane Centre "The Pore"
Delft University of Technology
Julianalaan 136, 2628 BL Delft, The Netherlands

ELSEVIER

Amsterdam – Boston – Heidelberg – London – New York – Oxford
Paris – San Diego – San Francisco – Singapore – Sydney – Tokyo

Elsevier
Radarweg 29, PO Box 211, 1000 AE Amsterdam, The Netherlands
Linacre House, Jordan Hill, Oxford OX2 8DP, UK

First edition 2007

**Library of Congress Cataloging-in-Publication Data**
A catalog record for this book is available from the Library of Congress

**British Library Cataloguing in Publication Data**
A catalogue record for this book is available from the British Library

ISBN: 978-0-444-53095-0

For information on all Elsevier publications
visit our website at books.elsevier.com

Printed and bound in The Netherlands

07 08 09 10 11    10 9 8 7 6 5 4 3 2 1

Working together to grow
libraries in developing countries
www.elsevier.com | www.bookaid.org | www.sabre.org

ELSEVIER    BOOK AID International    Sabre Foundation

# TABLE OF CONTENTS

| | |
|---|---|
| **ABW** | **BPH** |
| **ACO** | **BRE** |
| **AEI** | **CAN** |
| **AEL** | **CAS** |
| **AEN** | **CDO** |
| **AET** | **CFI** |
| **AFG** | **CGF** |
| **AFI** | **CGS** |
| **AFN** | **CHA** |
| **AFO** | **-CHI** |
| **AFR** | **-CLO** |
| **AFS** | **CON** |
| **AFT** | **CZP** |
| **AFX** | **DAC** |
| **AFY** | **DDR** |
| **AHT** | **DFO** |
| **ANA** | **DFT** |
| **APC** | **DOH** |
| **APD** | **DON** |
| **AST** | **EAB** |
| **ASV** | **EDI** |
| **ATN** | **EMT** |
| **ATO** | **EON** |
| **ATS** | **EPI** |
| **ATT** | **ERI** |
| **ATV** | **ESV** |
| **AWO** | **ETR** |
| **AWW** | **EUO** |
| **BCT** | **EZT** |
| **\*BEA** | **FAR** |
| **BEC** | **FAU** |
| **BIK** | **FER** |
| **BOG** | **FRA** |

| | |
|---|---|
| **SGT** | **TUN** |
| **SIV** | **UEI** |
| **SOD** | **UFI** |
| **SOS** | **UOZ** |
| **SSY** | **USI** |
| **STF** | **UTL** |
| **STI** | **VET** |
| **STT** | **VFI** |
| **SZR** | **VNI** |
| **TER** | **VSV** |
| **THO** | **WEI** |
| **TOL** | **-WEN** |
| **TON** | **YUG** |
| **TSC** | **ZON** |

Appendices

# PREFACE

The *Atlas of Zeolite Framework Types*[1] contains 168 topological distinct tetrahedral TO4 frameworks where T may be Si, Al, P, Ga, B, Be etc. The compiled framework types, characterized by Framework Type Codes consisting of three capital letters, do not depend on the composition, distribution of the various T atoms, cell dimensions or symmetry. Their frameworks exhibit such a diversity of four-connected three-dimensional nets that finite and infinite component units were introduced to describe their topologies.

Finite units were introduced by Meier[2,3] and Smith[4]. The secondary building units (SBUs) of Meier, e.g., 4-, 5- or 6-rings, are invariably non-chiral. This means that only one kind of SBU rather than enantiomeric pairs is needed to assemble the three-dimensional framework. The assemblage of the structure does not necessarily involve crystallographic symmetry operations.

The finite structural subunits (SSUs) developed by Smith are often of greater complexity (e.g., polyhedral cages). The SSUs represent a structural feature. They are not, however, SBUs in the sense just mentioned because very often the framework cannot be constructed from SSUs alone. Frequently, SSUs need to share corners, edges or faces to complete the framework.

The SBUs, as such, are not meant to describe precursors from which the zeolite grows. On the other hand, inspection of the systematic in existing framework types may give clues to choose targets for synthesis because equal segments in different frameworks, like (some of) the polyhedral cages, may play a role during crystal growth.

Several authors extensively discussed infinite units, e.g., chains and layers[5–8]. The 5-ring zeolites were described in terms of component chains[9] as well as in terms of layers[10].

In this "Compendium", each framework type is described using Periodic Building Units (PerBUs) defined in the next Section. Most framework types are covered using two pages for each framework type. The first page gives a pictorial description of how the framework can be built using PerBUs. The second page shows the larger cages, cavities and/or channels in the framework. All drawings are prepared using the ORTEP program of Carroll K. Johnson[11] using atomic coordinates of ideal TO2 frameworks in the highest possible symmetry. The atomic coordinates are obtained from http://www.iza-structure.org/databases/ and from references cited there. Only T atoms are drawn. Open circles are tetrahedral coordinated T atoms; bridging oxygen atoms, about midway between T atoms, are left out for clarity.

Three appendices are added. The first appendix gives a survey of cages as type characteristics. The second appendix summarizes the pore descriptors of those channels and cavities that appear in more than one framework type. Finally, the third appendix depicts the channel intersections and other cavities tabulated in Appendix 2.

The "Compendium" contains information that is complementary to the data listed in the Atlas[1] from which the topological symmetry, unit cell data and pore dimensions in a "real" structure with a particular framework type can be obtained. The entries on each page are described in more detail in the Introduction.

It does not seem possible to assemble such a compilation free of errors. The author will therefore be grateful for any additions and/or corrections for future updates.

Thanks are due to Christian Baerlocher and Lynne McCusker for many helpful discussions.

February 2007                                                                 Henk van Koningsveld

# INTRODUCTION: ENTRIES USED IN EACH FRAMEWORK TYPE DOCUMENT

## 1. The Periodic Building Unit (PerBU)

Crystal structures, which are periodically ordered in three dimensions, are ordered structures (regular crystalline solids). In this sense, chemical disorder (e.g., different cations on a particular site) and dynamic disorder (e.g., rotational disorder of template molecules) are excluded. Structural disorder within cavities of zeolite frameworks is also excluded. In this "Compendium", the frameworks are built from periodic 0-, 1-, or 2-dimensional structurally invariant Periodic Building Units (PerBUs). The PerBUs are built from smaller units composed of a limited number of T atoms by applying simple operation(s) to the smaller unit, e.g., translation, rotation. The zeolite framework types are analyzed in terms of these component PerBUs. The infinite PerBUs, like (multiple) chains, tubes and layers, and finite PerBUs, like (double) 4-rings, (double) 6-rings and cages, are far from unique. However, they are common to several zeolite framework types and allow an easy description of the frameworks. Infinite PerBUs and finite PerBUs can be used to build the zeolite frameworks. 6-Ring layers are frequently curled up to form tubes of 6-rings.

Many PerBUs can readily be constructed from (infinite) chains shown in Scheme 1. Three of these chains, with identity periods of $\sim m*2.5\,\text{Å}$, are referred to as zigzag chain, saw chain and crankshaft chain with $m = 2$, 3 and 4, respectively. The number of T atoms in the independent repeat unit along the

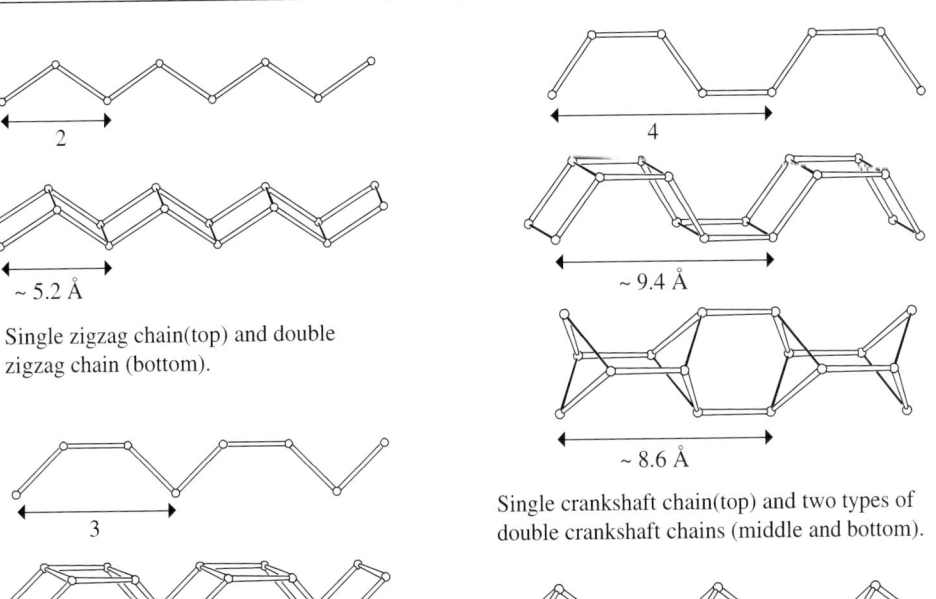

2

~ 5.2 Å

Single zigzag chain(top) and double zigzag chain (bottom).

3

~ 7.6 Å

Single saw chain(top) and double saw chain (bottom).

4

~ 9.4 Å

~ 8.6 Å

Single crankshaft chain(top) and two types of double crankshaft chains (middle and bottom).

5

~ 6.4 Å

Natrolite chain (or fibrous chain).

Scheme 1.    Some examples of frequently occurring chains in zeolite frameworks. The number of T atoms in the repeat unit (of the single chain) and the length of the identity period are indicated.

chain axis equals *m*. The fibrous zeolites can be built using the natrolite chain. The unit cell dimension in a certain direction very often reflects the presence of zigzag, saw or crankshaft chains in that direction.

A large number of framework types can be constructed using a hexagonal PerBU consisting of an array of non-connected 6-rings shown in Scheme 2. They all belong to the so-called ABC-6-family. In these framework types the unit cell dimension along the hexagonal axis is about $n \times 2.55\,\text{Å}$, where $n$ is the number of PerBUs connected along the hexagonal axis.

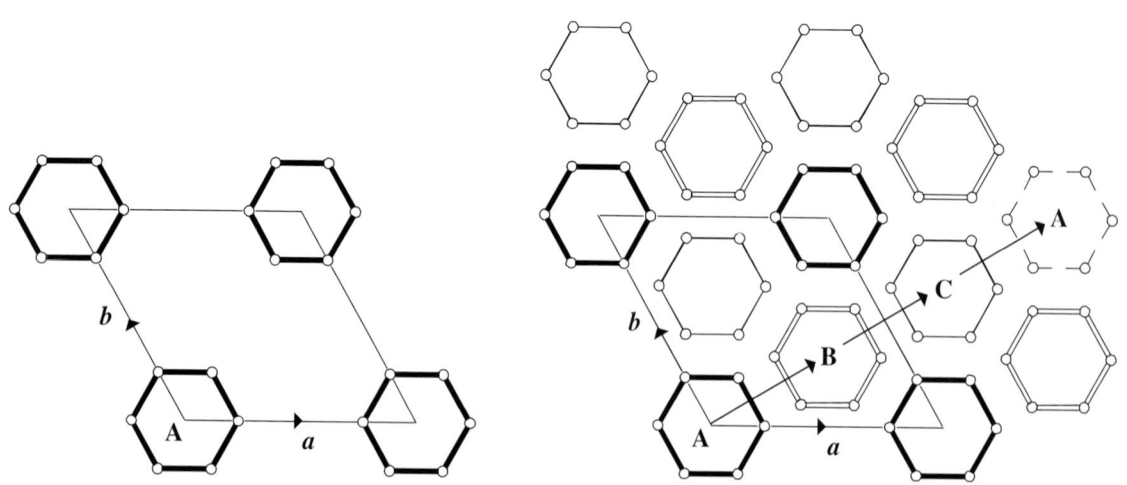

Scheme 2. The two-dimensional Periodic Building Unit (PerBU) in the ABC-6-family consists of a hexagonal array of non-connected planar 6-rings (in bold), which are related by pure translations along $a$ and $b$. The 6-rings are centered at (0,0) in the $ab$ layer. This position is usually called the **A** position. Neighboring PerBUs can be connected along the hexagonal $c$ axis through tilted 4-rings in three different ways:

(1) The next layer is shifted by $+(\frac{2}{3}a + \frac{1}{3}b)$ before connecting it to the first layer. The 6-rings in the second layer are centered at $(\frac{2}{3}, \frac{1}{3})$. This position is usually denoted as the **B** position (Scheme 2, right). The same connection mode can be repeated: a third PerBU is shifted with respect to the second layer by (again) $+(\frac{2}{3}a + \frac{1}{3}b)$. The 6-rings are now centered at $(\frac{4}{3}, \frac{2}{3})$ (or, equivalently, at $(\frac{1}{3}, \frac{2}{3})$). This position is called the **C** position. Adding a fourth layer with the same connection mode gives a shift with respect to the first layer of $(2a + b)$ [or zero] and an **A** position of the 6-rings is again obtained.

(2) the added layers are shifted by $-(\frac{2}{3}a + \frac{1}{3}b)$ before connecting them along $+c$ to the previous layer.

(3) the added layer has a zero lateral shift along $a$ and $b$.

The Scheme shows the PerBU in the ABC-6-family (left) and illustrates the definition of the 6-ring positions in neighboring PerBUs with respect to each other (right). The distance between two neighboring PerBUs is about 2.55 Å.

## 2. Connection mode

The relative orientation of neighboring PerBUs, or the connectivity of the PerBUs, can be described in terms of connection modes, which describe the symmetry elements (including lateral translation components given in fractions of the basis vectors of the invariant PerBU) that relate the PerBUs to each other. The connection mode (e.g., a rotation of 180° about an axis parallel to $a$ followed by a shift of $\frac{1}{2}a$) has been replaced with the shorter "a screw rotation of 180° about $a$". When neighboring PerBUs are "related by a shift of (e.g., $\frac{1}{2}(a+b)$)", all equivalent positions (obtained after pure translations along $a$ and $b$) are implied.

## 3. Channels and/or cages

The cages, cavities and channels (the type characteristics) are described using the pore descriptor. According to the IUPAC Recommendations 2001[12] the pore system is described with the general pore descriptor

$$\{D\ [n^m]_i\ [uvw]\ (W_{(eff)})\}$$

where

$D$ is the dimensionality of the pore system. For cages $D = 0$. For cavities (i.e., cages with at least one window large enough ($n > 6$) to allow diffusion of guest species) $D = 1$, 2 or 3. For channels, $D = 1$. For intersections of channels or systems of interconnected channels, $D = 2$ or 3;

$[n^m]_i$ is the shape of the pore, where $m$ is the number of $n$-rings (or windows) defining the faces of the polyhedral pore and $\Sigma m_i$ is the total number of faces; a polyhedron is called a cage when the largest ring size has $n = 6$.

$[uvw]$ is the direction of the channel. The term $[uvw]$ can be replaced by $<uvw>$ to indicate that all crystallographic equivalent directions are involved.

$(W_{(eff)})$ is the effective channel width. In topological description this is the smallest $n$-ring that determines the accessibility of the pore system to guest species along the dimension of infinite extension.

If more than one pore system is present, the descriptions are separated by a slash ( / ).

A survey of cages ($D = 0$) as type characteristics is added in Appendix 1. The polyhedrons are arranged in increasing number of T atoms ($n$) defining the smallest faces of the polyhedron and in increasing number of faces ($\Sigma m_i$). Appendix 2 and Appendix 3 summarizes those channels ($D = 1$) and cavities ($D \geq 1$) that appear in more than one framework type. An extensive compilation of polyhedral units in zeolites is published by Smith[13]. A list of those polyhedral units (or Composite Building Units) that are found in at least two different framework types is also published in the new edition of the Atlas[14].

## References

(1) Baerlocher, Ch., Meier, W.M. and Olson, D.H. *Atlas of Zeolite Framework Types*, 5th revised ed. (2001), Elsevier, London
(2) Meier, W.M. In *Molecular Sieves*, pp. 10–27 (1968), Soc. of Chem. and Ind., London
(3) Meier, W.M. and Olson, D.H. In *Atlas of Zeolite Structure Types*, p. 5, 2nd revised ed. (1987), Butterworths, London
(4) Smith, J.V. *Chem. Rev.*, **88**, 149 (1988)

(5) Liebau, F. In *Structural Chemistry of Silicates* (1985), Springer-Verlag, Berlin, Heidelberg

(6) Gottardi, G. and Galli, E. In *Natural Zeolites, Volume 18, Minerals and Rocks* (1987), Springer-Verlag, Berlin

(7) Meier, W.M. In *Natural Zeolites, Occurance, Properties and Use* (eds. L.B. Sand and F.A. Mumpton) pp. 99–103 (1978), Pergamon Press, Oxford

(8) van Koningsveld, H. In *Introduction to Zeolite Science and Practise* (eds. H. van Bekkum, E.M. Flanigen and J.C. Jansen) (1991), Elsevier, Amsterdam; *Stud. Surf. Sci. Catal.*, **58**, 35 (1991)

(9) van Koningsveld, H. *Zeolites*, **12**, 114 (1992)

(10) Akporiaye, D.E. *Z. Kristallogr.*, **188**, 103 (1989)

(11) Johnson, C.K. *ORTEP*. Report ORNL-3794, revised June 1970. Oak Ridge National Laboratory, TN, USA (1970)

(12) McCusker, L.B., Liebau, F. and Engelhardt, G. *Pure Appl. Chem.*, **73**, 381 (2001)

(13) Smith, J.V., In *Landolt-Boernstein*, New Series IV/14 (Subvolume A) (eds. W.H. Baur and R.X. Fischer) p. 81 (2000), Springer-Verlag, Berlin

(14) Baerlocher, Ch., McCusker, L.B. and Olson, D.H. *Atlas of Zeolite Framework Types,* 6th ed. (2007), Elsevier, Amsterdam

# BUILDING SCHEMES AND TYPE CHARACTERISTICS

(arranged by the framework type code in alphabetical order)

# ABW                          Building Scheme

## 1. Periodic Building Unit

**ABW** can be built using the zigzag chain (bold in Figure 1) running parallel to **b**. The two-dimensional PerBU is obtained when zigzag chains are connected along **c** into a layer of (fused) 6-ring chairs as shown in Figure 1.

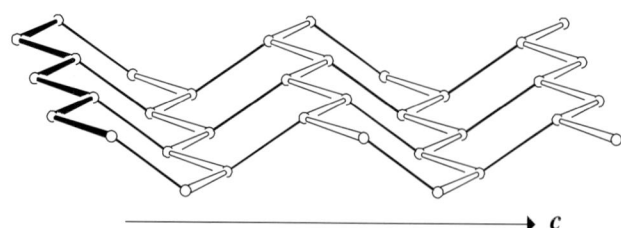

Fig. 1.   PerBU, constructed from zigzag chains, viewed along **b**.

## 2. Connection mode

Neighboring PerBUs, related along **a** by a shift of $\frac{1}{2}(\boldsymbol{a}+\boldsymbol{b}+\boldsymbol{c})$, are connected along **a** through 4-rings as depicted in Figure 2.

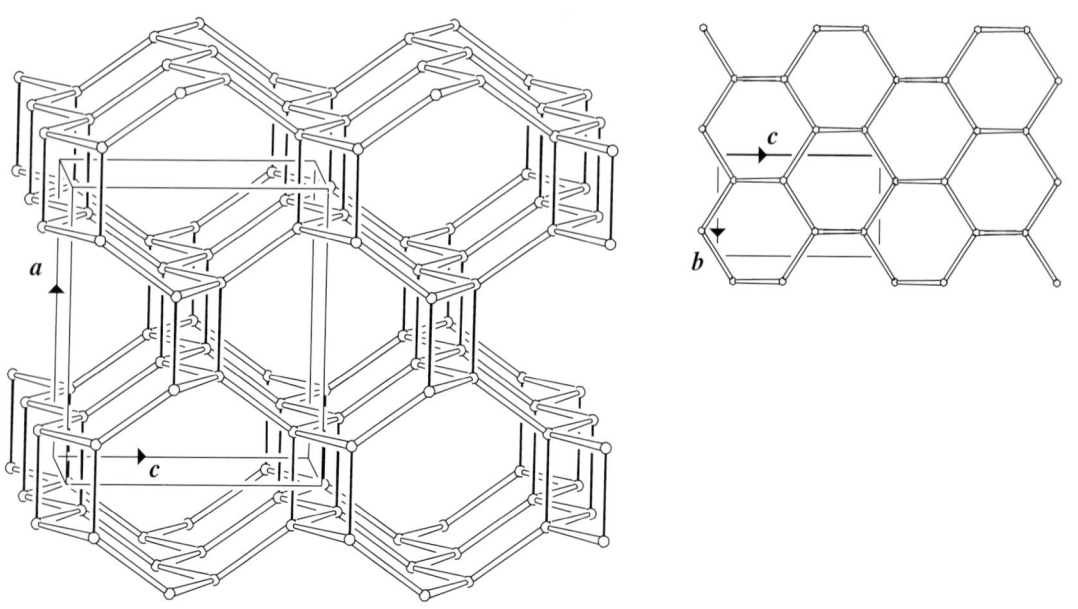

Fig. 2.   Connection mode and unit cell content viewed along **b** (left) and projection of the unit cell content along **a** (right).

### 3. Channels and/or cages

Non-interconnecting 8-ring channels (Figure 3) are parallel to **b**. The channel is topologically equivalent to the channel in **JBW**.

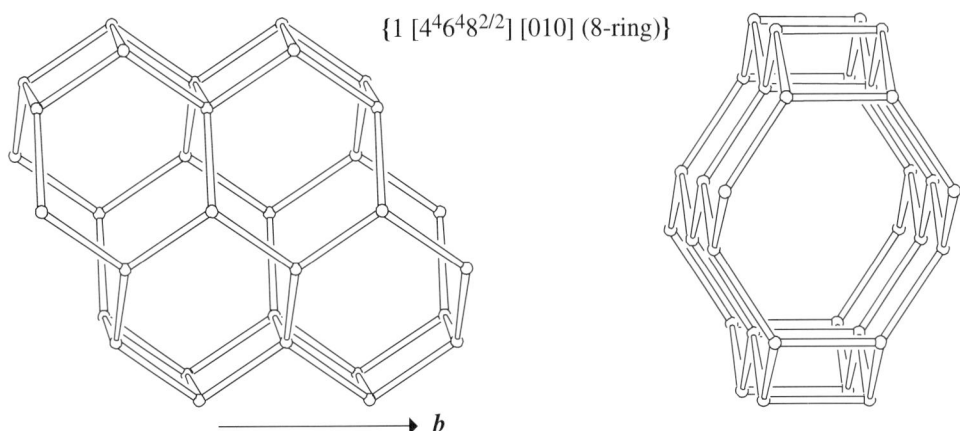

$\{1\ [4^4 6^4 8^{2/2}]\ [010]\ (8\text{-ring})\}$

Fig. 3.    Channel (with side-pockets) viewed perpendicular to the channel axis (left) and along the channel axis (right).

# ACO
# Building Scheme

## 1. Periodic Building Unit

Cubic **ACO** can be built using the double 4-ring (D4R) as zero-dimensional PerBU (bold in Figure 1).

## 2. Connection mode

Neighboring D4Rs are related by shifts of $\frac{1}{2}(a+b+c)$. All eight T-sites in a D4R are singly connected to other D4Rs. 8-Ring windows are formed.

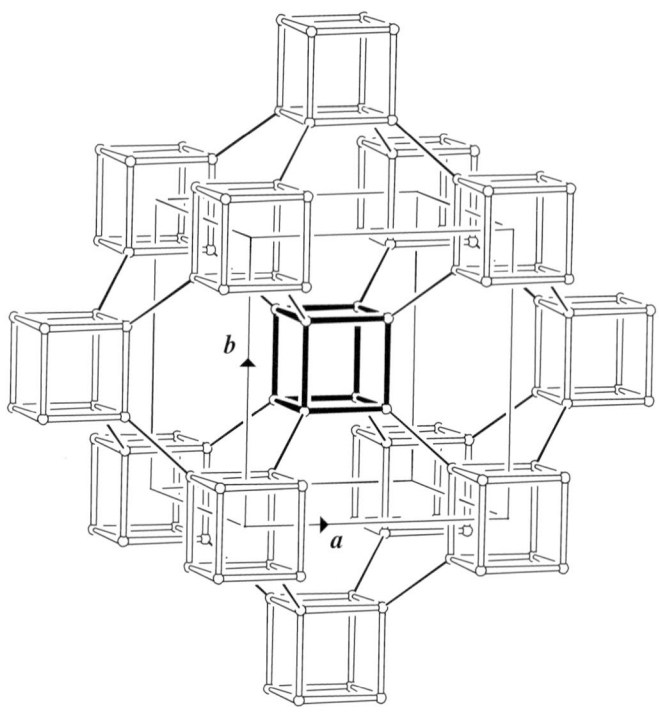

Fig. 1.   Connection mode and unit cell content viewed along a cube axis.

### 3. Channels and/or cages

Intersecting 8-ring channels (Figure 2) are parallel to $<100>$. The channel is topologically equivalent to the channel in **MER** and to one of the channels in **DFT**.

$\{[4^4 8^4 8^{2/2}] <100> \text{ (8-ring)}\}$

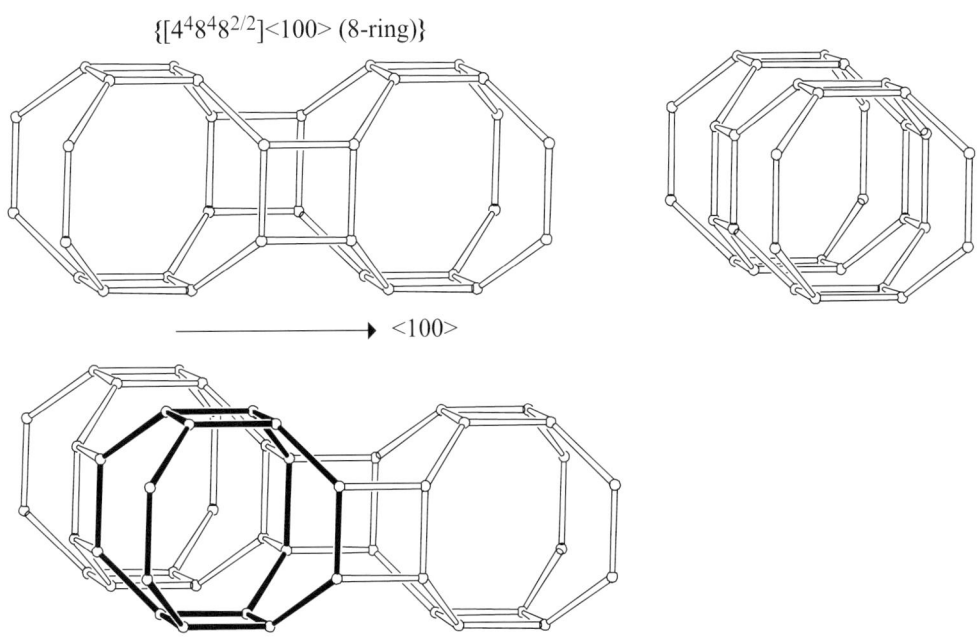

$\longrightarrow <100>$

Fig. 2. Top: 8-ring channel viewed perpendicular to a cube axis (left) and along that cube axis (right). Bottom: intersecting channels have $[4^2 8^4]$-cavities in common. One intersection in bold.

# AEI                    Building Scheme

## 1. Periodic Building Unit

The two-dimensional PerBU in **AEI** is the double 6-ring layer depicted in Figure 1. Double 6-rings, related along $(a+b)$ by a shift of $\frac{1}{2}(a+b)$, are connected into the **ab** layer through 4-rings. (See also **CHA**; compare with **SAV**.)

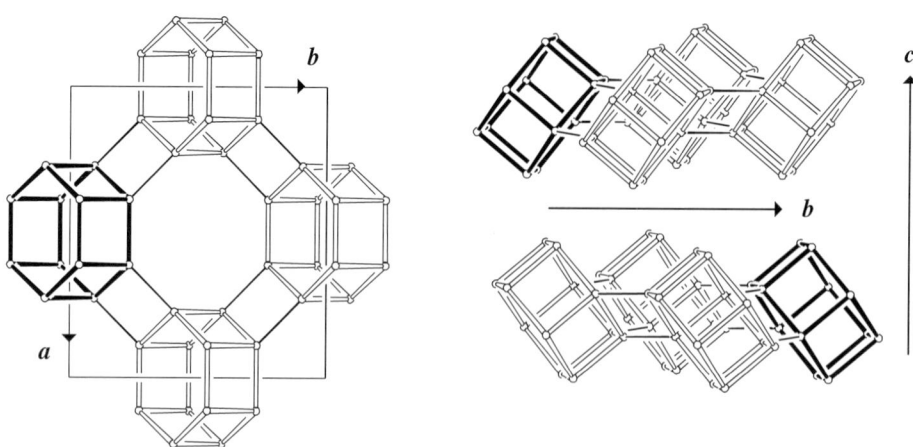

Fig. 1.    PerBU viewed along $c$ (left) and along $a$ (right). The PerBUs at the right are related by a rotation of 180° about $c$.

## 2. Connection mode

Neighboring PerBUs, related along $c$ by a screw rotation of 180° about $c$, are connected along $c$ through 4-rings as shown in Figure 2.

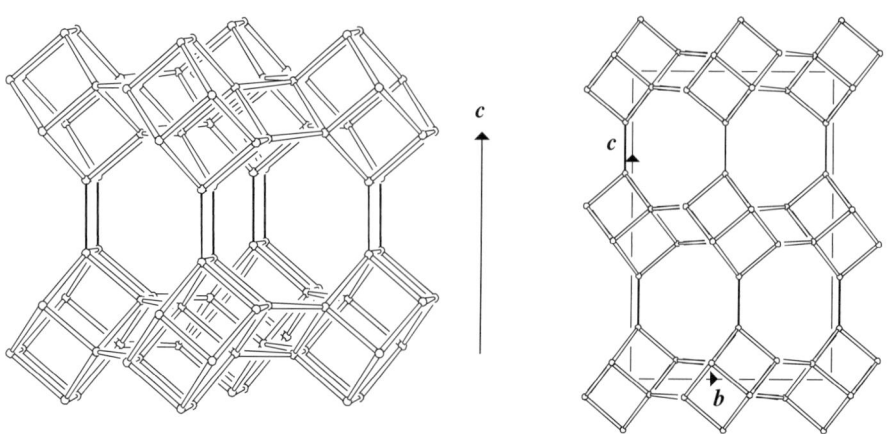

Fig. 2.    Connection mode (left) and unit cell content (right) viewed along $a$. (The projection of the cell content along $c$ is equal to Figure 1.)

## 3. Channels and/or cages

Intersecting 8-ring channels are parallel to *c*, *a* and <110>. The channel intersection and channels are depicted in Figure 3. The channel intersection is topologically equivalent to the ***chab*** cavity (see **AFT** and **CHA**).

{3 [$4^{12}6^{2}8^{6}$] [100] (8-ring), <110> (8-ring), [001](8-ring)}

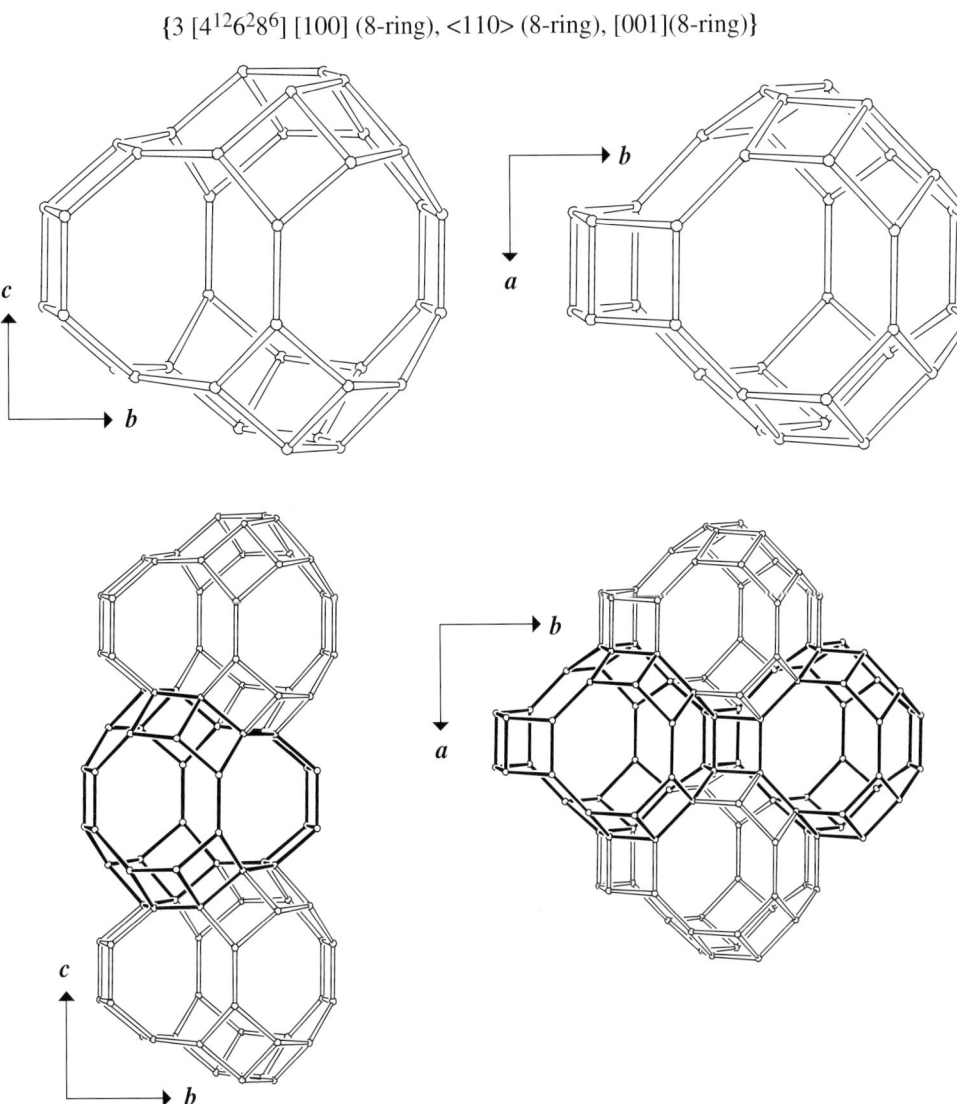

Fig. 3.  Top: channel intersection viewed along *a* (left) and along *c* (right). Bottom: channel intersections are connected along *c* (left) and along [110] and [1–10] (right) through common 8-rings into 8-ring channels parallel to *c* (left) and parallel to *a* and <110> (right).

# AEL     Building Scheme

## 1. Periodic Building Unit

**AEL** can be built using the crankshaft chain (bold in Figure 1, left) running parallel to $a$. A one-dimensional PerBU is obtained when five crankshaft chains are connected into a channel with a 10-ring aperture. The channel wall consists of fused 6-rings. The repeat unit of the PerBU is a cylindrical 6-ring band of 20 T atoms (bold in Figure 1, right).

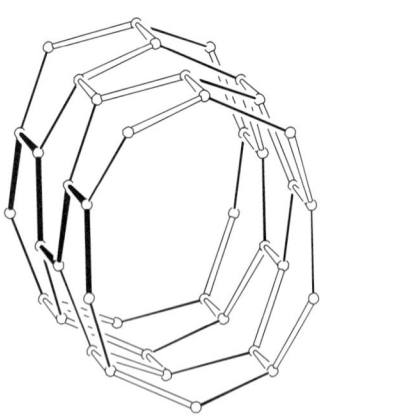 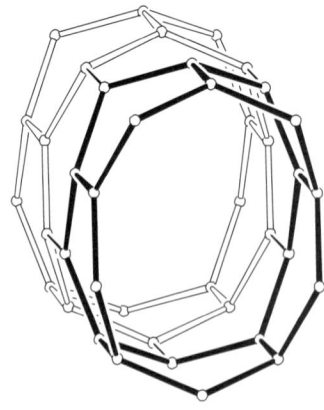

Fig. 1.   PerBU (viewed along $a$) constructed from five crankshaft chains (left) and from 6-ring bands (right).

## 2. Connection mode

Neighboring PerBUs, related along $c$ by a pure translation and along $b$ by a shift of $\frac{1}{2}(a+b+c)$, are connected through single- and double- and crankshaft chains, respectively (Figure 2).

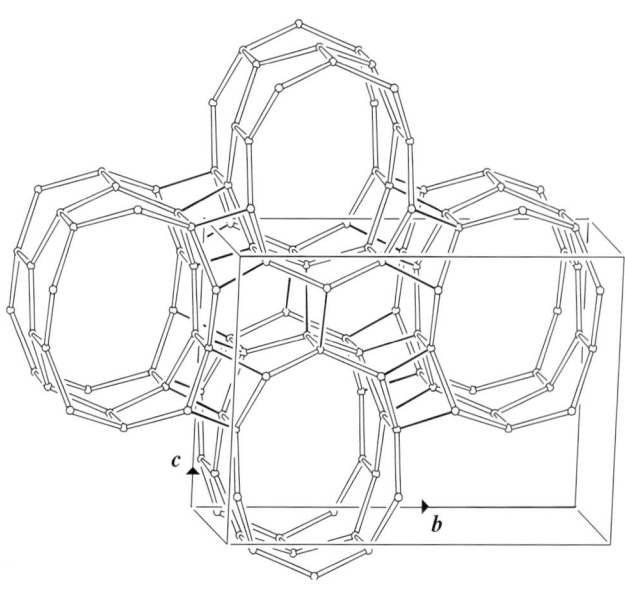

Fig. 2.   Connection mode and unit cell content viewed along $a$. For clarity, only $1\frac{1}{2}$ repeat unit along $a$ of each PerBU is drawn.

### 3. Channels and/or cages

Non-interconnecting 10-ring channels, parallel to $a$, are topologically equivalent to the channels in **AFO** and **AHT**. One channel is depicted in Figure 3.

{1 [$6^{10}10^{2/2}$] [100] (10-ring)}

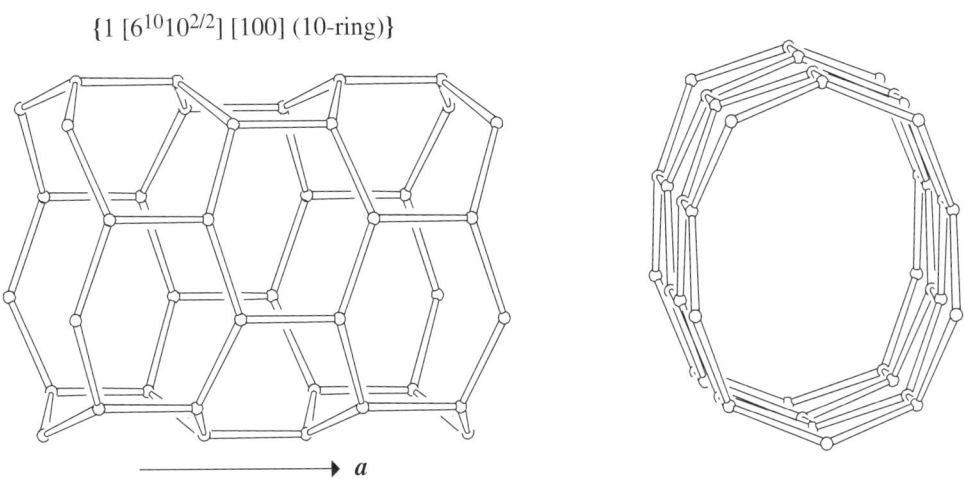

$\longrightarrow a$

Fig. 3.    Channel viewed perpendicular to the channel axis (left) and along the channel axis (right).

# AEN

## Building Scheme

### 1. Periodic Building Unit

**AEN** can be built using the T12-unit (bold in Figure 1) consisting of two strongly deformed 6-rings. The two 6-rings are 2-fold (1,3)-connected into three fused 6-rings. The two-dimensional PerBU is obtained when T12-units, related along **b** by a glide mirror plane parallel to the **bc** plane and along **c** by a pure translation, are connected into the **bc** layer shown in Figure 1.

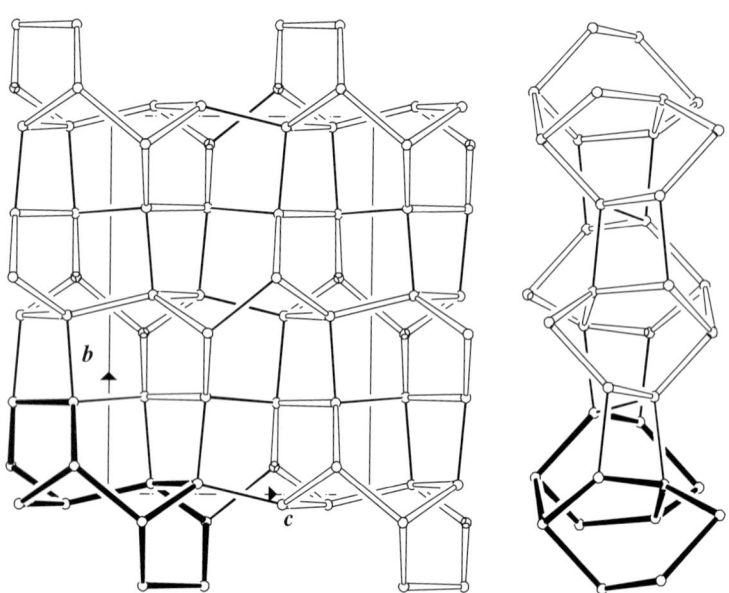

Fig. 1.   PerBU viewed along **a** (left) and along **c** (right). In the perspective drawing along **c** only one T12-unit along **c** is shown for clarity.

### 2. Connection mode

Neighboring PerBUs, related along **a** by a shift of $\frac{1}{2}(a+b)$, are connected through 4-rings (Figure 2).

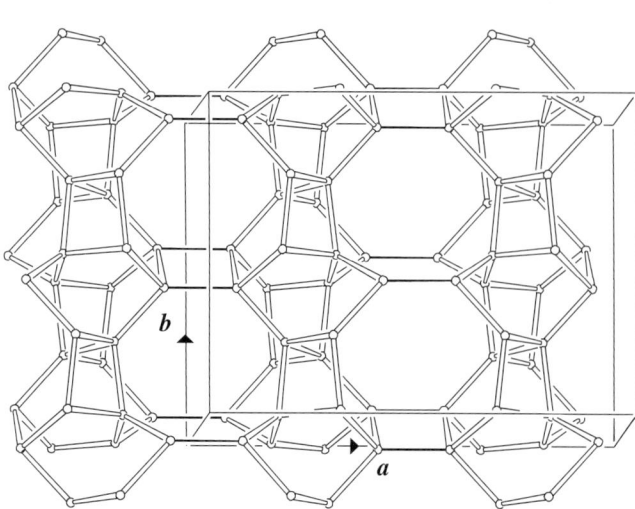

Fig. 2.   Connection mode and unit cell content viewed along **c**. Only one repeat unit along **c** is shown for clarity.

## 3. Channels and/or cages

Interconnecting sinusoidal 8-ring channels are parallel to **b** and **c**. The channels are shown in Figure 3.

{1 [$4^2 6^8 8^2 8^{2/2}$] [001] (8-ring)}

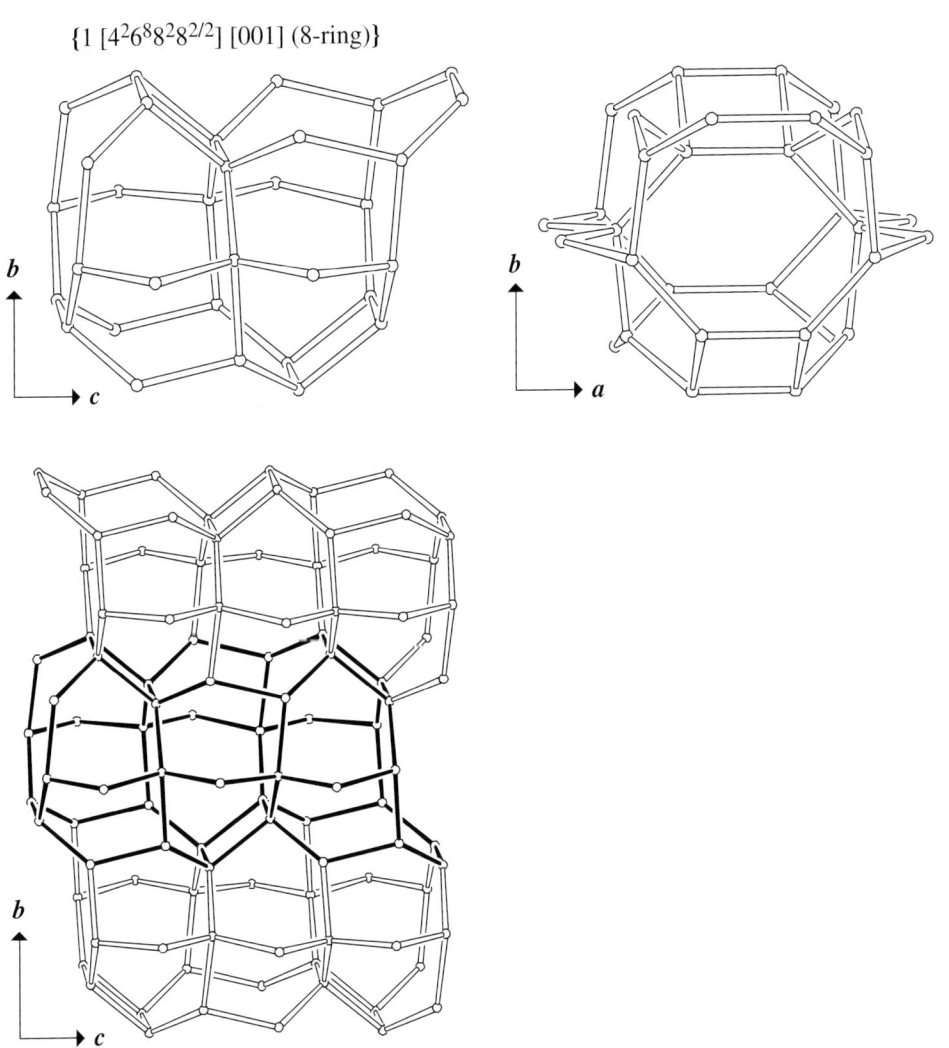

Fig. 3. Top: sinusoidal 8-ring channel parallel to **c** viewed perpendicular to the channel axis (left) and along the channel axis (right). Bottom: channels parallel to **c** (one channel in bold) are linked into sinusoidal channels parallel to **b**. View perpendicular to channel axes.

# AET

## Building Scheme

### 1. Periodic Building Unit

**AET** can be built using the crankshaft chain (bold in Figure 1, left) running parallel to *c*. A one-dimensional PerBU is obtained when seven crankshaft chains are linked into a channel with a 14-ring aperture with two additional chains as "handles".

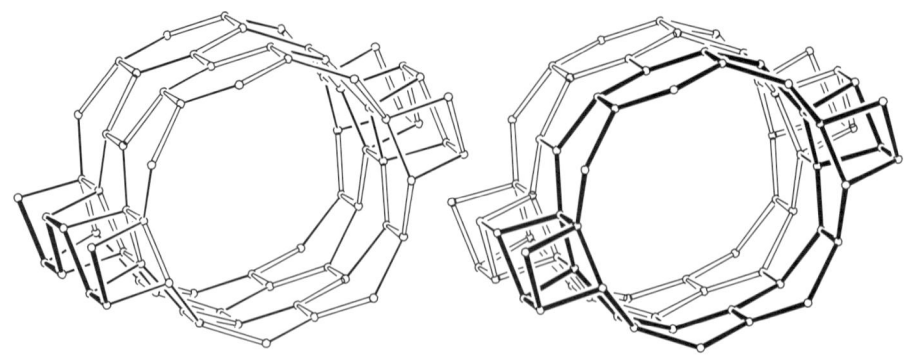

Fig. 1.   PerBU, constructed from crankshaft chains (left) and from 6-ring bands with "handles" (right), viewed along *c*. The channel wall consists of fused 6-rings.

### 2. Connection mode

Neighboring PerBUs, related along *b* by a pure translation and along *a* by a shift of $\frac{1}{2}(a+b)$, are connected through triple crankshaft chains and through 6-rings, respectively (Figure 2).

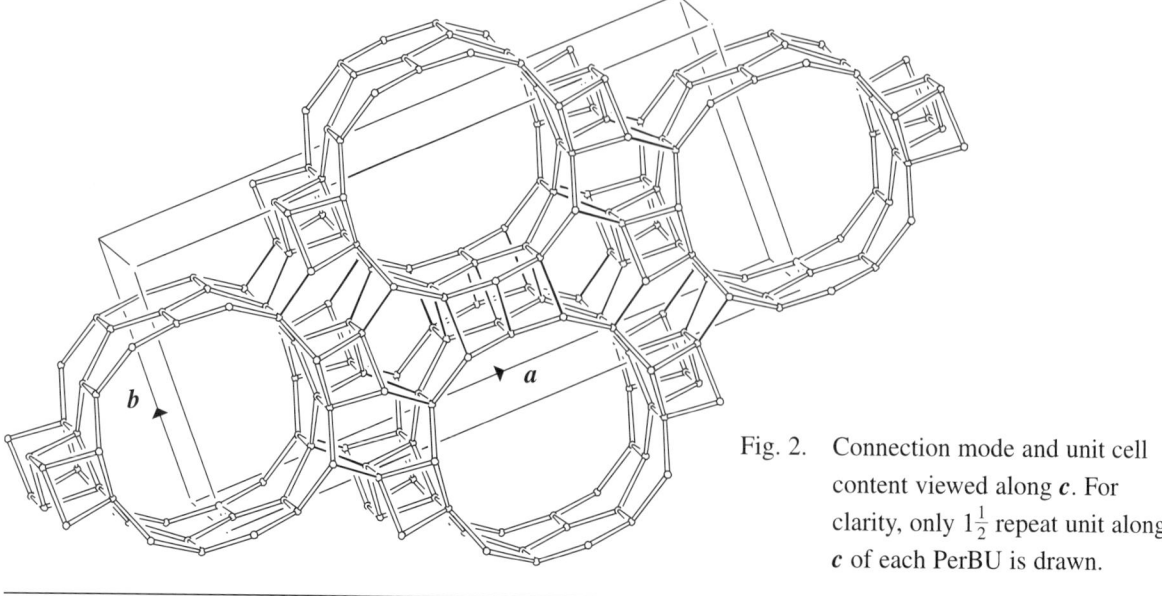

Fig. 2.   Connection mode and unit cell content viewed along *c*. For clarity, only $1\frac{1}{2}$ repeat unit along *c* of each PerBU is drawn.

## 3. Channels and/or cages

The non-interconnecting 14-ring channels in **AET**, parallel to $c$, are topologically equivalent to the channels in **DON**. One channel is depicted in Figure 3.

$\{1 \; [6^{14}14^{2/2}] \; [001] \; (14\text{-ring})\}$

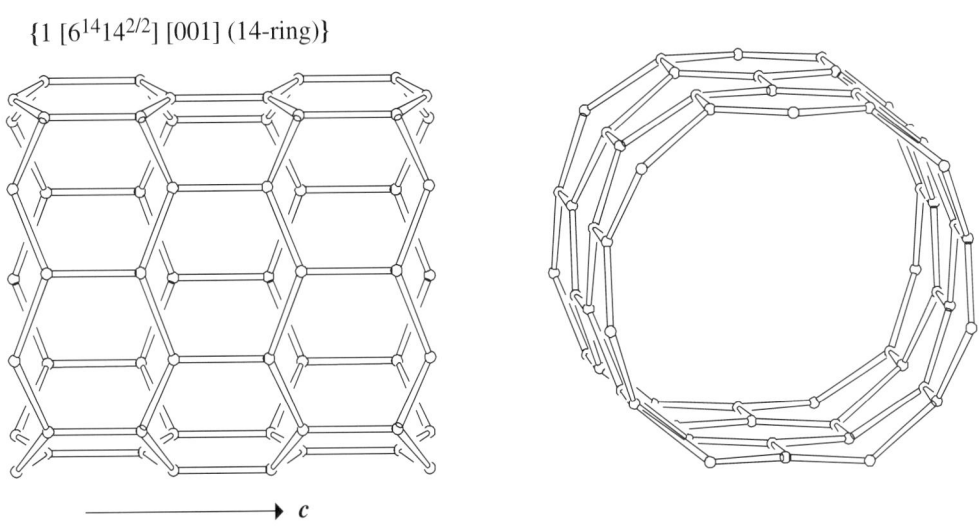

$\longrightarrow c$

Fig. 3.    Channel viewed perpendicular to the channel axis (left) and along the channel axis (right).

# AFG                    Building Scheme

## 1. Periodic Building Unit

The two-dimensional PerBU of **AFG** consists of a hexagonal array of non-connected planar 6-rings (bold in Figure 1), which are related along *a* and *b* by pure translations. The 6-rings are centered at (0,0) in the *ab* layer. This position is usually called the **A** position. **AFG** belongs to the ABC-6 family (see Introduction).

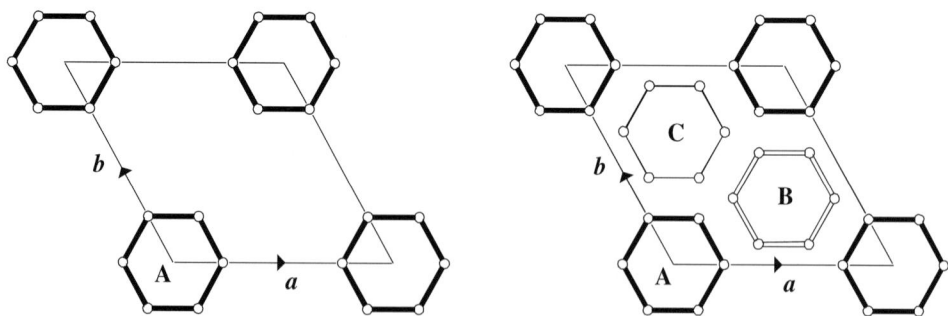

Fig. 1.   PerBU (left) and definition of 6-ring positions used in the stacking modes (right).

## 2. Connection mode

Neighboring PerBUs are connected along *c* through tilted 4-rings by connection modes **(1)** and **(2)** (see Introduction) as illustrated in Figure 2.

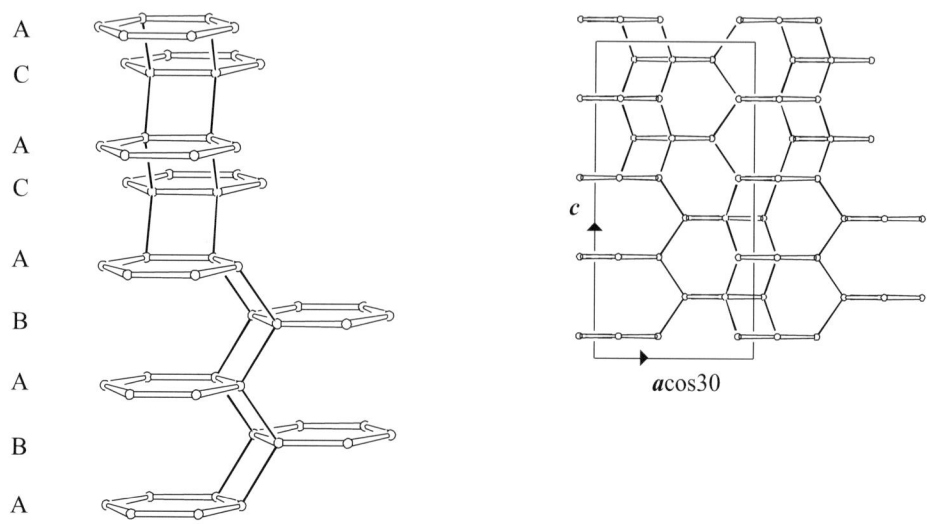

Fig. 2.   Connection mode (left) and unit cell content (right) viewed along *b*. The stacking sequence is given. In the perspective drawing, each PerBU is represented by one 6-ring only.

### 3. Channels and/or cages

The two types of cages in **AFG** are depicted in Figure 3. The *can* cage is also present in **CAN**, **ERI**, **FAR**, **FRA**, **GIU**, **LIO**, **LOS**, **LTL**, **LTN**, **MAR**, **MOZ**, **OFF**, **SAT**, **SBS**, **SBT**, **TOL** and **-WEN**. The *lio* cage is also found in **FAR**, **LIO**, **MAR** and **TOL**. Apertures of "channels" are formed by 6-rings only.

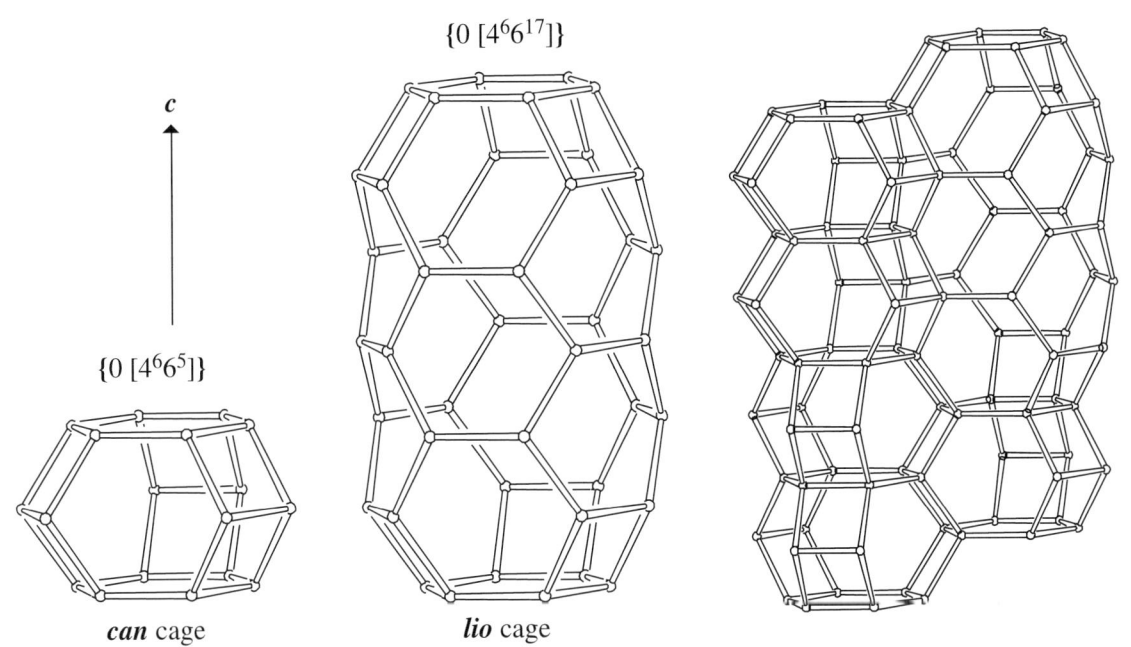

$\{0\,[4^6 6^{17}]\}$

$c$

$\{0\,[4^6 6^5]\}$

*can* cage      *lio* cage

Fig. 3.    *can* Cage (left), *lio* cage (middle) and connection of cages (right) viewed perpendicular to $c$.

# AFI                    Building Scheme

## 1. Periodic Building Unit

Hexagonal **AFI** can be built using the crankshaft chain (bold in Figure 1, left) running parallel to *c*. A one-dimensional PerBU is obtained when six crankshaft chains are connected into a channel with a 12-ring aperture. The channel wall consists of fused 6-rings. The repeat unit of the PerBU is a cylindrical 6-ring band of 24 T atoms (bold in Figure 1, right).

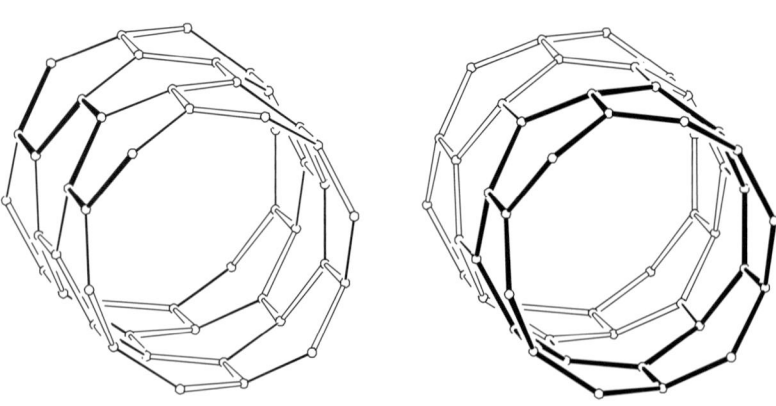

Fig. 1.   PerBU, constructed from six crankshaft chains (left) and from 6-ring bands (right), viewed along *c*.

## 2. Connection mode

Neighboring PerBUs, related along *a* and *b* by pure translations, are connected along *a* and *b* (and along (*a* + *b*)) through double crankshaft chains as shown in Figure 2. [6⁵]-Cages are formed.

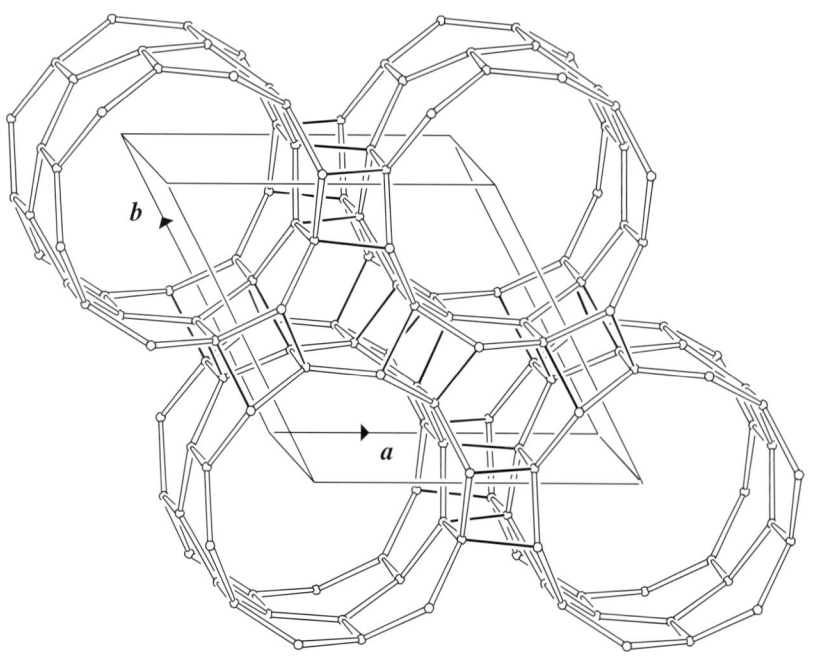

Fig. 2.   Connection mode and unit cell content viewed along *c*. For clarity, only $1\frac{1}{2}$ repeat unit along *c* of each PerBU is drawn.

## 3. Channels and/or cages

Non-interconnecting 12-ring channels are parallel to $c$. The channel is depicted in Figure 3. **AFI** can also be built from [$6^5$]-cages as can be seen from Figure 2. (See also **ATV**.)

{1 [$6^{12}12^{2/2}$] [001] (12-ring)}

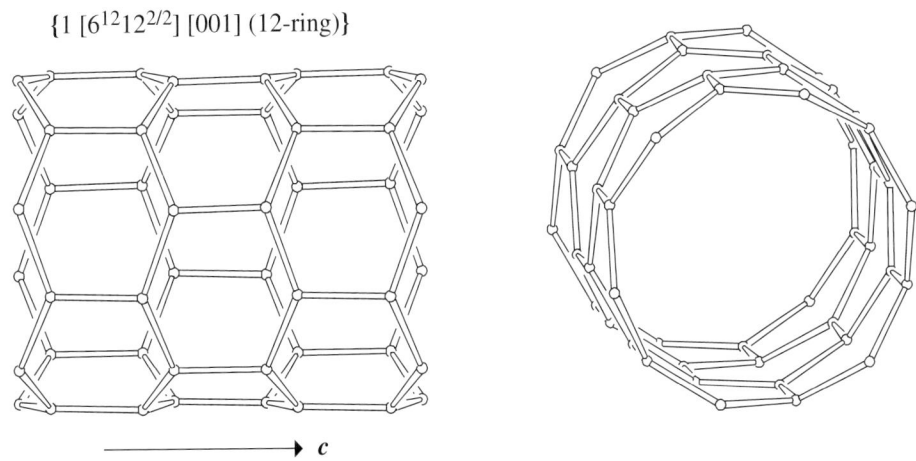

$\longrightarrow c$

Fig. 3.　Channel viewed perpendicular to the channel axis (left) and along the channel axis (right).

# AFN

# Building Scheme

## 1. Periodic Building Unit

**AFN** can be built using T8-units composed of three fused 4-rings (bold in Figure 1). T8-units, related along **a** by a rotation of 180° about **b**, are connected through (fused) 4-rings into chains along **a**. The two-dimensional PerBU is obtained when neighboring chains, related along **c** by pure translations, are connected along **c** through single T–T bonds into the **ac** layer depicted in Figure 1.

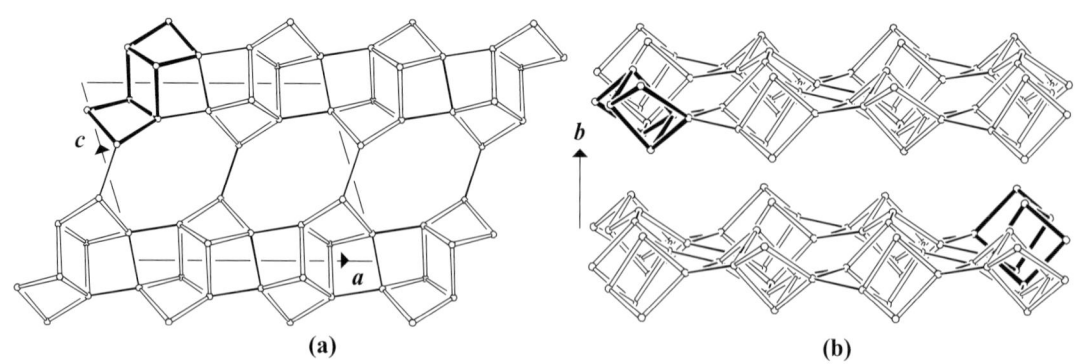

(a)  (b)

Fig. 1.  (a) PerBU seen along **b**; and (b) PerBUs viewed down [102]. The PerBUs depicted in (b) are identical and related by a rotation of 180° about **b**.

## 2. Connection mode

Neighboring PerBUs, related along **b** by a screw rotation of 180° about **b**, are connected along **b** through (fused) 4- and 6-rings as shown in Figure 2. 8-Rings are formed.

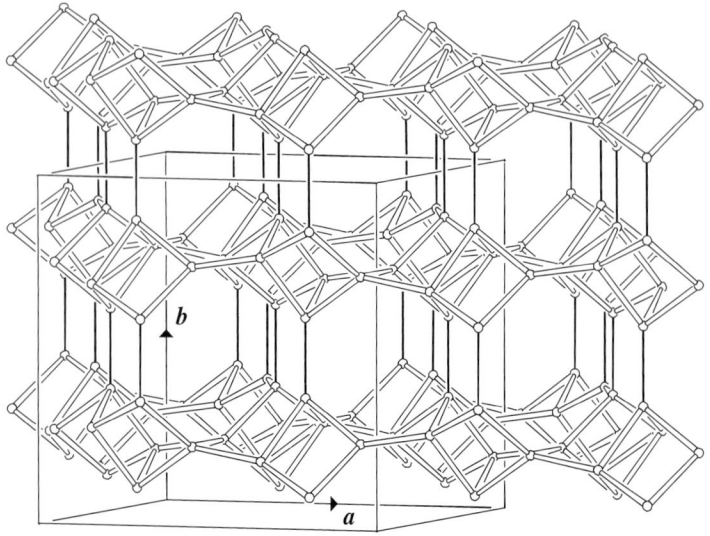

Fig. 2.  Connection mode and unit cell content viewed along [102].

# Cage/Channel AFN

## 3. Channels and/or cages

Intersecting 8-ring channels are parallel to *b*, [102] and <110>. Channel intersection and channels are depicted in Figure 3.

{3 [$4^6 8^8$] [010] (8-ring), [102] (8-ring), <110> (8-ring)}

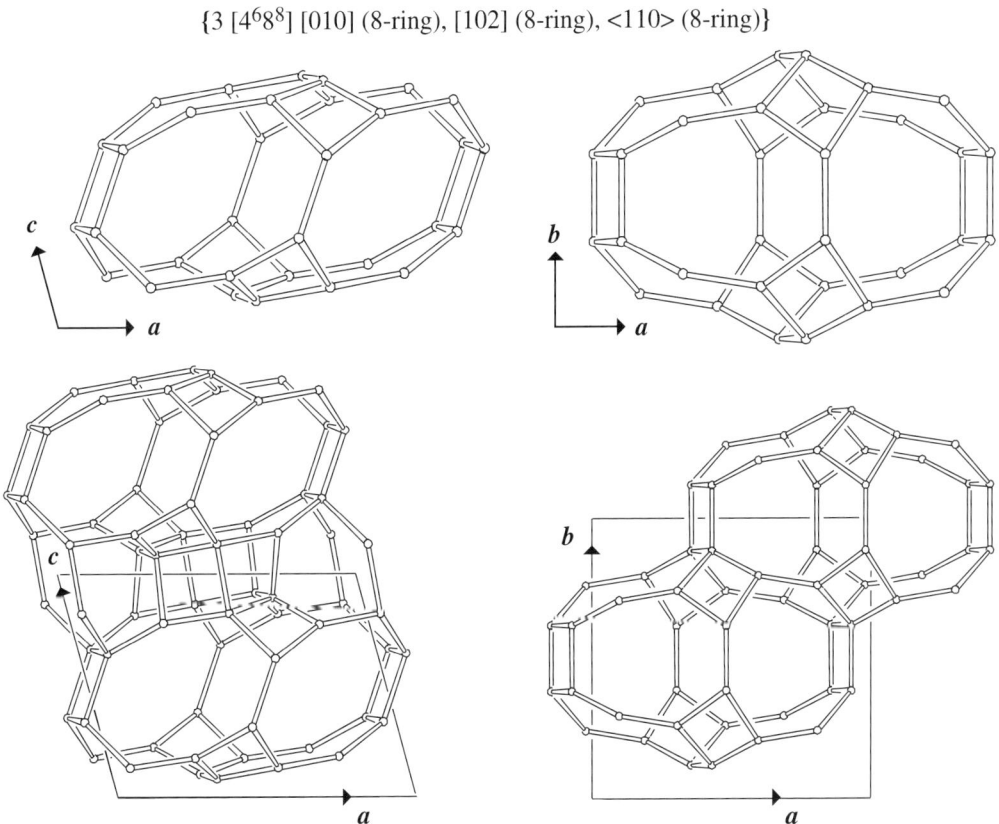

Fig. 3. Top: channel intersection viewed along *b* (left) and perpendicular to the *ab* plane (right). Bottom: channel intersections are connected along *c* (left) and along [110] (right) into 8-ring channels parallel to [102] (left) and parallel to *b* and [110] (and [1–10]) (right).

# AFO Building Scheme

## 1. Periodic Building Unit

**AFO** can be built using the crankshaft chain (bold in Figure 1, left) running parallel to $c$. A one-dimensional PerBU is obtained when five crankshaft chains are connected into a column of $[6^5]$-cages with two additional chains as "handles". The repeat unit of the PerBU contains 20 T atoms: a 3-fold (1,3,5) connected double 6-ring (or a $[6^5]$-cage) with "handles".

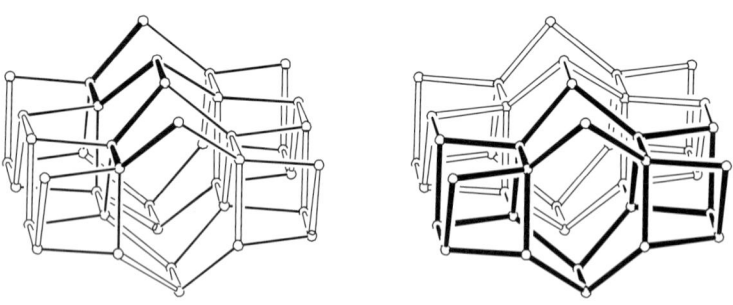

Fig. 1. PerBU, constructed from five crankshaft chains (left) and from T 20-units (right), viewed along $c$.

## 2. Connection mode

Neighboring PerBUs, related along $a$ by a pure translation and along $b$ by a shift of $\frac{1}{2}(a+b)$, are connected through single crankshaft chains. 10-Rings are formed as illustrated in Figure 2.

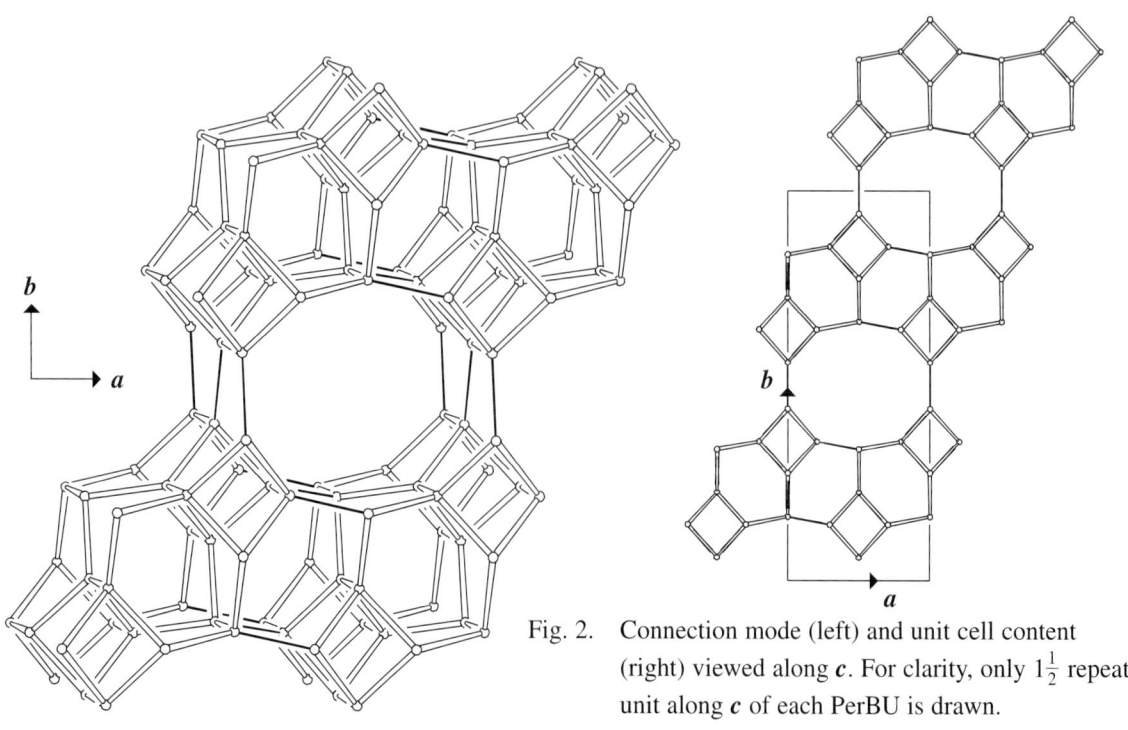

Fig. 2. Connection mode (left) and unit cell content (right) viewed along $c$. For clarity, only $1\frac{1}{2}$ repeat unit along $c$ of each PerBU is drawn.

## 3. Channels and/or cages

The non-interconnecting 10-ring channels, parallel to *c*, are topologically equivalent to the channels in **AEL** and **AHT**. One channel is depicted in Figure 3.

{1 [$6^{10}10^{2/2}$] [001] (10-ring)}

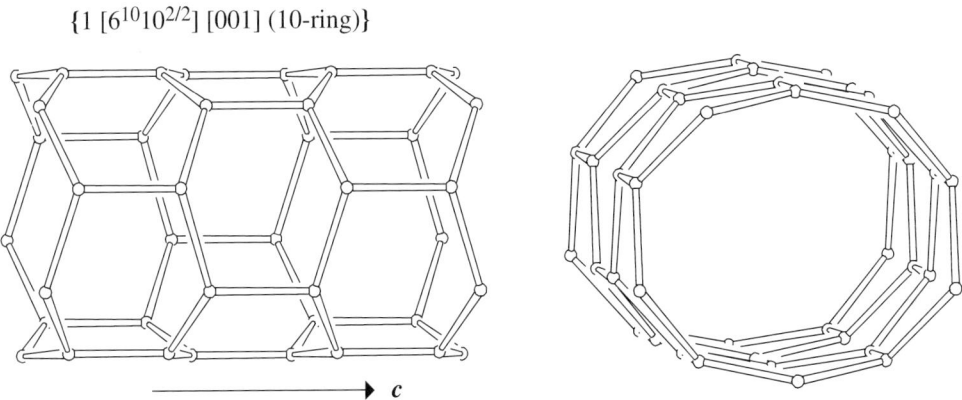

Fig. 3.    Channel viewed perpendicular to the channel axis (left) and along the channel axis (right).

# AFR                    Building Scheme

## 1. Periodic Building Unit

**AFR** can be built using double 4-rings with one disconnected edge (a $[4^46^1]$-cage; bold in Figure 1). These T8-units, related by a pure translation along $c$, are connected into chains parallel to $c$. Neighboring chains, related by a screw rotation of 180° about $b$, are linked into the undulating $bc$ layer. This two-dimensional PerBU is depicted in Figure 1. (See also **SFO**; compare with **ZON**.)

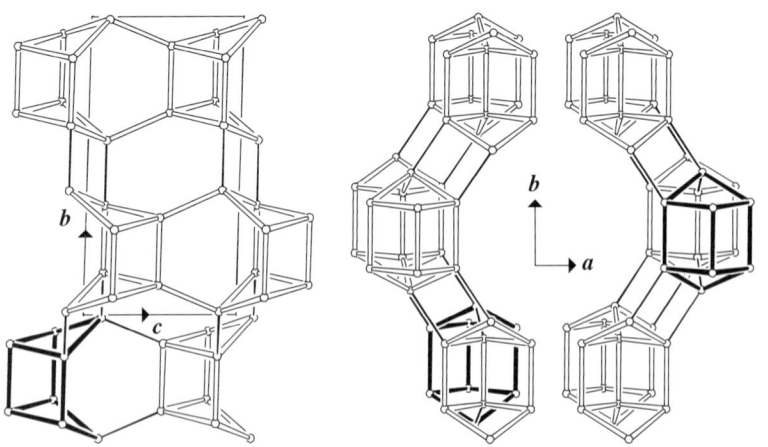

Fig. 1.  PerBU viewed along $a$ (left) and along $c$ (middle and right). The PerBUs at the right are related by a rotation of 180° about $a$ followed by a shift of $\frac{1}{2}b$.

## 2. Connection mode

Neighboring PerBUs, related along $a$ by a screw rotation of 180° about $a$ followed by a shift of $\frac{1}{2}b$, are connected along $a$ through 4-rings as illustrated in Figure 2.

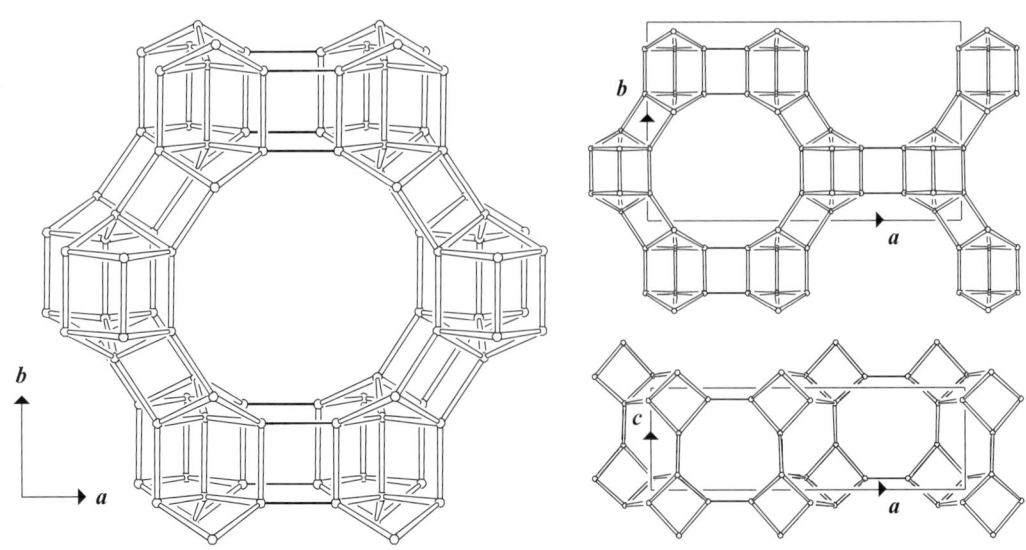

Fig. 2.  Connection mode viewed along $c$ (left) and unit cell content projected along $c$ (top) and along $b$ (bottom).

### 3. Channels and/or cages

Intersecting 8- and 12-ring channels are parallel to *b* and *c*, respectively. Channel intersection and channels are depicted in Figure 3. The channel intersection is topologically equivalent to the channel intersection in **SFO**.

$$\{2\ [4^{10}6^48^212^2]\ [010]\ (8\text{-ring}),\ [001]\ (12\text{-ring})\}$$

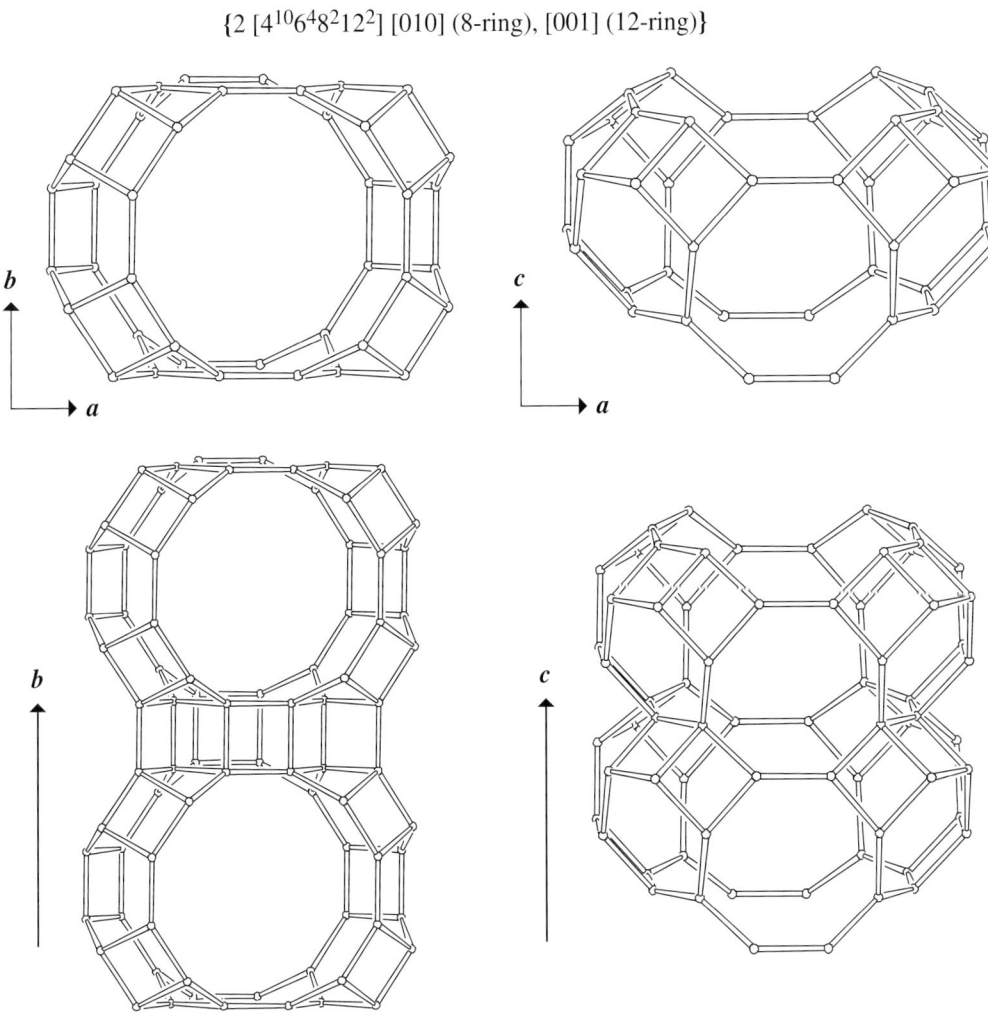

Fig. 3.  Top: channel intersection viewed along *c* (left) and along *b* (right). Bottom: interconnection of channel intersections into an 8-ring channel parallel to *b* viewed along *c* (left) and into a 12-ring channel parallel to *c* viewed along *b* (right).

# AFS                          Building Scheme

## 1. Periodic Building Unit

Hexagonal **AFS** can be built using units of 14 T atoms. The T14-unit consists of a 3-fold (1,3,5)-connected double 6-ring "capped" on each side by a single T atom (a $[4^6 6^3]$-cage; see inset Figure 1). The two-dimensional PerBU equals the layer obtained by connecting T14-units, related by a 3-fold axis and a 2-fold axis (parallel to [110]), through 4-rings as shown in Figure 1. (See also **BPH**.)

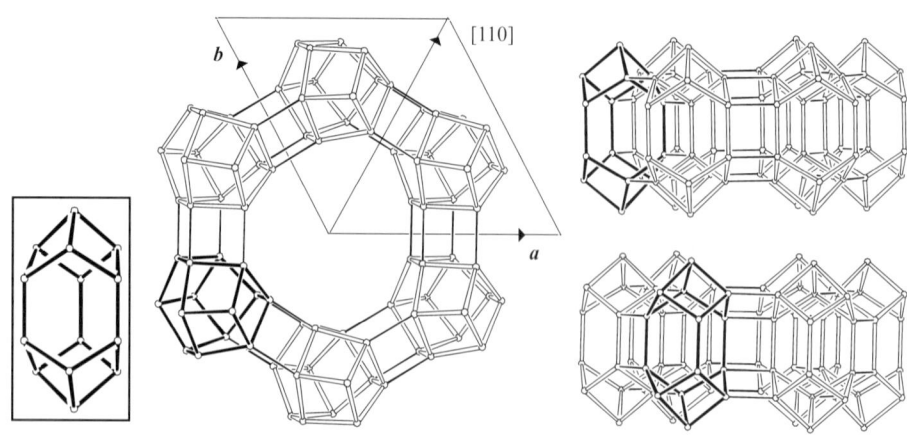

Fig. 1.   PerBU viewed along **c** (middle), along **b** (top right) and along [110] (bottom right). The PerBUs at the right are related by a rotation of 60° about **c**.

## 2. Connection mode

Neighboring PerBUs, related by a rotation of 60° about **c** followed by a shift of $\frac{1}{2}$**c** (i.e., related by a $6_3$-axis), are connected along **c** through 8-rings as depicted in Figure 2.

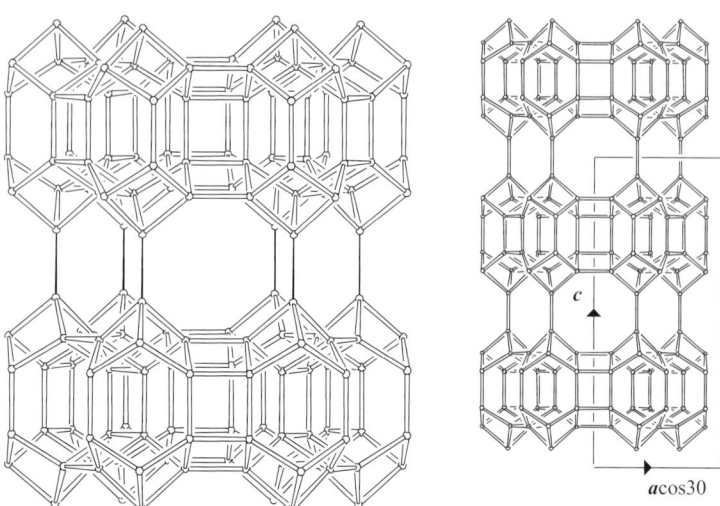

Fig. 2.   Connection mode (left) and unit cell content (right) viewed along **b.**

## 3. Channels and/or cages

Intersecting 8- and 12-ring channels are parallel to <100> and *c*, respectively (Figure 3). The channel intersection is topologically equivalent to the channel intersection in **AFY** and **BPH**.

{3 [4$^{18}$8$^6$12$^2$] <100> (8-ring), [001] (12-ring)}

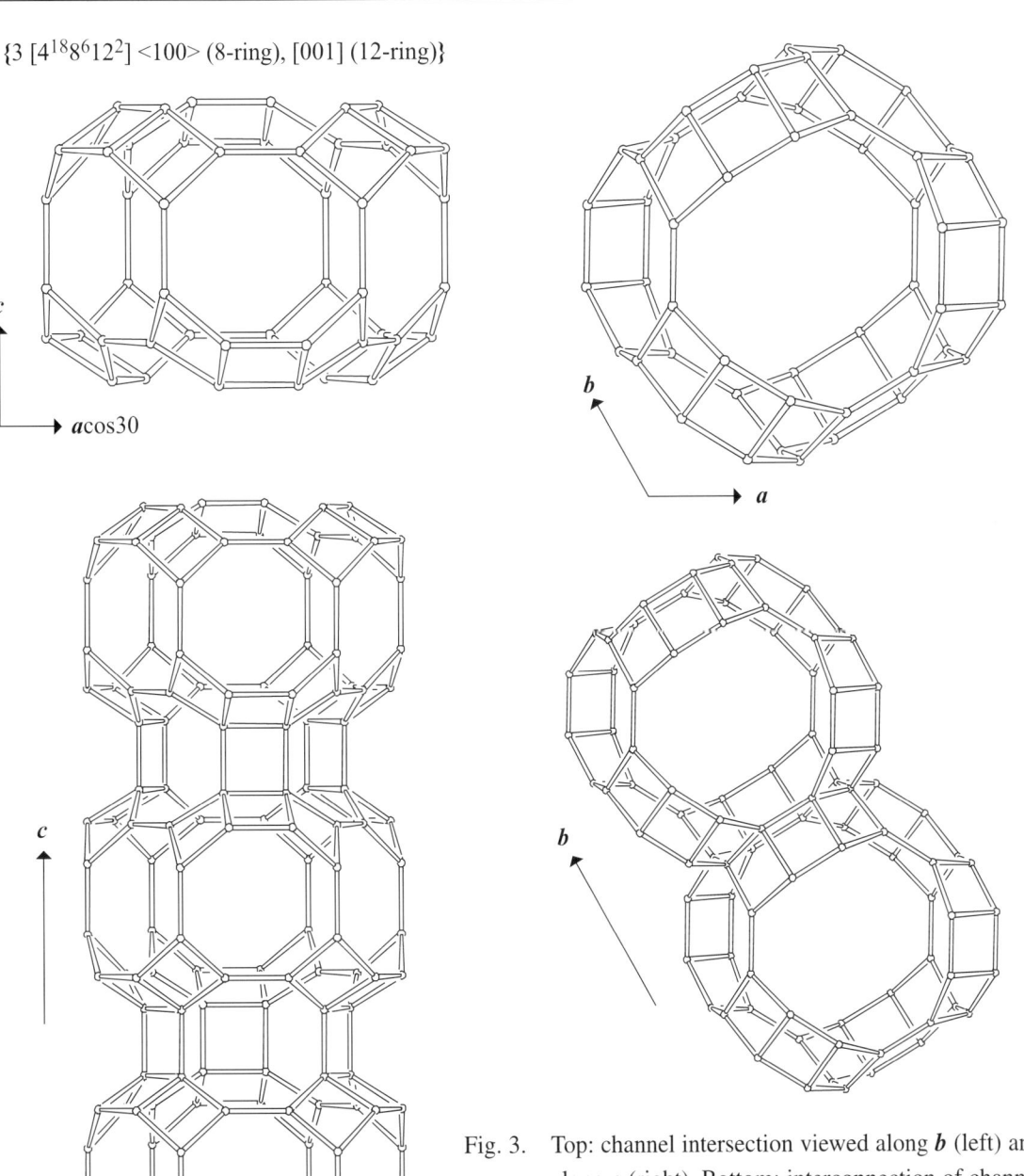

*c*

→ *a*cos30

*b*

→ *a*

*c*

*b*

Fig. 3.   Top: channel intersection viewed along *b* (left) and along *c* (right). Bottom: interconnection of channel intersections into a 12-ring channel parallel to *c* viewed along <100> (left) and into an 8-ring channel parallel to *b* (or *a*) viewed along *c* (right).

## 1. Periodic Building Unit

The two-dimensional PerBU of **AFT** consists of a hexagonal array of non-connected planar 6-rings (bold in Figure 1), which are related along *a* and *b* by pure translations. The 6-rings are centered at (0,0) in the *ab* layer. This position is usually called the **A** position. **AFT** belongs to the ABC-6 family (see Introduction).

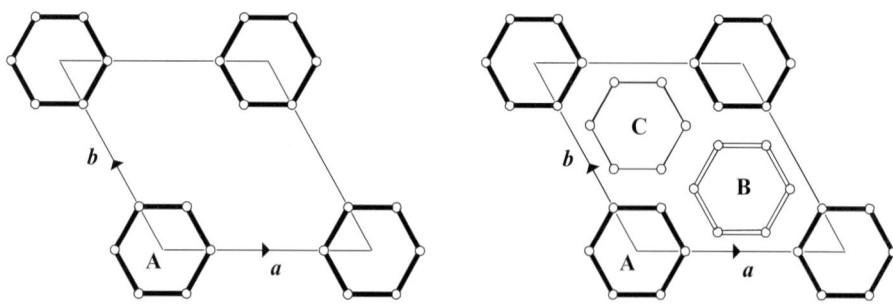

Fig. 1.　PerBU (left) and definition of 6-ring positions used in the stacking modes (right).

## 2. Connection mode

Neighboring PerBUs can be connected along *c* through tilted 4-rings in three different ways (see Introduction). In **AFT** all three connection modes between the PerBUs are observed. ***chab*** Cavities are formed (see Figure 2 and **CHA**).

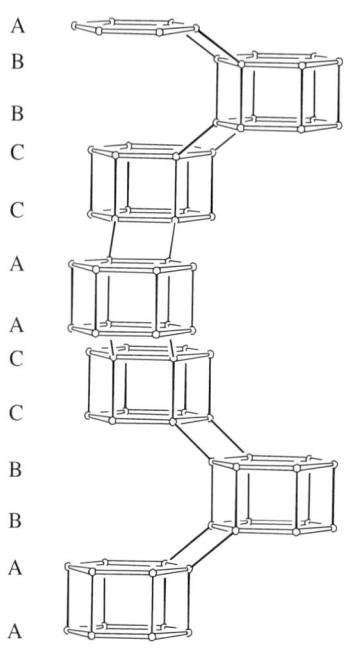

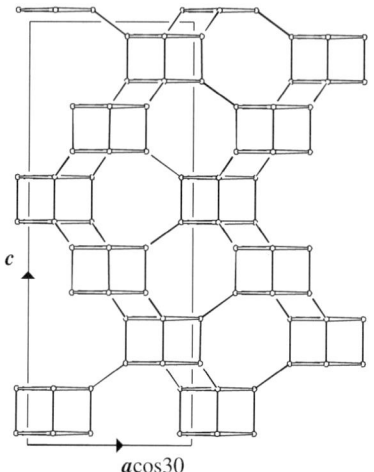

Fig. 2.　Connection mode (left) and unit cell content (right) viewed along *b*. The stacking sequence is given. In the perspective drawing, each PerBU is represented by one 6-ring only.

### 3. Channels and/or cages

The three types of cavities are depicted in Figure 3. **AFT** can be built from *chab* cavities alone (see also **CHA**). The *chab* cavity is topologically equivalent to the channel intersection in **AEI**. The *gmel* cavity is also present in **GME**, **AFX**, **EAB**, **EON**, **MAZ** and **OFF**. The *aft* cavity is also found in **AFX**.

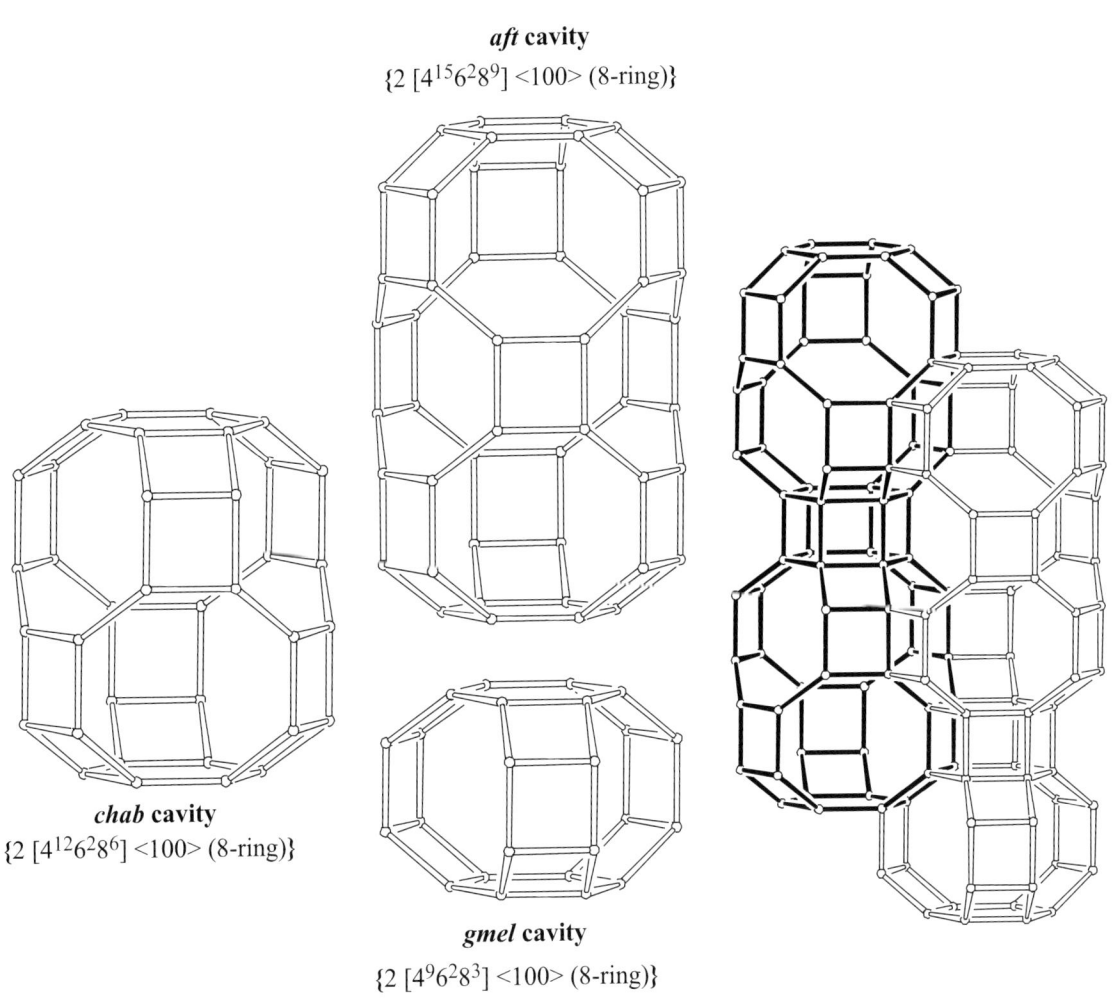

**aft cavity**
$\{2 \, [4^{15}6^{2}8^{9}] <100> (8\text{-ring})\}$

**chab cavity**
$\{2 \, [4^{12}6^{2}8^{6}] <100> (8\text{-ring})\}$

**gmel cavity**
$\{2 \, [4^{9}6^{2}8^{3}] <100> (8\text{-ring})\}$

Fig. 3.   Cavities viewed along <120> (left and middle). Two-dimensional 8-ring channels, perpendicular to *c*, are interconnected along *c* through 12-rings in the *chab* and *aft* cavities (*chab* cavities in bold) leading to a three-dimensional channel system (right).

# AFX

# Building Scheme

## 1. Periodic Building Unit

The two-dimensional PerBU of **AFX** consists of a hexagonal array of non-connected planar 6-rings (bold in Figure 1), which are related along **a** and **b** by pure translations. The 6-rings are centered at (0,0) in the **ab** layer. This position is usually called the **A** position. **AFX** belongs to the ABC-6 family (see Introduction).

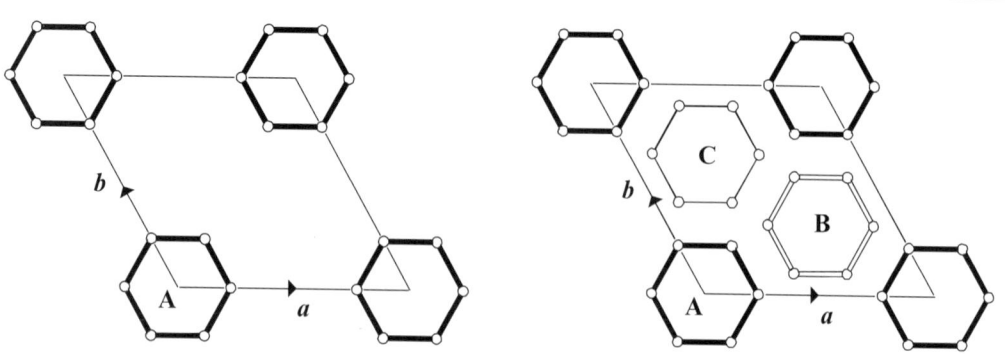

Fig. 1.   PerBU (left) and definition of 6-ring positions used in the stacking modes (right).

## 2. Connection mode

Neighboring PerBUs can be connected along **c** through tilted 4-rings in three different ways (see Introduction). In **AFX** all three connection modes between the PerBUs are observed. **gmel** Cavities are formed (see Figure 2 and **GME**).

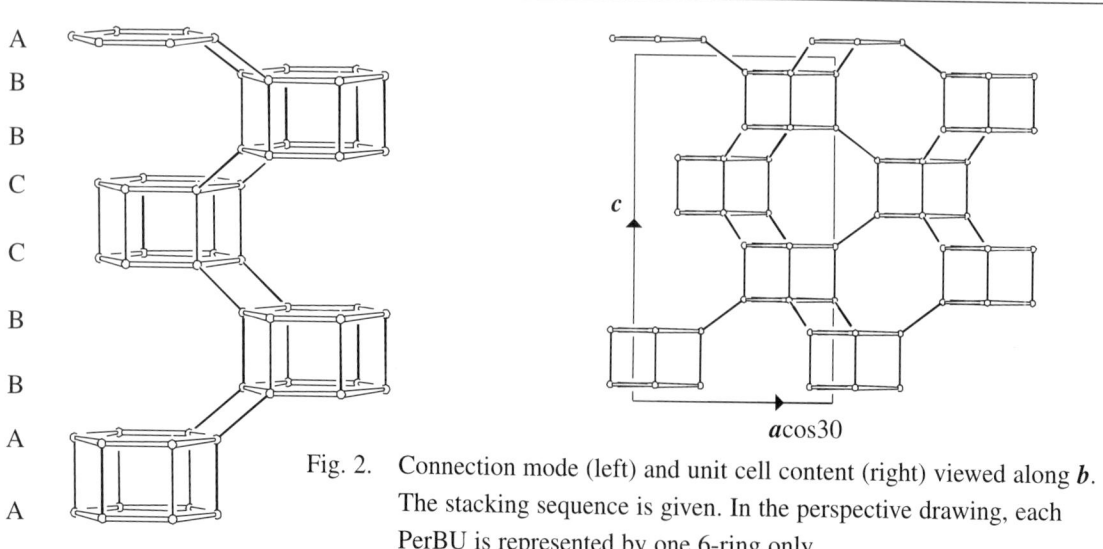

Fig. 2.   Connection mode (left) and unit cell content (right) viewed along **b**. The stacking sequence is given. In the perspective drawing, each PerBU is represented by one 6-ring only.

## 3. Channels and/or cages

The two types of cavities are depicted in Figure 3. **AFX** can be built from *gmel* cavities alone (see also **GME**). The *gmel* cavity is also present in **GME**, **AFT**, **EAB**, **EON**, **MAZ** and **OFF**. The *aft* cavity is also found in **AFT**.

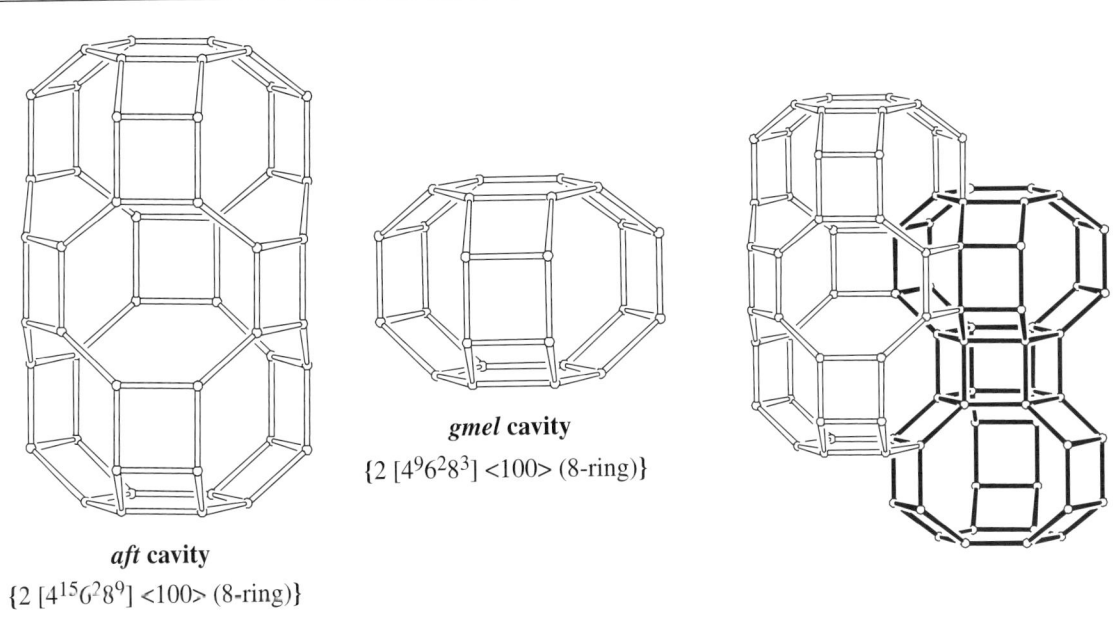

*gmel* **cavity**
$\{2 [4^9 6^2 8^3] <100> (8\text{-ring})\}$

*aft* **cavity**
$\{2 [4^{15} 6^2 8^9] <100> (8\text{-ring})\}$

Fig. 3.   Cavities viewed along <120> (left and middle). Two-dimensional 8-ring channels, perpendicular to *c*, are interconnected along *c* through 12-rings in the *aft* cavities leading to a three-dimensional channel system (*gmel* cavities in bold; right).

# AFY                              Building Scheme

## 1. Periodic Building Unit

**AFY** can be built using the double 4-ring (D4R) drawn bold in Figure 1. The PerBU equals the hexagonal layer obtained by connecting D4Rs through 4-rings around a 3-fold inversion axis as shown in Figure 1.

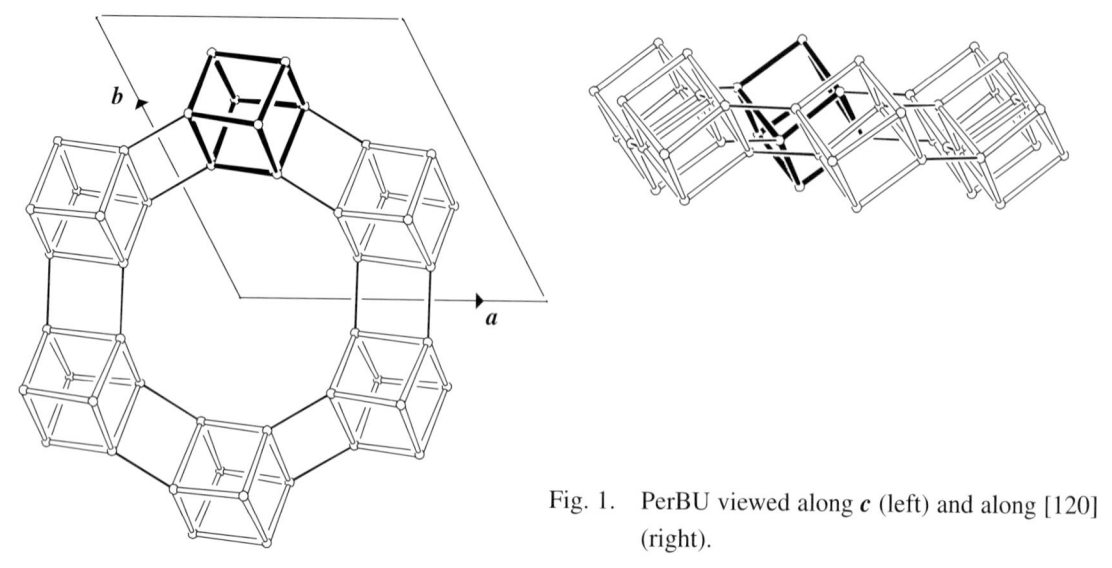

Fig. 1.   PerBU viewed along **c** (left) and along [120] (right).

## 2. Connection mode

Neighboring PerBUs, related along **c** by a pure translation, are connected along **c** through single T–T bonds. 8-Rings are formed (Figure 2).

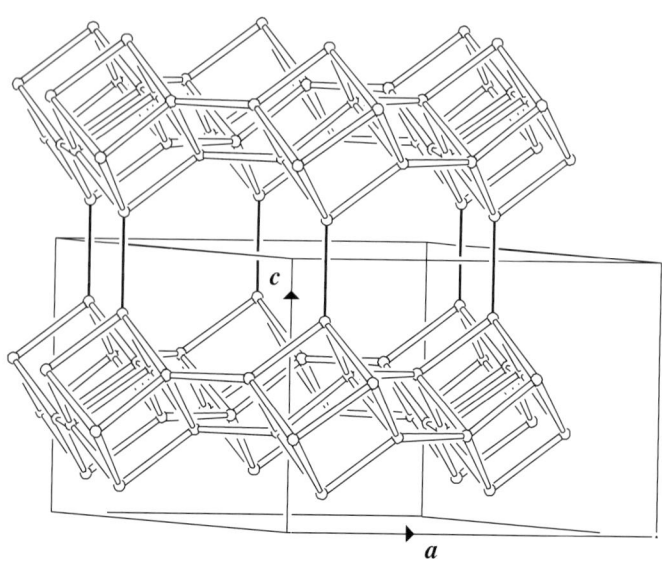

Fig. 2.   Connection mode and unit cell content viewed along [120].

### 3. Channels and/or cages

Intersecting 8- and 12-ring channels are parallel to <100> and **c**, respectively. The channel intersection, depicted in Figure 3, is topologically equivalent to the channel intersection in **AFS** and **BPH**.

{3 [4$^{18}$8$^{6}$12$^{2}$] <100> (8-ring), [001] (12-ring)}

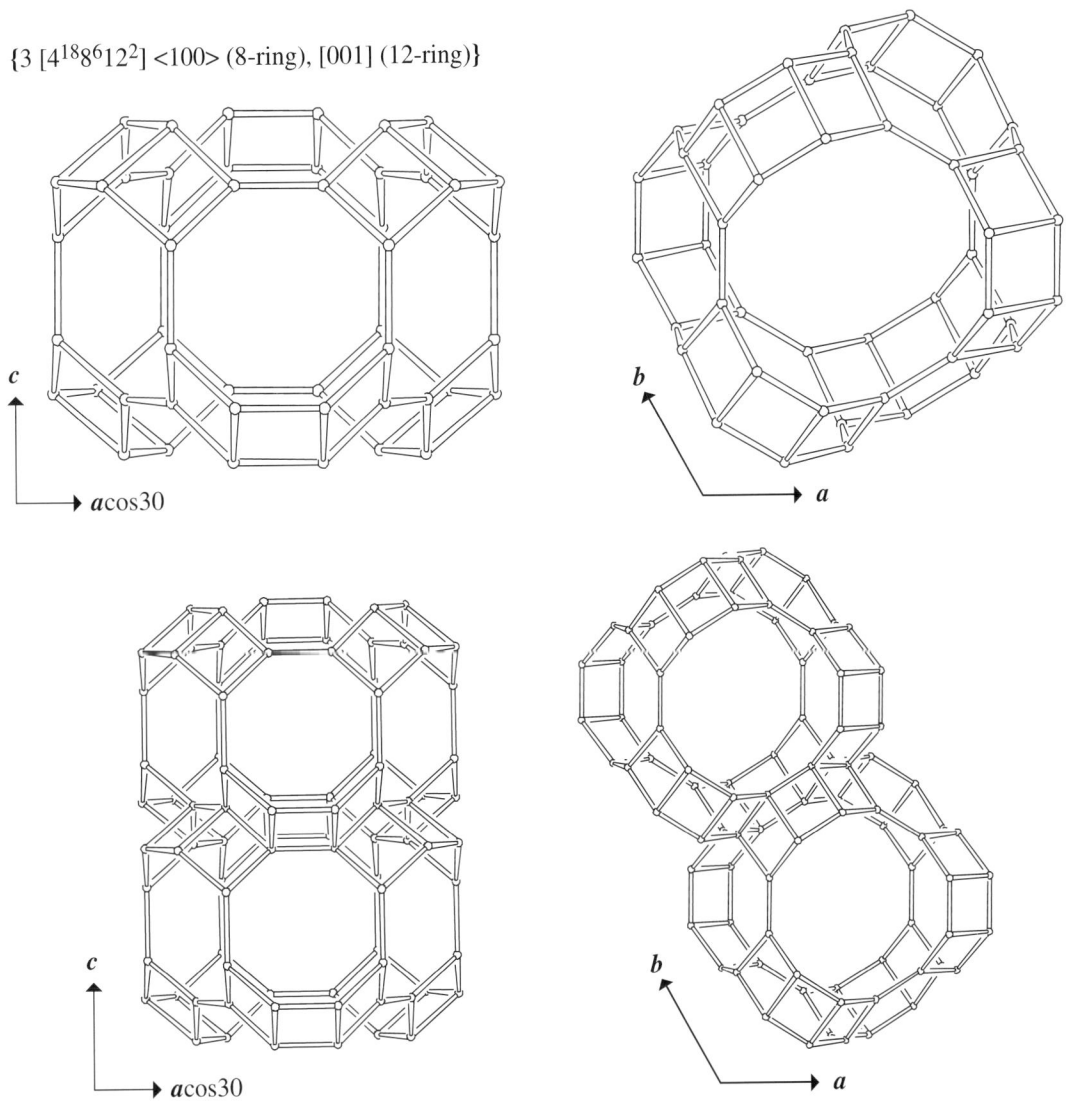

Fig. 3. Top: channel intersection viewed along **b** (left), and along **c** (right). Bottom: interconnection of channel intersections into a 12-ring channel parallel to **c** viewed along **b** (left) and into an 8-ring channel parallel to **b** (or **a**) viewed along **c** (right).

# AHT                    Building Scheme

## 1. Periodic Building Unit

**AHT** can be built using the crankshaft chain (bold in Figure 1) running parallel to *c*. The two-dimensional PerBU1 is obtained when crankshaft chains are connected into a layer of (fused) 6-rings as shown in Figure 1 (left). A one-dimensional PerBU2, built from three crankshaft chains is illustrated in Figure 1 (right).

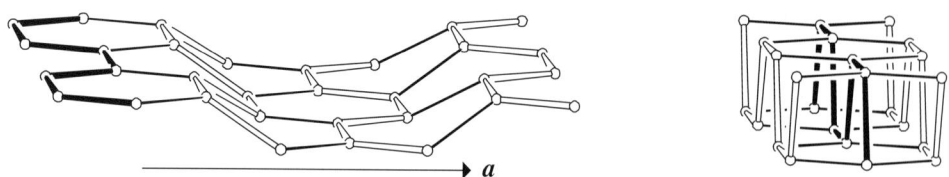

Fig. 1.    Crankshaft chains are connected into a 6-ring layer (PerBU1; left) or into a triple crankshaft chain (PerBU2; right). Views along *c*.

## 2. Connection mode

Neighboring PerBU1s, related along *b* by a shift of $\frac{1}{2}(a+b)$, are connected along *b* through triple crankshaft chains as illustrated in Figure 2. Channels with 10-ring apertures are formed. The channel wall consists of (fused) 6-rings.

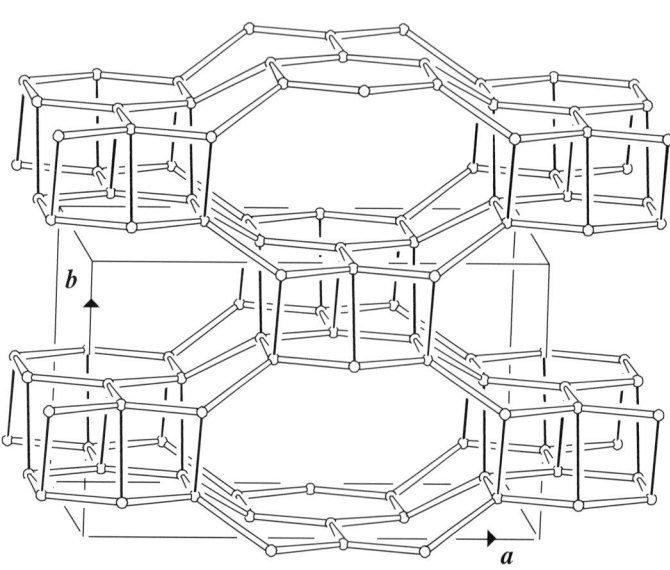

Fig. 2.    Connection mode and unit cell content viewed along *c*. For clarity, only $1\frac{1}{2}$ repeat unit along *c* of each PerBU is drawn.

## 3. Channels and/or cages

The non-interconnecting 10-ring channels, parallel to *c*, are topologically equivalent to those in **AEL** and **AFO**. One channel is depicted in Figure 3.

{1 [$6^{10}10^{2/2}$] [100] (10-ring)}

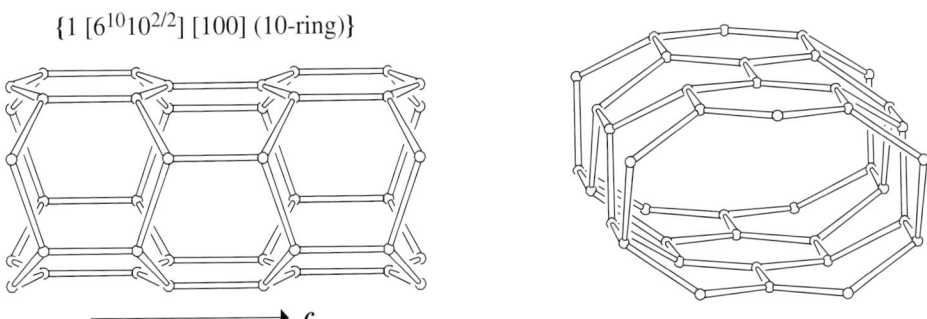

Fig. 3.   Channel viewed perpendicular to the channel axis (left) and along the channel axis (right).

# ANA        Building Scheme

## 1. Periodic Building Unit

Cubic **ANA** can be built using distorted 6-ring chairs (one in bold in Figure 1). 6-Rings, related along $b$ by a rotation of 180° about an axis parallel to $a$, are connected through (distorted) 4-rings into chains along $b$. A two-dimensional PerBU is obtained when chains, related along $a$ by a screw rotation of 180° about $b$, are connected along $a$ through (distorted) 4-rings as shown in Figure 1.

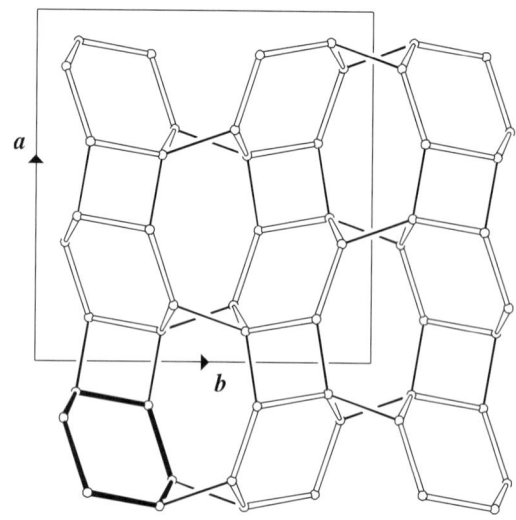

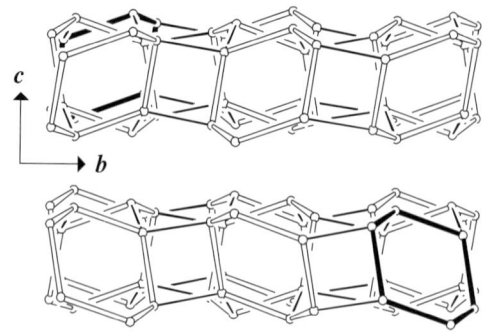

Fig. 1. PerBU viewed along the cube axis $c$ (left) and along the cube axis $a$ (right). The PerBUs at the right are related by a rotation of 180° about $c$.

## 2. Connection mode

Neighboring PerBUs, related along $c$ by a screw rotation of 180° about $c$ followed by a shift of $\frac{1}{2}b$, are connected along $c$ through (distorted) 4- and 8-rings as illustrated in Figure 2.

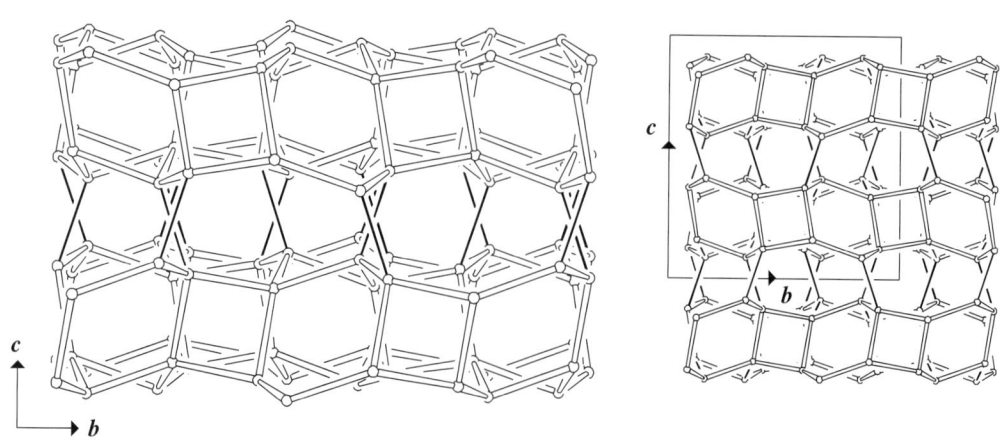

Fig. 2. Connection mode (left) and unit cell content (right) viewed along the cube axis $a$.

### 3. Channels and/or cages

Channel intersections are connected into "double" cavities that form irregular 8-ring channels as illustrated in Figure 3.

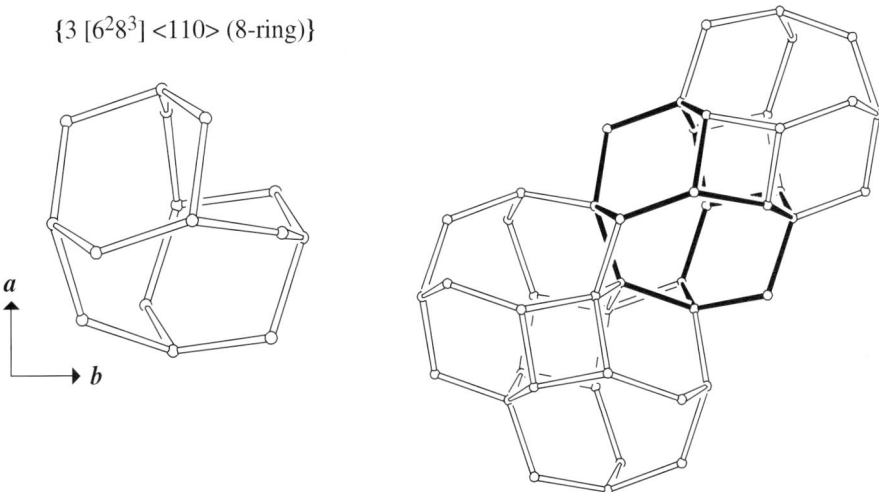

$\{3\ [6^2 8^3]\ <110>\ (8\text{-ring})\}$

Fig. 3.   Channel intersection (left) and irregular 8-ring channels (right) viewed along $c$.

# APC                    Building Scheme

## 1. Periodic Building Unit

**APC** can be built using the crankshaft chain (bold in Figure 1, left) running parallel to **a**. A one-dimensional PerBU is obtained when two crankshaft chains and 4-rings are connected into a channel with an 8-ring aperture. The channel wall consists of 4-, 6- and 10-rings. The repeat unit of the PerBU consists of a 4-fold (1,2,3,5)-connected double 8-ring (bold in Figure 1, right).

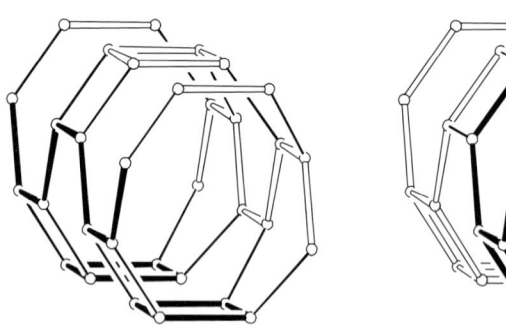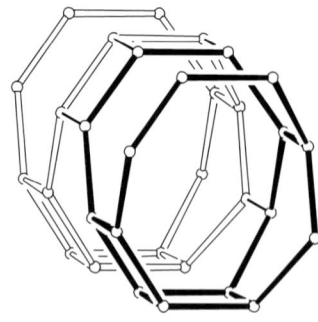

Fig. 1.   PerBU constructed from crankshaft chains and 4-rings (left) and from 4-fold connected double 8-rings (right) viewed along **a**.

## 2. Connection mode

Neighboring PerBUs, related along **c** by a pure translation and along **b** by a shift of $\frac{1}{2}(a+b)$, are connected along **b** and **c** through double-crankshaft chains as shown in Figure 2.

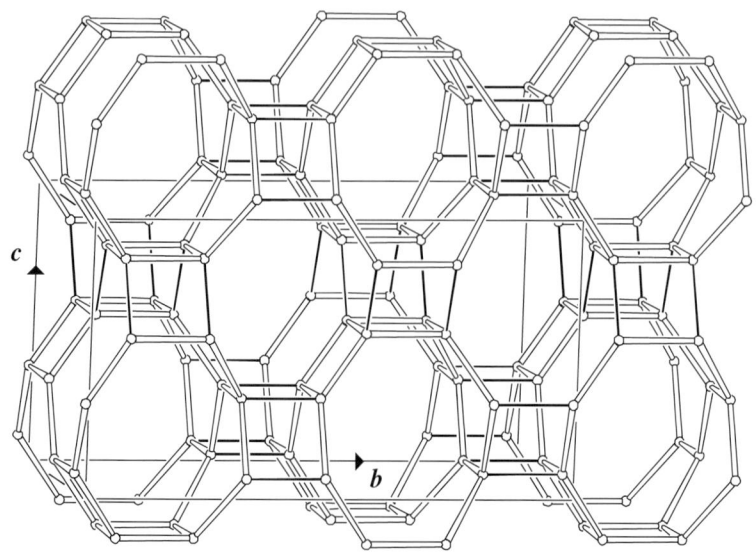

Fig. 2.   Connection mode and unit cell content viewed along **a**. For clarity, only $1\frac{1}{2}$ repeat unit along **a** of each PerBU is drawn.

### 3. Channels and/or cages

8-Ring channels are parallel to *a*. Pairs of interconnecting channels are formed when the channels are linked through common 8-rings as depicted in Figure 3.

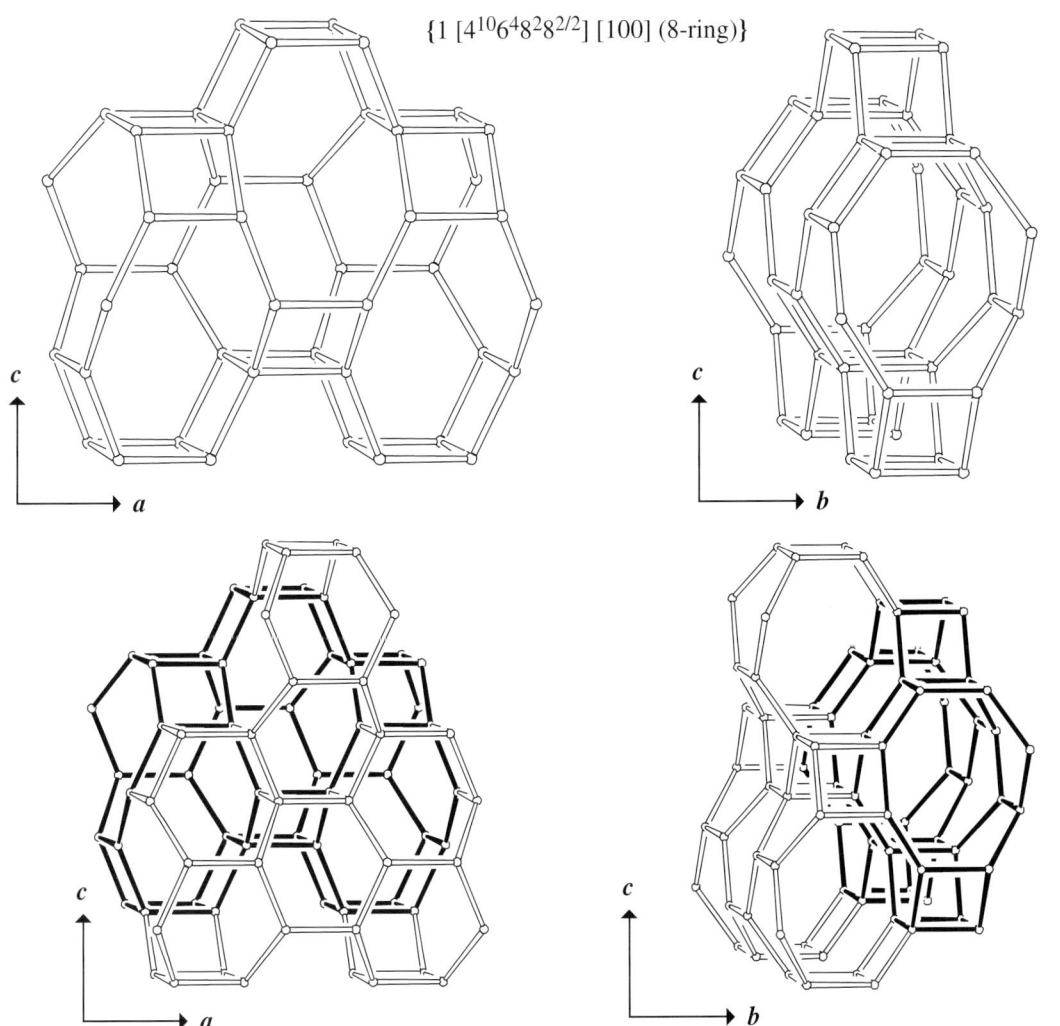

$\{1\ [4^{10}6^48^28^{2/2}]\ [100]\ (8\text{-ring})\}$

Fig. 3.  Top: 8-ring channel viewed perpendicular to the channel axis (left) and along the channel axis (right). Bottom: pairs of interconnecting channels viewed along *b* (left) and along *a* (right). One channel in bold.

# APD        Building Scheme

## 1. Periodic Building Unit

**APD** can be built using the crankshaft chain (bold in Figure 1, left) running parallel to **a**. A one-dimensional PerBU is obtained when two crankshaft chains and 4-rings are connected into a channel with an 8-ring aperture. The channel wall consists of 4-, 6- and 8-rings. The repeat unit of the PerBU consists of a 4-fold (1,2,4,6)-connected double 8-ring (bold in Figure 1, right).

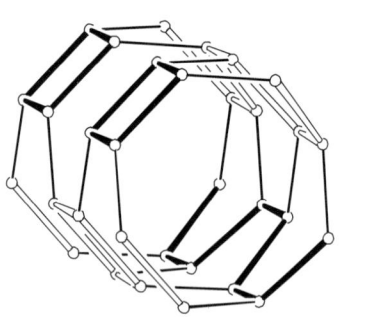 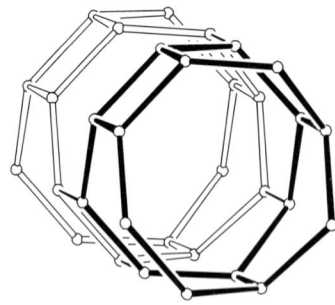

Fig. 1.   PerBU constructed from crankshaft chains and 4-rings (left) and from 4-fold connected double 8-rings (right) viewed along **a**.

## 2. Connection mode

Neighboring PerBUs, related along **c** by a pure translation and along **b** by a shift of $\frac{1}{2}(a+b)$, are connected along **c** and **b** through double-crankshaft chains as shown in Figure 2.

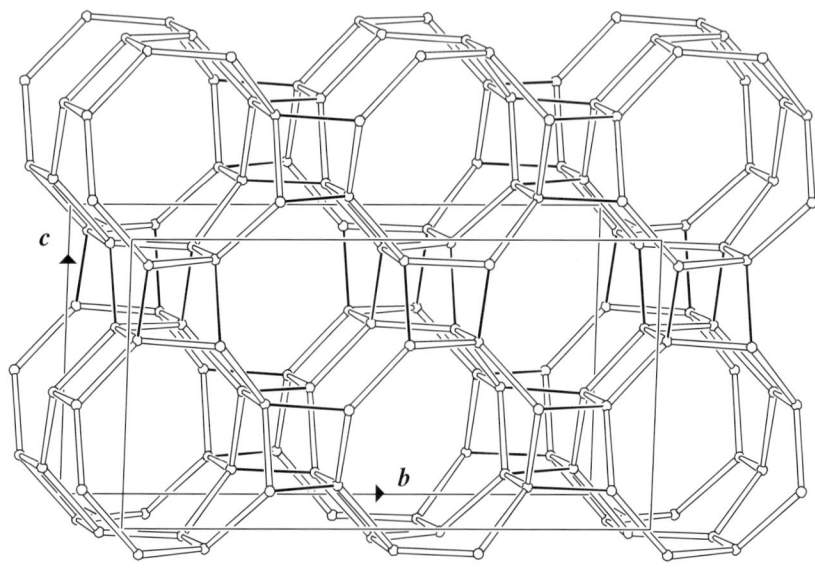

Fig. 2.   Connection mode and unit cell content viewed along **a**. For clarity, only $1\frac{1}{2}$ repeat unit along **a** of each PerBU is drawn.

# Cage/Channel — APD

## 3. Channels and/or cages

Pairs of interconnecting 8-ring channels are parallel to *a*. One channel and pairs of interconnecting channels are depicted in Figure 3.

$$\{1\ [4^2 6^4 8^2 8^{2/2}]\ [100]\ (8\text{-ring})\}$$

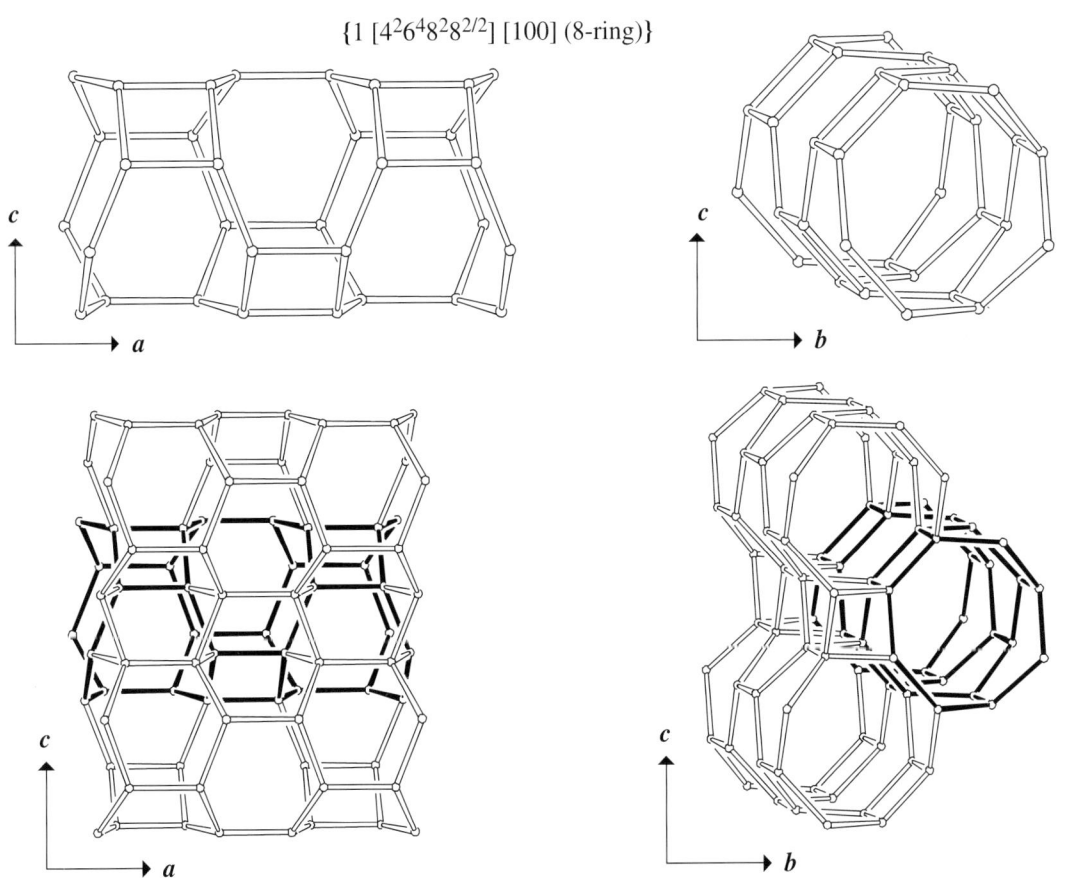

Fig. 3.   Top: 8-ring channel viewed perpendicular to the channel axis (left) and along the channel axis (right).
Bottom: pairs of interconnecting channels viewed along *b* (left) and along *a* (right). One channel in bold.

# AST         Building Scheme

## 1. Periodic Building Unit

Cubic **AST** can be built using T10-units consisting of a double 4-ring with two "dangling" T atoms (bold in Figure 1). T10-units, related by translations along half a face diagonal, are connected through the "dangling" T atoms into a one-dimensional PerBU shown in Figure 1. (Compare with **ASV** and **UOZ**.)

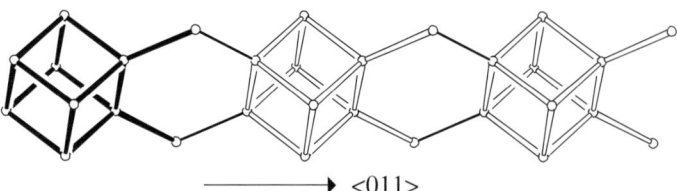

$\longrightarrow$ <011>

Fig. 1. PerBU viewed perpendicular to <011>.

## 2. Connection mode

Neighboring PerBUs, related by shifts of half a face diagonal, are connected through 6-rings as illustrated in Figure 2. [$4^6 6^{12}$]-Cages are formed.

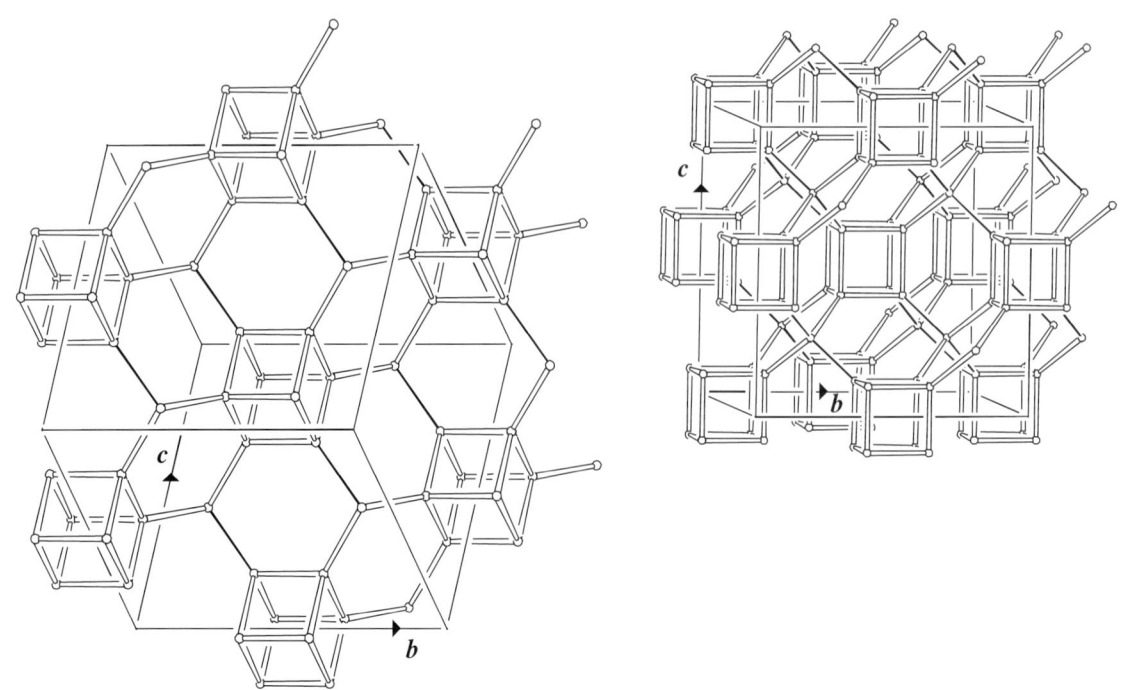

Fig. 2. Connection mode in the (22-2) plane (left) and unit cell content viewed along **a** (right).

## 3. Channels and/or cages

The cage in **AST** is depicted in Figure 3. The apertures are formed by 6-rings only.

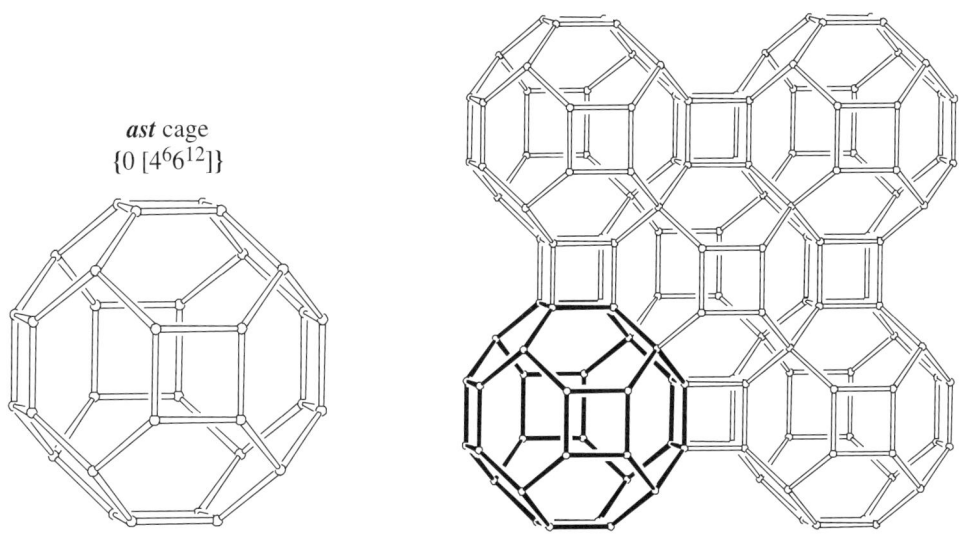

*ast* cage
$\{0\ [4^6 6^{12}]\}$

Fig. 3.　Cage (left) and connected cages (right) viewed along ***a***. One cube face is shown.

# ASV                    Building Scheme

## 1. Periodic Building Unit

Tetragonal **ASV** can be built using the T20-unit consisting of two double 4-ring with two "dangling" T atoms (bold in Figure 1). Neighboring T20-units, related along *a* and *c* by pure translations, are connected along *a* and *c* through distorted (fused) 6-rings as shown in the drawing of the two-dimensional PerBU in Figure 1. (Compare with **AST** and **UOZ**.)

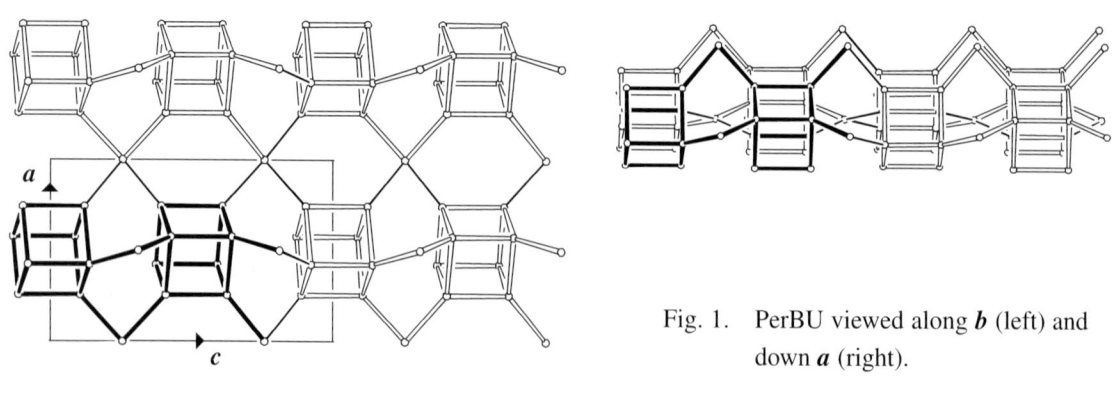

Fig. 1.    PerBU viewed along *b* (left) and down *a* (right).

## 2. Connection mode

Neighboring PerBUs, related along *b* by a pure translation, are connected along *b* through the "dangling" T atoms. A 12-ring channel parallel to *c* is formed (Figure 2).

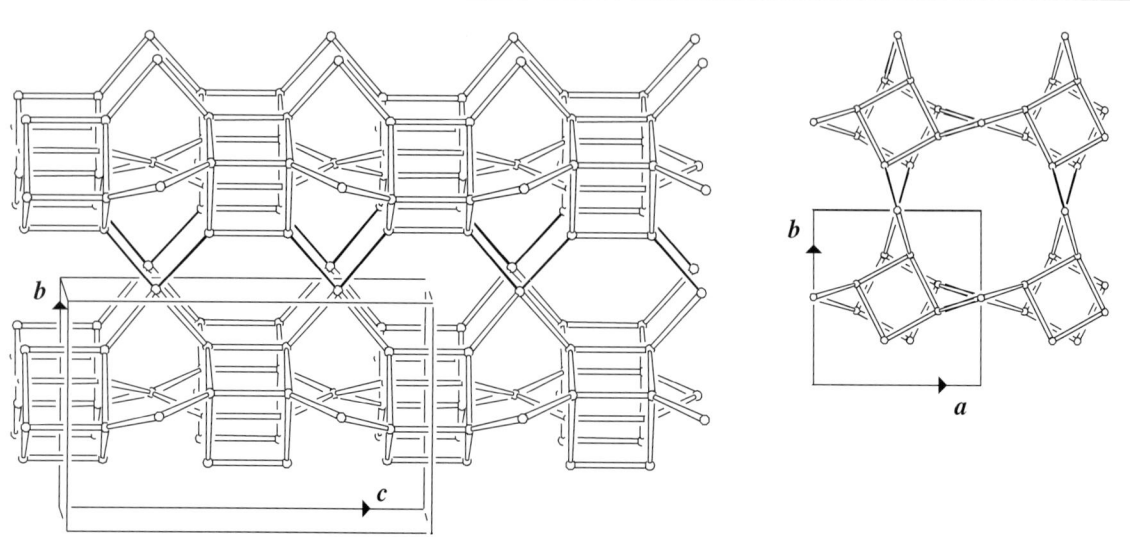

Fig. 2.    Connection mode (and unit cell content) viewed along *a* (left) and unit cell content projected along *c* (right).

## 3. Channels and/or cages

Non-interconnecting 12-ring channels are parallel to *c* as depicted in Figure 3.

$\{1\ [4^8 6^{16} 12^{2/2}]\ [001]\ (12\text{-ring})\}$

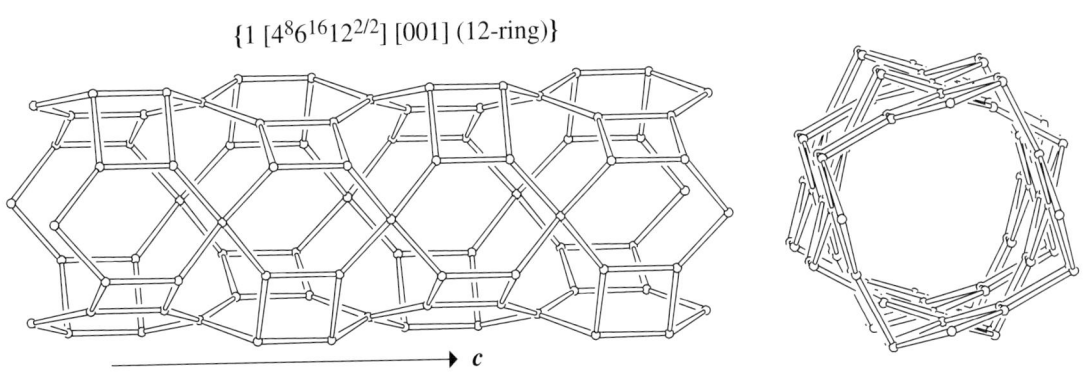

Fig. 3.   Channel viewed perpendicular to the channel axis (left) and along the channel axis (right).

# ATN    Building Scheme

## 1. Periodic Building Unit

Tetragonal **ATN** can be built using the zigzag chain (bold in Figure 1, left) running parallel to $c$. The one-dimensional PerBU is obtained when four zigzag chains are connected into a channel with an 8-ring aperture. The repeat unit of the PerBU is an 8-ring (bold in Figure 1, right). The channel wall consists of fused 6-rings. (See also **BCT**.)

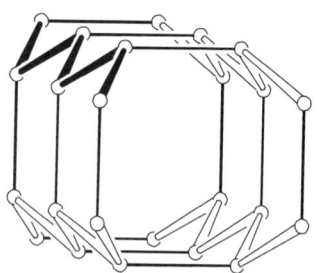 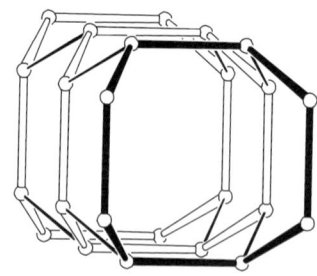

Fig. 1.  PerBU, constructed from four zigzag chains (left) and from 8-rings (right), viewed along $c$.

## 2. Connection mode

Neighboring PerBUs, related along $a$ (and $b$) by a shift of $\frac{1}{2}(a+b+c)$, are connected along $a$ (and $b$) through double zigzag chains which form (fused) **atn** cavities (see Figure 2).

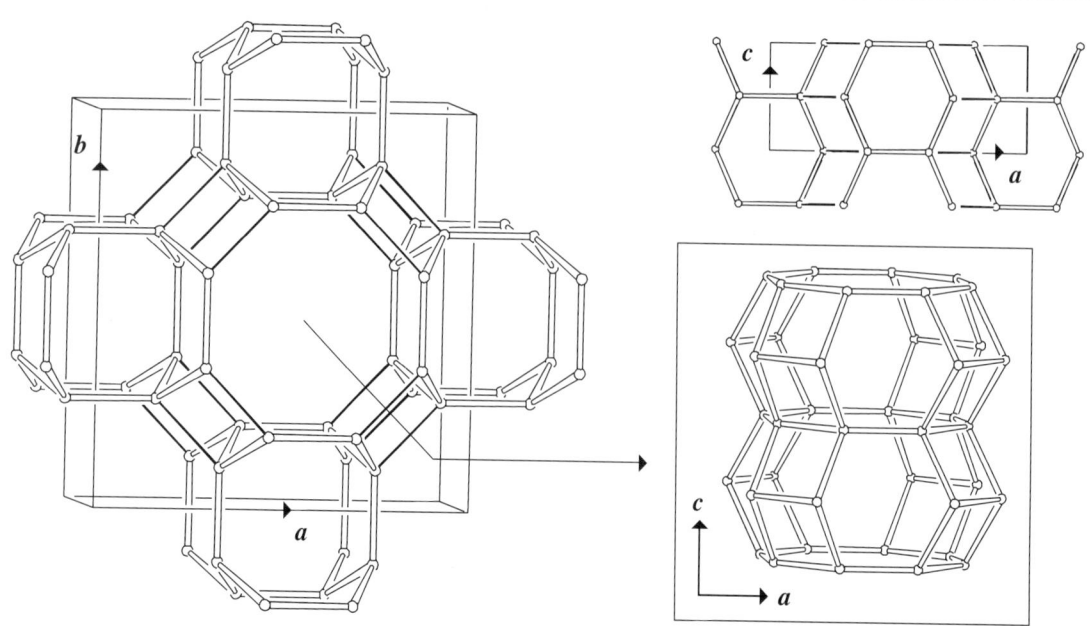

Fig. 2.  Connection mode (and unit cell content) viewed along $c$ (left) and unit cell content projected along $b$ (or $a$; top right). For clarity, only two repeat units of the PerBUs are drawn. The inset shows the (fused) **atn** cavities formed.

### 3. Channels and/or cages

Two types of non-interconnecting 8-ring channels are parallel to *c*. The first type equals the PerBU and is topologically equivalent to the 8-ring channel in **BCT** and **GON**. The second type is obtained when *atn* cavities (also present in **SBE**) are connected through common 8-rings (Figure 3).

Type 1: {1 [$6^4 8^{2/2}$] [001] (8-ring)}

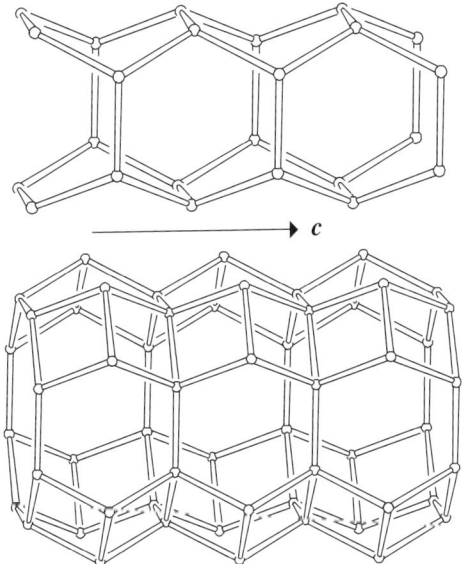
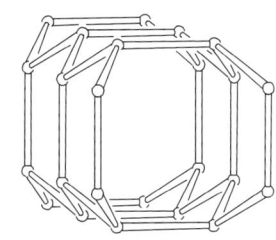
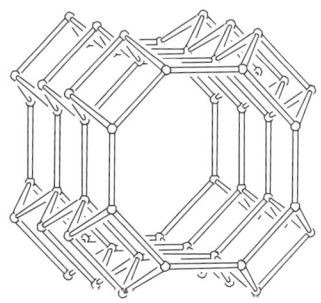

Type 2: {1 [$4^8 6^4 8^{2/2}$] [001] (8-ring)}

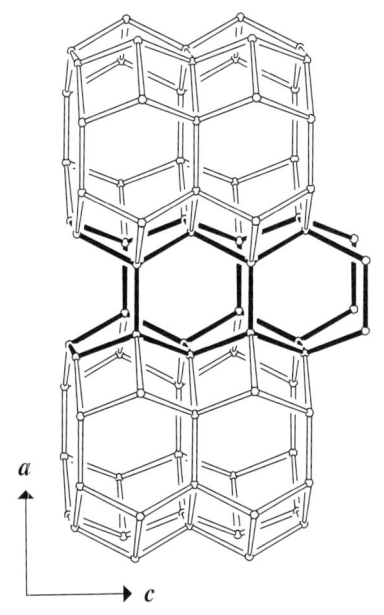

Fig. 3.  Top and middle: channels viewed normal to channel axis (left) and along the channel axis (right). Bottom: connection of both types of 8-ring channels (Type 1 channel in bold).

# ATO
# Building Scheme

## 1. Periodic Building Unit

Trigonal **ATO** can be built using the zigzag chain (bold in Figure 1, left) running parallel to $c$. The one-dimensional PerBU is obtained when six zigzag chains are connected into a channel with a 12-ring aperture. The repeat unit of the PerBU is a 12-ring (bold in Figure 1, right). (See also **CAN**.)

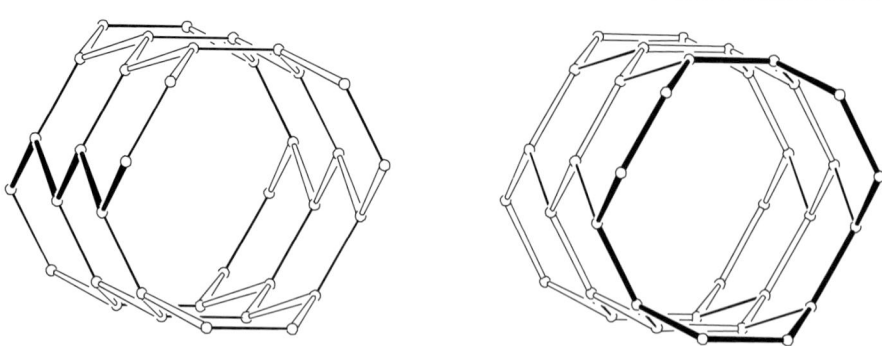

Fig. 1.   PerBU constructed from six zigzag chains (left) and from 12-rings (right) viewed along $c$.

## 2. Connection mode

Neighboring PerBUs, related by a rotation of 120° about $c$ followed by a shift of $\frac{1}{3}c$ (i.e., related by a $3_1$ axis), are connected through 4-rings as illustrated in Figure 2.

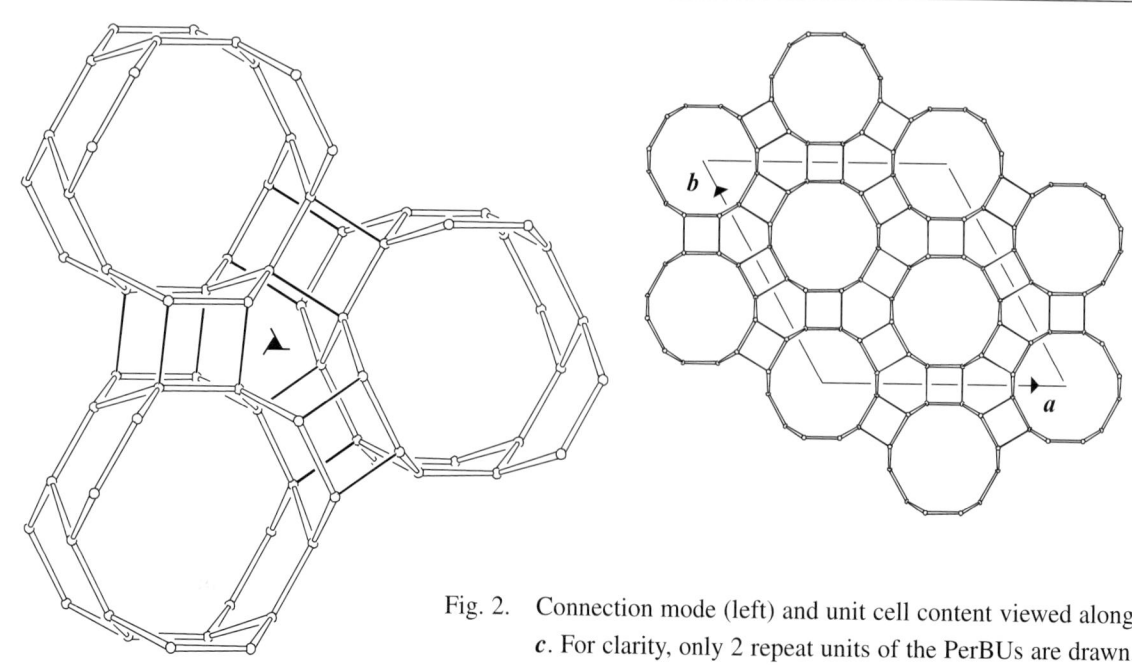

Fig. 2.   Connection mode (left) and unit cell content viewed along $c$. For clarity, only 2 repeat units of the PerBUs are drawn.

## 3. Channels and/or cages

Non-interconnecting 12-ring channels are parallel to $c$. The channel wall consists of fused 6-rings as depicted in Figure 3. The channel is topologically equivalent to the 12-ring channels in **CAN** and **NPO**.

{1 [$6^6 12^{2/2}$] [001] (12-ring)}

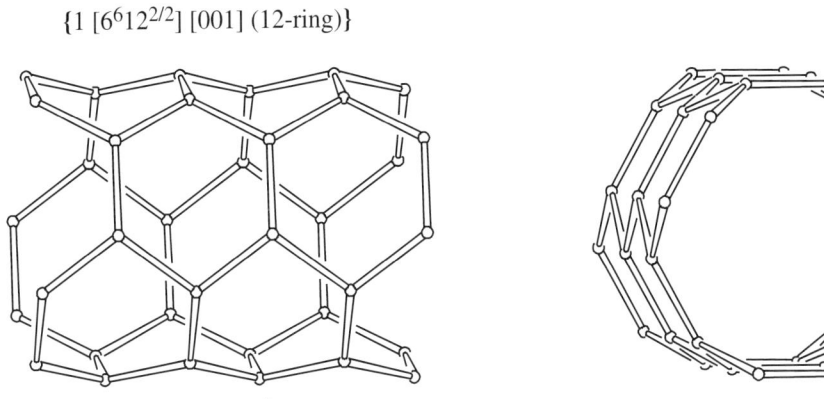

Fig. 3.   Channel viewed perpendicular to the channel axis (left) and along the channel axis (right).

# ATS                     Building Scheme

## 1. Periodic Building Unit

**ATS** can be built using the zigzag chain (bold in Figure 1, left) running parallel to $c$. The one-dimensional PerBU is obtained when six zigzag chains are connected as shown in Figure 1. The repeat unit of the PerBU contains 12 T atoms and consists of two doubly connected 6-ring boats (bold in Figure 1, right).

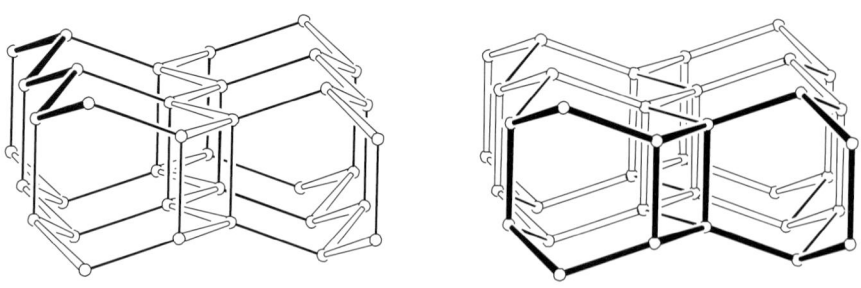

Fig. 1.   PerBU, constructed from six zigzag chains (left) and from doubly connected 6-ring boats (right), viewed along $c$.

## 2. Connection mode

Neighboring PerBUs, related along $a$ (and $b$) by a shift of $\frac{1}{2}(a+b)$, are connected along $a$ (and $b$) through double zigzag chains (Figure 2).

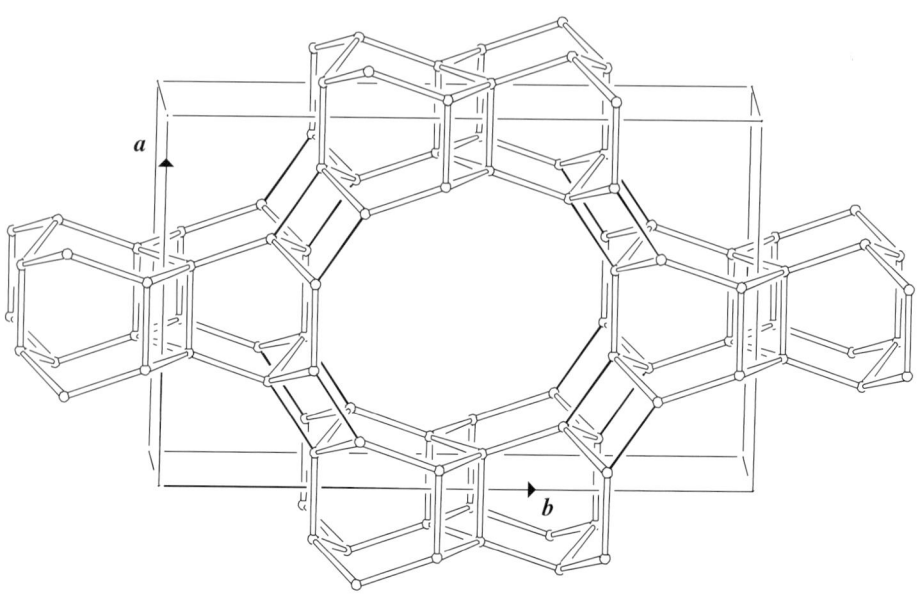

Fig. 2.   Connection mode and cell content viewed along $c$. For clarity, only two repeat units in each PerBU are drawn.

## 3. Channels and/or cages

Non-interconnecting 12-ring channels with side-pockets are parallel to $c$ as depicted in Figure 3.

$\{1\ [4^86^612^{2/2}]\ [001]\ (12\text{-ring})\}$

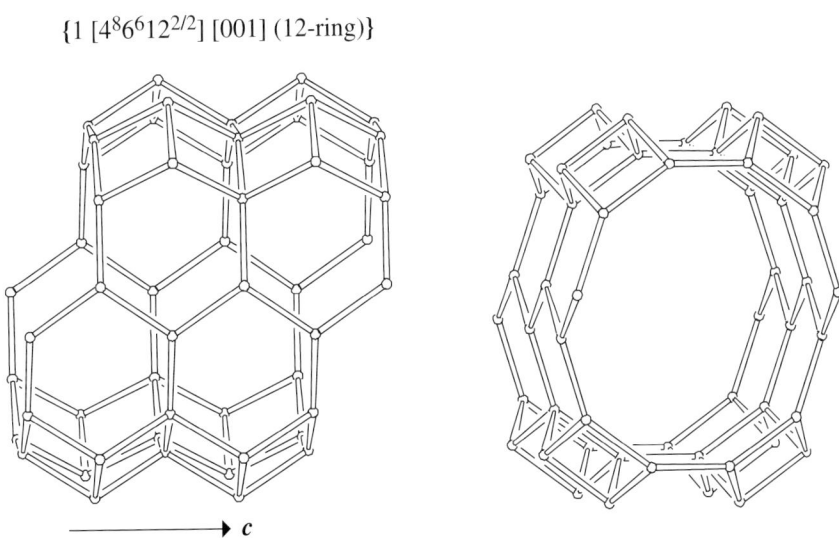

$\longrightarrow c$

Fig. 3.   Channel viewed perpendicular to the channel axis (left) and along the channel axis (right).

# ATT                    Building Scheme

## 1. Periodic Building Unit

**ATT** can be built using crankshaft chains (running parallel to **a**) and T–T dimers (bold in Figure 1, left). A one-dimensional PerBU is obtained when two crankshaft chains and T–T dimers are connected in such a way that a tube with a 6-ring aperture is formed. The repeat unit of the PerBU consists of a 3-fold (1,2,3)-connected double 6-ring (bold in Figure 1, right).

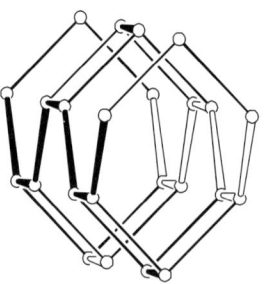 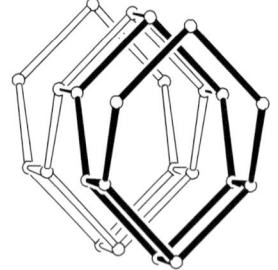

Fig. 1.   PerBU, constructed from crankshaft chains and dimers (left) and from three-fold connected double 6-rings (right), viewed along **a**.

## 2. Connection mode

Neighboring PerBUs, related along **b** and **c** by pure translations, are connected along **b** through double crankshaft chains and along **c** through single crankshaft chains as shown in Figure 2.

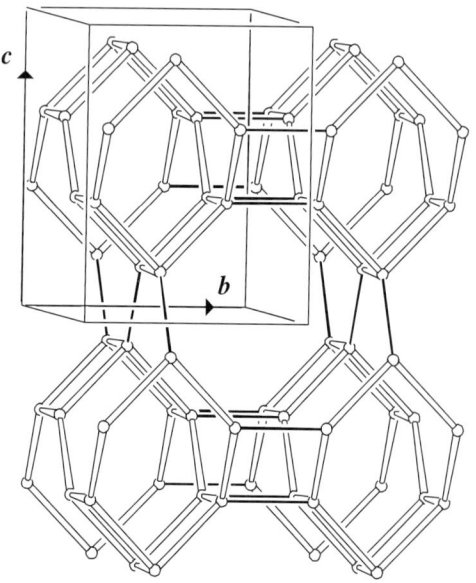 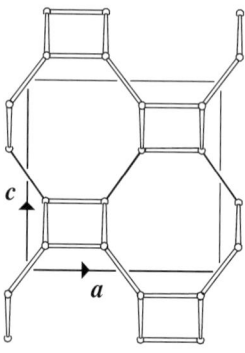

Fig. 2.   Connection mode (and unit cell content) viewed **a** (left) and unit cell content projected along **b** (right). For clarity, only $1\frac{1}{2}$ repeat unit along **a** of each PerBU is drawn.

## 3. Channels and/or cages

Intersecting 8-ring channels (of different type) are parallel to *a* and *b*. The channel parallel to *a* is (topologically) equivalent to (one of) the 8-ring channels in **GIS**, **PHI** and **SIV**. The channels are shown in Figure 3.

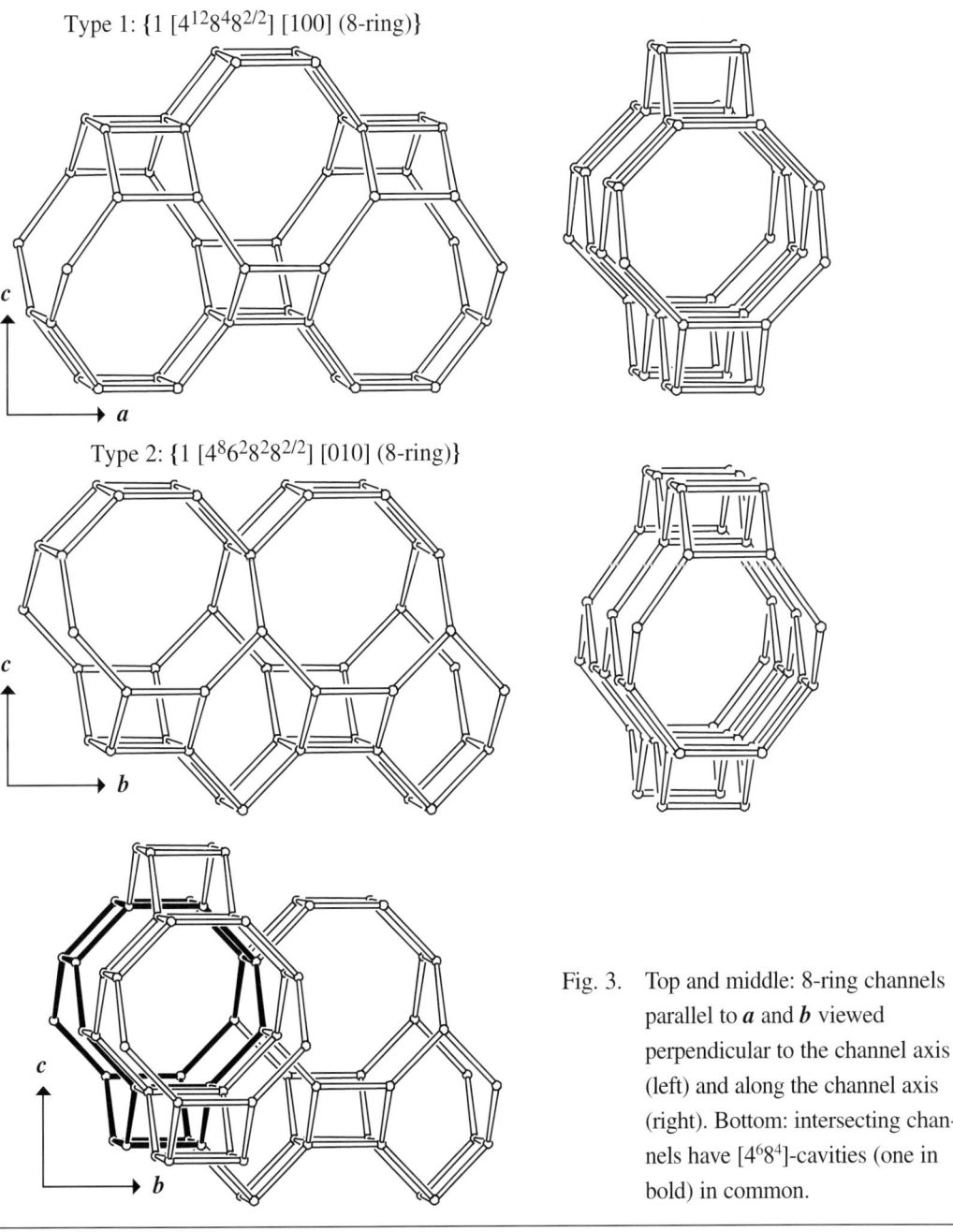

Type 1: {1 [$4^{12}8^48^{2/2}$] [100] (8-ring)}

Type 2: {1 [$4^86^28^28^{2/2}$] [010] (8-ring)}

Fig. 3.    Top and middle: 8-ring channels parallel to *a* and *b* viewed perpendicular to the channel axis (left) and along the channel axis (right). Bottom: intersecting channels have [$4^68^4$]-cavities (one in bold) in common.

# ATV            Building Scheme

## 1. Periodic Building Unit

**ATV** can be built using the crankshaft chain (bold in Figure 1, left) running parallel to $a$. A one-dimensional PerBU is obtained when three crankshaft chains are connected in such a way that a tube with a 6-ring aperture is formed. The tube wall consists solely of 6-rings. The repeat unit of the PerBU consists of a 3-fold (1,3,5)-connected double 6-ring (or [$6^5$]-cages; bold in Figure 1, right). (See also **AFI**.)

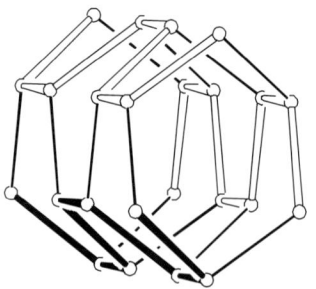 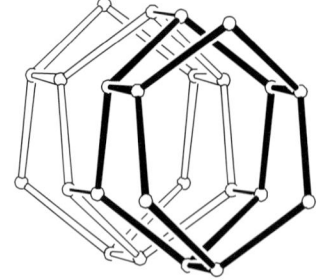

Fig. 1.  PerBU, constructed from crankshaft chains (left) and from [$6^5$]-cages (right), viewed along $a$.

## 2. Connection mode

Neighboring PerBUs, related along $b$ by a shift of $\frac{1}{2}a$ and along $c$ by a pure translation, are connected along $b$ through double crankshaft chains and along $c$ through single crankshaft chains as shown in Figure 2.

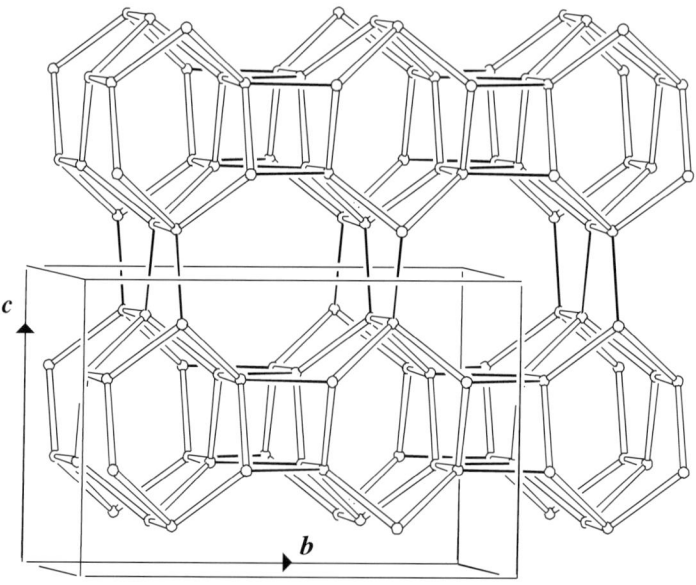

Fig. 2.  Connection mode and unit cell content viewed along $a$. For clarity, only $1\frac{1}{2}$ repeat unit along $a$ of each PerBU is drawn.

# **ATV**

## 3. Channels and/or cages

Non-interconnecting 8-ring channels are parallel to $a$. The channel wall consists of fused 6-rings as shown in Figure 3.

$\{1\ [6^8 8^{2/2}]\ [100]\ (8\text{-ring})\}$

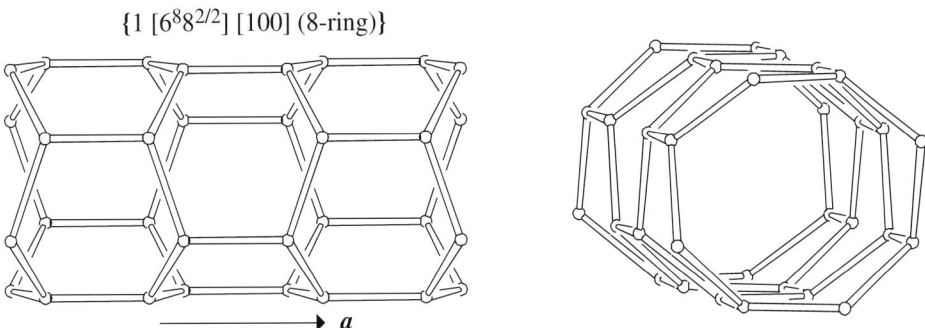

Fig. 3.   Channel viewed perpendicular to the channel axis (left) and along the channel axis (right).

# AWO                    Building Scheme

## 1. Periodic Building Unit

**AWO** can be built using the crankshaft chain (bold in Figure 1, left) running parallel to *a*. A one-dimensional PerBU is obtained when three crankshaft chains are connected into a tube with a 6-ring aperture. The tube wall consists of 4-, 6- and 8-rings. The repeat unit of the PerBU consists of a 3-fold (1,2,4)-connected double 6-ring (bold in Figure 1, right). [See also **UEI**]

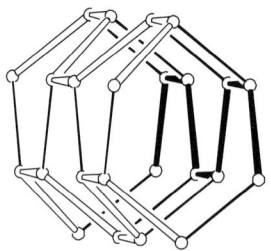 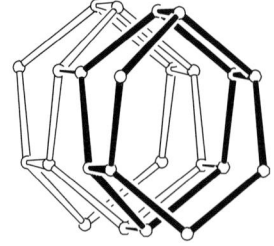

Fig. 1.   PerBU, constructed from crankshaft chains (left) and from 3-fold connected double 6-rings (right), viewed along *a*.

## 2. Connection mode

Neighboring PerBUs, related along *b* by a shift of $\frac{1}{2}(a+b)$ and along *c* by a screw rotation of 180° about *c*, are connected along *b* and *c* through double crankshaft chains and 4-rings, respectively (Figure 2).

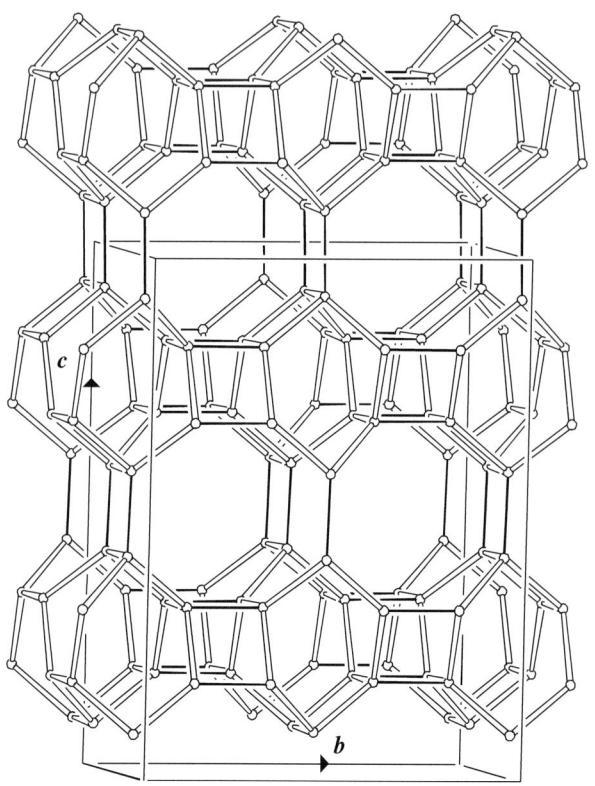

Fig. 2.   Connection mode and unit cell content viewed along *a*. For clarity, only $1\frac{1}{2}$ repeat unit along *a* of each PerBU is drawn.

### 3. Channels and/or cages

Interconnecting 8-ring channels are parallel to *a* (Figure 3).

{1 [$4^{12}6^210^28^{2/2}$] [100] (8-ring)}

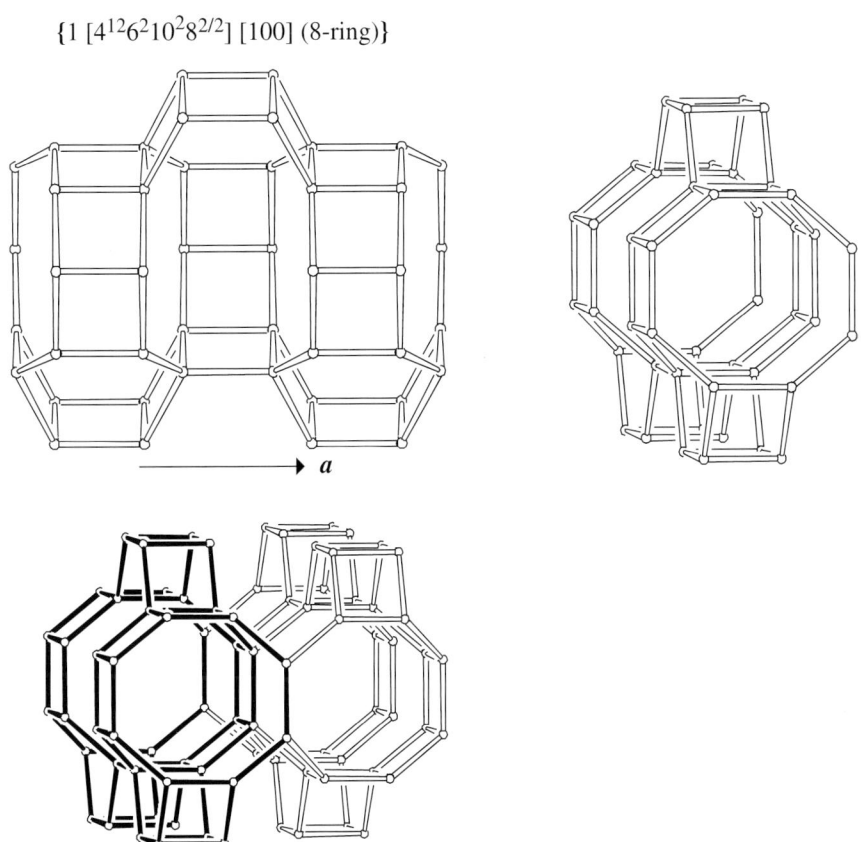

Fig. 3.  Top: 8-ring channel viewed perpendicular to the channel axis (left) and along the channel axis (right). Bottom: interconnected channels viewed along *a* (one channel in bold).

# AWW                    Building Scheme

## 1. Periodic Building Unit

The one-dimensional PerBU in tetragonal **AWW** is equal to a chain parallel to $c$ and is depicted in Figure 1. The repeat unit of the PerBU consists of a 4-fold (1,2,3,4)-connected double 6-ring (bold in Figure 1). The 4-fold connected double 6-rings in the chain are related along $c$ by a pure translation.

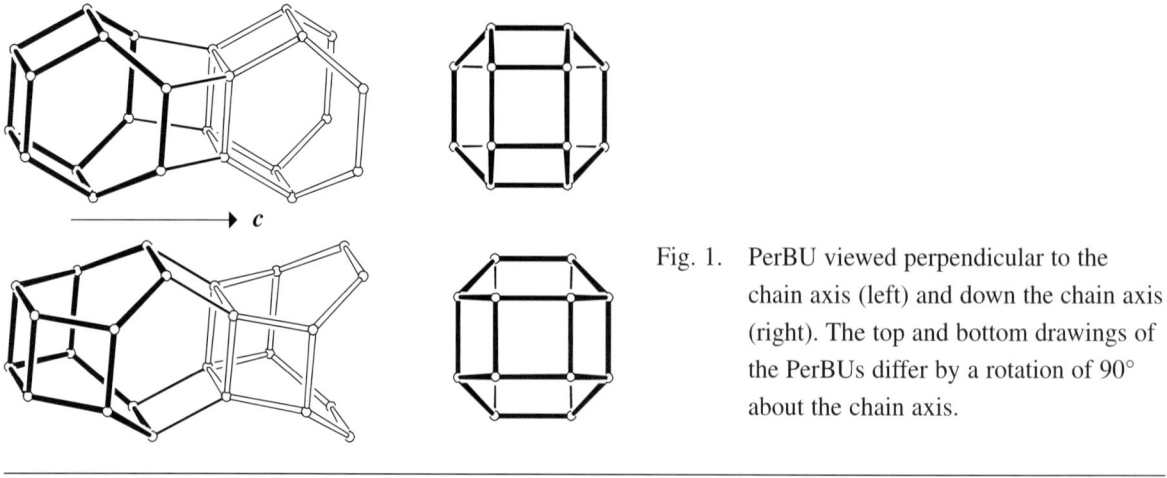

Fig. 1.   PerBU viewed perpendicular to the chain axis (left) and down the chain axis (right). The top and bottom drawings of the PerBUs differ by a rotation of 90° about the chain axis.

## 2. Connection mode

Neighboring PerBUs, related along $a$ (and $b$) by a rotation of 90° about the chain axis and a shift of $\frac{1}{2}(a+b)$, are connected along $a$ (and $b$) through (fused) 4- and 6-rings as shown in Figure 2.

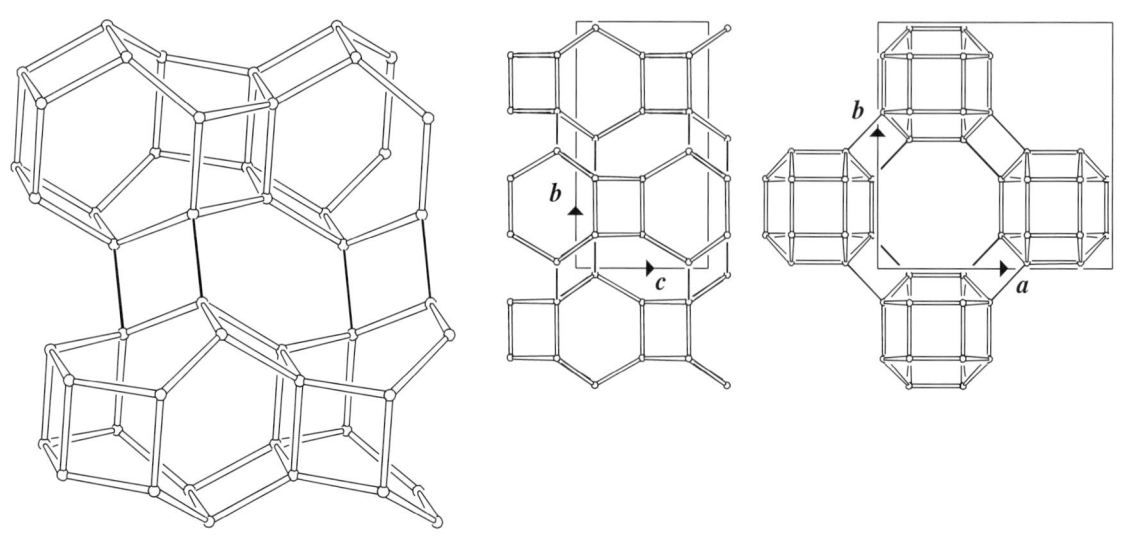

Fig. 2.   Connection mode viewed along $a$ (left) and unit cell content projected along $a$ (middle) and $c$ (right). The unit cell is obtained by connecting four PerBUs around a 4-fold rotation axis parallel to $c$.

### 3. Channels and/or cages

Non-interconnecting 8-ring channels, consisting of *aww* cavities, are parallel to *c* as illustrated in Figure 3. (Compare with the 8-ring channel in **-CLO**.)

$\{1 \ [4^8 6^8 8^{2/2}] \ [001] \ (8\text{-ring})\}$

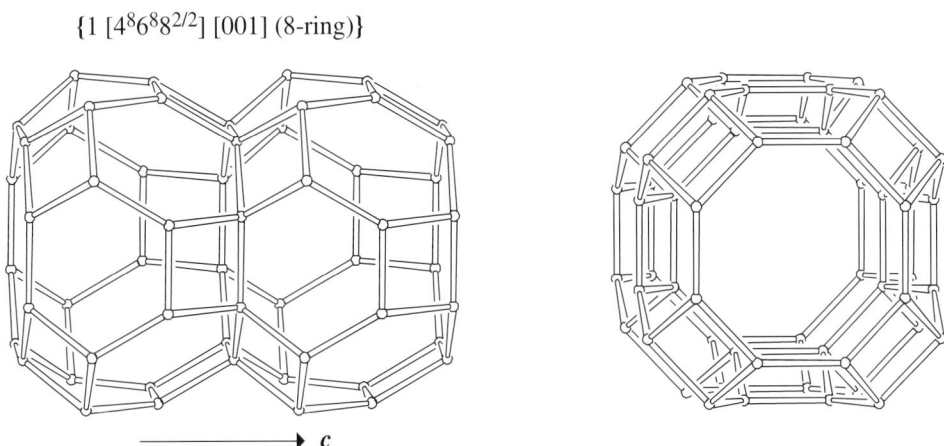

→ *c*

Fig. 3.   Channel viewed perpendicular to the channel axis (left) and along the channel axis (right).

# BCT                    Building Scheme

## 1. Periodic Building Unit

Tetragonal **BCT** can be built using the zigzag chain (bold in Figure 1, left) running parallel to $c$. The one-dimensional PerBU is obtained when four zigzag chains are connected into a channel with an 8-ring aperture. The repeat unit of the PerBU is an 8-ring (bold in Figure 1, right). The cylinder wall consists of fused 6-rings. (See also **ATN**.)

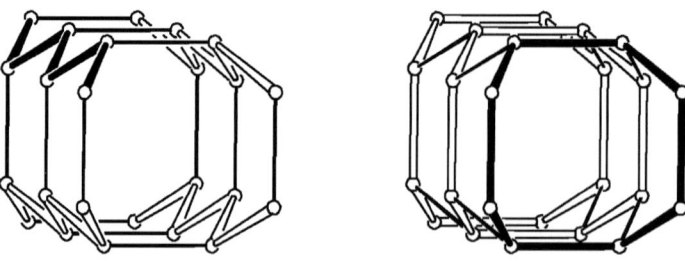

Fig. 1.   PerBU, constructed from four zigzag chains (left) and from 8-rings (right), viewed along $c$.

## 2. Connection mode

Neighboring PerBUs, related along $a$ and $b$ by pure translations, are connected along $a$ and $b$ through 4-rings as depicted in Figure 2.

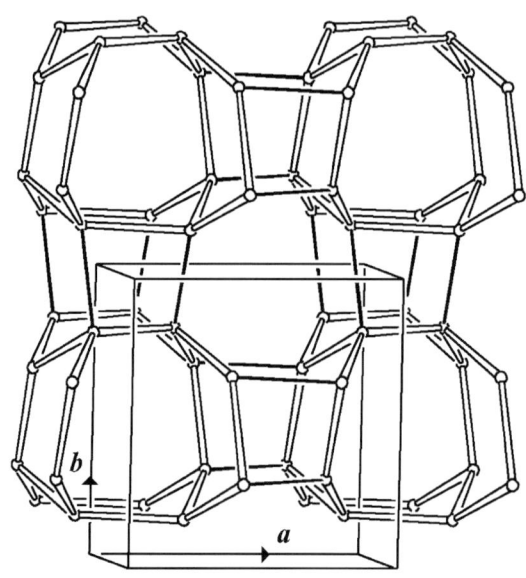

Fig. 2.   Connection mode and unit cell content viewed along $c$. For clarity, only two repeat units of each PerBU are drawn.

## 3. Channels and/or cages

Non-interconnecting 8-ring channels are parallel to $c$. The channel, depicted in Figure 3, is topologically equivalent to the 8-ring channel in **ATN** and **GON**.

$\{1\ [6^{4}8^{2/2}]\ [001]\ (8\text{-ring})\}$

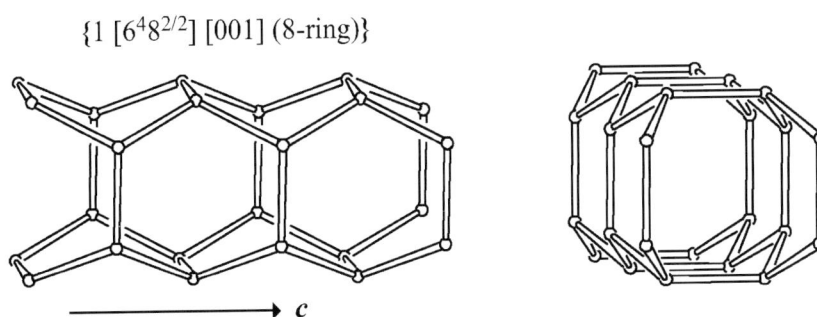

$\longrightarrow c$

Fig. 3.    Channel viewed perpendicular to the channel axis (left) and along the channel axis (right).

# *BEA                           **Building Scheme**

## 1. Periodic Building Unit

The two-dimensional PerBU in tetragonal *BEA equals the beta layer shown in Figure 1. The PerBU is composed of T16-units (four fused 6-rings or eight fused 5-rings (in bold)) related along the cell edges *a* and *b* by pure translations. (See also **BEC**; compare with **ISV**.)

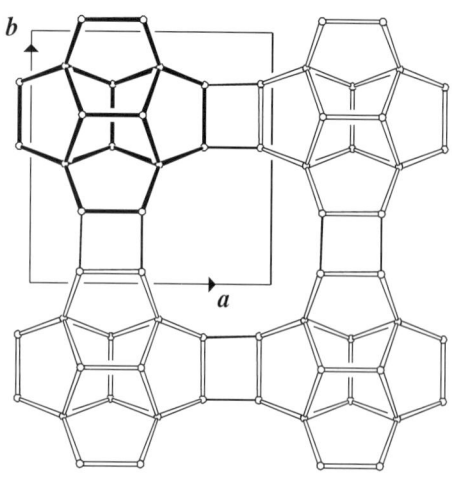
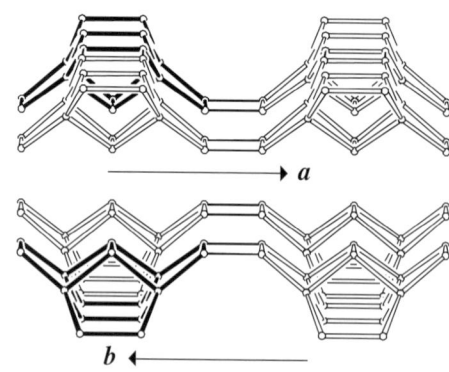

Fig. 1.   PerBU viewed along *c* (left), along *b* (top right) and down *a* (bottom right). The PerBUs at the right are related by a rotation of 90° about *c*.

## 2. Connection mode

Neighboring PerBUs, related by a rotation of 90° about *c* followed by a shift of $\frac{1}{4}c$ (i.e. related by a $4_1$-axis) and a lateral shift of $\pm\frac{1}{3}a$ or $\pm\frac{1}{3}b$ (denoted as $\pm(1/3,0)$ or $\pm(0,1/3)$), are connected through 6-5-5-ring sequences (Figure 2).

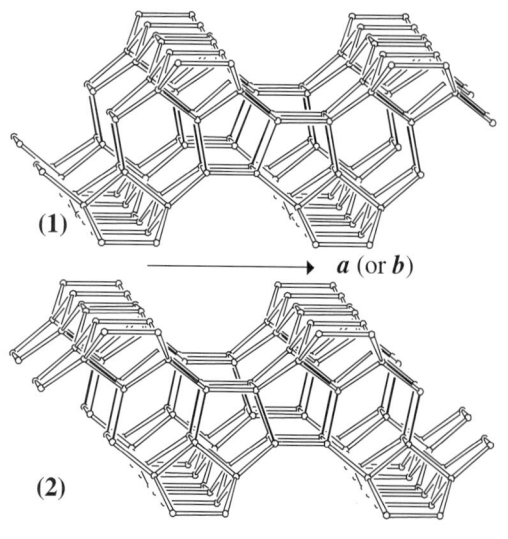

(1)

⟶ *a* (or *b*)

(2)

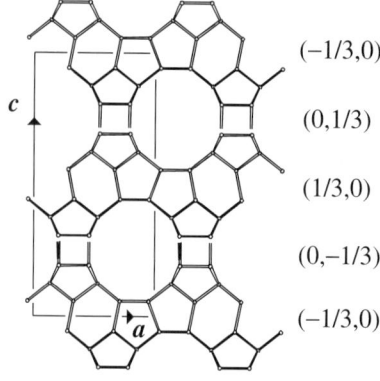

(−1/3,0)

(0,1/3)

(1/3,0)

(0,−1/3)

(−1/3,0)

Fig. 2.   Connection modes viewed perpendicular to *c* (left) and unit cell projected along <010> (right). The lateral shifts are added. The enantiomorph (not drawn) has opposed lateral shifts.

### 3. Channels and/or cages

Interconnecting 12-ring channels are parallel to <100> and *c*. The channels parallel to <100>, depicted in Figure 3, are topologically equivalent to the 12-ring channels in **CON**. The sinusoidal 12-ring channel parallel to *c* is also illustrated in Figure 3.

$\{1 \ [4^2 5^8 6^2 12^2 12^{2/2}] <100> \ (12\text{-ring})\}$

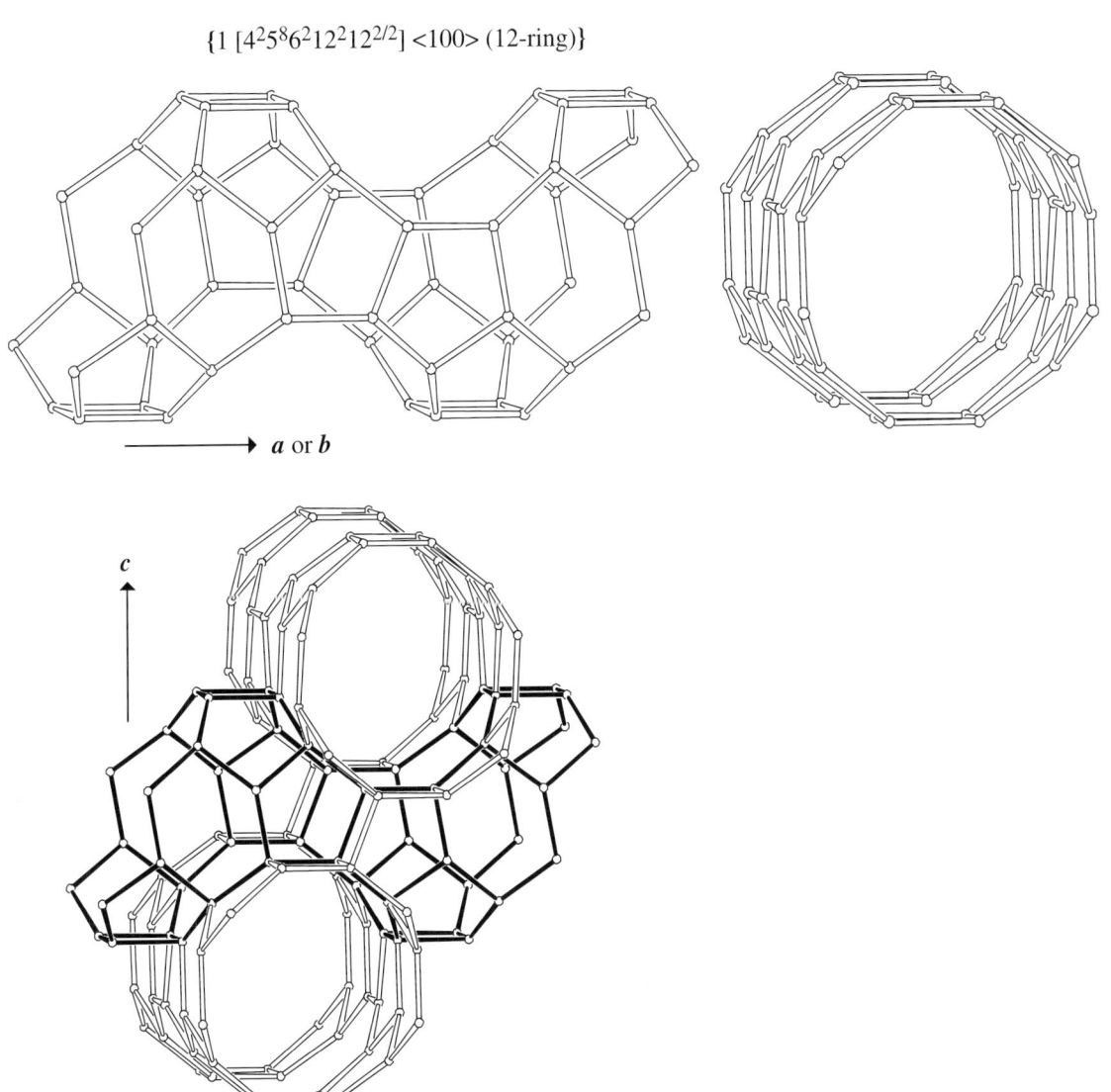

*a* or *b*

*c*

Fig. 3.   Top: 12-ring channel parallel to <100> viewed perpendicular to channel axis (left) and down the channel axis (right). Bottom: (part of) the sinusoidal 12-ring channel parallel to *c* viewed perpendicular to the tetragonal $4_1$-axis.

# BEC                    Building Scheme

## 1. Periodic Building Unit

The two-dimensional PerBU in tetragonal **BEC** equals the beta layer shown in Figure 1. The PerBU is composed of T16-units (four fused 6-rings or eight fused 5-rings (in bold)) related along the cell edges *a* and *b* by pure translations. (See also ***BEA**; compare with **ISV**.)

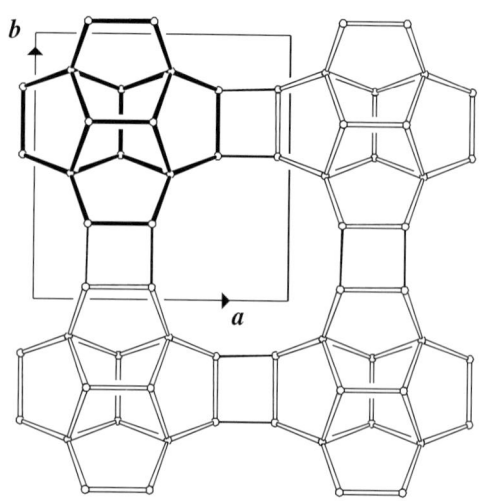
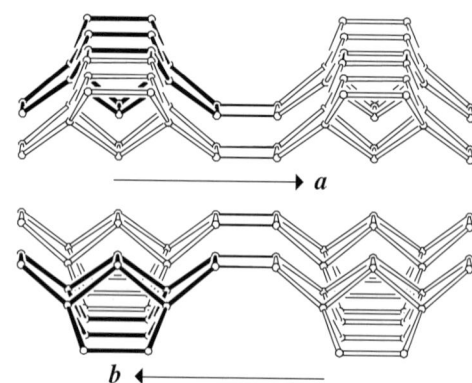

Fig. 1.   PerBU viewed along *c* (left), along *b* (top right) and down *a* (bottom right). The PerBUs at the right are related by a rotation of 90° about *c*.

## 2. Connection mode

Neighboring PerBUs, related by a rotation of 90° about *c* followed by a shift of $\frac{1}{2}c$ (i.e. related by a $4_2$-axis) and no lateral shifts along *a* and *b* (denoted as (0,0)), are connected through (fused) 4-6-6-ring sequences (Figure 2).

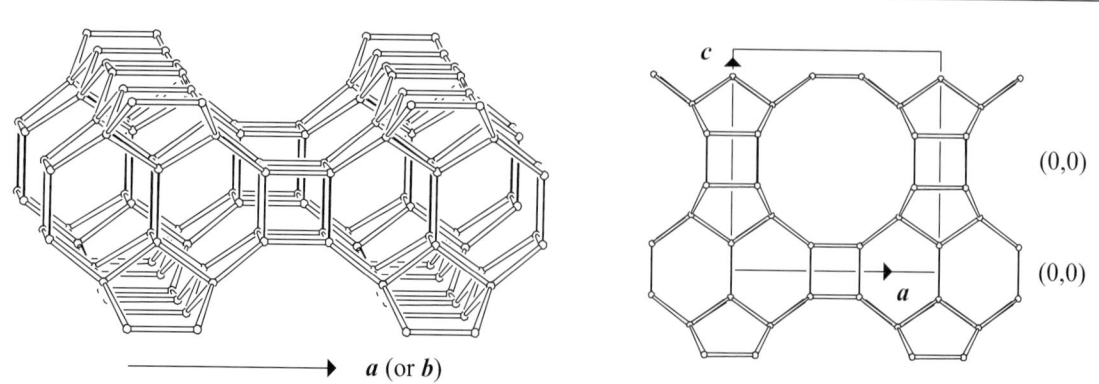

Fig. 2.   Connection mode viewed perpendicular to the tetragonal axis (left) and unit cell projected along *b* (or *a*; right). The lateral shifts are added.

## 3. Channels and/or cages

Interconnecting 12-ring channels are parallel to <100> and $c$. The channels parallel to <100>, depicted in Figure 3, are topologically equivalent to the 12-ring channels in **IWR**. The straight 12-ring channel parallel to $c$ is also illustrated in Figure 3.

$\{1 \ [4^4 5^4 6^4 12^2 12^{2/2}] <100> \ (12\text{-ring})\}$

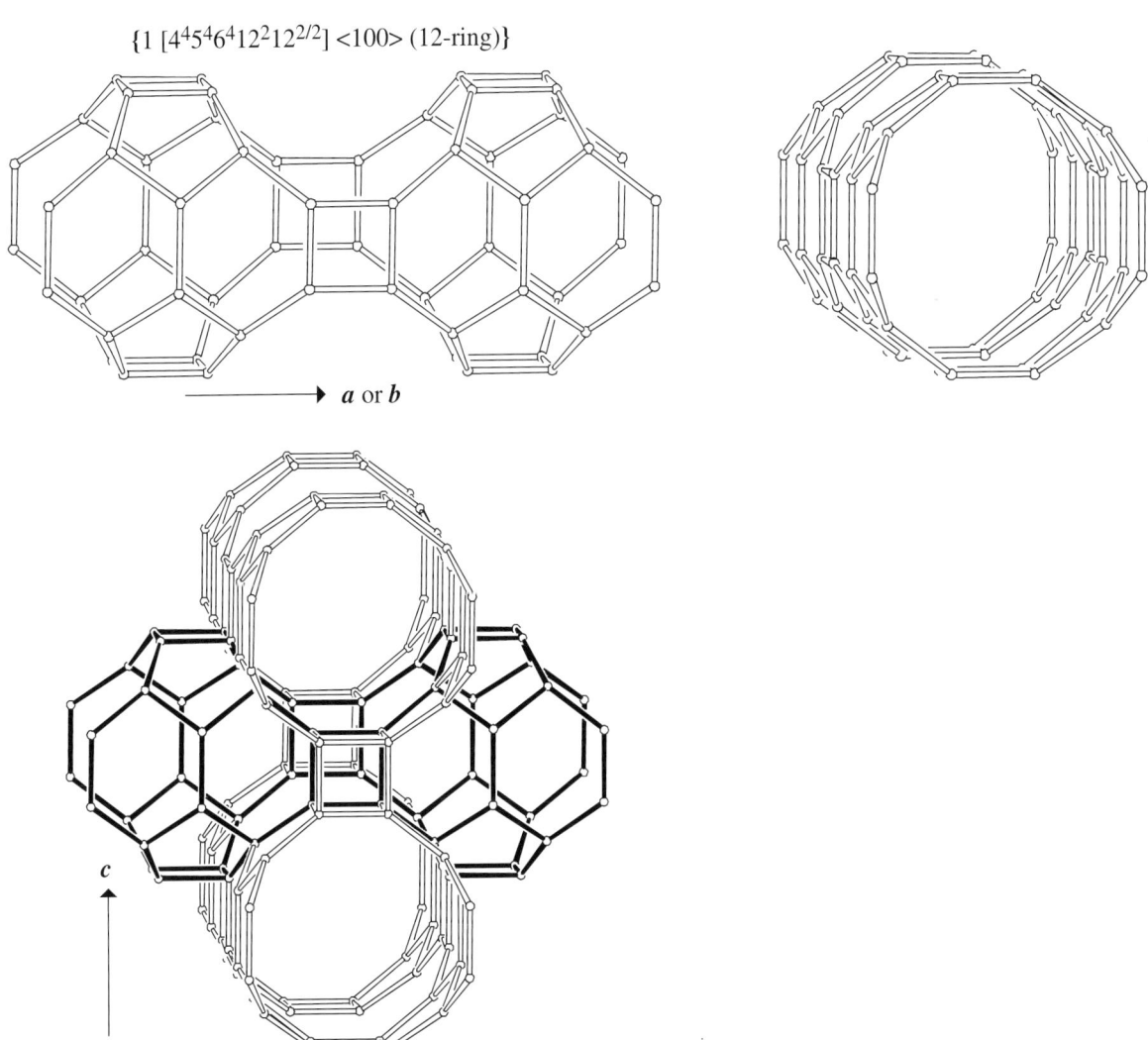

Fig. 3.  Top: 12-ring channel parallel to <100> viewed perpendicular to the channel axis (left) and down the channel axis (right). Bottom: straight 12-ring channel parallel to $c$ viewed perpendicular to the tetragonal $4_2$-axis.

# BIK                                        Building Scheme

## 1. Periodic Building Unit

**BIK** can be built using the zigzag chain (bold in Figure 1) running parallel to *c*. Three zigzag chains are connected into an infinite building unit. A two-dimensional PerBU is obtained when infinite building units, related along [110] by a shift of $\frac{1}{2}(\boldsymbol{a}+\boldsymbol{b})$, are connected along [110] into a layer of (fused) 6-ring chairs decorated with additional zigzag chains as shown in Figure 1 (compare with **CAS** and **NSI**).

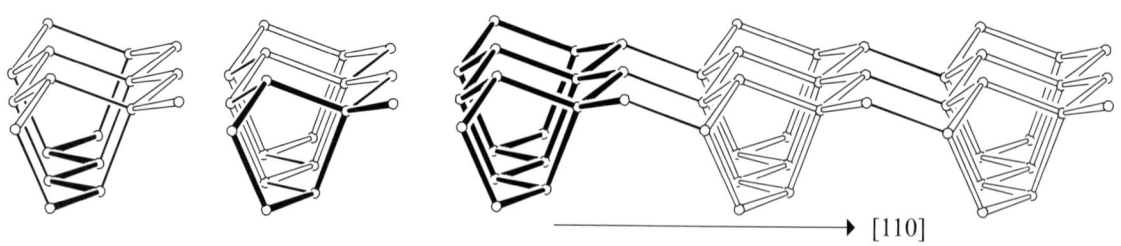

Fig. 1.    Infinite building unit constructed from three zigzag chains (left) and from 5–1 repeat units (middle) viewed along the chain axis *c*. Right: PerBU viewed along *c*.

## 2. Connection mode

Neighboring PerBUs, related along *a* by a pure translation, are connected along *a* through (fused) 5- and 6-rings as illustrated in Figure 2.

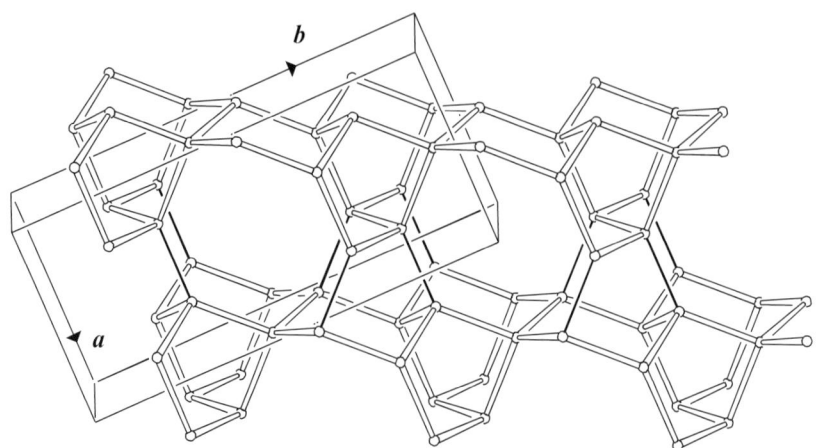

Fig. 2.    Connection mode and unit cell content viewed along *c*. For clarity, only two repeat units of each PerBU are drawn.

# Cage/Channel

### 3. Channels and/or cages

Non-interconnecting 8-ring channels are parallel to *c*. The 8-ring channel, depicted in Figure 3, is topologically equivalent to the 8-ring channel in **NSI**.

$$\{1 \ [6^6 8^{2/2}] \ [001] \ (8\text{-ring})\}$$

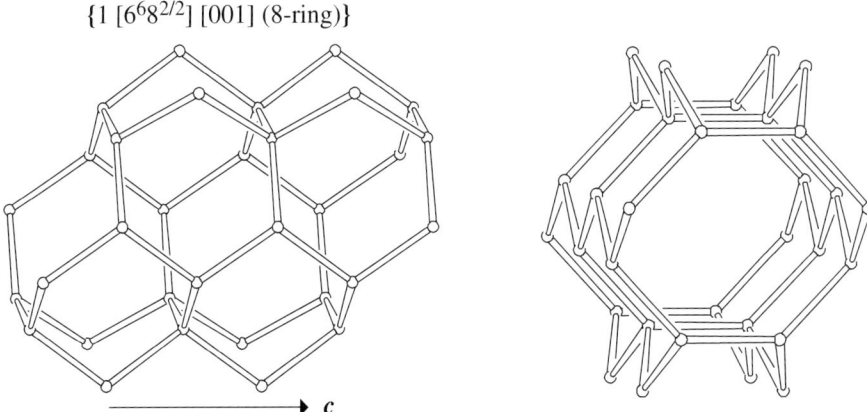

Fig. 3.   Channel viewed perpendicular to the channel axis (left) and along the channel axis (right).

# BOG

**Building Scheme**

## 1. Periodic Building Unit

**BOG** can be built using the T12-unit (a $[5^4 6^2]$-cage; see inset) consisting of two 4-fold connected 6-ring chairs. Neighboring T12-units, related by a screw rotation of 180° about $c$, and a shift along $\pm b$ are connected into left- and right-handed chains along $c$, respectively. A two-dimensional PerBU is obtained when left- and right-handed chains are connected along $b$ through double zigzag chains (Figure 1).

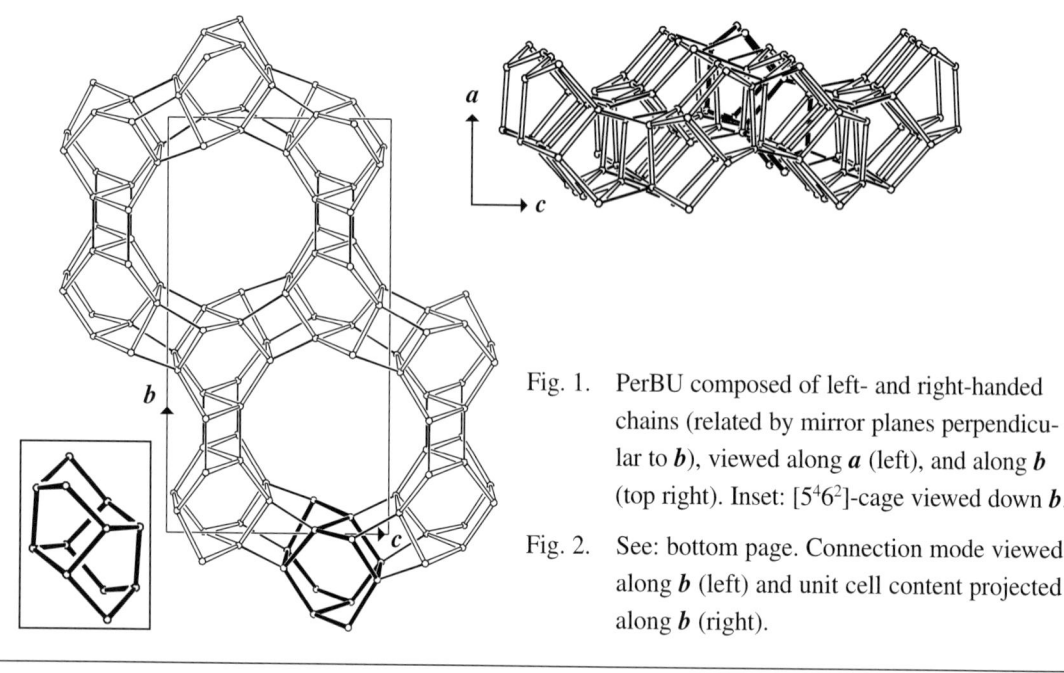

Fig. 1.  PerBU composed of left- and right-handed chains (related by mirror planes perpendicular to $b$), viewed along $a$ (left), and along $b$ (top right). Inset: $[5^4 6^2]$-cage viewed down $b$.

Fig. 2.  See: bottom page. Connection mode viewed along $b$ (left) and unit cell content projected along $b$ (right).

## 2. Connection mode

Neighboring PerBUs, related along $a$ by a shift of $\frac{1}{2}(a+b+c)$, are connected as shown in Figure 2.

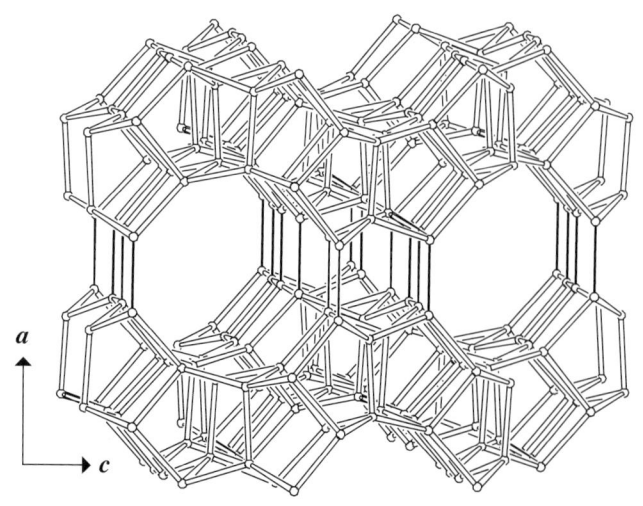

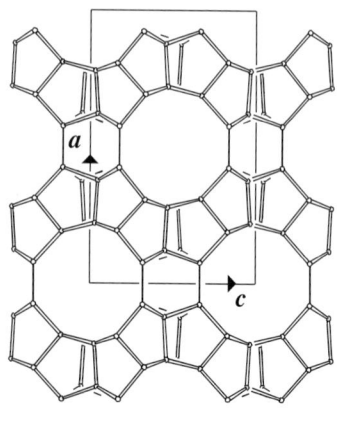

Fig. 2.  Legend: see above.

## 3. Channels and/or cages

Interconnecting 12- and 10-ring channels are parallel to **a** and **b**, respectively. The channels are shown in Figure 3.

{1 [$4^{16}5^{16}6^410^412^{2/2}$] [100] (12-ring)}

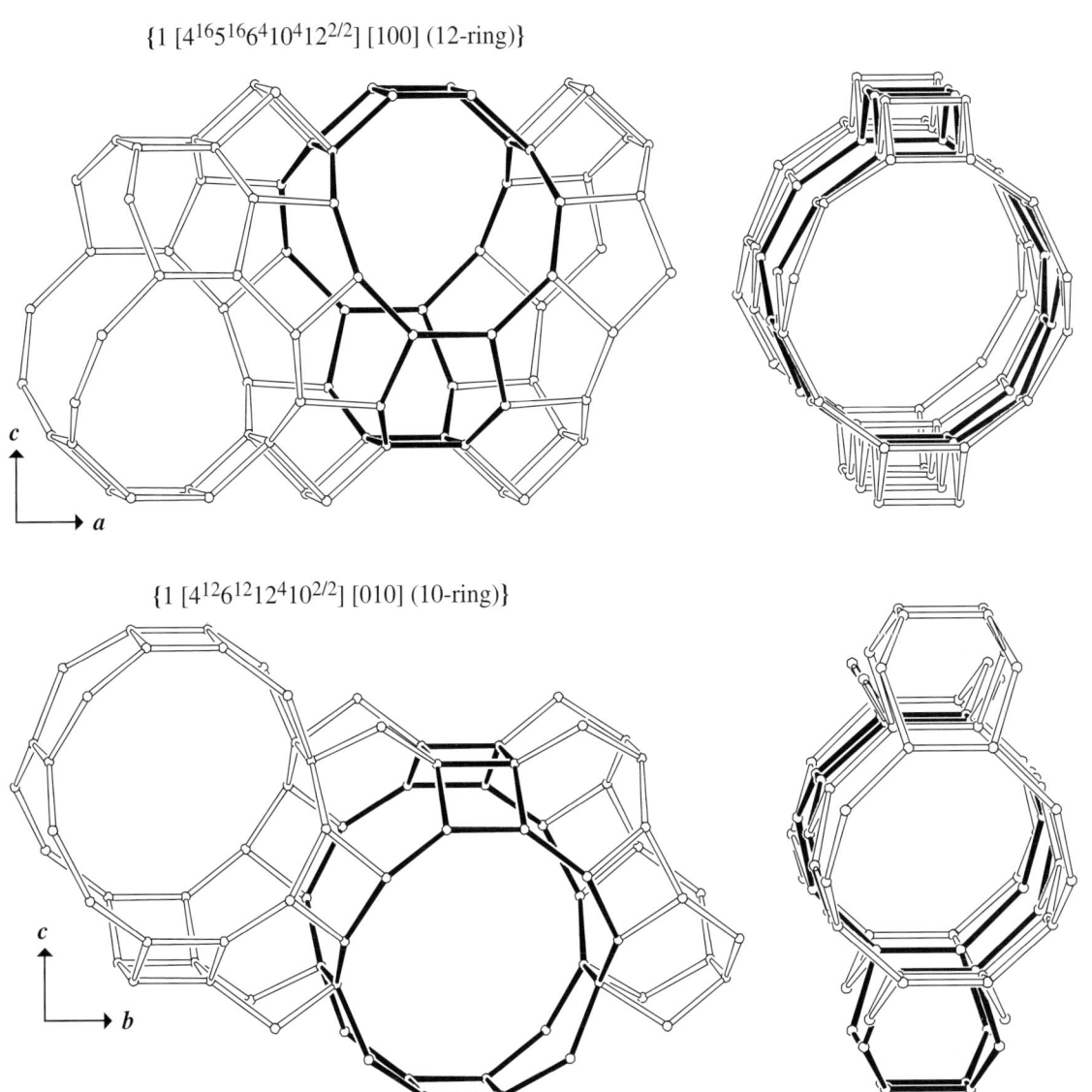

{1 [$4^{12}6^{12}12^410^{2/2}$] [010] (10-ring)}

Fig. 3.   10-Ring channel (top) and 12-ring channel (bottom) viewed perpendicular to the channel axis (left) and along the channel axis (right). The channels have [$4^46^210^212^4$]-cavities (one in bold) in common.

# BPH                    Building Scheme

## 1. Periodic Building Unit

Hexagonal **BPH** can be built using units of 14 T atoms. The T14-unit consists of a 3-fold (1,3,5)-connected double 6-ring "capped" on each side by a single T atom (a $[4^6 6^3]$-cage; see inset Figure 1). The two-dimensional PerBU equals the layer obtained by connecting T14-units related by a 3-fold axis and a 2-fold axis (parallel to [110]) through 4-rings as shown in Figure 1. (See also **AFS**.)

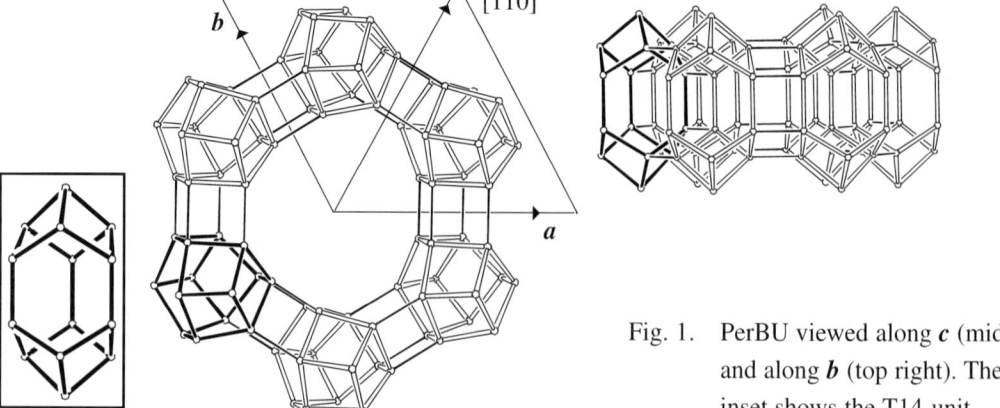

Fig. 1.   PerBU viewed along *c* (middle) and along *b* (top right). The inset shows the T14-unit.

## 2. Connection mode

Neighboring PerBUs, related along *c* by a pure translation, are connected along *c* through 8-rings as depicted in Figure 2.

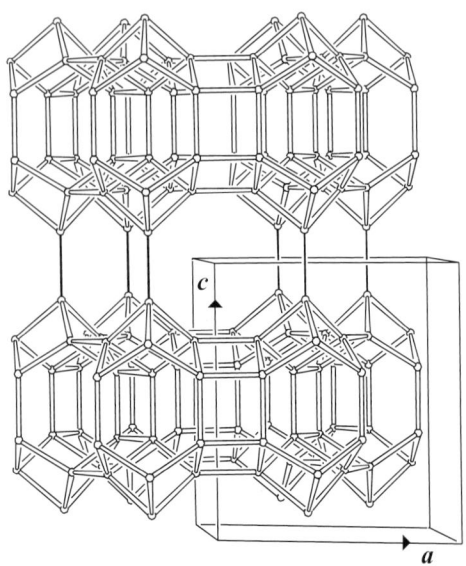

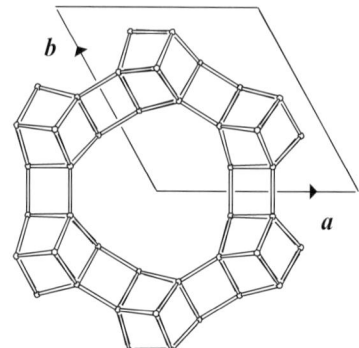

Fig. 2.   Connection mode (and unit cell content) viewed along *b* (left) and unit cell content projected along *c* (right).

## 3. Channels and/or cages

Intersecting 8- and 12-ring channels are parallel to <100> and **c**, respectively (Figure 3). The channel intersection is topologically equivalent to the channel intersection in **AFS** and **AFY**.

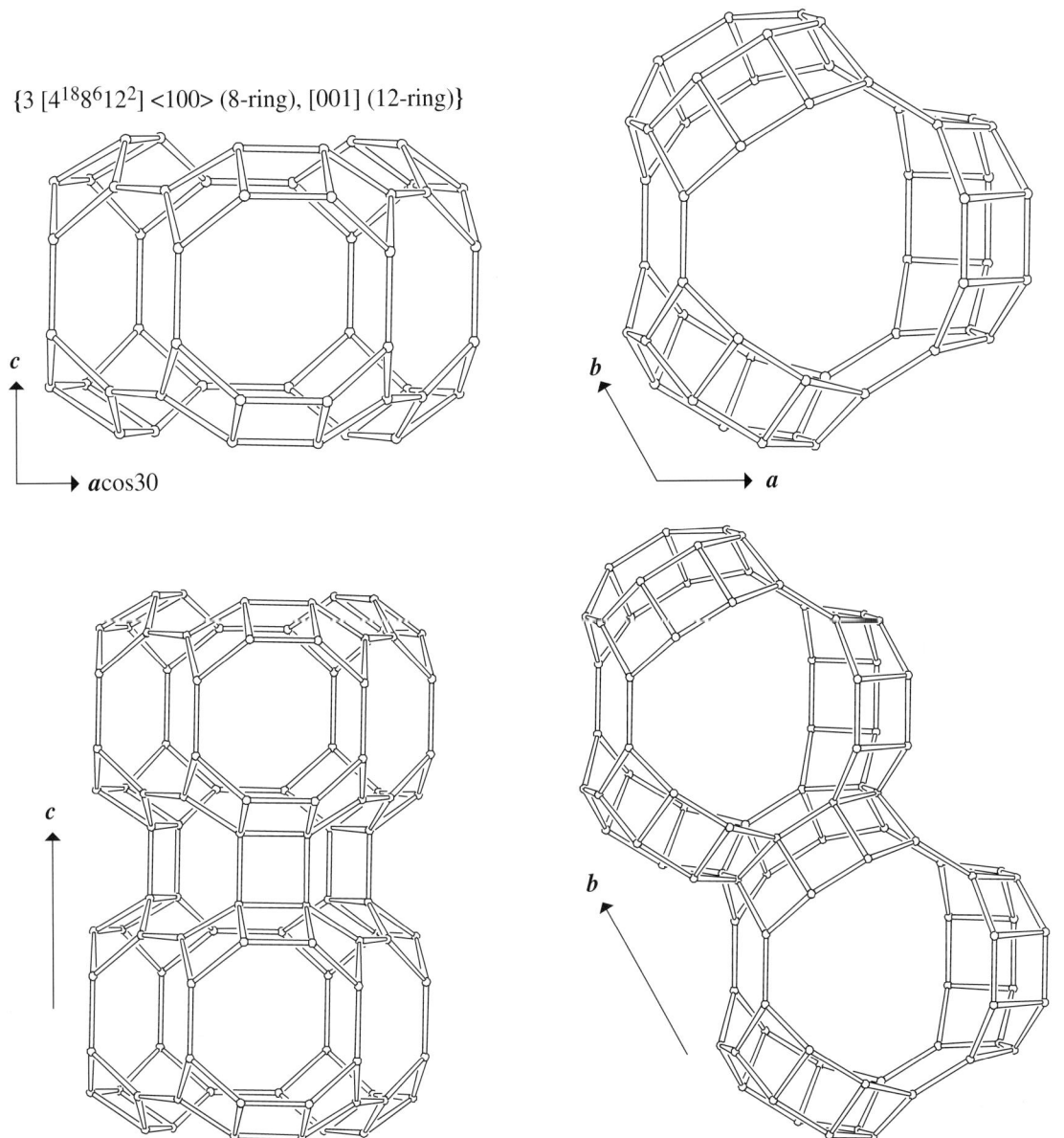

{3 [$4^{18}8^612^2$] <100> (8-ring), [001] (12-ring)}

Fig. 3. Top: channel intersection viewed along <100> (left) and along **c** (right). Bottom: interconnection of channel intersections into a 12-ring channel parallel to **c** viewed along <010> (left) and into an 8-ring channel parallel to **b** (or **a**) viewed along **c** (right).

# BRE           Building Scheme

## 1. Periodic Building Unit

The PerBU in **BRE** is the layer depicted in Figure 1. Twofold (1,3)-connected double 4-rings (one in bold), related along *a* and *c* by pure translations, are connected through (fused) 5- and 4-rings, respectively.

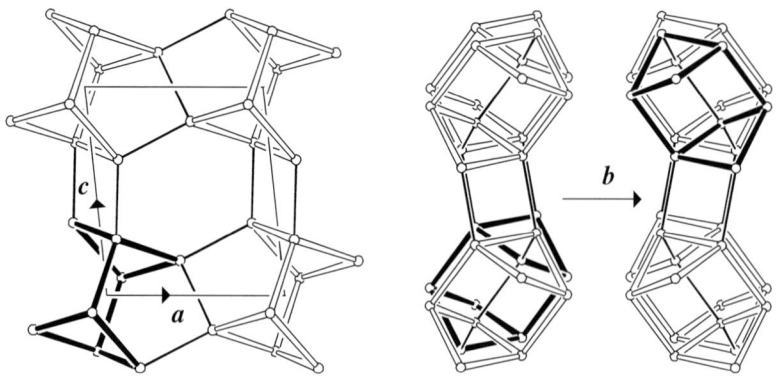

Fig. 1.    PerBU viewed along *b* (left) and along *a* (right). The PerBUs at the right are related by a rotation of 180° about *b*.

## 2. Connection mode

Neighboring PerBUs, related by a screw rotation of 180° about *b*, are connected along *b* through single T–T bonds as shown in Figure 2. 8-Rings perpendicular to *a* and *b* are formed.

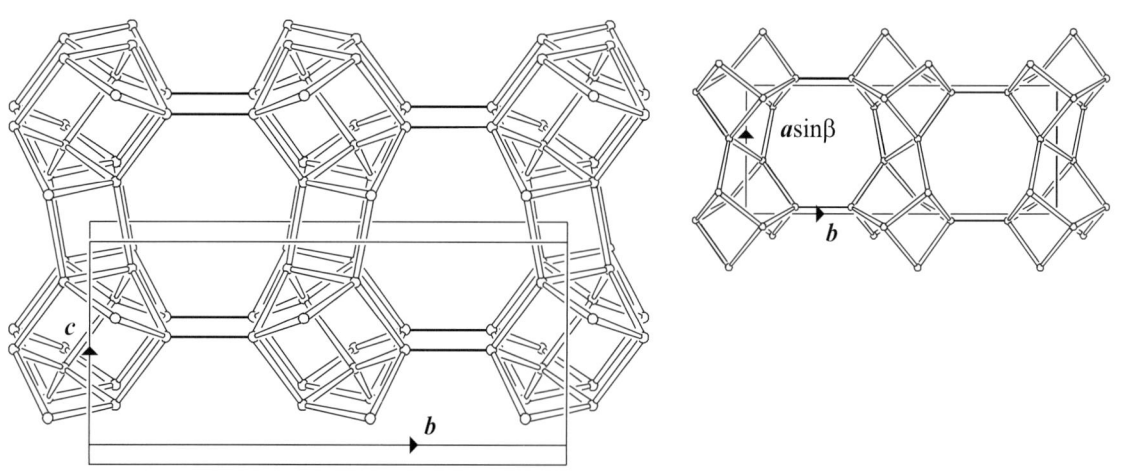

Fig. 2.    Connection mode (and unit cell content) viewed along *a* (left) and unit cell content projected along *c* (right).

## 3. Channels and/or cages

Intersecting 8-ring channels are parallel to *a* and *c*. The channel intersection is shown in Figure 3 together with the linkage of intersections.

{2 [$4^4 5^4 6^2 8^4$] [100] (8-ring), [001] (8-ring)}

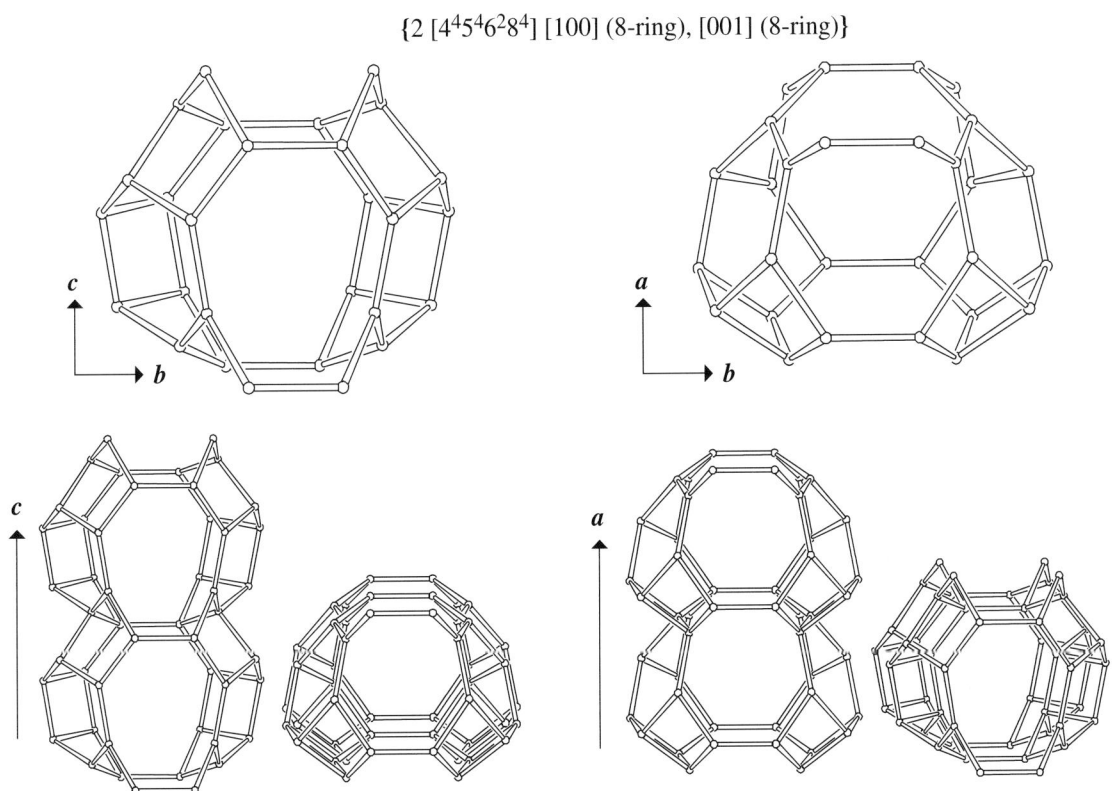

Fig. 3. Top: intersection of channels viewed along *a* (left) and along *c* (right). Bottom: interconnection of channel intersections into 8-ring channels along *c* (left) and along *a* (right) viewed perpendicular to the channel axis and along the channel axis.

# CAN                  Building Scheme

## 1. Periodic Building Unit

The two-dimensional PerBU of **CAN** consists of a hexagonal array of non-connected planar 6-rings (bold in Figure 1), which are related along **a** and **b** by pure translations. The 6-rings are centered at (0,0) in the **ab** layer. This position is usually called the **A** position. **CAN** belongs to the ABC-6 family (see Introduction).

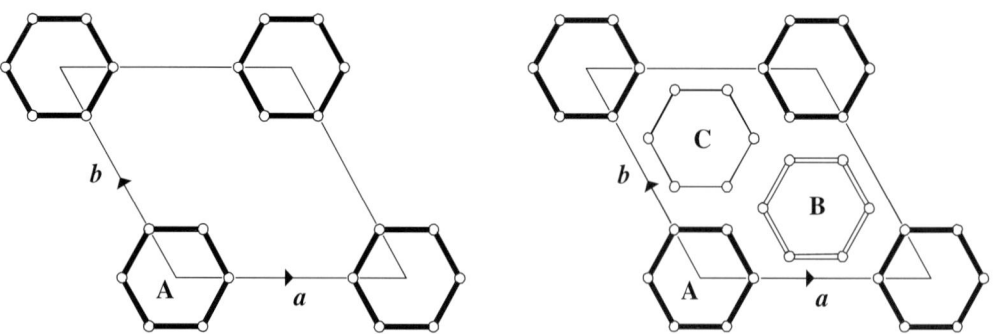

Fig. 1.    PerBU (left) and definition of 6-ring positions used in the stacking modes (right).

## 2. Connection mode

Neighboring PerBUs are connected along **c** through tilted 4-rings by connection modes **(1)** and **(2)** (see Introduction) as illustrated in Figure 2. 12-Ring channels are formed. (See Figure 2 and **ATO**.)

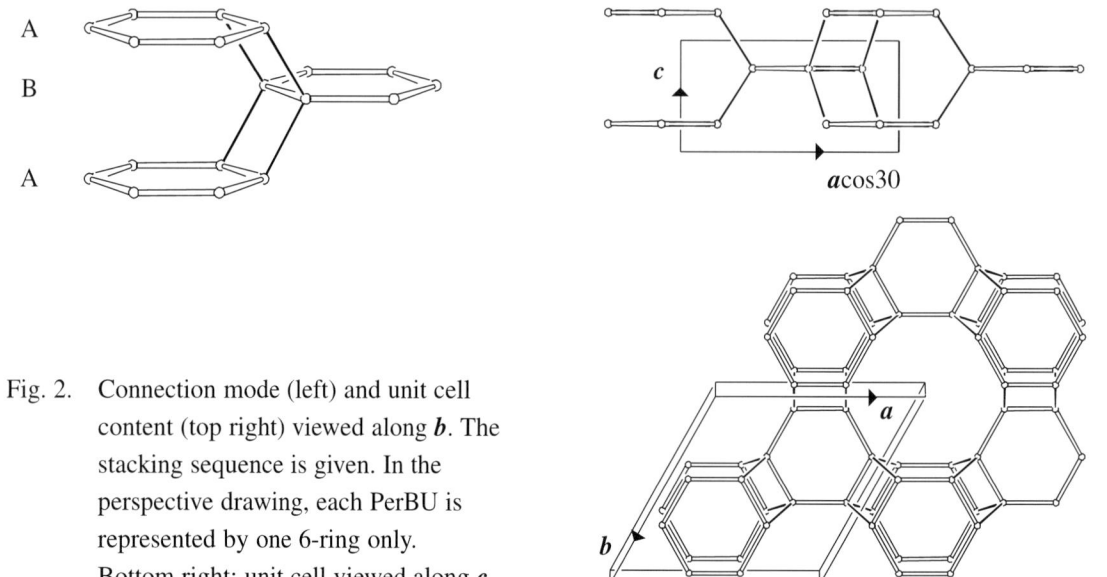

Fig. 2.    Connection mode (left) and unit cell
content (top right) viewed along **b**. The
stacking sequence is given. In the
perspective drawing, each PerBU is
represented by one 6-ring only.
Bottom right: unit cell viewed along **c**.

## 3. Channels and/or cages

Non-interconnecting 12-ring channels are parallel to *c*. The channel is topologically equivalent to the 12-ring channels in **ATO** and **NPO**. **CAN** can as well be constructed using the 12-ring channels as building unit. A column of (fused) *can* cages is formed when channels are connected around a 3-fold axis. 12-Ring channel and *can* cage(s) are depicted in Figure 3. The *can* cage is also present in **AFG**, **ERI**, **FAR**, **FRA**, **GIU**, **LIO**, **LOS**, **LTL**, **LTN**, **MAR**, **MOZ**, **OFF**, **SAT**, **SBS**, **SBT**, **TOL** and **-WEN**.

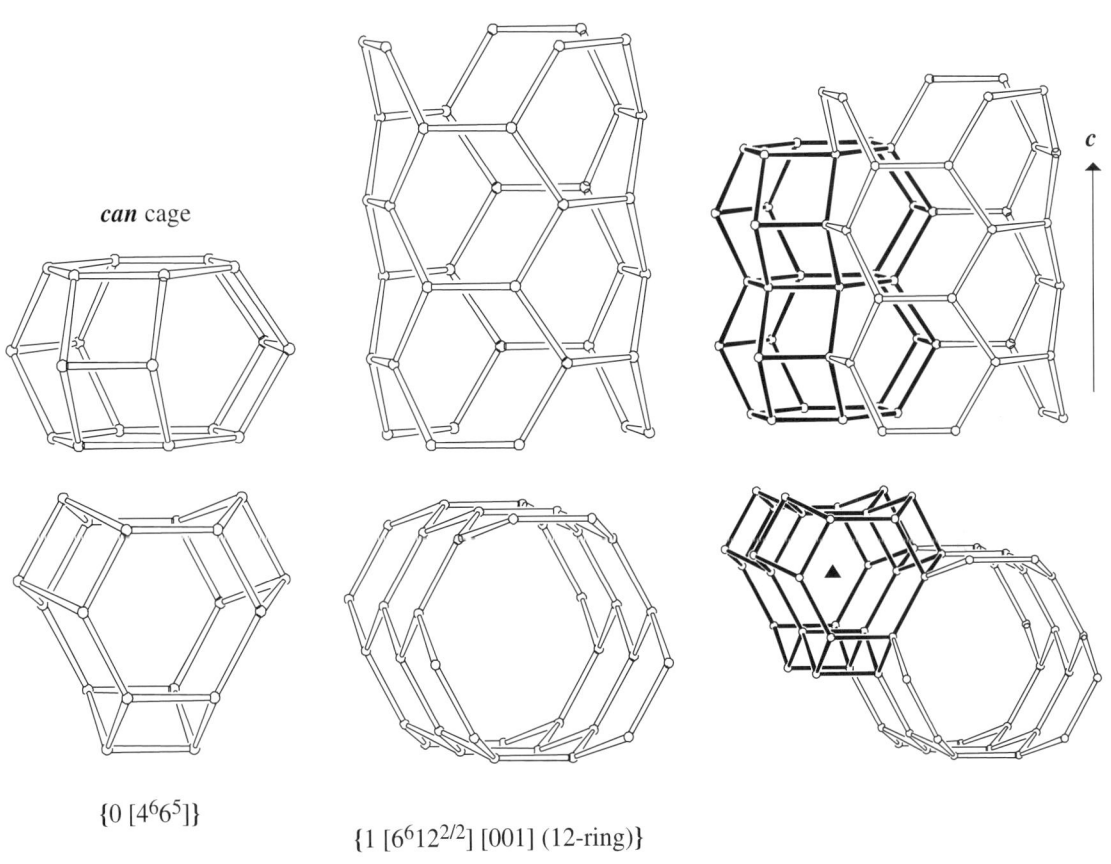

*can* cage

*c*

{0 [4⁶6⁵]}

$\{0\ [4^6 6^5]\}$

$\{1\ [6^6 12^{2/2}]\ [001]\ (12\text{-ring})\}$

Fig. 3.   *can* Cage (left), 12-ring channel (middle) and fused *can* cages and channel (right) viewed perpendicular to *c* (top) and along the 3-fold axis parallel to *c* (bottom).

# CAS

## Building Scheme

### 1. Periodic Building Unit

**CAS** can be built using the zigzag chain (bold in Figure 1) running parallel to *a*. Three zigzag chains are connected into an infinite building unit. A two-dimensional PerBU is obtained when infinite building units, related by a 2-fold axis parallel to the plane of the paper (pointing from top to bottom), are connected along *c* into a layer of (fused) 6-ring chairs and 6-ring boats decorated with additional zigzag chains as shown in Figure 1. (Compare with **BIK** and **NSI**.)

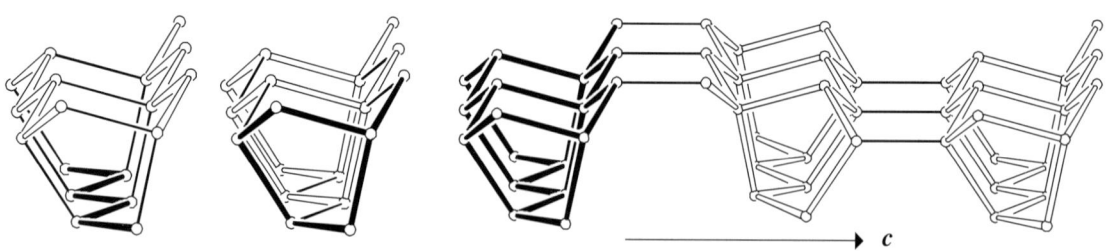

Fig. 1.   Infinite building unit constructed from three zigzag chains (left) and from 5-1 units (middle) viewed along the chain axis *a*. Right: PerBU viewed along *a*.

### 2. Connection mode

Neighboring PerBUs, related along *b* by a shift of $\frac{1}{2}(a+b)$, are connected along *b* through 5-rings as can be seen from Figure 2.

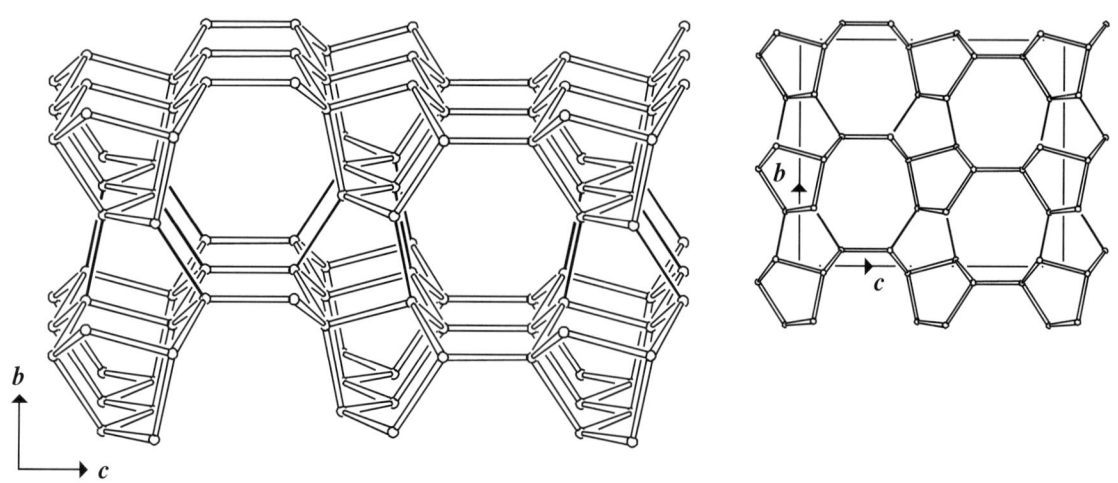

Fig. 2.   Connection mode (left) and unit cell content (right) viewed along *a*.

## 3. Channels and/or cages

Non-interconnecting 8-ring channels (Figure 3) are parallel to *a*.

{1 [$6^6 8^{2/2}$] [100] (8-ring)}

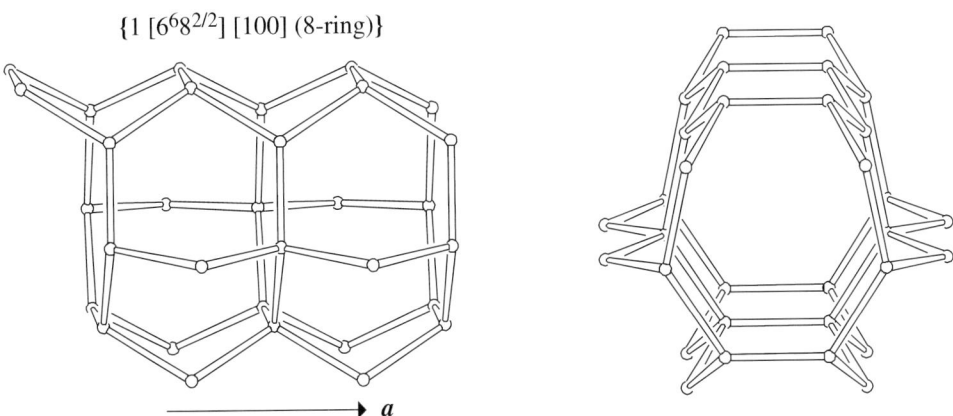

Fig. 3.    Channel viewed perpendicular to the channel axis (left) and along the channel axis (right).

# CDO

## Building Scheme

### 1. Periodic Building Unit

**CDO** can be built using the saw chain (bold in Figure 1) parallel to *a*. Three saw chains are connected into an infinite building unit. A two-dimensional PerBU is obtained when infinite building units, related along *c* by a rotation of 180° about *c* and a shift of $\frac{1}{2}$*a*, are connected along *c* into the *ac* layer depicted in Figure 1 (right). (See also **FER**.)

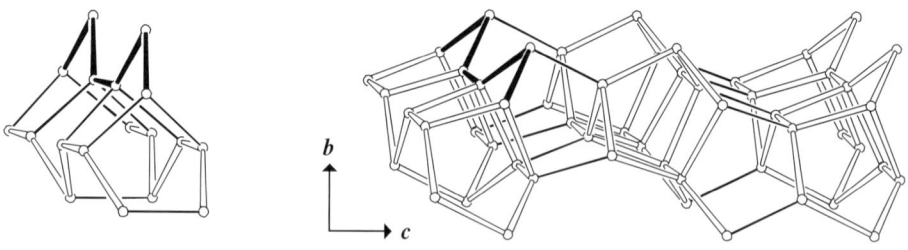

Fig. 1. Infinite building unit (left) and PerBU (right) viewed along *a*.

### 2. Connection mode

Neighboring PerBUs, related along *b* by a shift of $\frac{1}{2}(a+b)$, are connected along *b* through 8-rings (Figure 2).

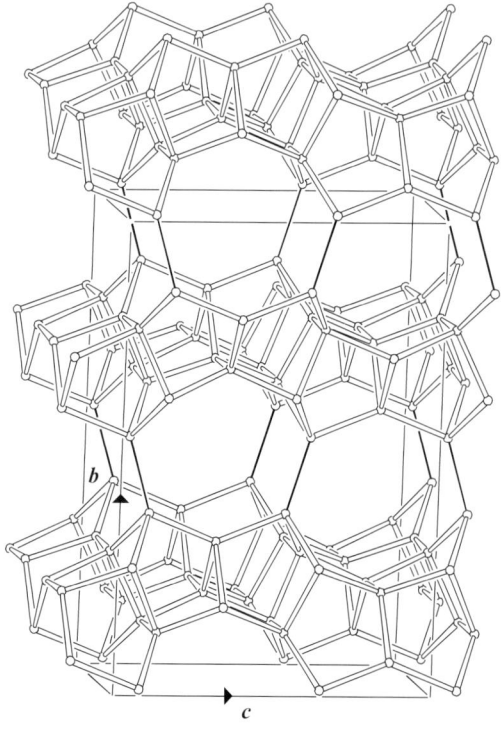

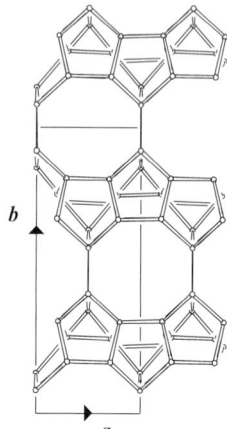

Fig. 2. Connection mode (and unit cell content) viewed along *a* (left) and unit cell content projected along *c* (right).

### 3. Channels and/or cages

Intersecting 8-ring channels are parallel to *a* and *c*. The channels are shown in Figure 3.

{1 [$5^8 6^6 8^4 8^{2/2}$] [001] (8-ring)}

{1 [$5^6 6^4 8^2 8^{2/2}$] [100] (8-ring)}

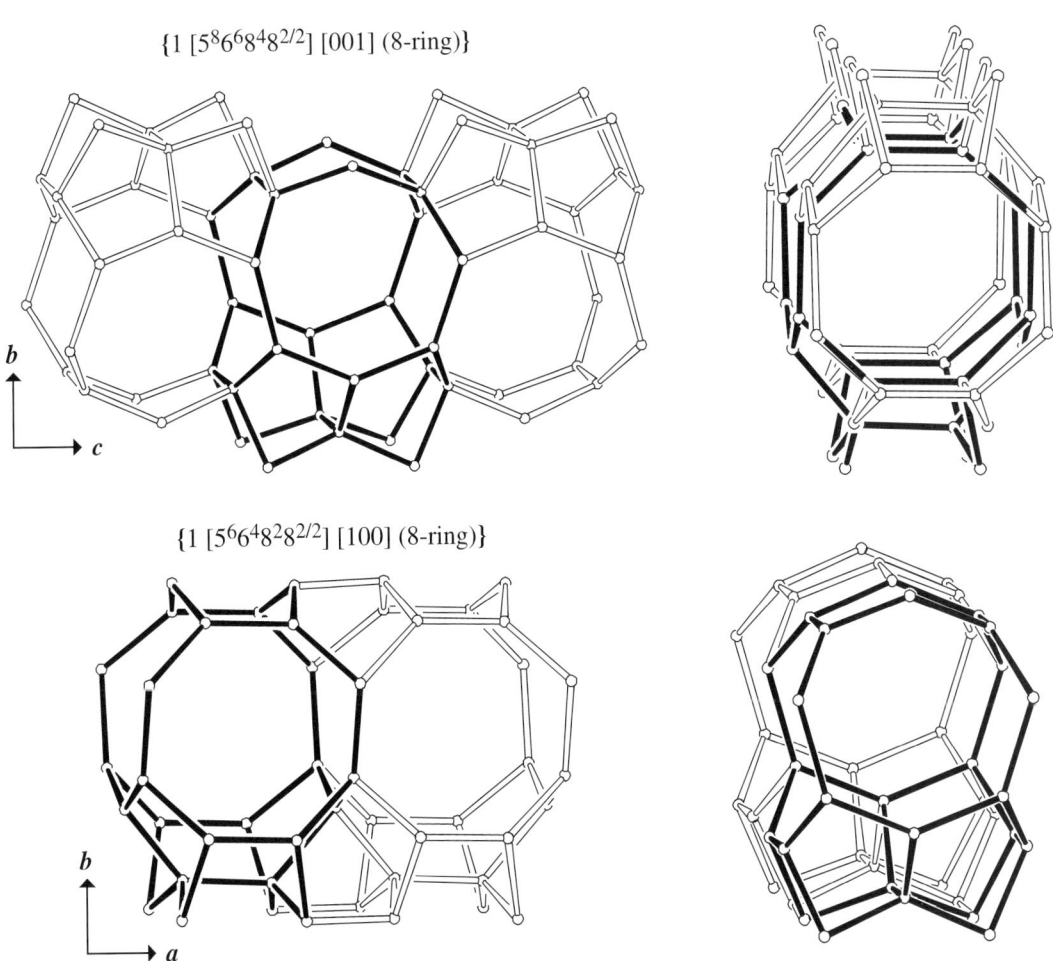

Fig. 3.　8-Ring channels viewed perpendicular to the channel axis (left) and along the channel axis (right). The channels have [$5^4 6^3 8^4$]-cavities in common. One intersection in bold.

# CFI                    Building Scheme

## 1. Periodic Building Unit

**CFI** can be built using the zigzag chain (bold in Figure 1, left) running parallel to **b**. The one-dimensional PerBU is obtained when eight zigzag chains are connected into a channel with a 14-ring aperture. The channel wall consists of fused 6-rings.

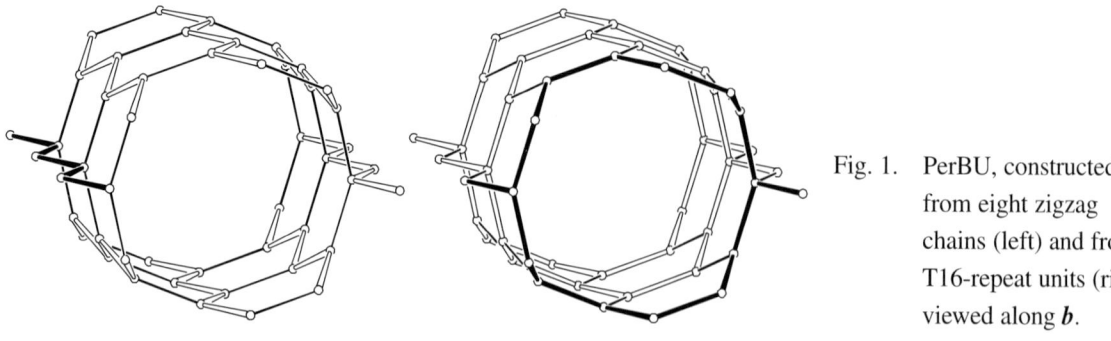

Fig. 1.   PerBU, constructed from eight zigzag chains (left) and from T16-repeat units (right), viewed along **b**.

## 2. Connection mode

Neighboring PerBUs, related along **a** by a pure translation and along **c** by a shift of $\frac{1}{2}(a+b+c)$, are connected along **a** through double zigzag chains and along **c** through (fused) 5- and 6-rings as illustrated in Figure 2.

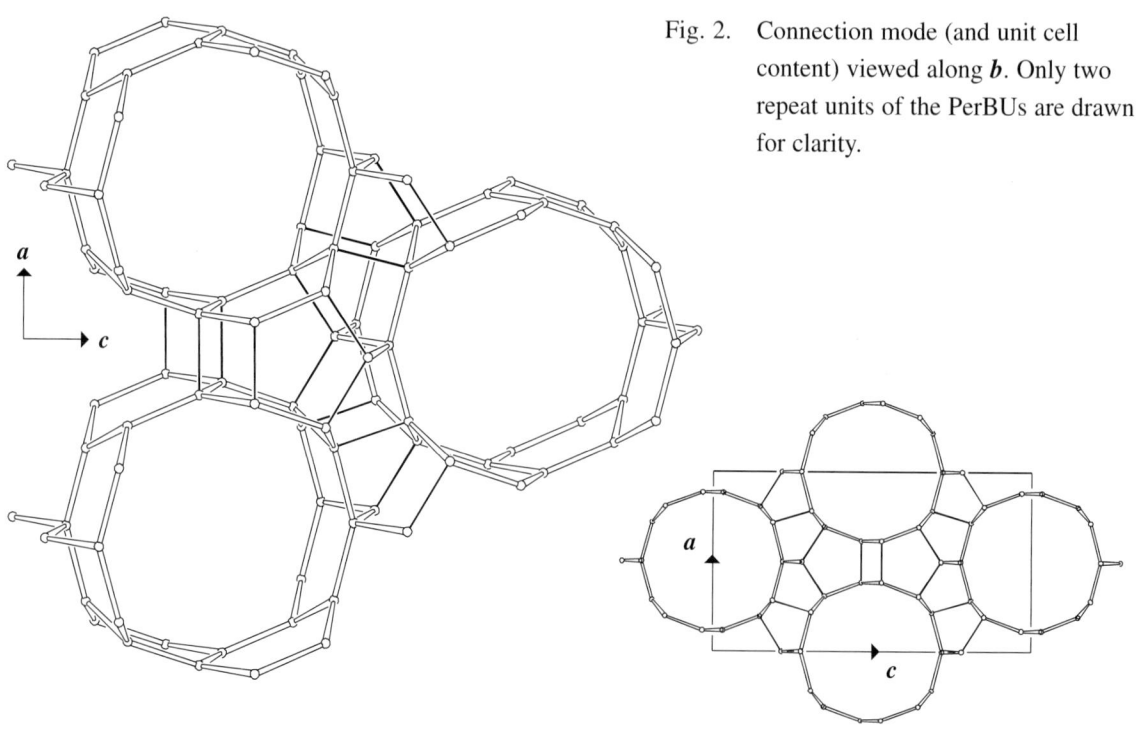

Fig. 2.   Connection mode (and unit cell content) viewed along **b**. Only two repeat units of the PerBUs are drawn for clarity.

## 3. Channels and/or cages

Non-interconnecting 14-ring channels are parallel to **b**. The channel wall consists of fused 6-rings as depicted in Figure 3.

{1 [$6^8 14^{2/2}$] [010] (14-ring)}

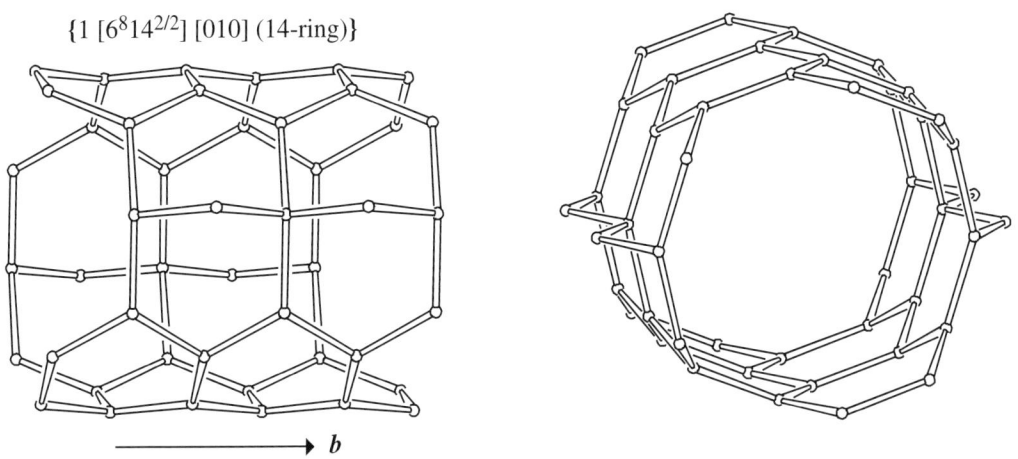

Fig. 3.    Channel viewed perpendicular to the channel axis (left) and along the channel axis (right).

# CGF

## Building Scheme

### 1. Periodic Building Unit

The two-dimensional PerBU in **CGF** equals the layer shown in Figure 1. The PerBU is composed of T9-units (in bold): a 2-fold (1,3)-connected double 4-ring and a single T atom. The T9-units are related along **c** by pure translations and along **a** by a rotation of 180° about **b**.

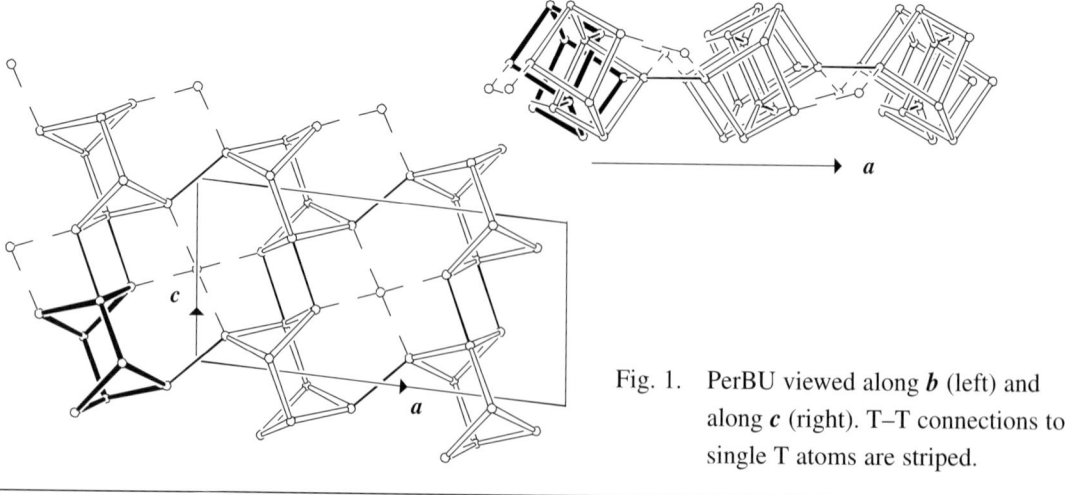

Fig. 1. PerBU viewed along **b** (left) and along **c** (right). T–T connections to single T atoms are striped.

### 2. Connection mode

Neighboring PerBUs, related along **b** by a shift of $\frac{1}{2}(a+b)$, are connected along **b** through single T–T bonds as illustrated in Figure 2.

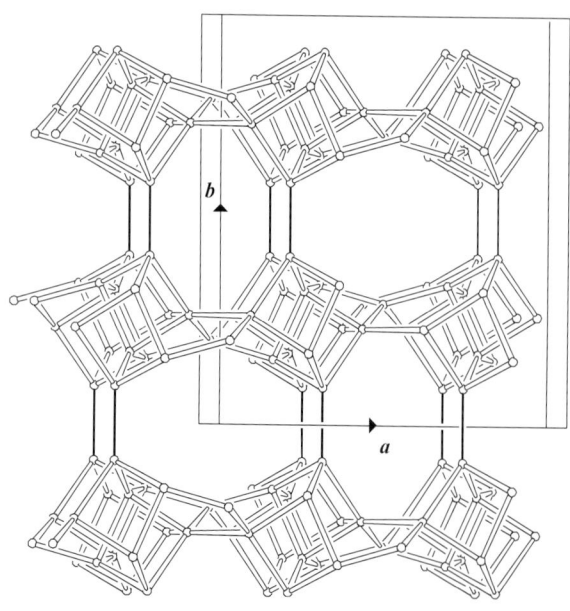

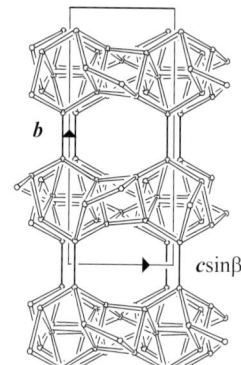

Fig. 2. Connection mode (and unit cell content) viewed along **c** (left) and unit cell content projected along **a** (right).

### 3. Channels and/or cages

Interconnecting 10- and 8-ring channels parallel to *c* form 8-ring channels parallel to *a*. The channels are shown in Figure 3.

$\{1 \, [4^4 6^4 8^2 10^{2/2}] \, [001] \, (10\text{-ring})\}$

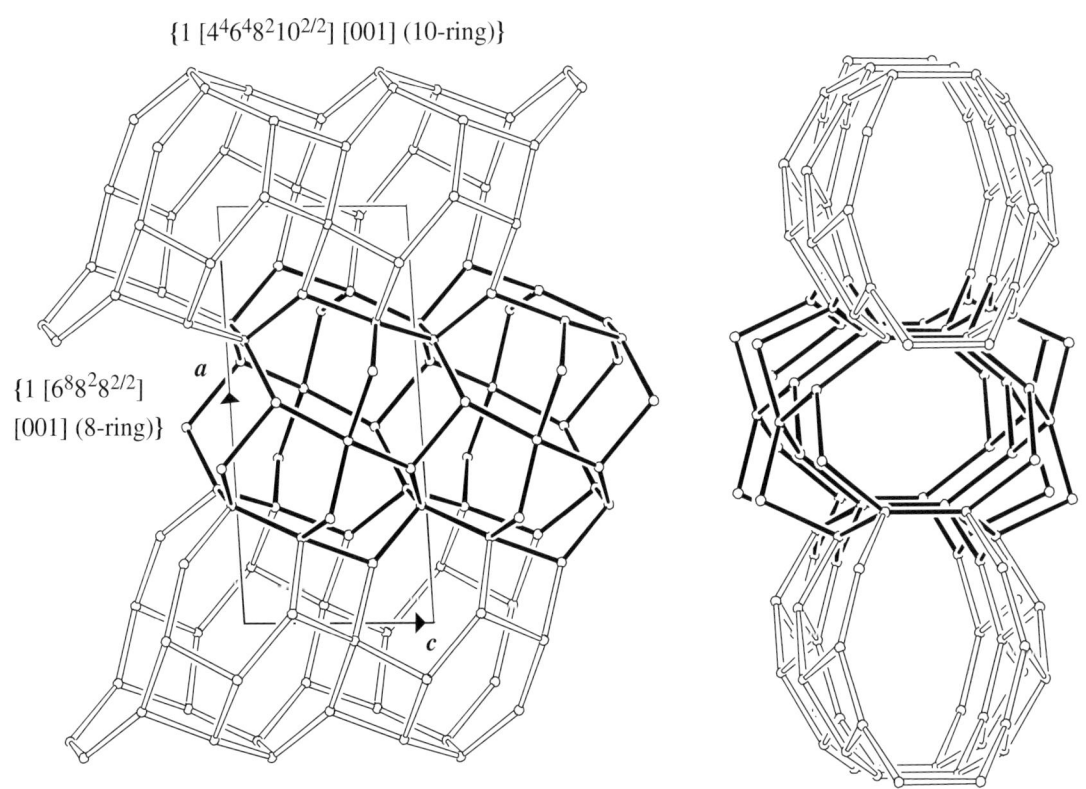

$\{1 \, [6^8 8^2 8^{2/2}] \, [001] \, (8\text{-ring})\}$

Fig. 3.   10-Ring channel(s) and 8-ring channel (bold) viewed perpendicular to the channel axis (left) and along the channel axis (right). Another type of 8-ring channel is parallel to *a*.

# CGS

# Building Scheme

## 1. Periodic Building Unit

**CGS** can be built using the chain depicted in Figure 1 as one-dimensional PerBU. The chain is composed of units of 16 T atoms (in bold) consisting of two 2-fold (1,2)-connected double 4-rings. The T16-units, related by pure translations along **a**, are connected along **a** through 4-rings. (See also **ETR**.)

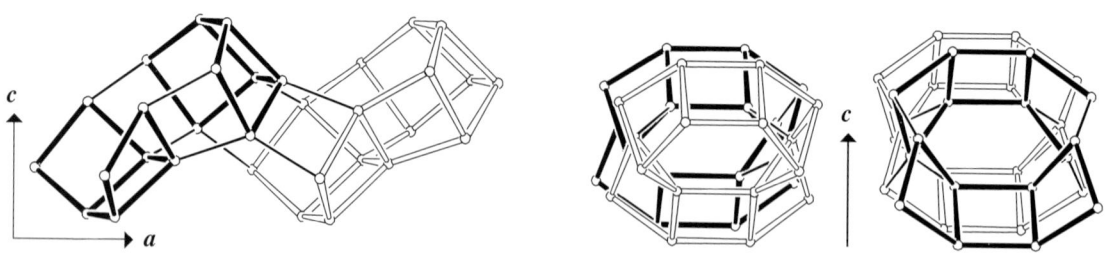

Fig. 1.  PerBU viewed along **b** (left) and along **a** (right). The PerBUs at the right are related by a rotation of 180° about **c**.

## 2. Connection mode

Neighboring PerBUs, related along **c** by a screw rotation of 180° about **c** followed by a shift of $\frac{1}{2}$**b**, are connected along **c** through 4-rings as illustrated in Figure 2.

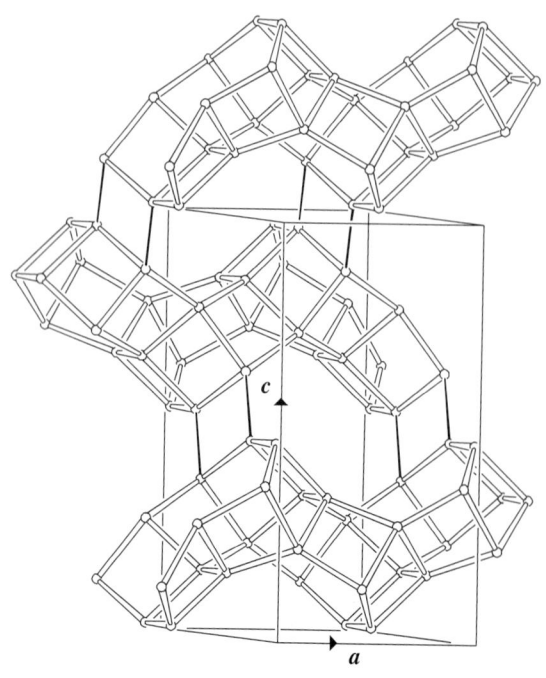

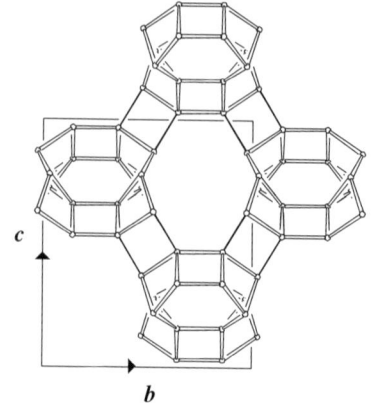

Fig. 2.  Connection mode (and unit cell content) viewed along **b** (left) and unit cell content projected along **a** (right).

## 3. Channels and/or cages

Interconnecting 10-ring channels are parallel to *a*. 8-Ring channels are parallel to *b*. The 10- and 8-ring channels intersect. The channel intersection is shown in Figure 3 together with the linkage of channel intersections into channels.

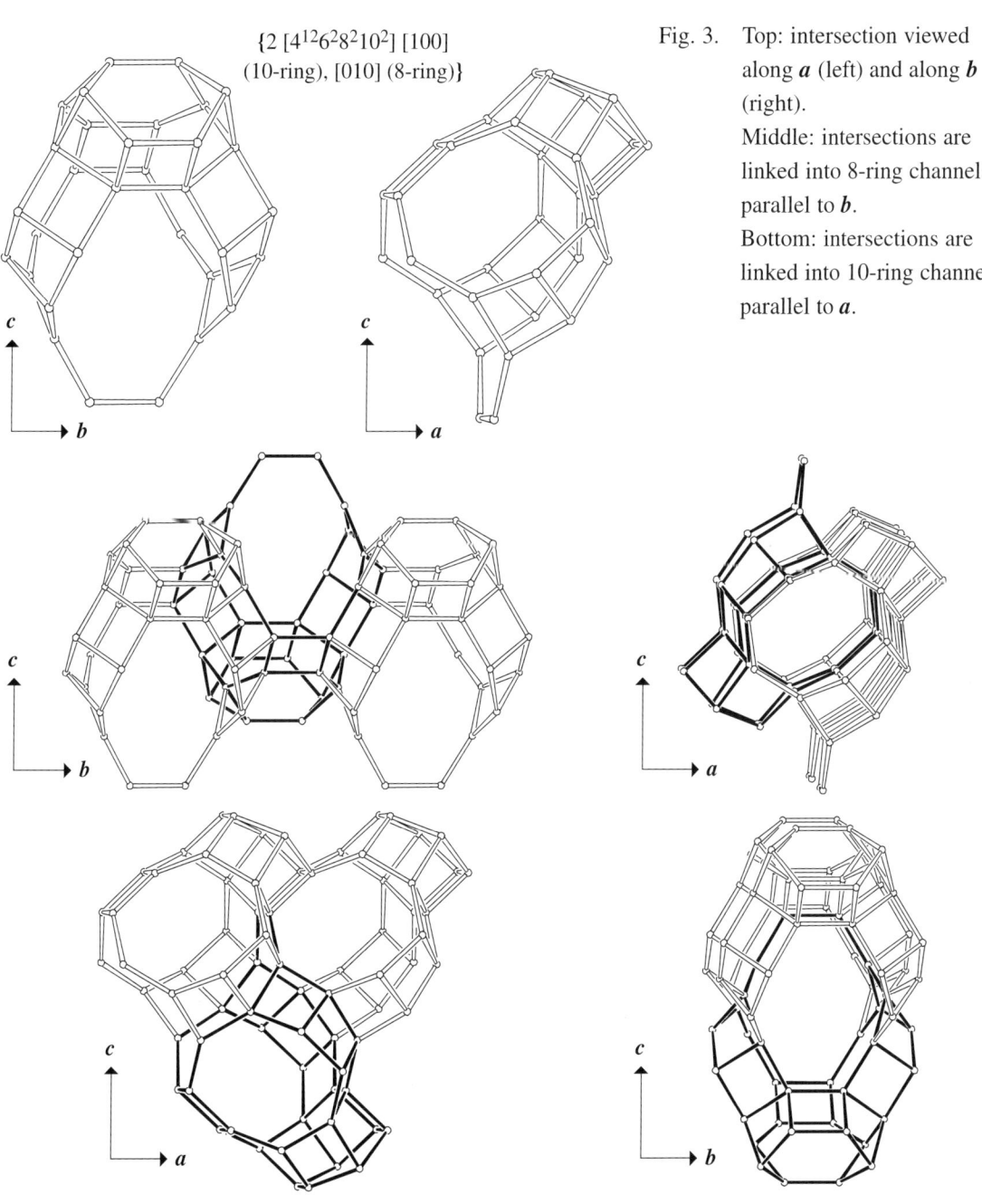

{2 [$4^{12}6^28^210^2$] [100] (10-ring), [010] (8-ring)}

Fig. 3. Top: intersection viewed along *a* (left) and along *b* (right).
Middle: intersections are linked into 8-ring channels parallel to *b*.
Bottom: intersections are linked into 10-ring channels parallel to *a*.

# CHA                    Building Scheme

## 1. Periodic Building Unit

The two-dimensional PerBU of **CHA** consists of a hexagonal array of non-connected planar 6-rings (bold in Figure 1), which are related along *a* and *b* by pure translations. The 6-rings are centered at (0,0) in the *ab* layer. This position is usually called the **A** position. **CHA** belongs to the ABC-6 family (see Introduction).

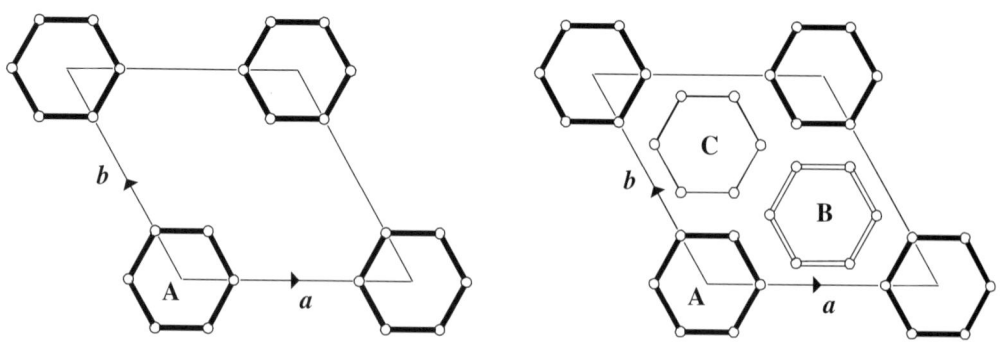

Fig. 1.    PerBU (left) and definition of 6-ring positions used in the stacking modes (right).

## 2. Connection mode

Neighboring PerBUs can be connected along *c* through tilted 4-rings in three different ways (see Introduction). In **CHA** all three connection modes between the PerBUs are observed. Double 6-ring layers are stacked along [201] (see Figure 2 and **AEI**). *chab* Cavities are formed. (See Figure 2 and **AFT**.)

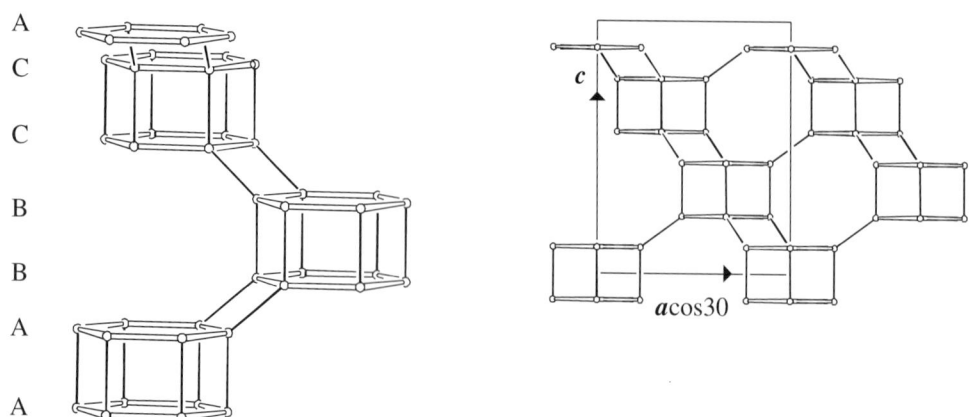

Fig. 2.    Connection mode (left) and unit cell content (right) viewed along *b*. The stacking sequence is given. In the perspective drawing, each PerBU is represented by one 6-ring only.

### 3. Channels and/or cages

Two-dimensional 8-ring channel systems perpendicular to *c* intersect with 8-ring channels parallel to <111>. The channel intersection (or *chab* cavity), topologically equivalent to the intersection in **AEI**, is depicted in Figure 3. **CHA** can be built from *chab* cavities alone (see also **AFT**). A three-dimensional channel system is obtained by connecting the cavities through common 8-rings.

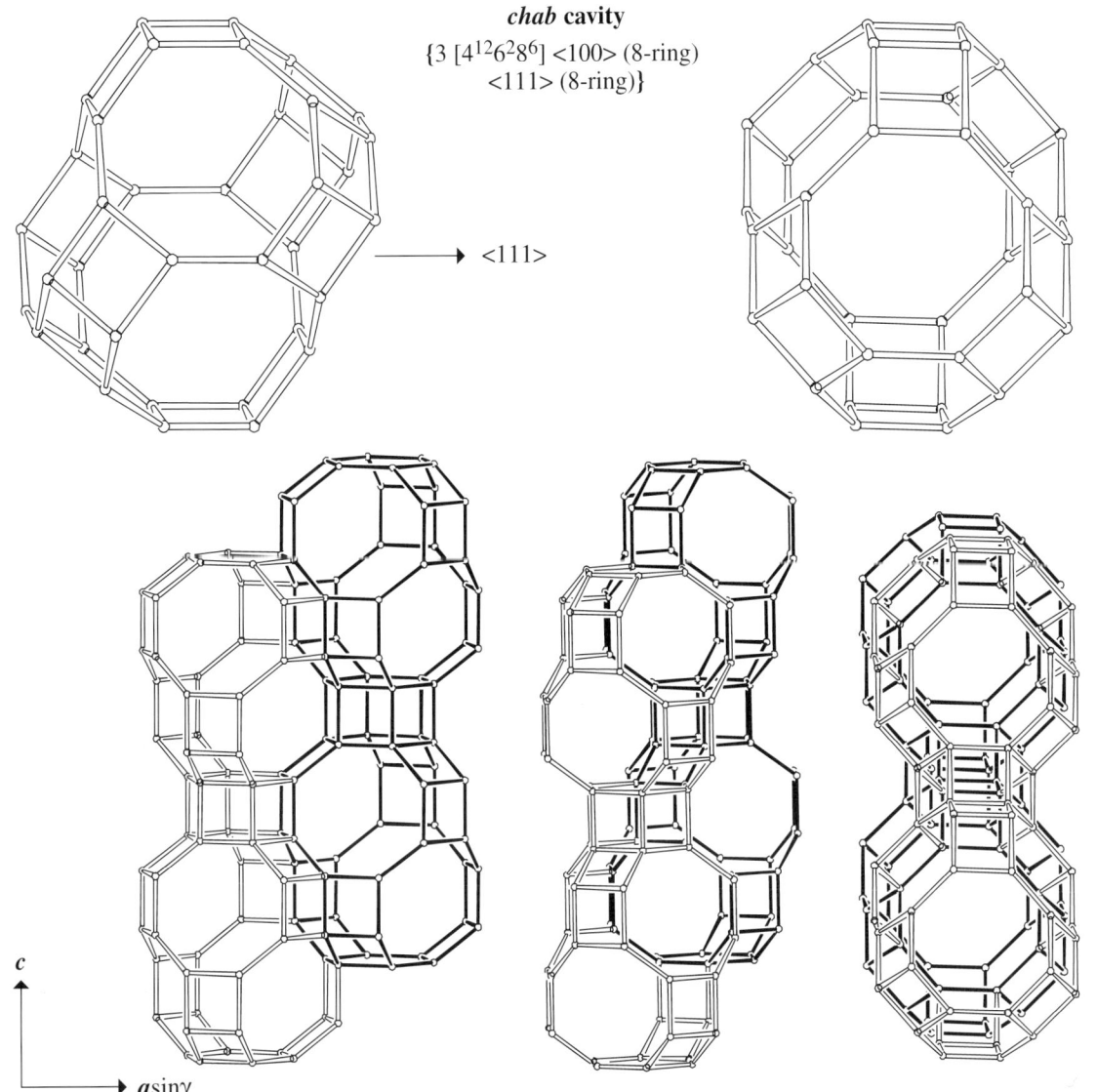

*chab* **cavity**

$\{3\ [4^{12}6^28^6]\ <100>\ (8\text{-ring})$
$<111>\ (8\text{-ring})\}$

$\longrightarrow\ <111>$

*c*

$a\sin\gamma$

Fig. 3.    Top: *chab* cavity viewed approximately along [010] (left) and along [111] (right). Bottom: *chab* cavities are connected into 8-ring channels parallel to <100> and <111>. The two-dimensional 8-ring channels parallel to <100> are interconnected along *c* through 12-rings in the *chab* cavities leading to a three-dimensional channel system. View along [010] (left), along [110] (middle) and along [111] (right).

# -CHI                                   Building Scheme

## 1. Periodic Building Unit

The interrupted framework of **-CHI** can be built using the zigzag chain (bold in Figure 1) parallel to *a*, and eight additional T atoms. The one-dimensional PerBU is obtained when three zigzag chains (connected into distorted 4-rings) are linked to trimers and pentamers. 5-Rings and 9-rings are formed (Figure 1). Terminal oxygen atoms are bonded to two T atoms in each 5- and 9-ring.

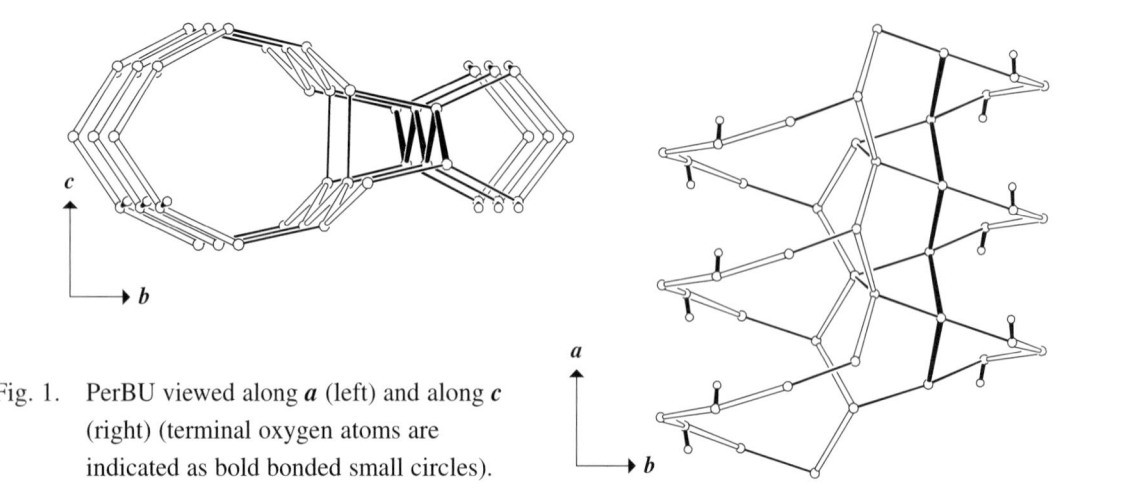

Fig. 1.   PerBU viewed along *a* (left) and along *c*
          (right) (terminal oxygen atoms are
          indicated as bold bonded small circles).

## 2. Connection mode

Neighboring PerBUs, related along *c* by a pure translation and along *b* by a screw rotation of 180° about *a*, are connected through zigzag chains and 4-rings. A backbone of 4-rings having one T atom in common parallel to *c* is formed (Figure 2).

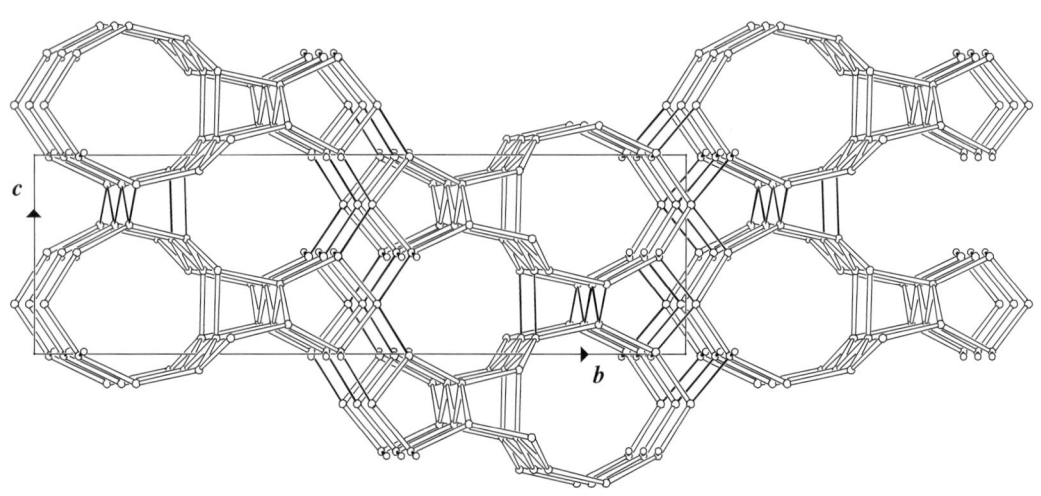

Fig. 2.   Connection mode and unit cell content viewed along *a*.

## 3. Channels and/or cages

9-Ring channels are parallel to *a*. Terminal oxygen atoms seriously hamper the diffusion between the 9-ring channels. The channel is depicted in Figure 3.

$$\{1\ [6^3 12^1 9^{2/2}]\ [100]\ (9\text{-ring})\}$$

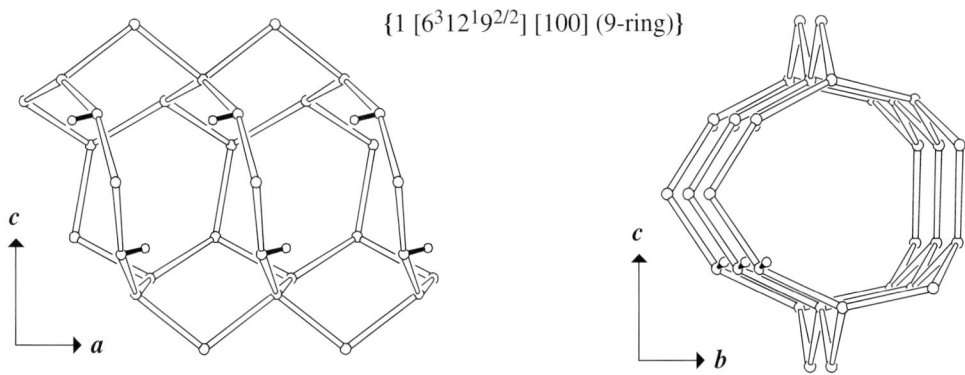

Fig. 3.    Channel viewed perpendicular to the channel axis (left) and along the channel axis (right).

# -CLO                    Building Scheme

## 1. Periodic Building Unit

The cubic interrupted framework of **-CLO** can be built using two types of building units: BU1 and BU2. BU1 equals the *rho* cavity (see **RHO**) in which all 12 4-rings are replaced by double 4-rings (D4Rs; Figure 1a). BU2 equals a slice of BU1 and consists of four D4Rs (Figure 1b). Each D4R is singly connected to two neighboring D4Rs. Each D4R bears two terminal oxygen atoms where the framework is not fully connected. A zero-dimensional PerBU is obtained when one BU1 and three BU2s are connected through 6-rings as shown in Figure 1c. *aww* Cavities are formed.

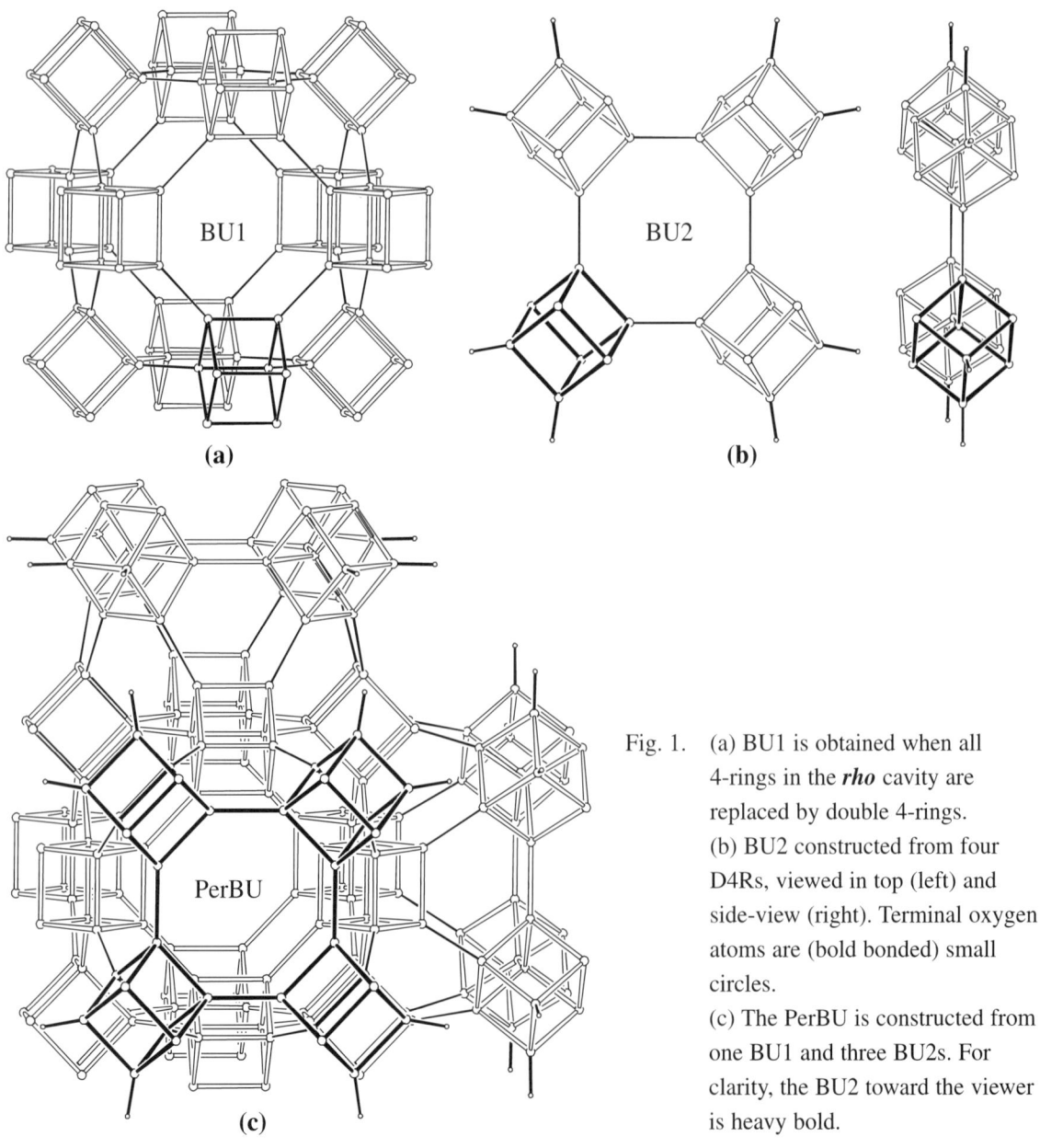

Fig. 1.  (a) BU1 is obtained when all 4-rings in the *rho* cavity are replaced by double 4-rings.
(b) BU2 constructed from four D4Rs, viewed in top (left) and side-view (right). Terminal oxygen atoms are (bold bonded) small circles.
(c) The PerBU is constructed from one BU1 and three BU2s. For clarity, the BU2 toward the viewer is heavy bold.

## 2. Connection mode

Neighboring PerBUs, related by pure translations along the cube axes, are connected into the three-dimensional framework. The connection mode within one cube face of the framework is illustrated in Figure 2.

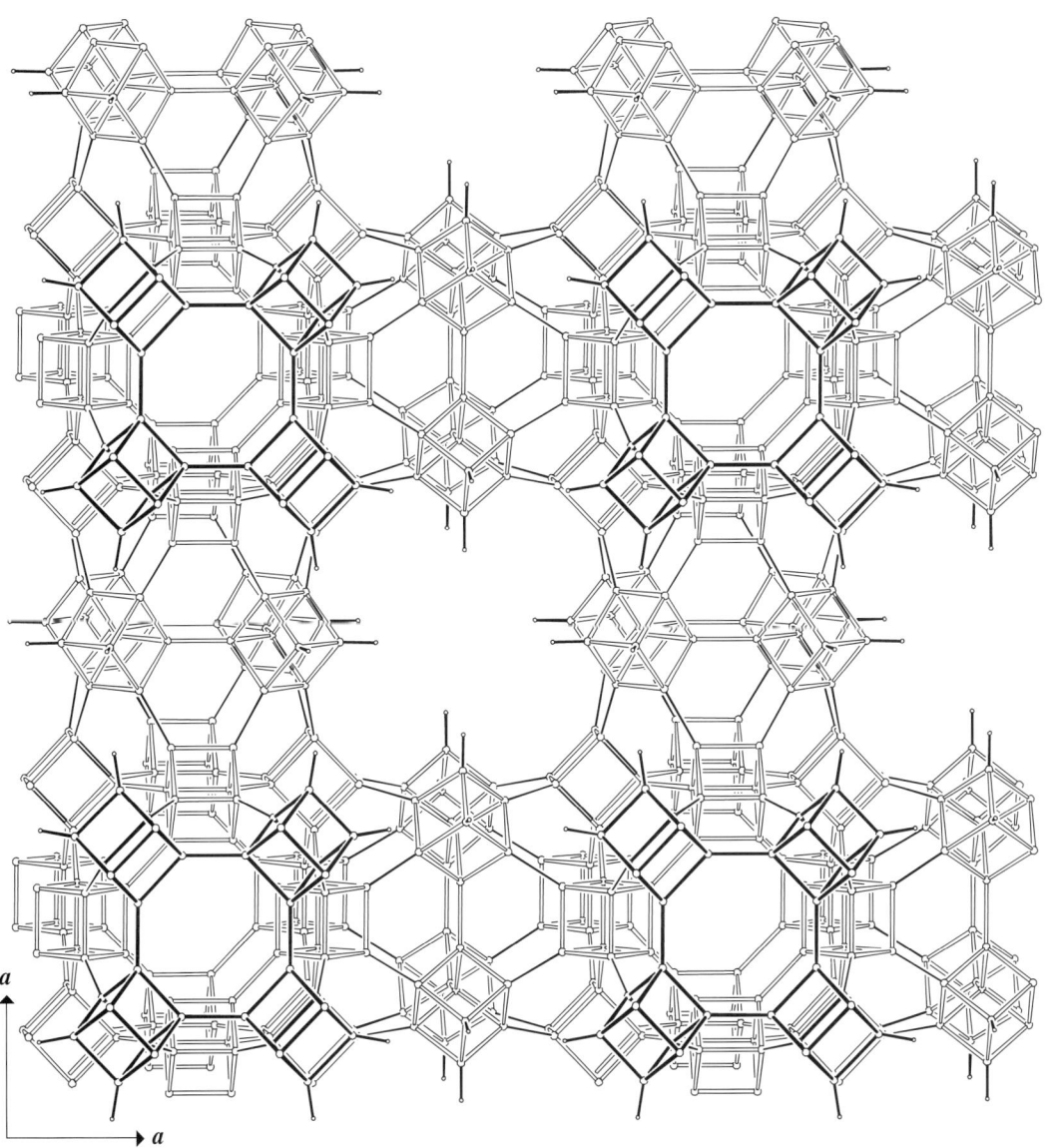

Fig. 2a.   Connection mode in the **aa** plane. The connection mode in the other cube faces is the same. The projection of the unit cell content is on next page.

Fig. 2b.   The cell content projected along a cube axis is shown at the bottom of the page number 97.

# -CLO

## Cage/Channel

### 3. Channels and/or cages

Intersecting 8-ring channels and intersecting 20-ring channels are parallel to the cube axes. There is no direct access from the 8-ring channels to the 20-ring channels. The 8-ring channels are composed of *rho* and *aww* cavities. The *rho* cavity is also present in **KFI**, **LTA**, **LTN**, **PAU**, **RHO**, **TSC** and **UFI**. The *aww* cavity is also found in **AWW**. One 8-ring channel and the 20-ring channel intersection (or *clo* cavity) are shown in Figure 3.

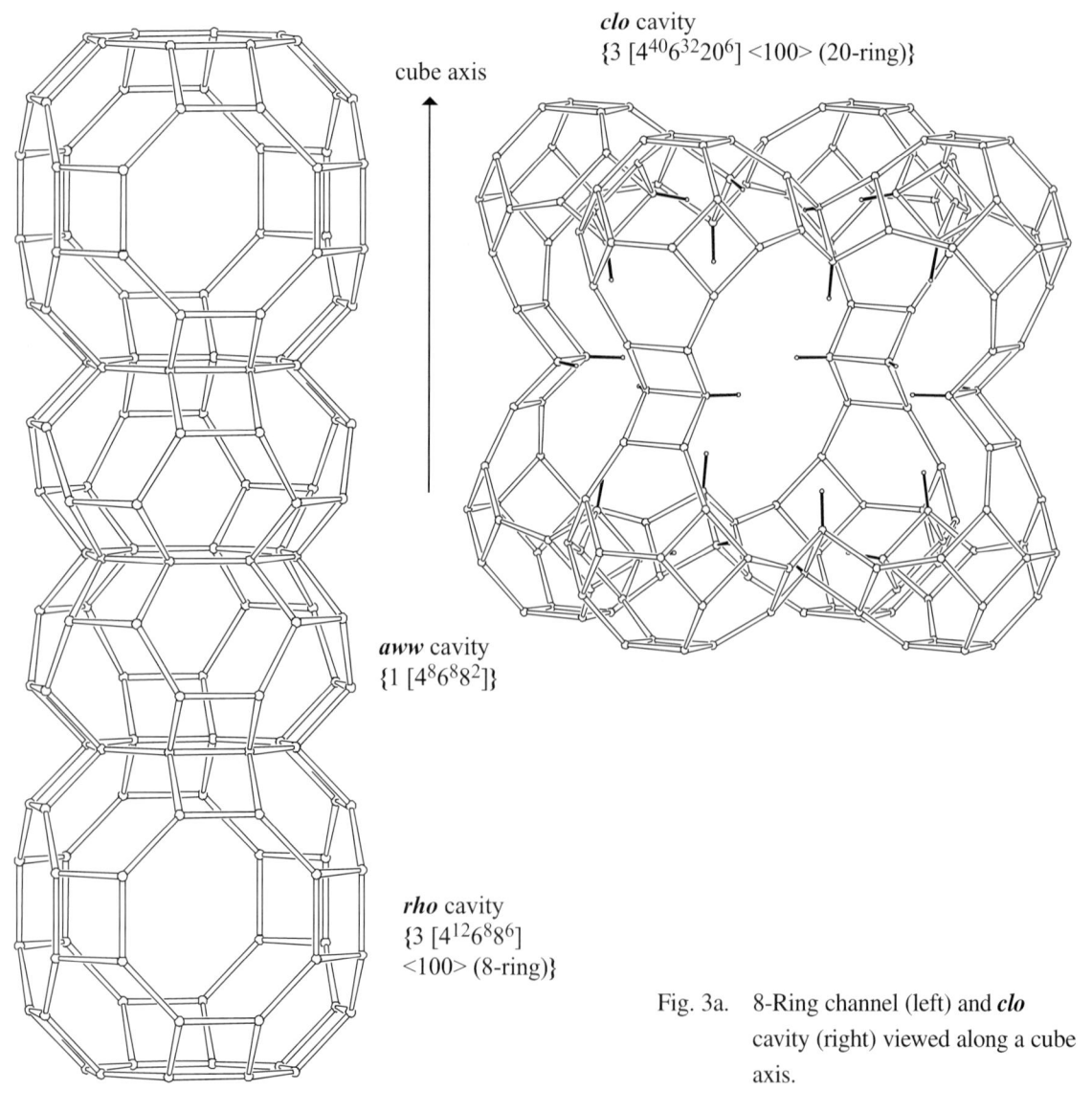

cube axis

*clo* cavity
{3 [$4^{40}6^{32}20^6$] <100> (20-ring)}

*aww* cavity
{1 [$4^86^88^2$]}

*rho* cavity
{3 [$4^{12}6^88^6$]
<100> (8-ring)}

Fig. 3a.   8-Ring channel (left) and *clo* cavity (right) viewed along a cube axis.

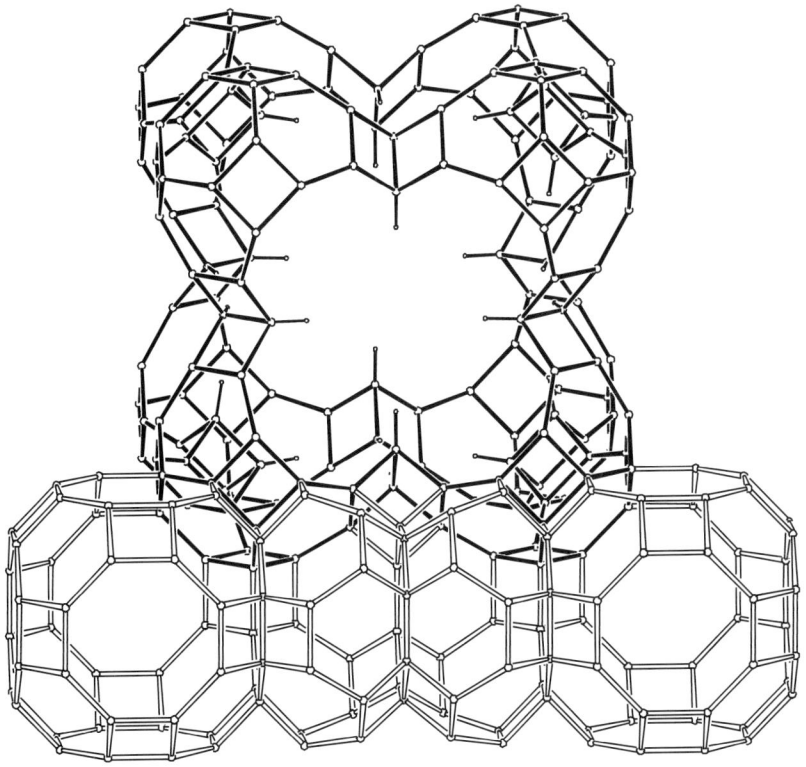

Fig. 3b.   Connection between 8-ring channel and *clo* cavity viewed along a cube axis.

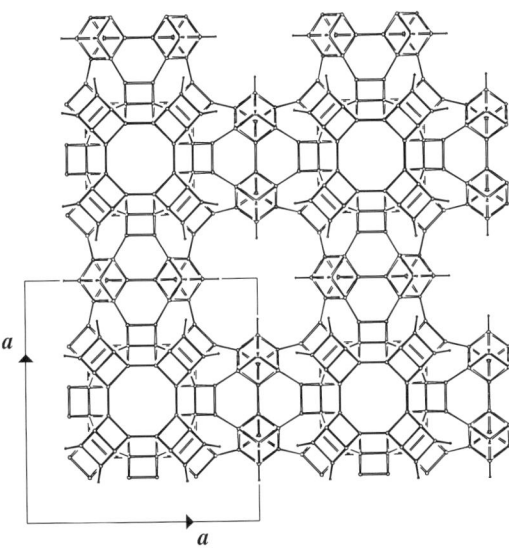

Fig. 2b.   Unit cell content projected along a cube axis.

# CON                          Building Scheme

## 1. Periodic Building Unit

**CON** can be built using chains parallel to **c** constructed from units of 14 T atoms (a $[4^1 5^2 6^4]$-cage, or two 1-5-1 units; bold in Figure 1) that are related along **c** by pure translations. A two-dimensional PerBU is obtained when parallel chains, related along **b** by a screw rotation of 180° about **b**, are connected into the **bc** layer through 4- and 5-rings (Figure 1). (See also **IWR**; compare with **ITH**.)

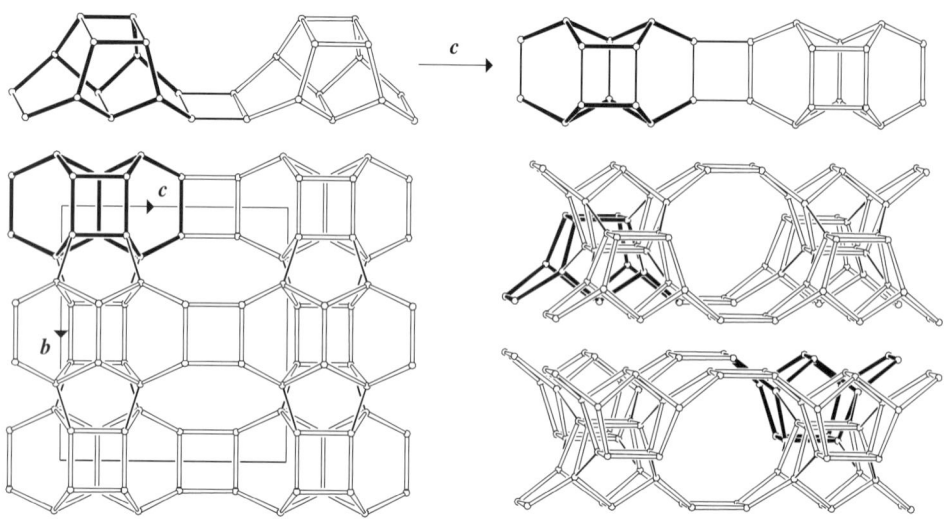

Fig. 1.   Chain of T14-units (top). The chain at the top right is rotated over 90° about **c** with respect to the left one. PerBU (bottom) viewed along **a** (left) and along **b** (right). The PerBUs depicted at the right are related by a rotation of 180° about **b**.

## 2. Connection mode

Neighboring PerBUs, related along **a** by a rotation of 180° about **b** (or related along **c** by a shift of $(\frac{1}{2}\mathbf{b} + \frac{1}{3}\mathbf{c})$), are connected along **a** through 4-rings as shown in Figure 2.

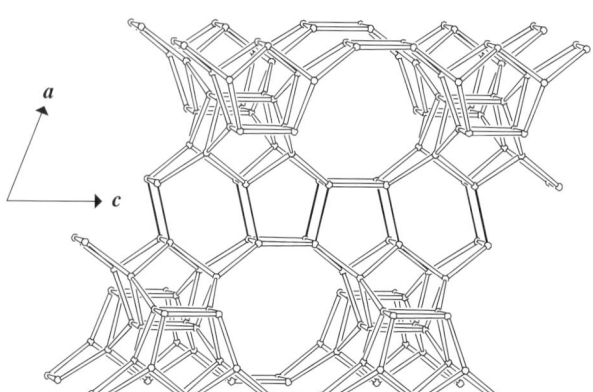

Fig. 2a.   Connection mode viewed along **b**. Fused 5-5-6-ring sequences are formed.

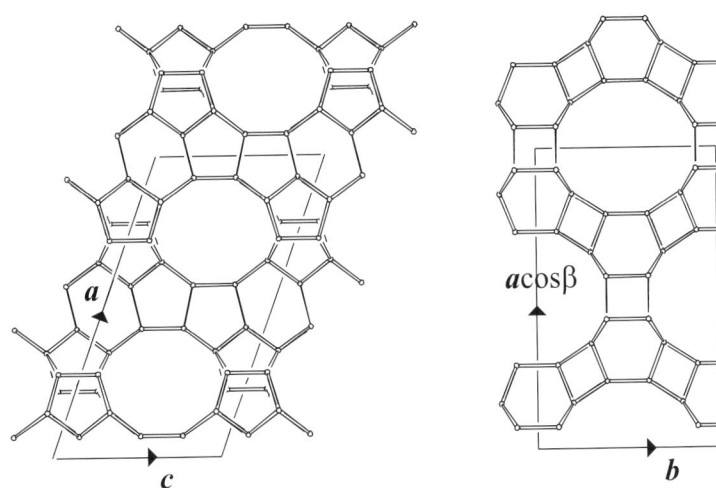

Fig. 2b. Unit cell content projected along *b* (left) and along *c* (right).

## 3. Channels and/or cages

Interconnecting straight 10- and 12-ring channels are parallel to *b* and *c*, respectively. The 10-ring and 12-ring channels are topologically equivalent to the 10-ring channel in **IWR** and the 12-ring channel in ***BEA**. An additional sinusoidal 12-ring channel, intersecting with the straight channels, is parallel to *a*. The channels are depicted in Figure 3.

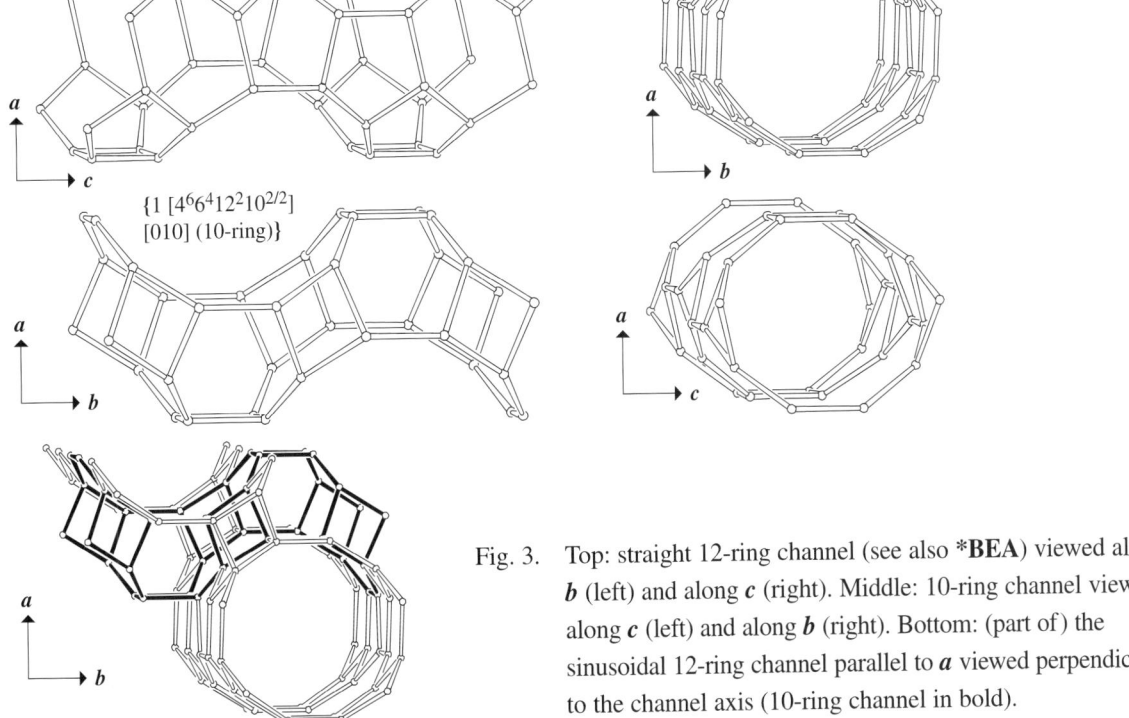

Fig. 3. Top: straight 12-ring channel (see also ***BEA**) viewed along *b* (left) and along *c* (right). Middle: 10-ring channel viewed along *c* (left) and along *b* (right). Bottom: (part of) the sinusoidal 12-ring channel parallel to *a* viewed perpendicular to the channel axis (10-ring channel in bold).

# CZP                    Building Scheme

## 1. Periodic Building Unit

Hexagonal **CZP** can be built using units of 12 T atoms (five fused 4-rings and an additional T atom; bold in Figure 1). T12-units, related along **a** and **b** by pure translations, are connected into the two-dimensional PerBU depicted in Figure 1. Deformed 7- and 12-rings are formed.

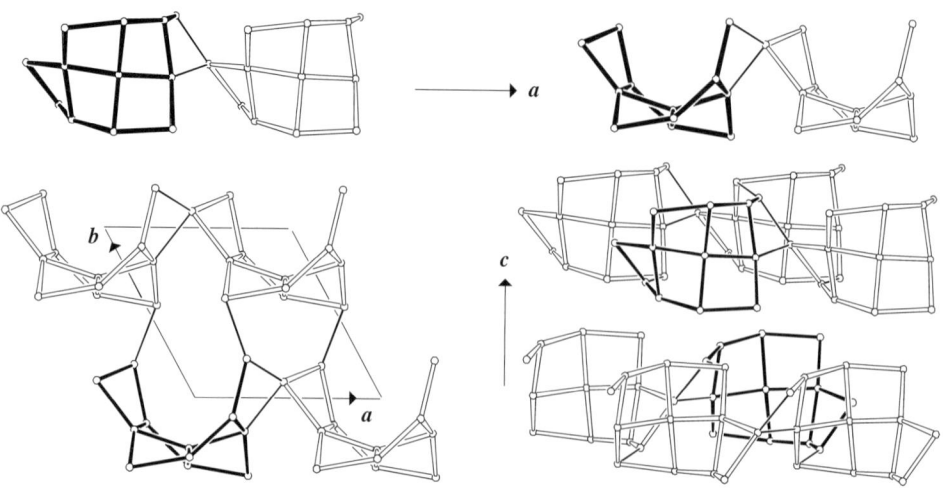

Fig. 1.   Top: T12-units are connected into a chain along **a**. The chains at left and right are related by a rotation of 90° about **a**. Bottom: PerBU viewed along **c** (left) and perpendicular to the **ac** plane (right). The PerBUs shown at the right differ by a rotation of 180° about **c**.

## 2. Connection mode

Neighboring PerBUs, related along **c** by a screw rotation of 180° about **c**, are connected along **c** through 4-rings as can be seen from Figure 2.

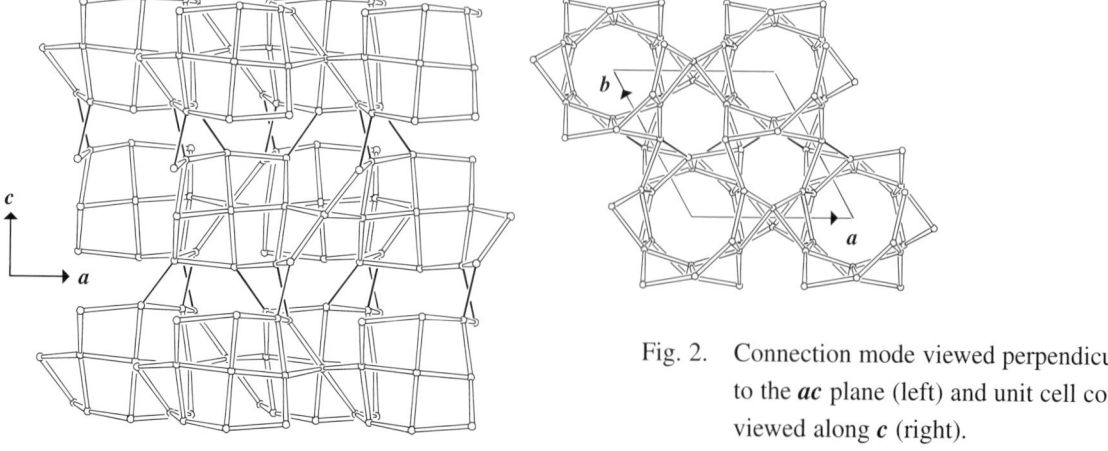

Fig. 2.   Connection mode viewed perpendicular to the **ac** plane (left) and unit cell content viewed along **c** (right).

# Cage/Channel

## 3. Channels and/or cages

Highly distorted chiral sinusoidal 12-ring channels are parallel to *c* as depicted in Figure 3.

$$\{1\ [4^{18}8^{12}12^{2/2}]\ [001]\ (12\text{-ring})\}$$

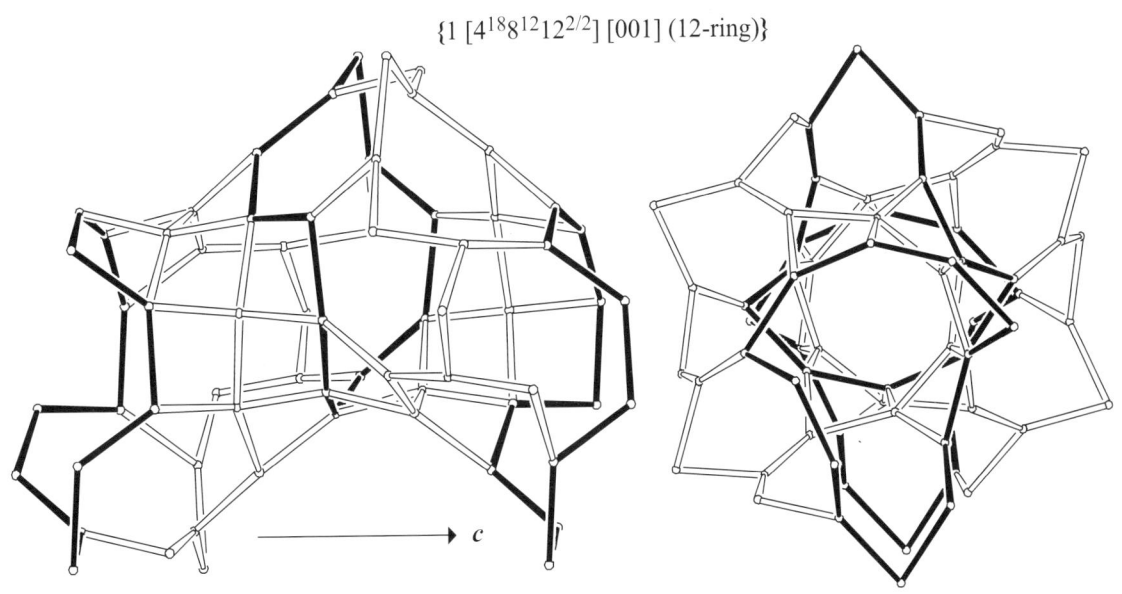

Fig. 3. Channel viewed perpendicular to the channel axis (left) and along the channel axis (right). Three 12-rings, related by a rotation of 180° about *c*, are in bold for clarity.

# DAC

## Building Scheme

### 1. Periodic Building Unit

Finite building units of 12 T atoms are composed of two 5-1 units (bold in Figure 1). The two-dimensional PerBU is obtained when these T12-units, related along $b$ and $c$ by pure translations, are connected into a layer with a rectangular repeat unit. Saw chains along $b$ and zigzag chains along $c$ are formed. (See also **EPI** and **MOR**.)

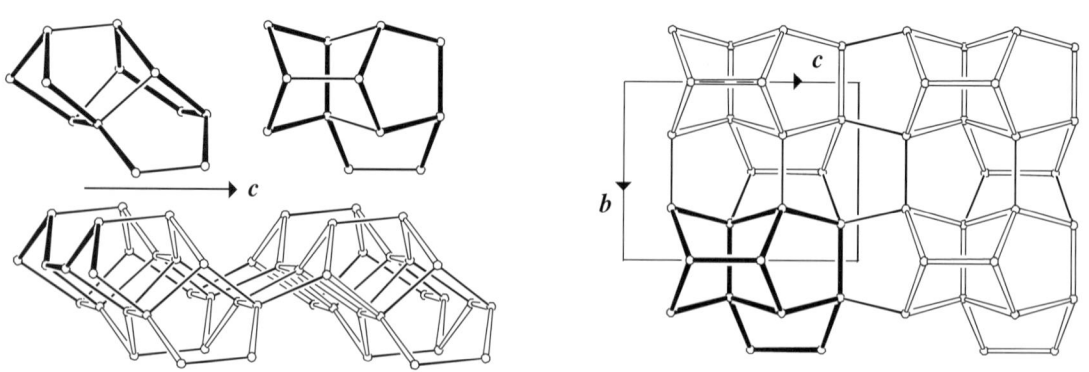

Fig. 1.    Top: T12-unit viewed along $b$ (left) and along $a$ (right). Right: PerBU constructed from T12-units (one T12-unit in bold) viewed along $a$. Bottom: PerBU built from saw chains (one saw chain in bold) viewed along $b$.

### 2. Connection mode

Neighboring PerBUs, related along $a$ by a shift of $\frac{1}{2}(a+b)$, are connected through 4-rings (Figure 2).

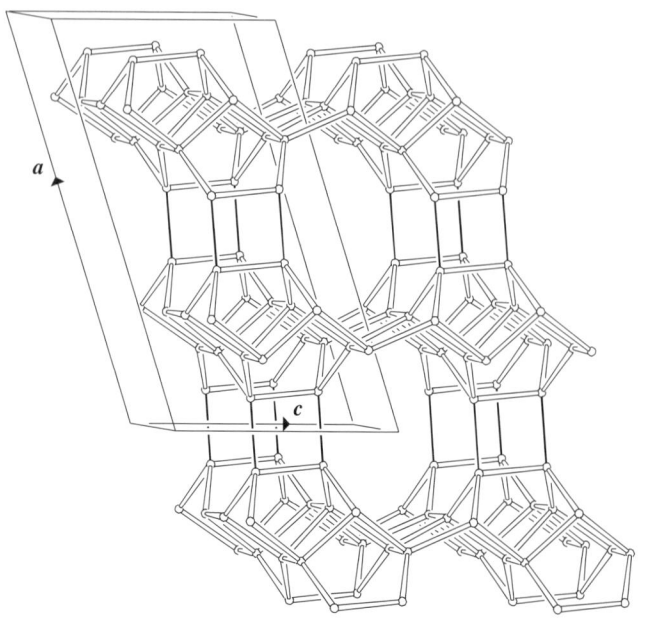

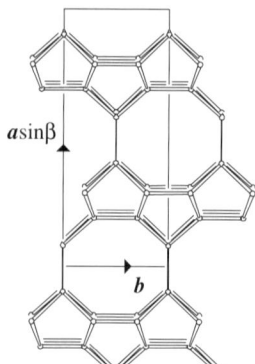

Fig. 2.    Connection mode (and unit cell content) viewed along $b$ (left) and unit cell content projected along $c$ (right).

## 3. Channels and/or cages

Intersecting 8- and 10-ring channels (Figure 3) are parallel to **c** and **b**, respectively. The 10-ring channel is topologically equivalent to the 10-ring channel in **FER**.

{1 [$5^4 6^2 8^2 10^{2/2}$] [010] (10-ring)}

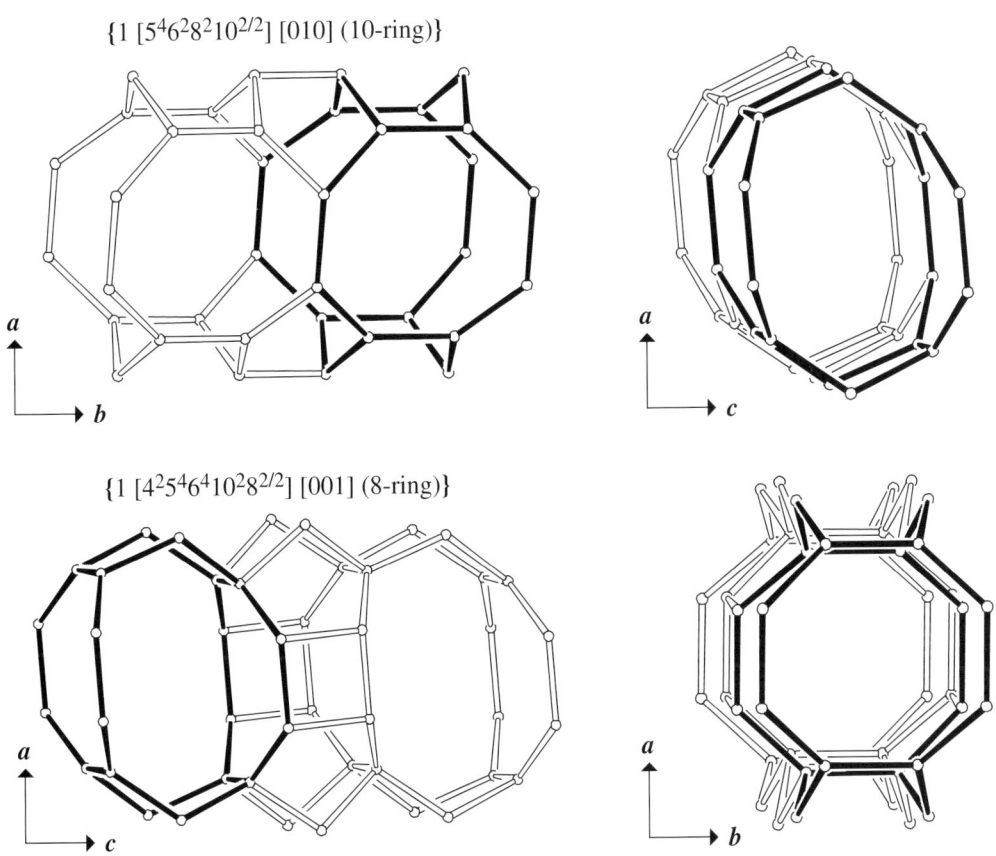

{1 [$4^2 5^4 6^4 10^2 8^{2/2}$] [001] (8-ring)}

Fig. 3.    8-Ring channel (top) and 12-ring channel (bottom) viewed perpendicular to the channel axis (left) and along the channel axis (right). The channels have [$6^2 8^2 10^2$]-cavities in common. One intersection in bold.

# DDR                         Building Scheme

## 1. Periodic Building Unit

In **DDR**, units of 30T atoms (two 12-rings and a 6-ring) are connected into a hexagonal layer. The "empty" spaces between the T30-units are filled with T2-dimers. $[5^{12}]$-Cages are generated. The T30-units are centered at (0,0) in the **ab** layer. This position is usually called the **A** position. At sites **B** and **C**, $[5^{12}]$-cages share faces. The two-dimensional PerBU is obtained when 6-rings from an additional 6-ring layer (bold in Figure 1) are stacked on top of the **C** sites. $[4^3 5^6 6^1]$-Cages are generated. The repeat unit of the PerBU consists of 40T atoms: the T30-unit, two "space-filling" dimers and three dimers (of the additional 6-ring). (See also **DOH**, **MEP** and **MTN**.)

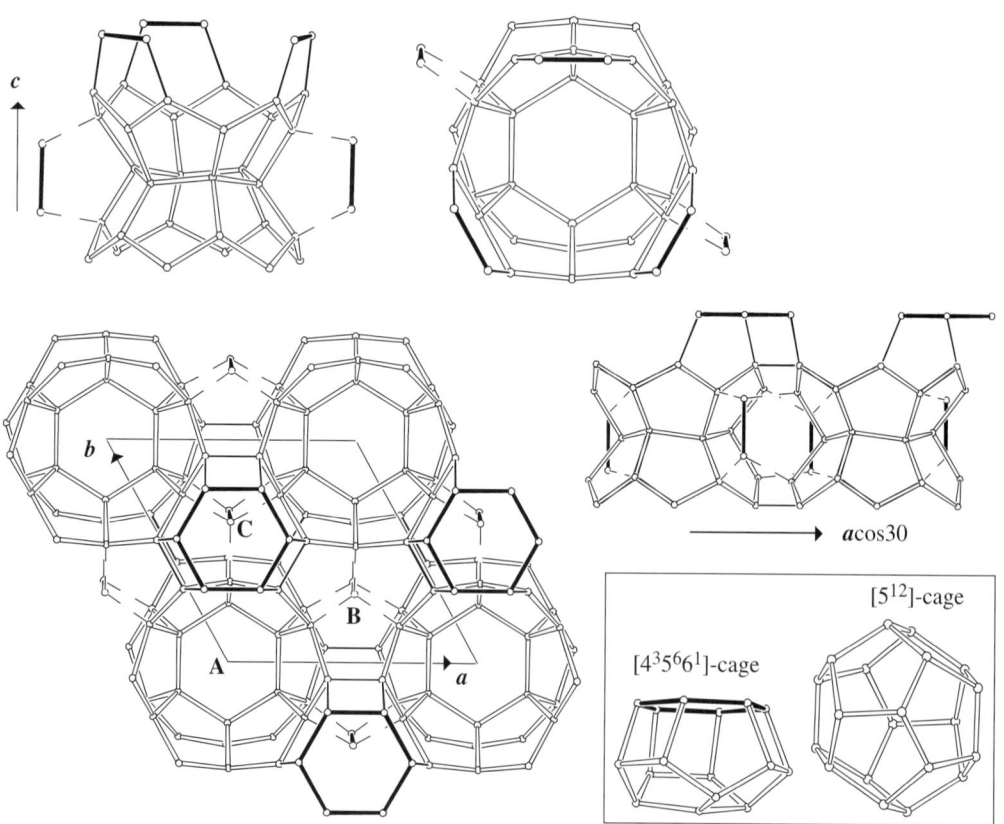

Fig. 1.  T40-unit (top) viewed along [110] (left) and along **c** (right). Hexagonal PerBU viewed along **c** (bottom left) and along **b** (top right). Connections to the space-filling dimers (in bold) are dashed. The additional 6-ring layer is also in bold. The inset shows the $[5^{12}]$- and $[4^3 5^6 6^1]$-cages.

## 2. Connection mode

Neighboring PerBUs, related by a lateral shift of $+(\frac{2}{3}\boldsymbol{a} + \frac{1}{3}\boldsymbol{b})$, are connected along **c** as shown in Figure 2 (compare this connection mode with mode (**1**) in the ABC-6 family; see Introduction).

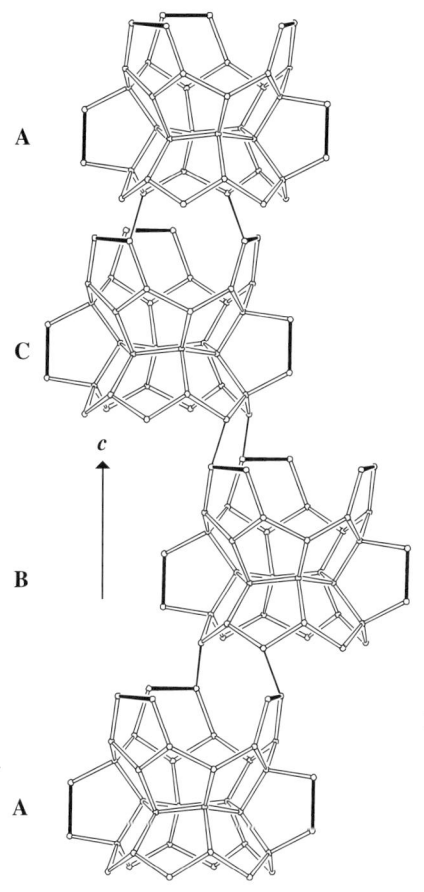

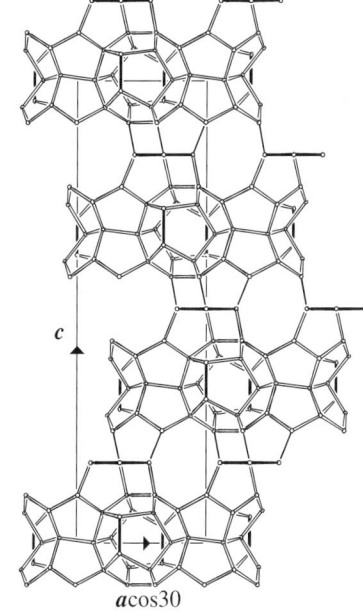

Fig. 2. Connection mode viewed along [110] (left) and unit cell content projected along **b** (right). The stacking sequence is given. In the perspective drawing, each PerBU is represented by one T40-unit only. (Parts of) the additional 6-rings are in bold.

## 3. Channels and/or cages

The [$5^{12}$]-cage, also present in **DOH**, **MEP** and **MTN**, and [$4^3 5^6 6^1$]-cage are shown in Figure 1. Intersecting two-dimensional 8-ring channel systems are perpendicular to **c**. The channel intersection is depicted in Figure 3.

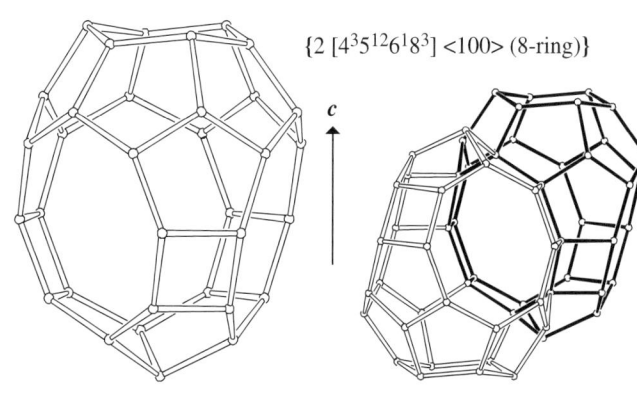

{2 [$4^3 5^{12} 6^1 8^3$] <100> (8-ring)}

Fig. 3. Channel intersection viewed along <100> (left) and linkage of intersections through common 8-rings into a two-dimensional 8-ring channel system perpendicular to **c** (right).

# DFO Building Scheme

## 1. Periodic Building Unit

Hexagonal **DFO** can be built using a building unit of 22 T atoms (bold in Figure 1) that consists of two 4-1 and two 1-4-1 units. T22-units, related along $c$ by pure translations, are connected through 4-rings into chains along $c$. Chains are connected along a 6-fold axis parallel to $c$ into the one-dimensional tubular PerBU shown in Figure 1 at the bottom right.

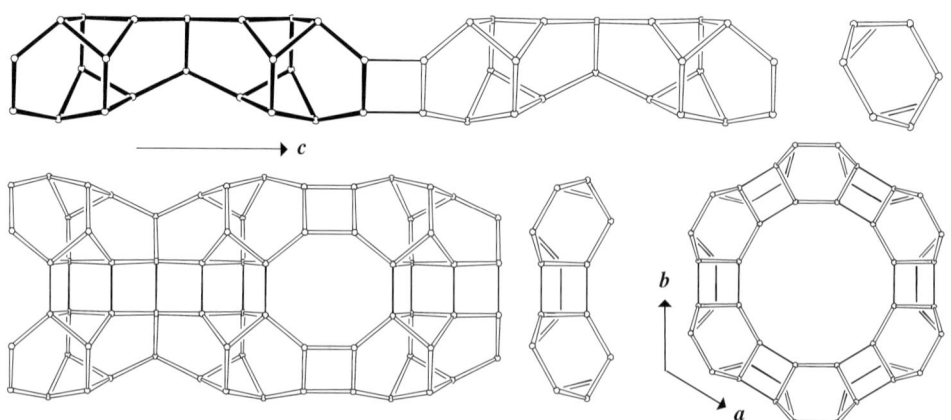

Fig. 1. Top: chain viewed perpendicular to $c$ (left) and down $c$ (right). Bottom left: connection mode of two chains in the PerBU viewed perpendicular to the 6-fold axis parallel to $c$. 8-Rings are formed. The chain length along $c$ has been limited to $1\frac{1}{2}$ of the repeat unit for clarity. Bottom right: PerBU projected along the 6-fold axis.

## 2. Connection mode

Neighboring PerBUs, related along $a$ and $b$ by pure translations, are connected along $a$ and $b$ (and along $(a+b)$) through 4- and 6-rings as shown in Figure 2.

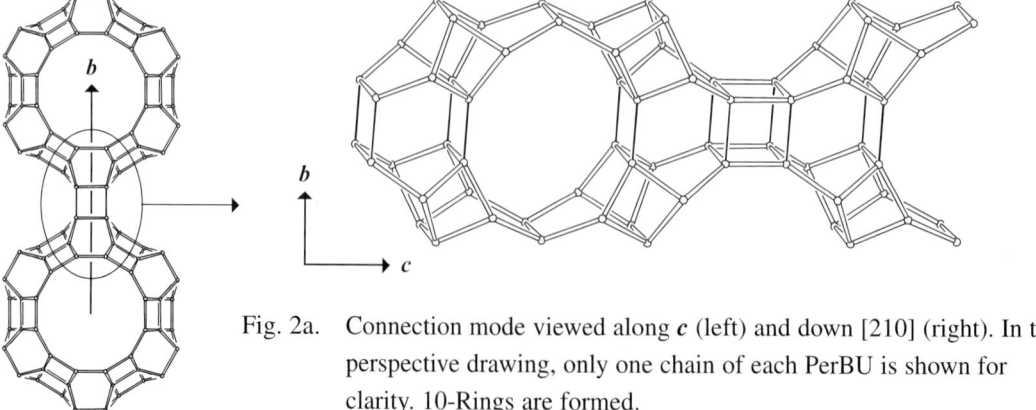

Fig. 2a. Connection mode viewed along $c$ (left) and down [210] (right). In the perspective drawing, only one chain of each PerBU is shown for clarity. 10-Rings are formed.

# Cage/Channel                    **DFO**

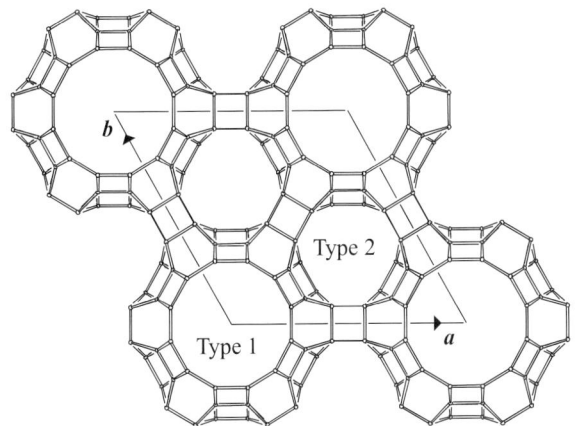

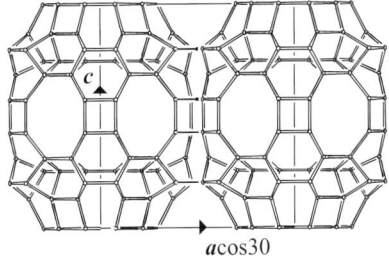

Fig. 2b.   Unit cell content projected along
           *c* (left) and along <010> (right).

## 3. Channels and/or cages

Two types of 12-ring channels are parallel to *c*. 12-Ring channels of Type 1 are interconnecting to 12-ring channels of Type 2 through common 8-rings (Figure 3). 12-Ring channels of Type 2 are connected through 10-rings. Two-dimensional 8- and 10-ring channels are perpendicular to *c*.

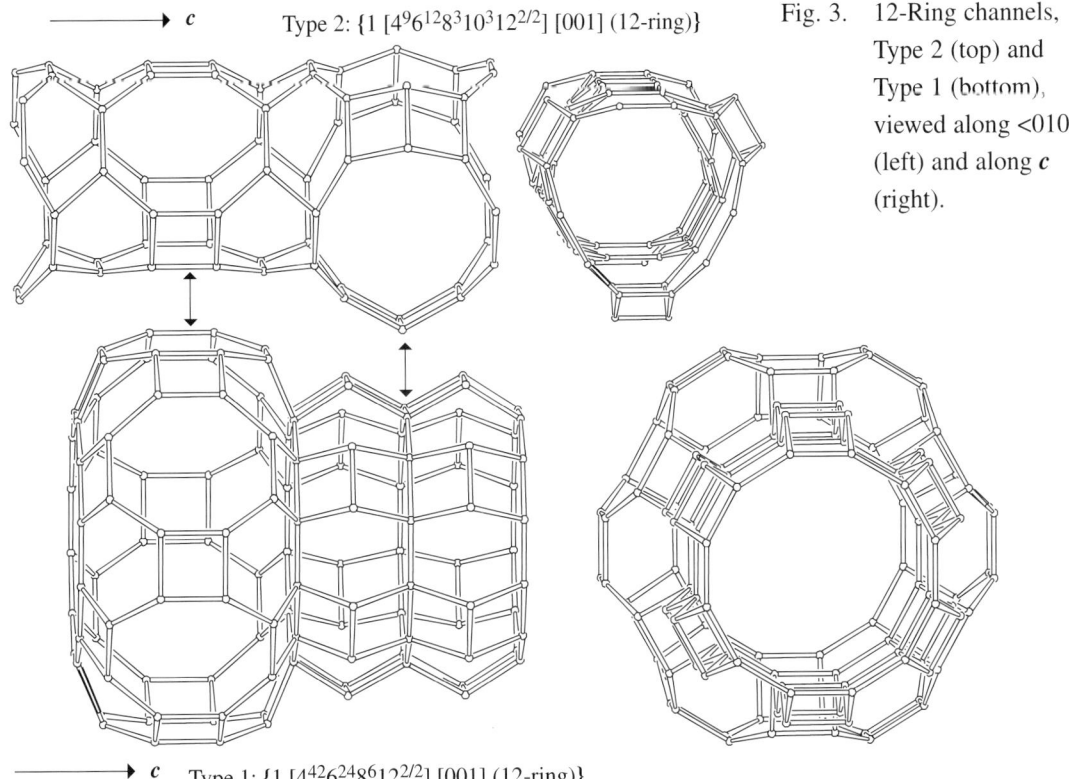

Type 2: $\{1\,[4^9 6^{12} 8^3 10^3 12^{2/2}]\,[001]\,(12\text{-ring})\}$

Type 1: $\{1\,[4^{42} 6^{24} 8^6 12^{2/2}]\,[001]\,(12\text{-ring})\}$

Fig. 3.   12-Ring channels,
          Type 2 (top) and
          Type 1 (bottom),
          viewed along <010>
          (left) and along *c*
          (right).

# DFT                         Building Scheme

## 1. Periodic Building Unit

Tetragonal **DFT** can be built using the crankshaft chain (bold in Figure 1, left) parallel to $c$. Two of these chains are connected into the one-dimensional PerBU (a double crankshaft chain). The repeat unit of the PerBU is a 2-fold (1,3)-connected double 4-ring (or a $[4^2 6^2]$-cage; bold in Figure 1, right).

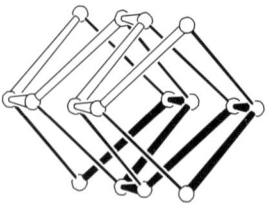 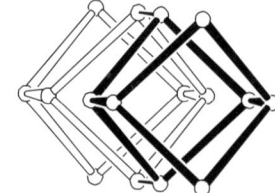

Fig. 1.   PerBU, constructed from two crankshaft chains (left) and from 2-fold (1,3)-connected double 4-rings (right), viewed along $c$.

## 2. Connection mode

Neighboring PerBUs, related along $a$ and $b$ by pure translations, are connected along $a$ and $b$ through 4-rings. 8-Rings perpendicular to $a$, $b$ and $c$ are formed (Figure 2).

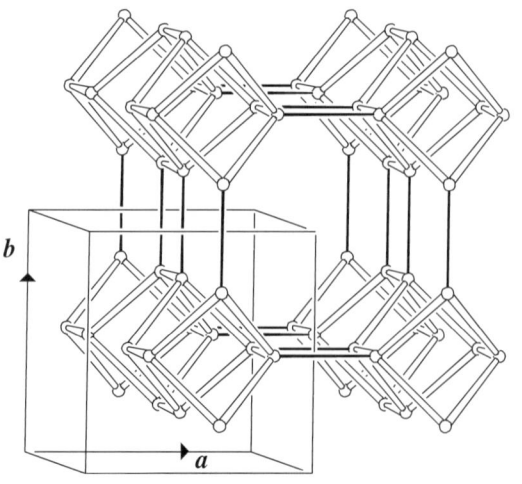 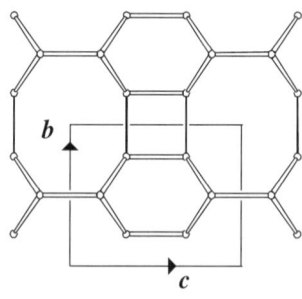

Fig. 2.   Connection mode (and unit cell content) viewed along $c$ (left), and unit cell content projected along $a$ (or along $b$; right). In the perspective drawing, only $1\frac{1}{2}$ repeat unit along $c$ of each PerBU is drawn for clarity.

## 3. Channels and/or cages

Intersecting 8-ring channels (of different type) are parallel to $c$ and <100>. The channel parallel to $c$ is topologically equivalent to the channel in **ACO** and **MER**. The channel parallel to <100> is topologically equivalent to one of the channels in **LOV**, **LTL**, **MOZ** and **RSN**. Channels and linkage of channels are illustrated in Figure 3.

{1 [$4^4 8^4 8^{2/2}$] [001] (8-ring)}

{1 [$4^2 6^2 8^2 8^{2/2}$] <100> (8-ring)}

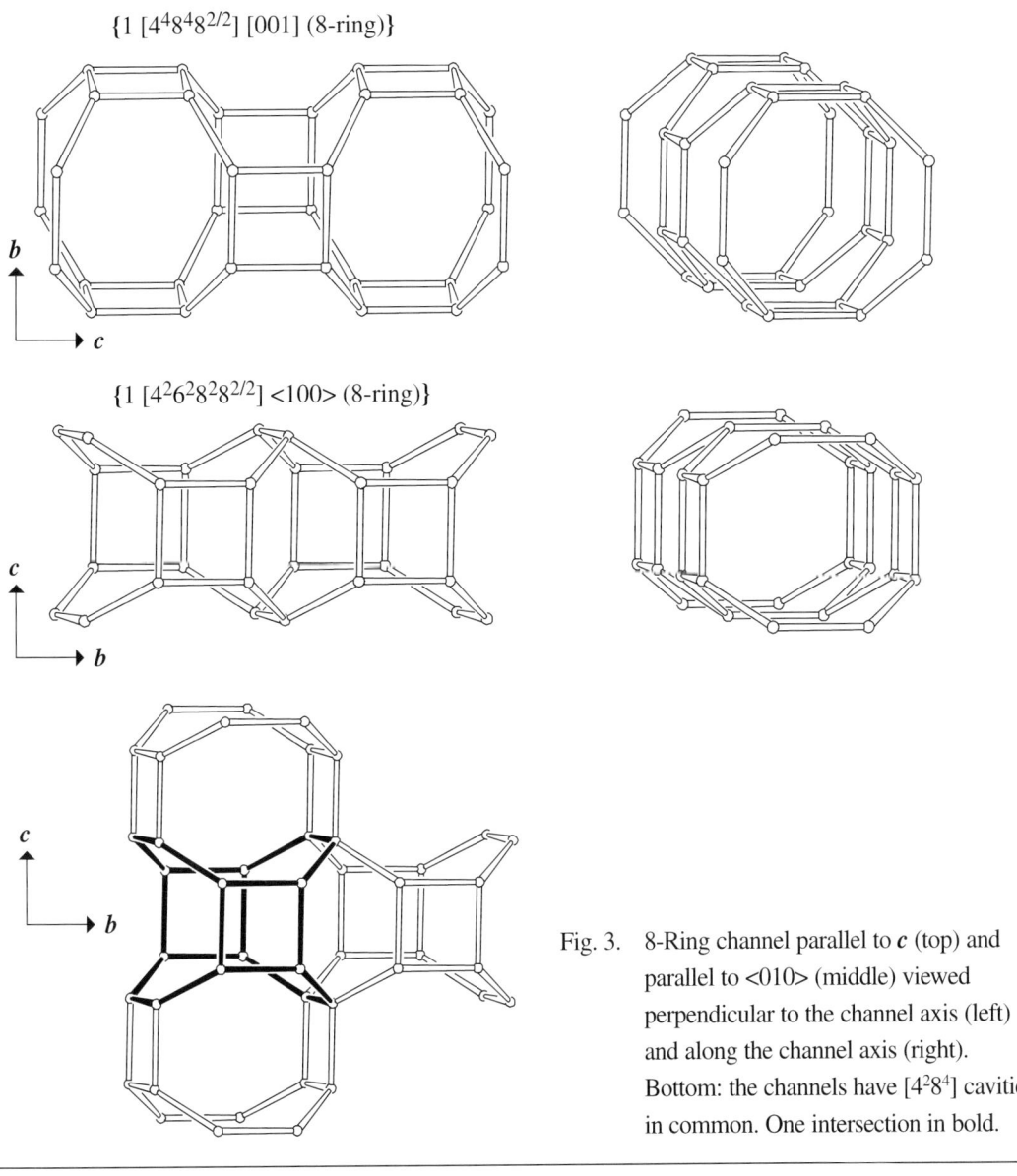

Fig. 3.　8-Ring channel parallel to $c$ (top) and parallel to <010> (middle) viewed perpendicular to the channel axis (left) and along the channel axis (right). Bottom: the channels have [$4^2 8^4$] cavities in common. One intersection in bold.

# DOH Building Scheme

## 1. Periodic Building Unit

In **DOH**, units of 30 T atoms (two 12-rings and a 6-ring) are connected into a hexagonal layer. The two-dimensional PerBU is obtained when the "empty" spaces between the T30-units are filled with T2-dimers. $[5^{12}]$-Cages are generated. The T30-units are centered at (0,0) in the **ab** layer. This position is usually called the **A** position. At sites **B** and **C**, $[5^{12}]$-cages share faces (see Figure 1). The repeat unit of the PerBU consists of 34 T atoms: the T30-unit and two "space-filling" dimers. (See also **DDR**, **MEP** and **MTN**.)

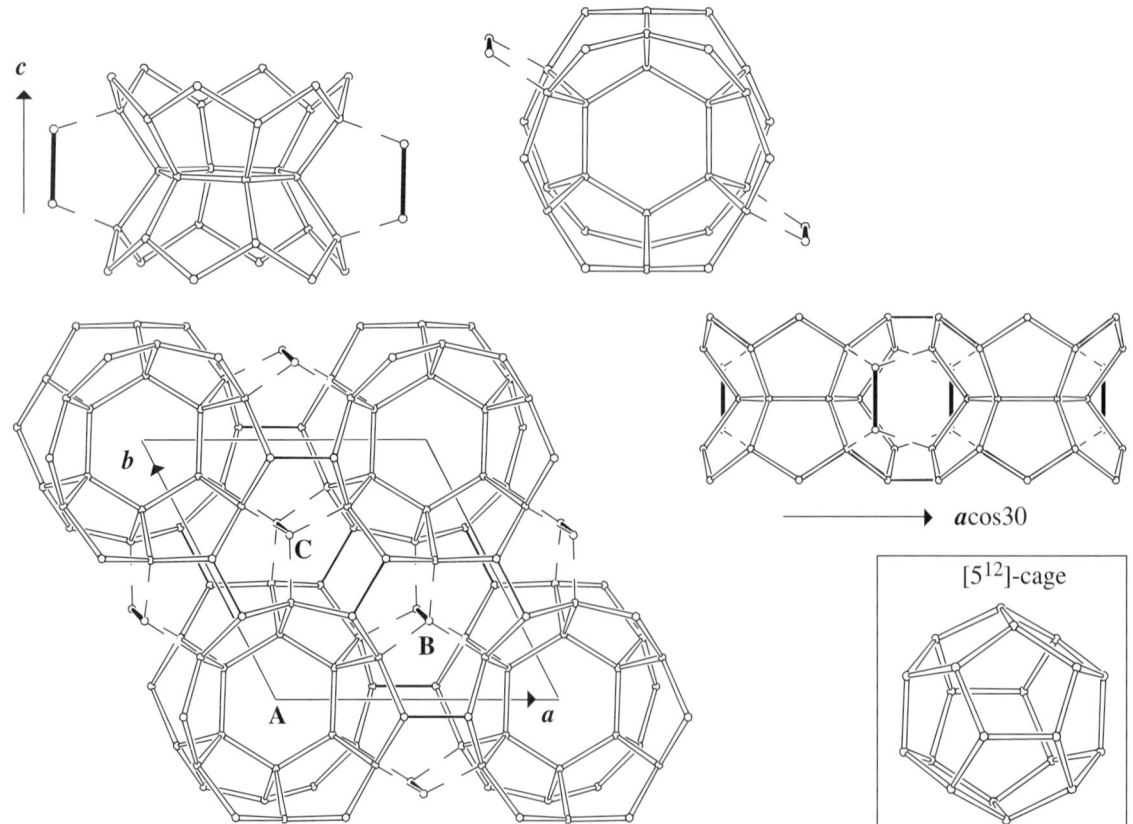

Fig. 1. T34-unit (top) viewed along [110] (left) and along **c** (right). Hexagonal PerBU viewed along **c** (bottom left) and along **b** (top right). Connections to the space-filling dimers (in bold) are dashed. The inset shows the $[5^{12}]$-cage.

## 2. Connection mode

Neighboring PerBUs are related along **c** by a pure translation as shown in Figure 2 (compare this connection mode with mode (**3**) in the ABC-6 family; see Introduction).

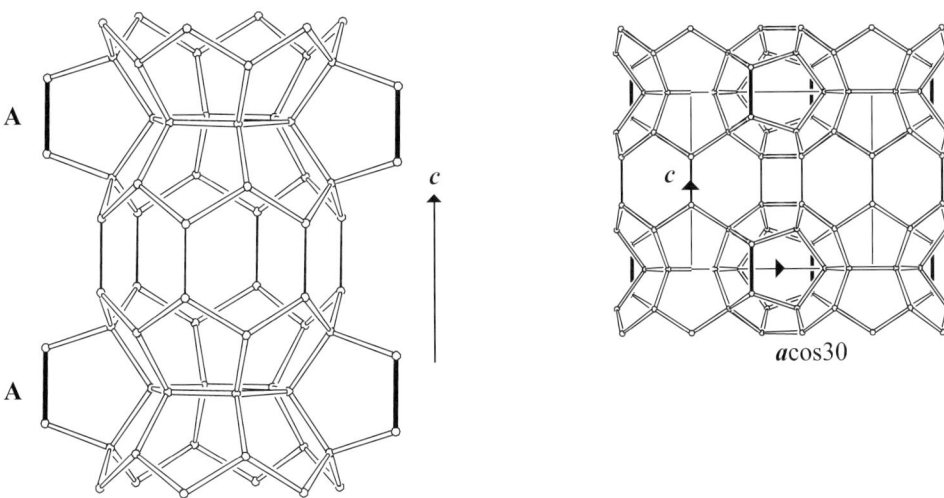

Fig. 2.     Connection mode viewed along [110] (left) and unit cell content projected along **b** (right). The stacking
sequence is given. In the perspective drawing, each PerBU is represented by one T34-unit only.

## 3. Channels and/or cages

The three types of cages are shown in Figure 3. The [$5^{12}$]-cage is also present in **DDR**, **MTN** and
**MEP**. Apertures of "channels" are formed by 6-rings only.

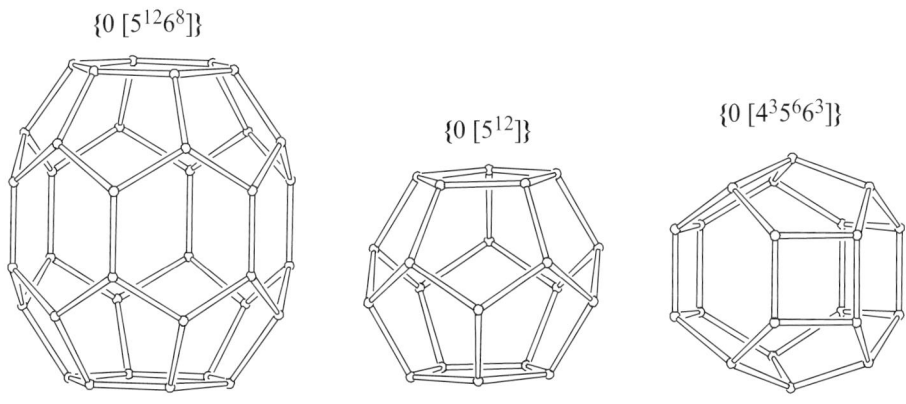

Fig. 3.     Cages viewed along <100>.

# DON

## Building Scheme

### 1. Periodic Building Unit

**DON** can be built using the crankshaft chain (bold in Figure 1) parallel to $c$. A one-dimensional PerBU is obtained when eight crankshaft chains are connected into a 14-ring channel with "handles". The channel wall consists of (fused) 6-rings.

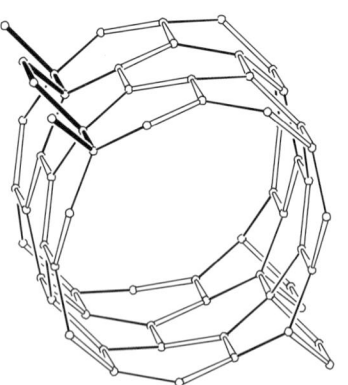 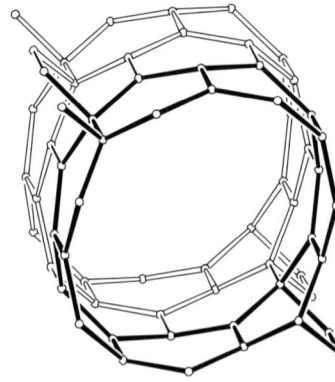

Fig. 1.  PerBU (seen along $c$) constructed from eight crankshaft chains (left) and from repeat units (of 32 T atoms) consisting of a 6-ring band decorated with four additional T atoms (right).

### 2. Connection mode

Neighboring PerBUs, related along $a$ by a pure translation and along $b$ by a shift of $\frac{1}{2}(a+b)$, are connected through single and double crankshaft chains as illustrated in Figure 2.

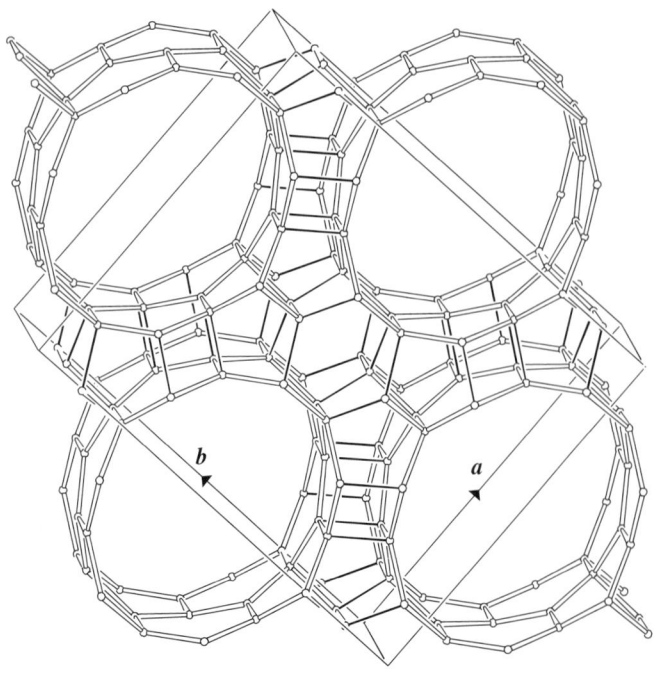

Fig. 2.  Connection mode and unit cell content viewed along $c$. Only $1\frac{1}{2}$ repeat unit along $c$ of each PerBU is drawn for clarity.

## 3. Channels and/or cages

Non-interconnecting 14-ring channels, parallel to *c*, are topologically equivalent to the channels in **AET**. One channel is depicted in Figure 3.

{1 [6$^{14}$14$^{2/2}$] [001] (14-ring)}

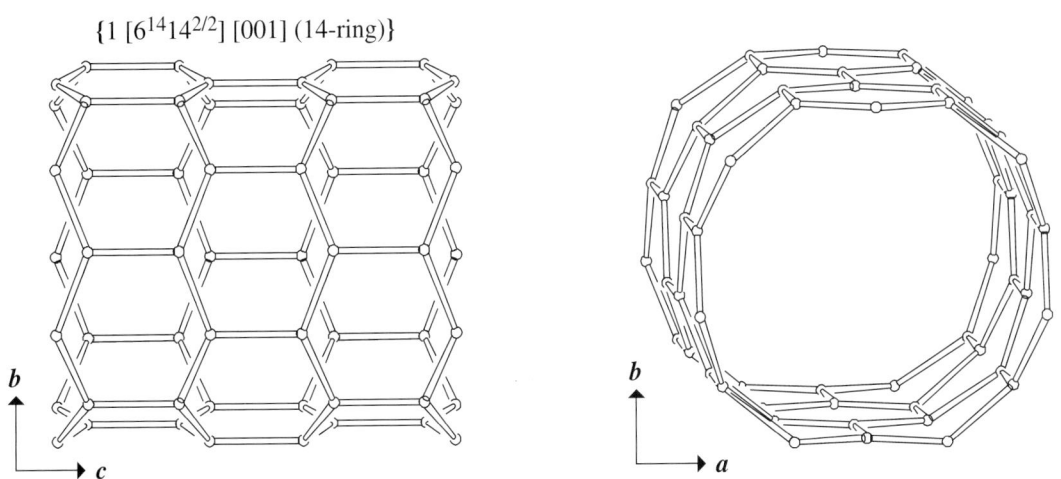

Fig. 3.    Channel viewed perpendicular to the channel axis (left) and along the channel axis (right).

# EAB

## Building Scheme

### 1. Periodic Building Unit

The two-dimensional PerBU of **EAB** consists of a hexagonal array of non-connected planar 6-rings (bold in Figure 1), which are related along **a** and **b** by pure translations. The 6-rings are centered at (0,0) in the **ab** layer. This position is usually called the **A** position. **EAB** belongs to the ABC-6 family (see Introduction).

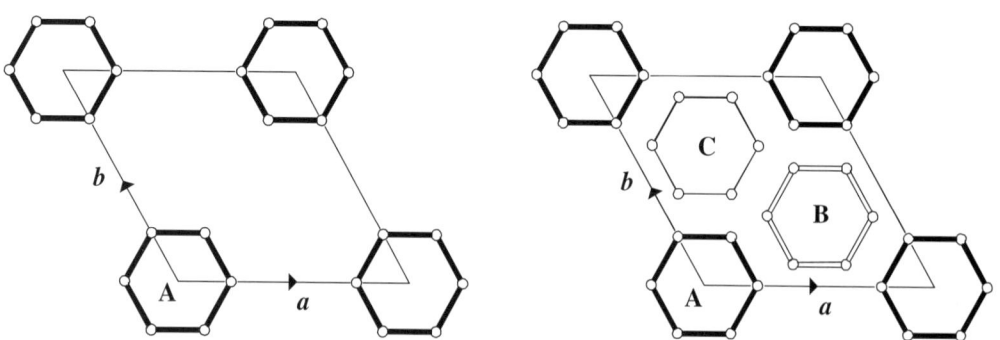

Fig. 1.   PerBU (left) and definition of 6-ring positions used in the stacking modes (right).

### 2. Connection mode

Neighboring PerBUs can be connected along **c** through tilted 4-rings in three different ways (see Introduction). In **EAB** all three connection modes between the PerBUs are observed (Figure 2).

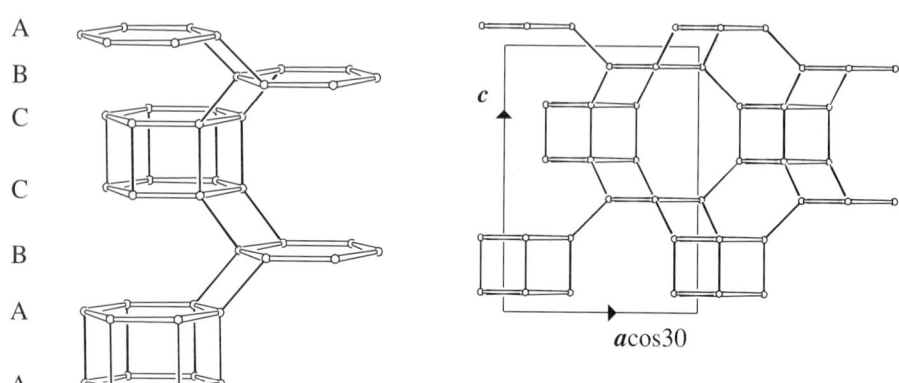

Fig. 2.   Connection mode (left) and unit cell content (right) viewed along **b**. The stacking sequence is given. In the perspective drawing, each PerBU is represented by one 6-ring only.

## 3. Channels and/or cages

The two types of cavities are depicted in Figure 3. The ***gmel*** cavity is also present in **GME**, **AFT**, **AFX**, **EON**, **MAZ** and **OFF**.

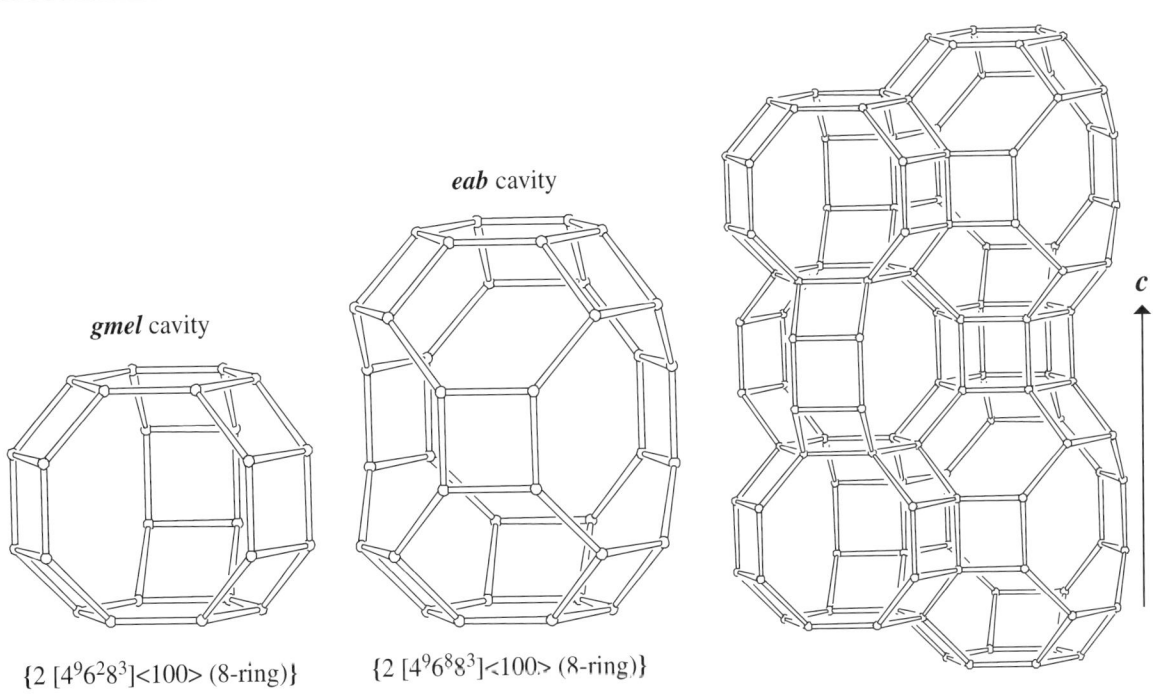

*gmel* cavity                    *eab* cavity

$\{2\ [4^9 6^2 8^3] {<}100{>}\ \text{(8-ring)}\}$         $\{2\ [4^9 6^8 8^3] {<}100{>}\ \text{(8-ring)}\}$

Fig. 3.    Cavities viewed along <120> (left and middle). Two equal non-interconnecting two-dimensional 8-ring channel systems are perpendicular to *c* (right).

# EDI                    Building Scheme

## 1. Periodic Building Unit

Tetragonal **EDI** can be built using the fibrous chain (or natrolite-chain) as one-dimensional PerBU. The chain is composed of units of 5 T atoms (bold in Figure 1). These T5-units ([4³]-cages or 4 = 1 units) are related along **c** by pure translations. (See also **NAT** and **THO**.)

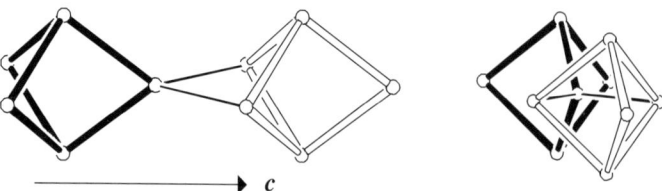

Fig. 1.    PerBU viewed perpendicular to the fibrous chain axis (left) and down the chain axis (right).

## 2. Connection mode

Neighboring PerBUs are related along **a** and **b** by pure translations as shown in Figure 2.

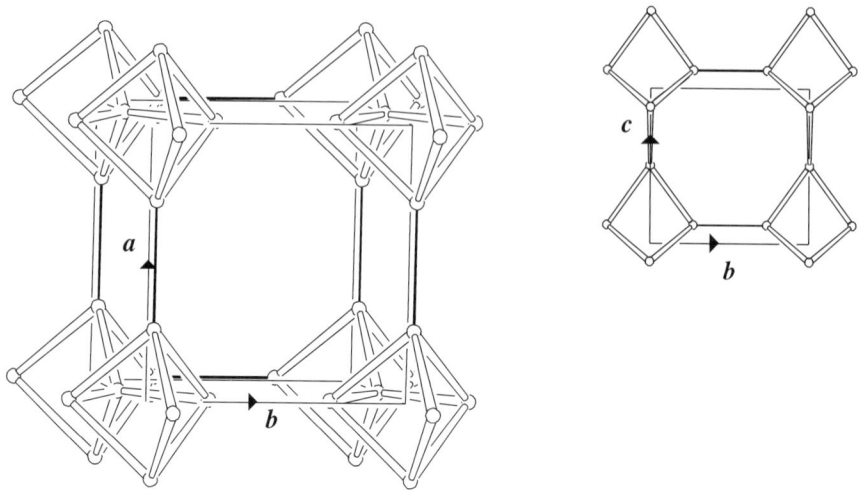

Fig. 2.    Connection mode (and unit cell content) viewed along the fibrous chain axis **c** (left) and unit cell content projected along **a** (or along **b**; right).

## 3. Channels and/or cages

Intersecting 8-ring channels are parallel to <100> and $c$ as shown in Figure 3.

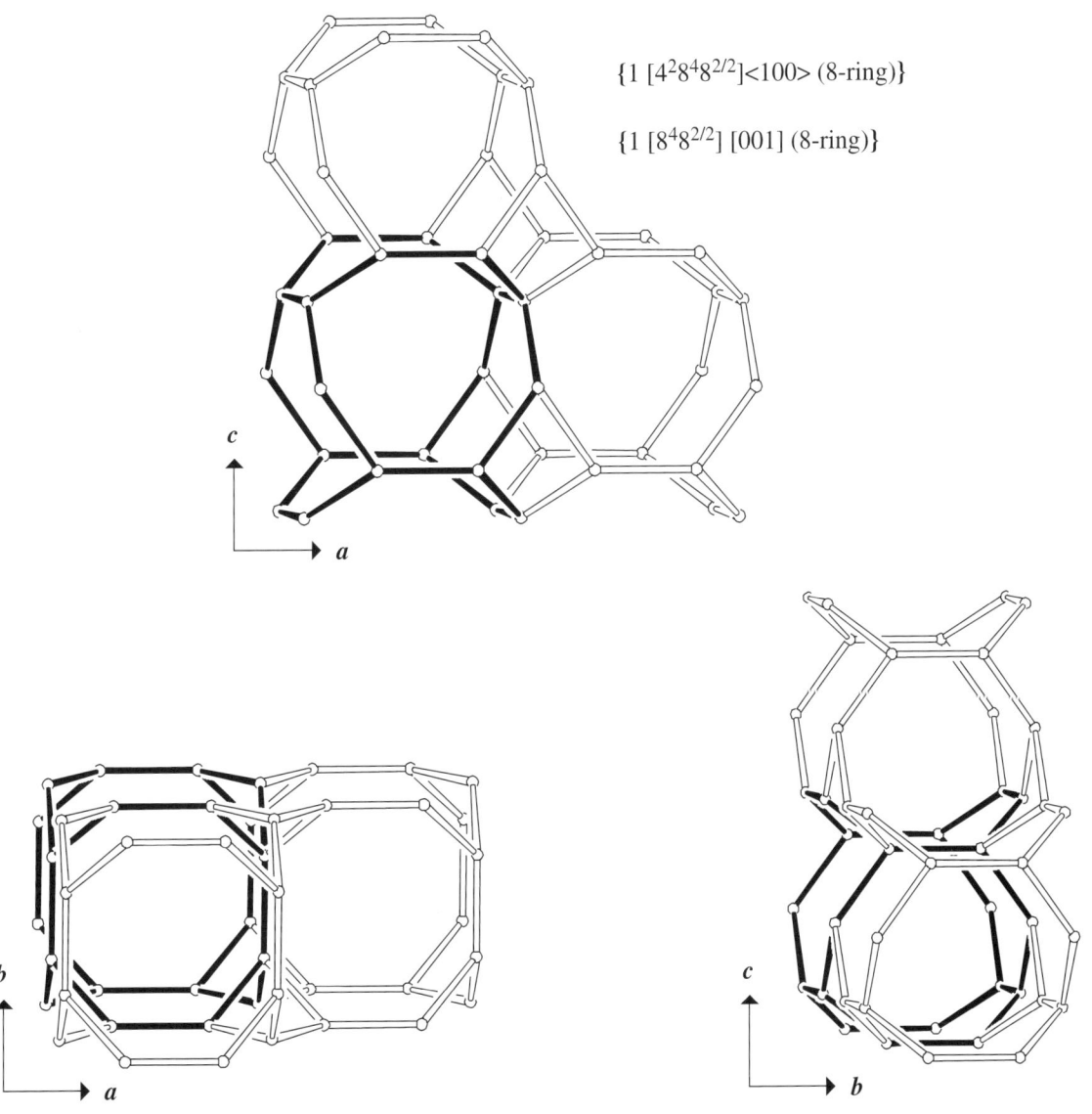

$\{1 \ [4^2 8^4 8^{2/2}] <100> \ (8\text{-ring})\}$

$\{1 \ [8^4 8^{2/2}] \ [001] \ (8\text{-ring})\}$

Fig. 3.  Intersecting 8-ring channels, parallel to <100> and $c$, viewed perpendicular to the channel axes (top), along the channel axis parallel to $c$ (bottom left) and along the channel axis parallel to <100> (bottom right). The channels have [8⁴]-cavities in common. One intersection in bold.

# EMT                    Building Scheme

## 1. Periodic Building Unit

Hexagonal **EMT** can be built using the sodalite cage (or *sod* cage) consisting of 24 T atoms (six 4-rings or four 6-rings; Figure 1a). The two-dimensional PerBU is obtained when *sod* cages are linked through double 6-rings into the hexagonal (faujasite) layer depicted in Figure 1b. (See also **FAU**; compare with **LTA**.)

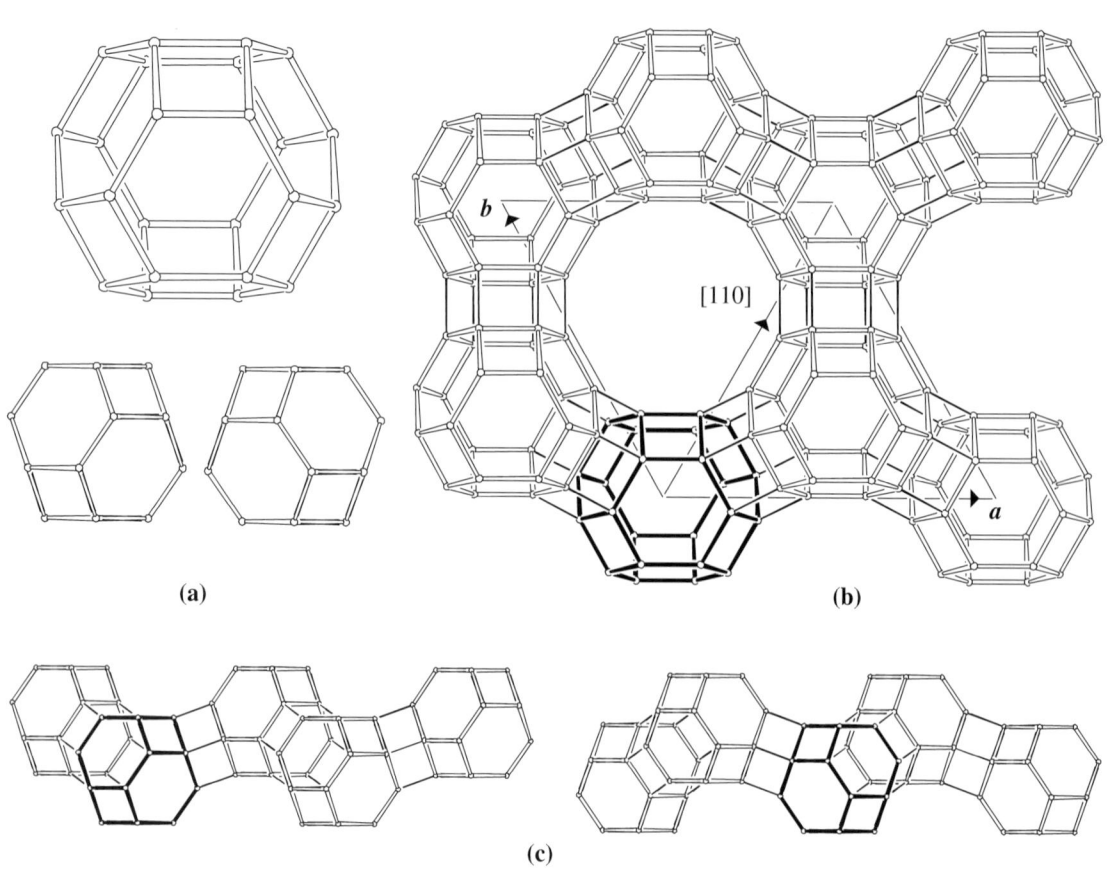

(a)                                      (b)

(c)

Fig. 1.  (a) *sod* Cage viewed along the hexagonal *c* axis (top) and projections of the cage along *b* (different scale; bottom). The two projections are related by a rotation of 60° about *c*; (b) PerBU (one *sod* cage in bold), viewed along *c*; and (c) PerBUs projected along *b*. The two PerBUs are related by a rotation of 60° about *c* or by a mirror plane perpendicular to *c*.

## 2. Connection mode

Neighboring PerBUs, related by a rotation of 60° about *c* followed by a shift of $\frac{1}{2}c$ (i.e. related by a $6_3$-axis) and a lateral shift of $\frac{1}{3}(-a+b)$, are connected along *c* through double 6-rings as shown in Figure 2 on next page.

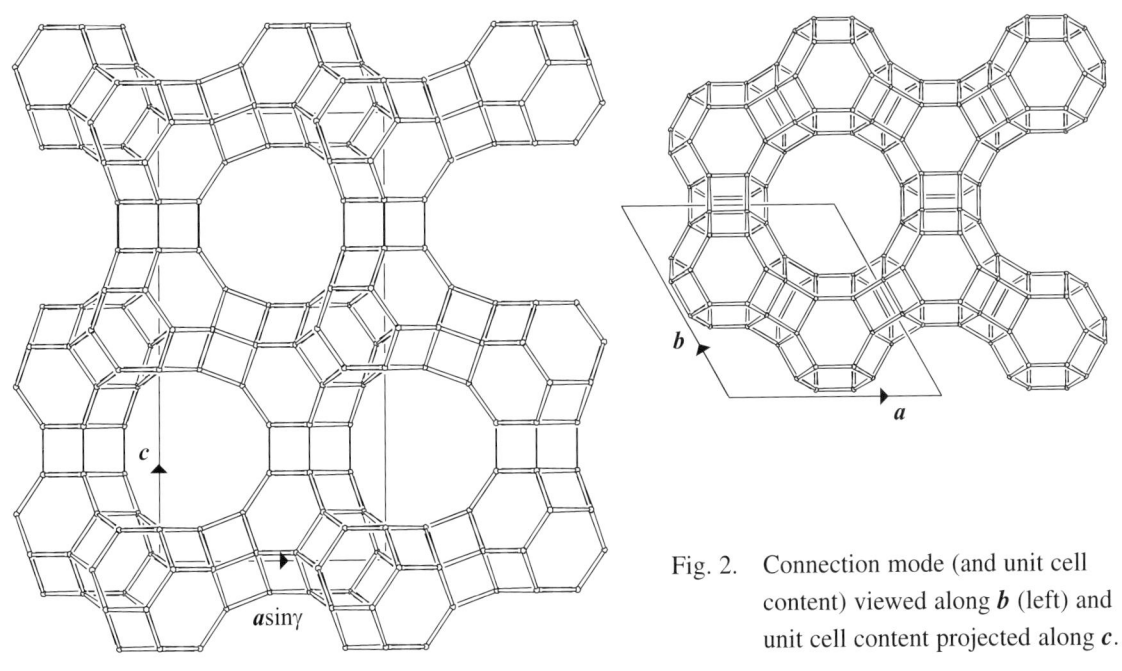

Fig. 2.   Connection mode (and unit cell content) viewed along **b** (left) and unit cell content projected along **c**.

## 3. Channels and/or cages

Intersecting two-dimensional 12-ring channel systems, parallel to <100>, are connected into straight 12-ring channels parallel to **c**. The channel intersection and the 12-ring channel parallel to **c** are depicted in Figure 3.

{3 [4²¹6⁶12⁵] <100> (12-ring), [001] (12-ring)}

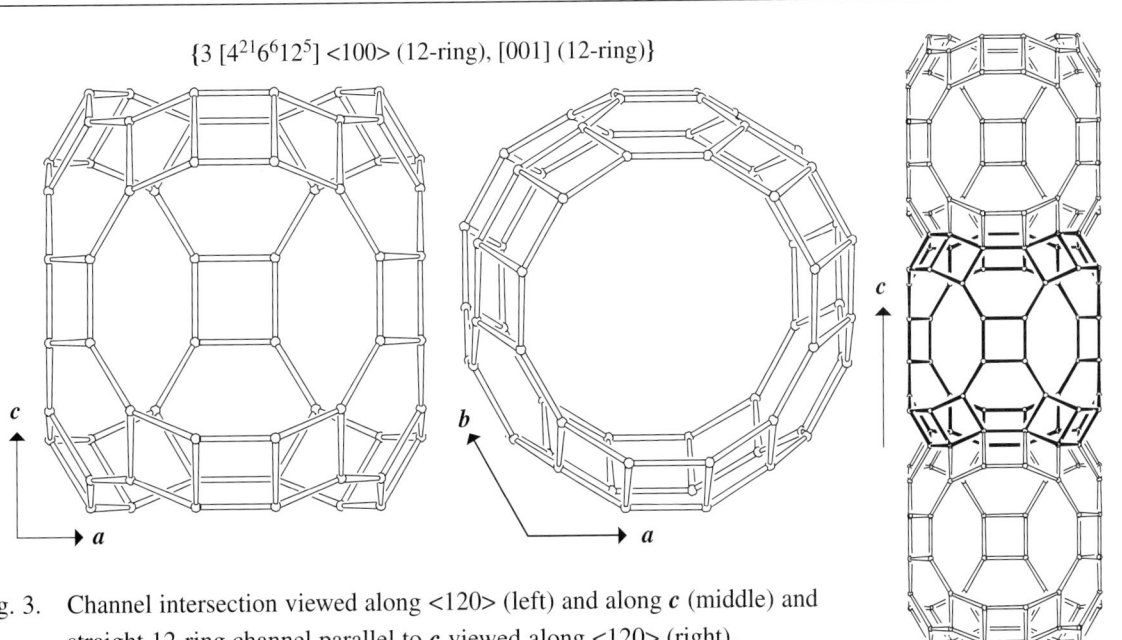

Fig. 3.   Channel intersection viewed along <120> (left) and along **c** (middle) and straight 12-ring channel parallel to **c** viewed along <120> (right).

# EON

## Building Scheme

### 1. Periodic Building Unit

**EON** can be built using the saw chain (bold in Figure 1) parallel to *a*. Six saw chains are connected into infinite building unit 1 consisting of a column of *gmel* cavities connected through common 6-rings (Figure 1, left). Four saw chains are connected into infinite building unit 2 shown in Figure 1 (right). Building units 1, related along *b* by a screw rotation of 180° about *b* and a shift of $\frac{1}{2}a$, are connected through 5-rings into PerBU1 (Figure 1, left). PerBU1 equals the (010) layer in **MAZ**. Building units 2, related along *b* by a rotation of 180° about *c*, are connected along *b* through 4-rings into PerBU2 (Figure 1, right). PerBU2 equals the (010) layer in **MOR**.

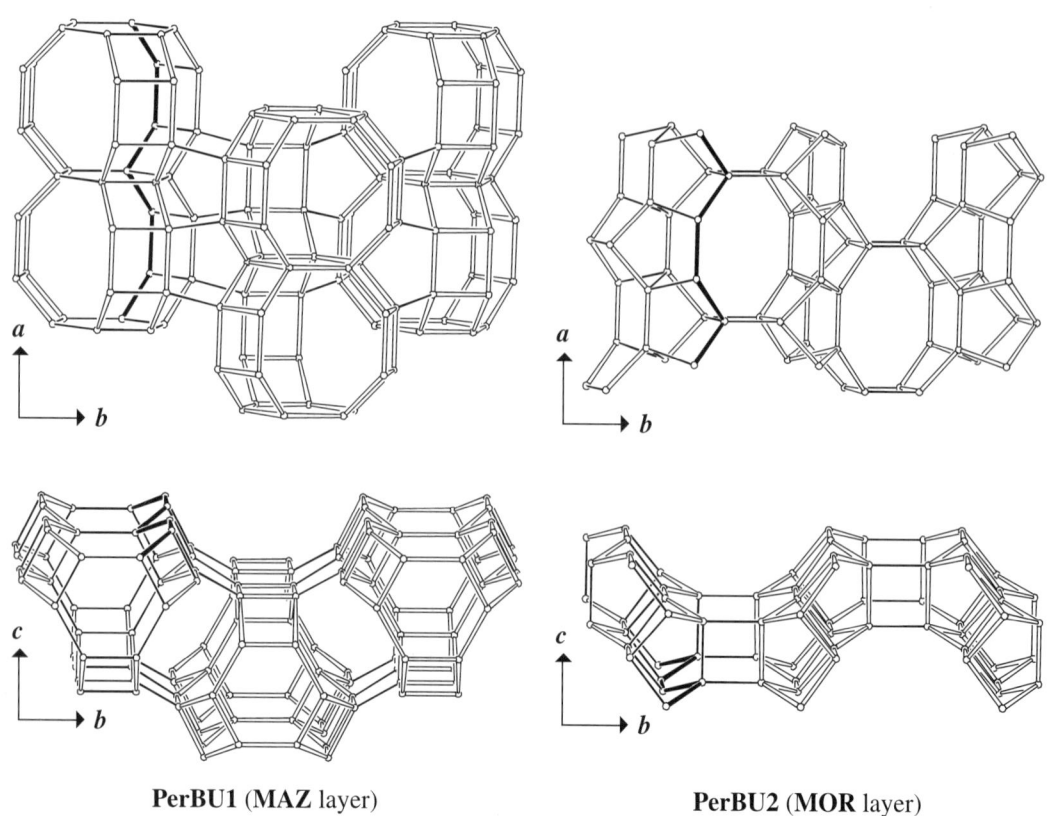

**PerBU1 (MAZ** layer)  **PerBU2 (MOR** layer)

Fig. 1.  Left: PerBU1 (or **MAZ** layer) composed of columns of (fused) *gmel* cavities viewed along *c* (top) and along the saw chain direction parallel to *a* (bottom). Right: PerBU2 (or **MOR** layer) viewed along *c* (top) and along the saw chain direction parallel to *a* (bottom).

# Building Scheme  **EON**

## 2. Connection mode

**MAZ** and **MOR** layers alternate along $c$. They are connected along $c$ through crankshaft chains. 12-Ring and 8-ring channels parallel to $a$ are formed as shown in Figure 2. The 8-ring channel formed between the layers is equal to the 8-ring channel in the **MAZ** layer.

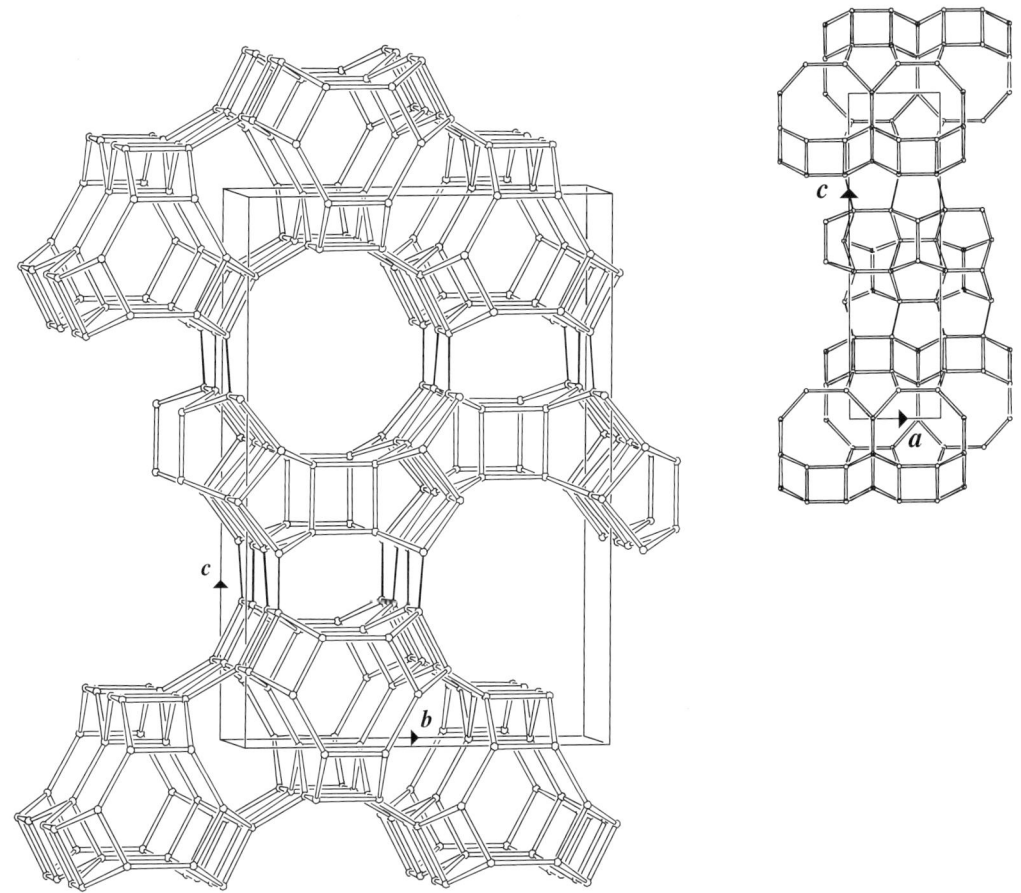

Fig. 2.  Connection mode (and unit cell content) viewed along $a$ (left) and unit cell content projected along $b$ (right).

## 3. Channels and/or cages

Interconnecting 12-ring and 8-ring channels are parallel to *a* as shown in Figure 3. The 8-ring channel is topologically equivalent to (one of) the 8-ring channels in **MAZ**, **MON**, **MOR**, **RSN** and **VSV**. Columns of *gmel* cavities are interconnected along *b* by the same type of 8-ring channels, leading to a rather complicated three-dimensional channel system with 8-ring apertures. The linkage between channels and *gmel* cavity is also shown in Figure 3. The *gmel* cavity is also present in **AFT**, **AFX**, **EAB**, **GME**, **MAZ** and **OFF**.

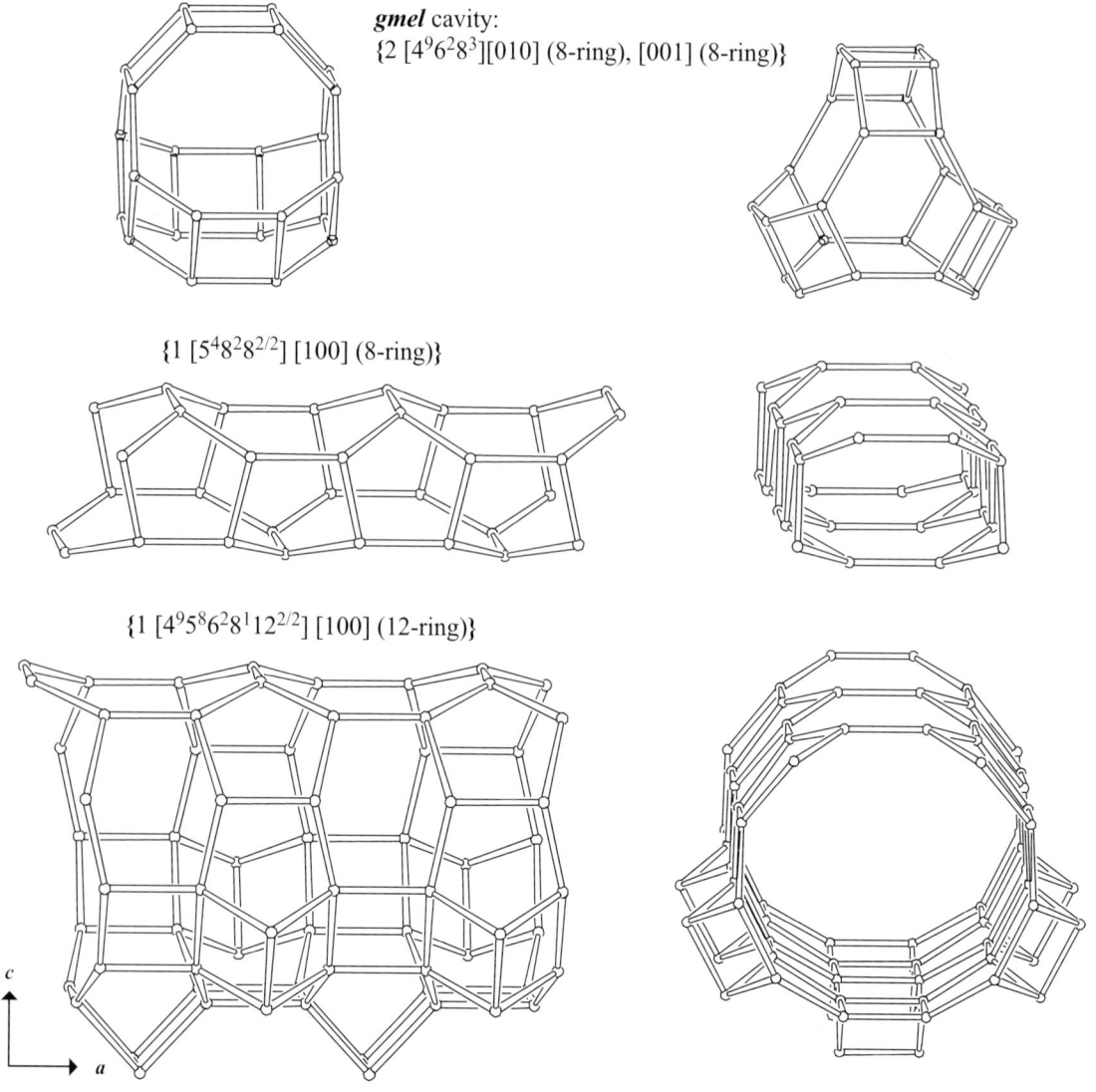

*gmel* cavity:
$\{2\ [4^9 6^2 8^3][010]\ (8\text{-ring}),\ [001]\ (8\text{-ring})\}$

$\{1\ [5^4 8^2 8^{2/2}]\ [100]\ (8\text{-ring})\}$

$\{1\ [4^9 5^8 6^2 8^1 12^{2/2}]\ [100]\ (12\text{-ring})\}$

Fig. 3a.   *gmel* Cavity (top), 8-ring channel (middle), 12-ring channel (bottom) viewed along *b* (left) and along *a* (right).

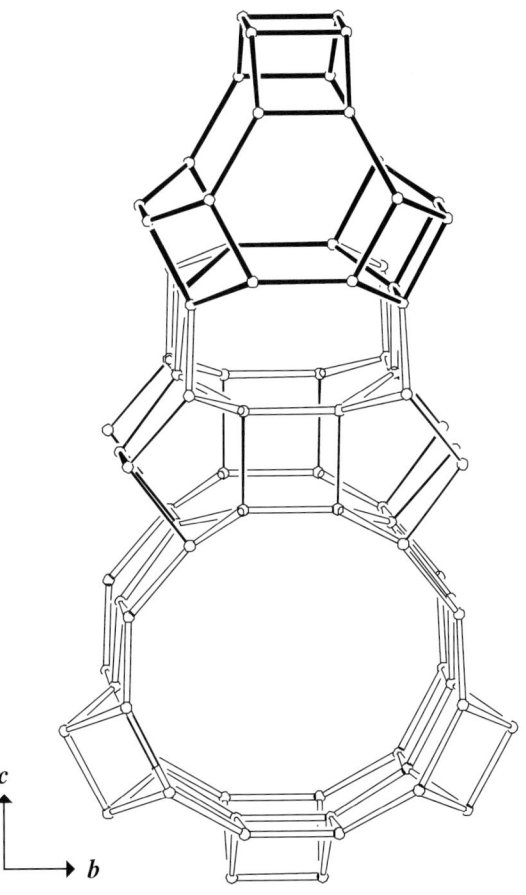

Fig. 3b.   Linkage of **gmel** cavity (in bold), 8-ring and 12-ring channels viewed along **a**.

# EPI                Building Scheme

## 1. Periodic Building Unit

Finite building units of 12 T atoms are composed of two 5-1 units (bold in Figure 1). The two-dimensional PerBU is obtained when T12-units, related along **a** and **c** by pure translations, are connected into a layer with an oblique repeat unit. Twisted saw chains along **bxc** and zigzag chains along **c** are formed. (See also **DAC** and **MOR**.)

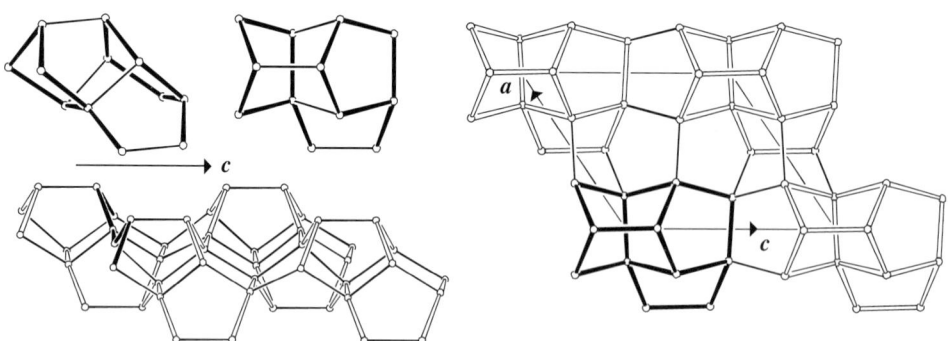

Fig. 1.   Top: T12-unit viewed along **bxc** (left) and along **b** (right). Right: PerBU viewed along **b** (one T12-unit in bold). Bottom left: PerBU built from twisted saw chains (one twisted saw chain in bold) viewed along **bxc**.

## 2. Connection mode

Neighboring PerBUs, related by a shift of $\frac{1}{2}(a+b)$, are connected along **b** as shown in Figure 2.

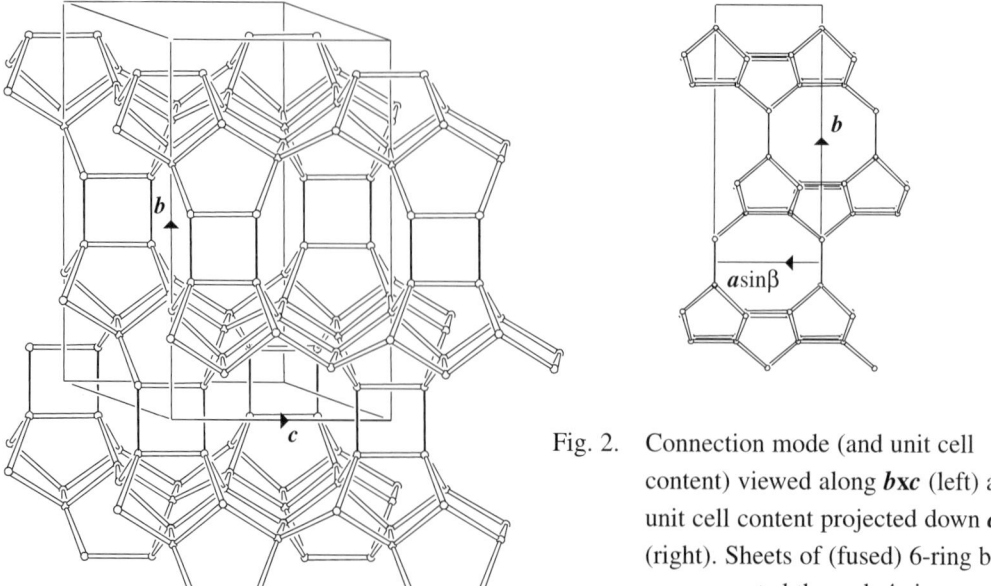

Fig. 2.   Connection mode (and unit cell content) viewed along **bxc** (left) and unit cell content projected down **c** (right). Sheets of (fused) 6-ring boats are connected through 4-rings.

## 3. Channels and/or cages

Intersecting 8-ring channels are parallel to *a*, *c* and [101]. The channel intersection is depicted in Figure 3 together with the linkage of intersections.

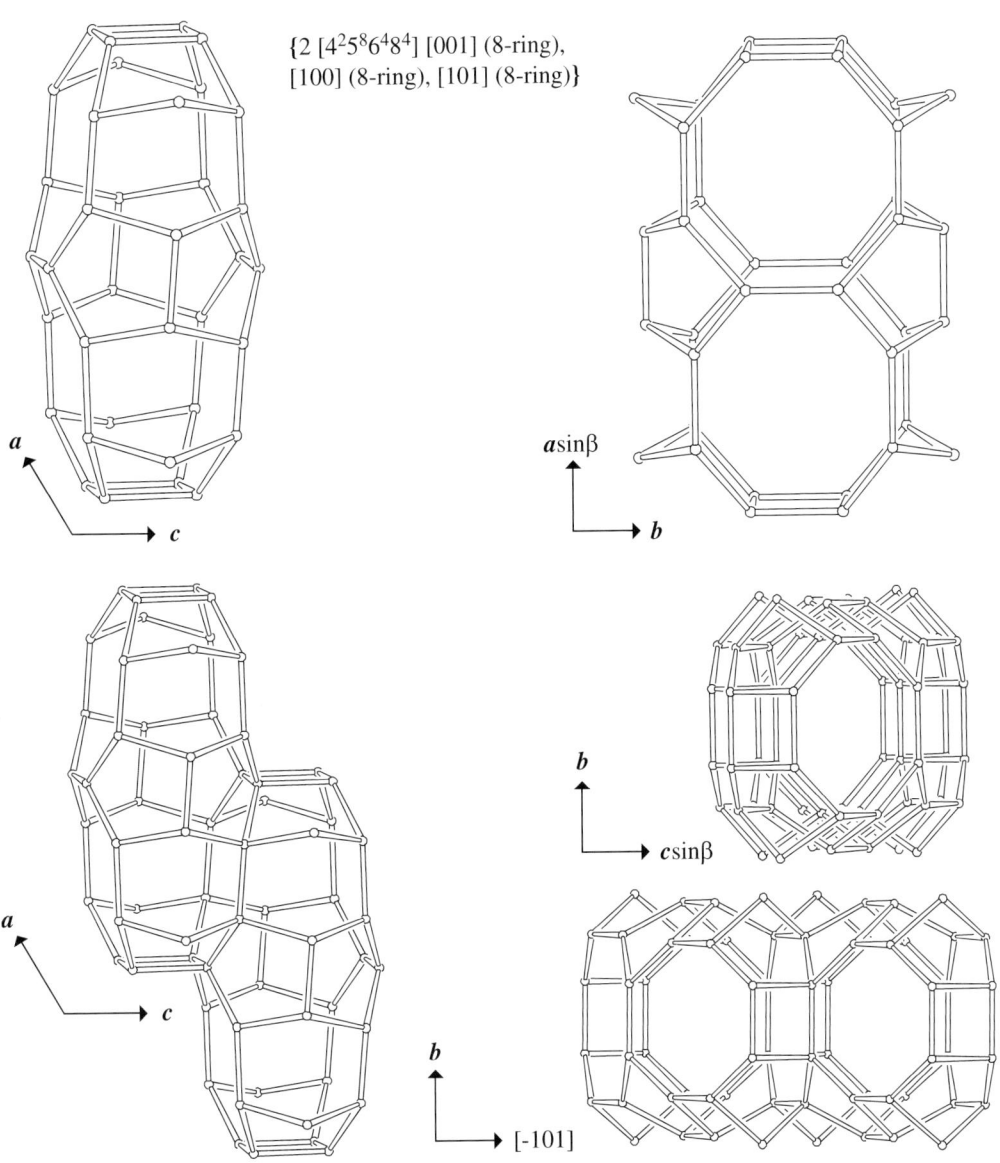

{2 [$4^2 5^8 6^4 8^4$] [001] (8-ring), [100] (8-ring), [101] (8-ring)}

Fig. 3. Top: channel intersection viewed along *b* (top left) and along the 8-ring channel axis parallel to *c* (right). The 8-ring channels parallel to *c* are interconnecting through 10-rings. Bottom: channel intersections are connected through common 8-rings. View along *b* (left), along the channel axis parallel to *a* (top right) and along the channel axis parallel to [101] (bottom right).

# ERI   Building Scheme

## 1. Periodic Building Unit

The two-dimensional PerBU of **ERI** consists of a hexagonal array of non-connected planar 6-rings (bold in Figure 1), which are related along *a* and *b* by pure translations. The 6-rings are centered at (0,0) in the *ab* layer. This position is usually called the **A** position. **ERI** belongs to the ABC-6 family (see Introduction).

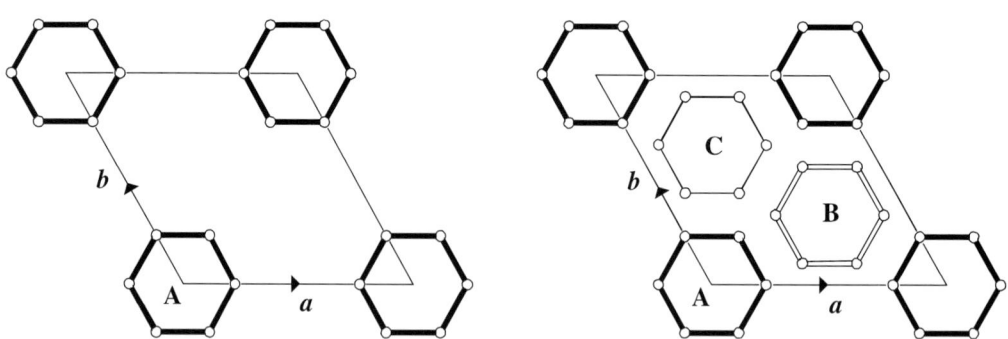

Fig. 1.   PerBU (left) and definition of 6-ring positions used in the stacking modes (right).

## 2. Connection mode

Neighboring PerBUs can be connected along *c* through tilted 4-rings in three different ways (see Introduction). In **ERI** all three connection modes between the PerBUs are observed. *can* Cages are formed (see Figure 2). (See also **OFF**.)

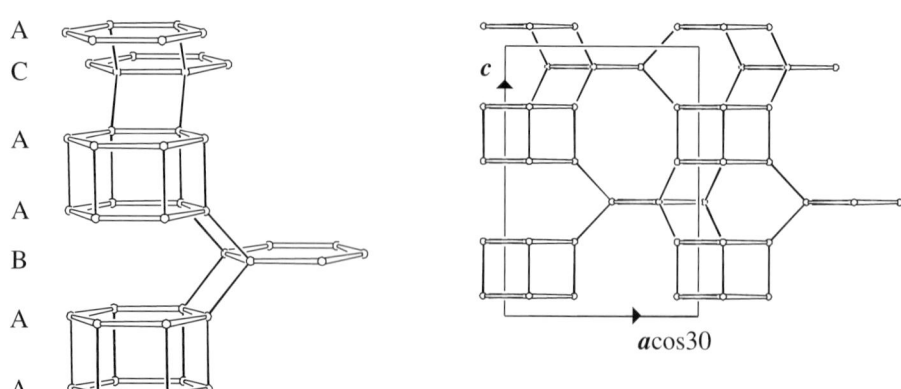

Fig. 2.   Connection mode (left) and unit cell content (right) viewed along *b*. The stacking sequence is given. In the perspective drawing, each PerBU is represented by one 6-ring only.

## 3. Channels and/or cages

The *can* cage and *eri* cavity in **ERI** are depicted in Figure 3. **ERI** can also be built from *can* cages (see also **LTL**, **MOZ** and **OFF**). The *can* cage is also present in **AFG**, **CAN**, **FAR**, **FRA**, **GIU**, **LIO**, **LOS**, **LTL**, **LTN**, **MAR**, **MOZ**, **OFF**, **SAT**, **SBS**, **SBT**, **TOL** and **-WEN**. A three-dimensional 8-ring channel system is obtained when *eri* cavities are connected through common 8-rings.

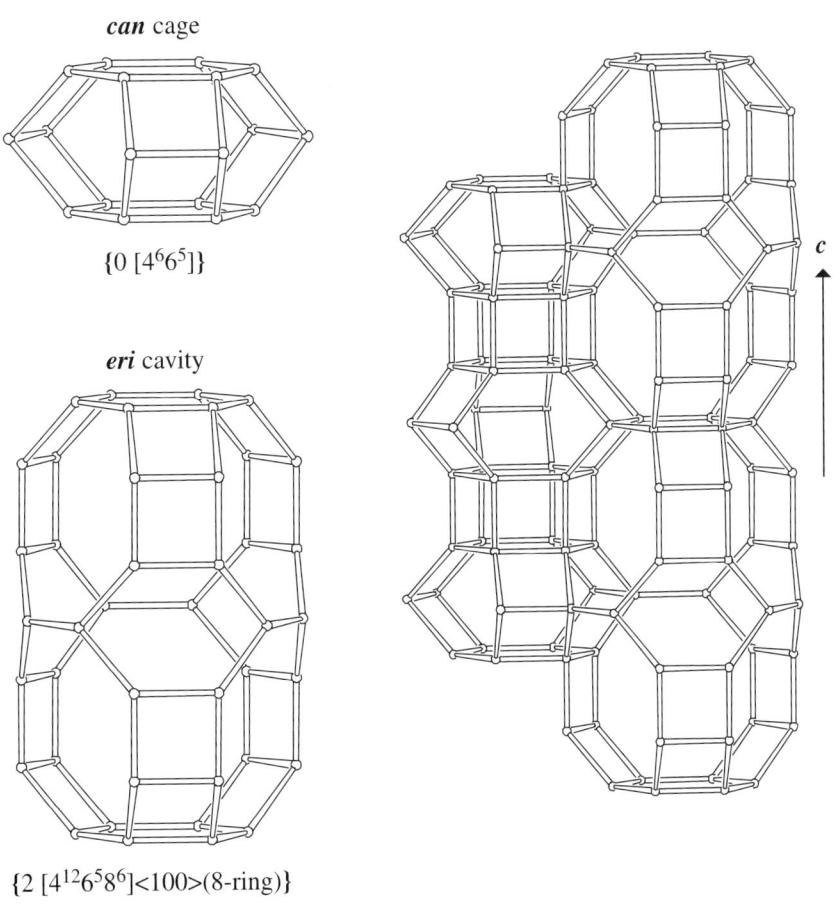

*can* cage

$\{0 \, [4^6 6^5]\}$

*eri* cavity

$\{2 \, [4^{12} 6^5 8^6] <100> (8\text{-ring})\}$

Fig. 3.   *can* Cage (top left) and *eri* cavity (bottom left) viewed along <120>. Linkage between cages and cavity viewed along <120> (right). Two-dimensional 8-ring channels, perpendicular to *c*, are interconnected along *c* through 12-rings in the *eri* cavities leading to a three-dimensional channel system.

# ESV          Building Scheme

## 1. Periodic Building Unit

**ESV** can be built using units composed of two 5-1 units (one T12-unit bold in Figure 1). Neighboring T12-units, related along *c* by a screw rotation of 180° about *c* and along *a* by a pure translation, are connected through (fused) 4- and 5-rings into the two-dimensional PerBU shown in Figure 1.

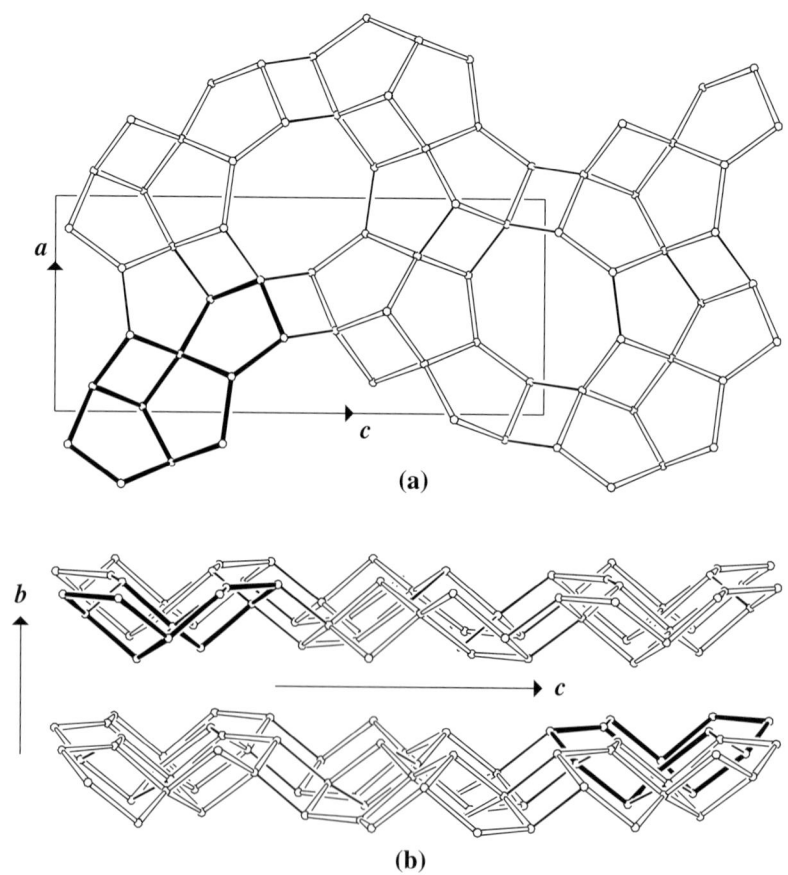

**(a)**

**(b)**

Fig. 1.  (a) PerBU viewed along *b*; (b) PerBUs viewed along *a*. The PerBUs are related by a rotation of 180° about *b*.

## 2. Connection mode

Neighboring PerBUs, related along *b* by a screw rotation of 180° about *b* and a lateral shift of $\frac{1}{2}(a + c)$, are connected along *b* through 4- and 6-rings as shown in Figure 2 on next page.

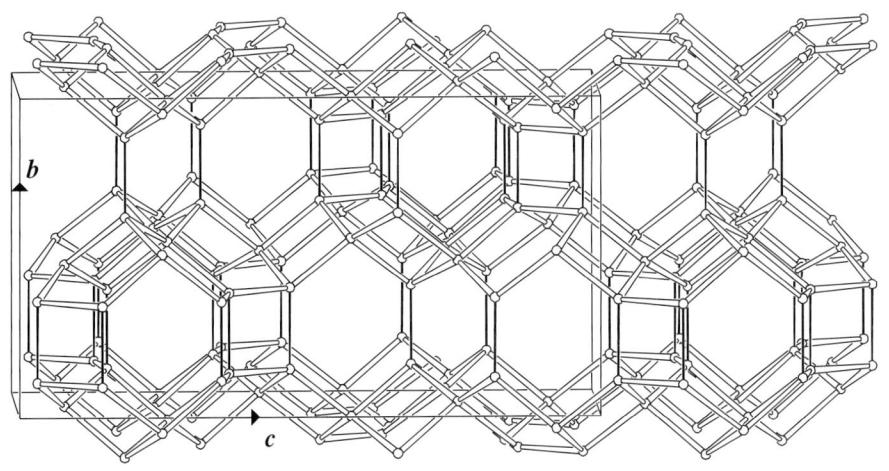

Fig. 2. Connection mode and unit cell content viewed along *a*. The projection of the unit cell content along *b* is equal to Figure 1a.

## 3. Channels and/or cages

Small cages are connected, by sharing 6-rings, into rows parallel to *a*. Cavities, related along *b* by 2-fold screw axes along *b*, are connected through common 8-rings into 8-ring channels parallel to *b* as illustrated in Figure 3.

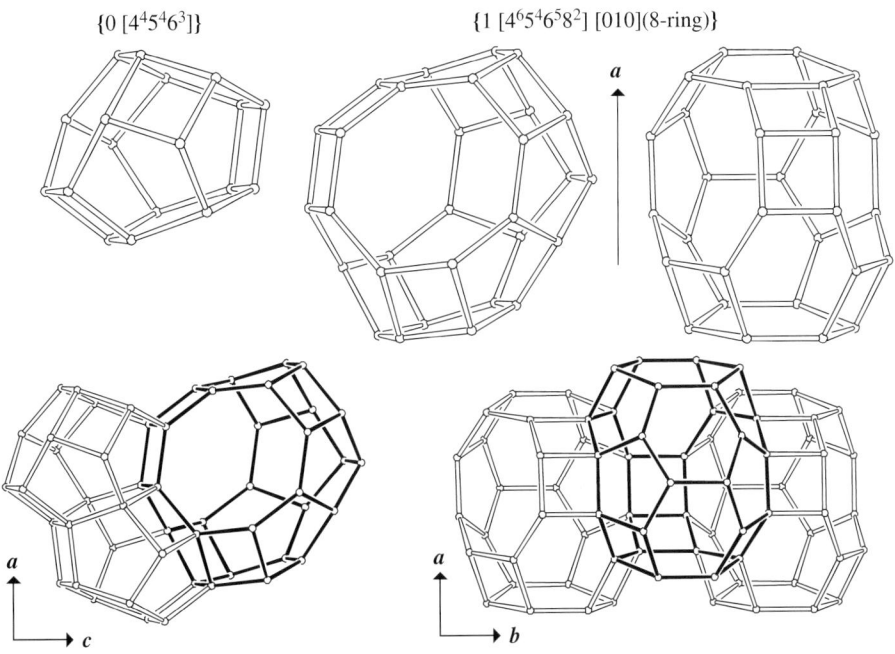

Fig. 3. Top: cage (left) and cavity (middle) viewed along *b* and cavity viewed along *c* (right).
Bottom: linkage of cages and cavity (in bold) along *a* and *c* viewed along *b* (left) and linkage of cavities into an 8-ring channel parallel to *b* viewed along *c* (right).

# ETR

# Building Scheme

## 1. Periodic Building Unit

Hexagonal **ETR** can be built using the chain depicted in Figure 1 as one-dimensional PerBU. The chain is composed of units of 16 T atoms (one in bold) and consists of two 2-fold (1,2)-connected double 4-rings. The T16-units, related along **a** by pure translations, are connected along **a** through 4-rings. (See also **CGS**.)

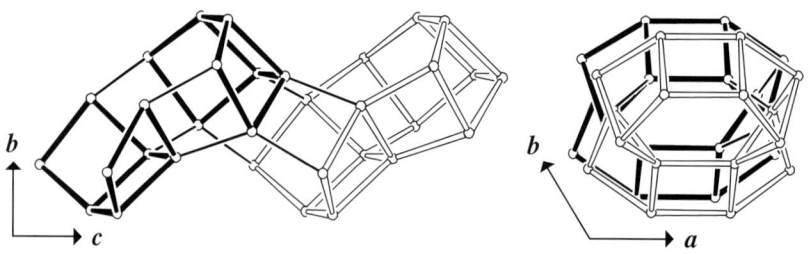

Fig. 1.   PerBU viewed along **b** (left) and along **a** (right).

## 2. Connection mode

Neighboring PerBUs, related by a rotation of 120° about **c**, are connected around 3-fold axes through 4-rings as illustrated in Figure 2.

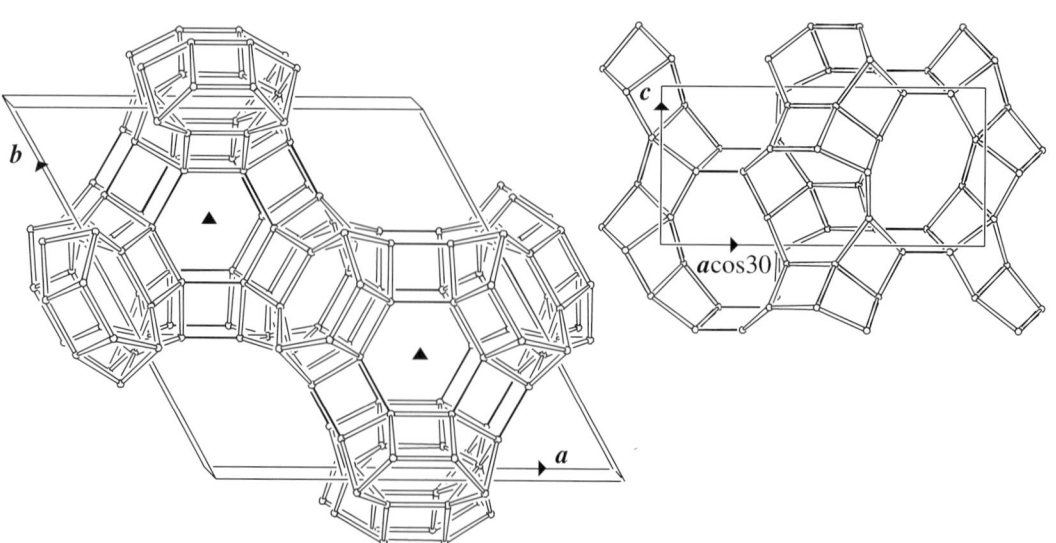

Fig. 2.   Connection mode (and unit cell content) viewed along **c** (left) and unit cell content projected along **b** (right).

## 3. Channels and/or cages

Two types of cavities are present. The first cavity is the intersection in a two-dimensional 8-ring channel system perpendicular to *c*. The second cavity is the intersection of an 18-ring channel parallel to *c* with this 8-ring channel system. The channel intersections are shown in Figure 3 together with their linkage into channels.

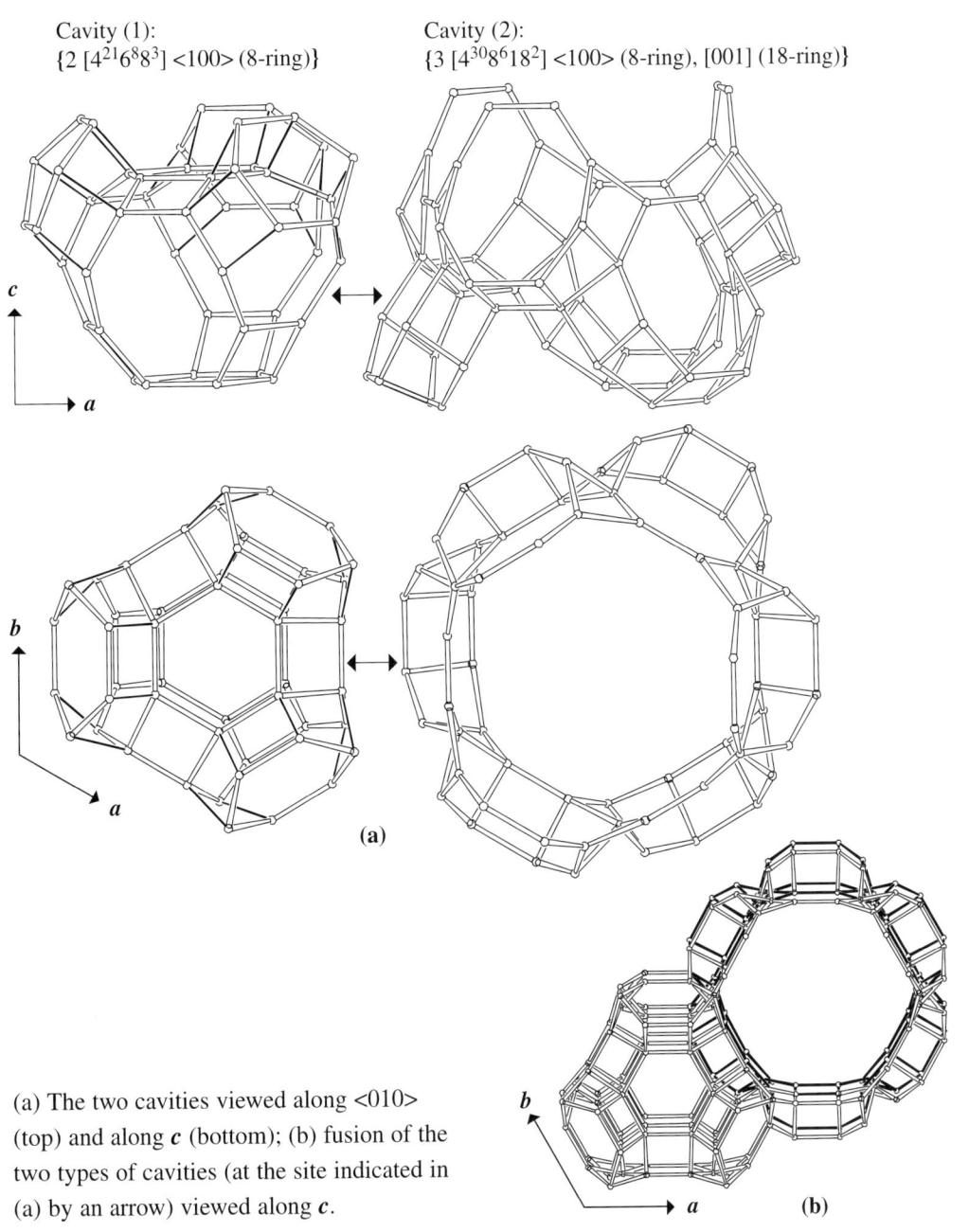

Cavity (1):
$\{2\ [4^{21}6^88^3] <100> (8\text{-ring})\}$

Cavity (2):
$\{3\ [4^{30}8^618^2] <100> (8\text{-ring}), [001] (18\text{-ring})\}$

(a)

(b)

Fig. 3.   (a) The two cavities viewed along <010> (top) and along *c* (bottom); (b) fusion of the two types of cavities (at the site indicated in (a) by an arrow) viewed along *c*.

# EUO                                    Building Scheme

## 1. Periodic Building Unit

**EUO** can be built using building units composed of 14 T atoms: two 1-5-1 units (one T14-unit bold in Figure 1). The two-dimensional PerBU is obtained when T14-units, related along *a* by a mirror glide plane perpendicular to *b* and along *b* by a rotation of 180° about *c*, are connected into the *ab* layer as shown in Figure 1. (Compare with the PerBUs in **IWV**, **NES** and **NON**.)

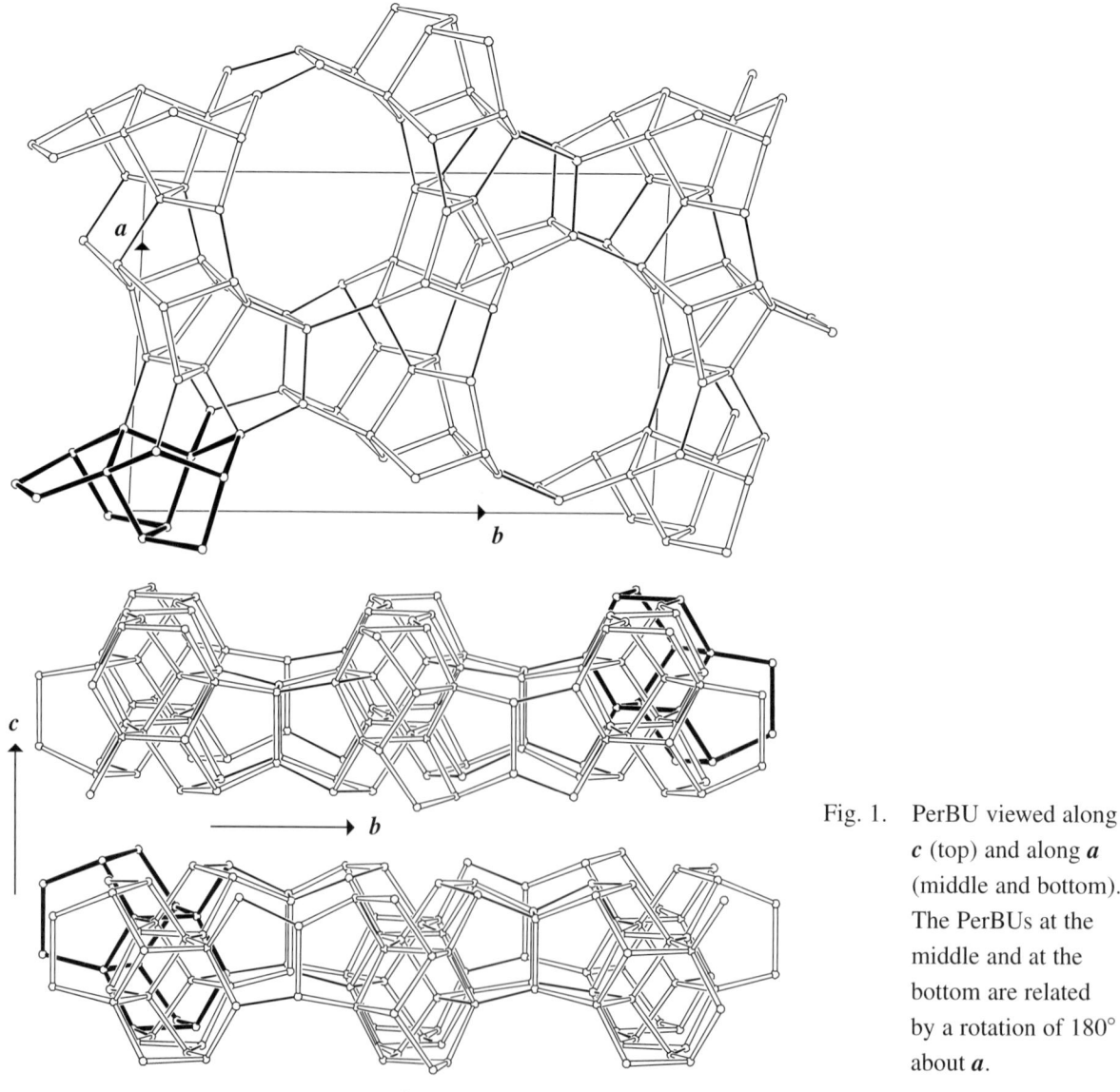

Fig. 1. PerBU viewed along *c* (top) and along *a* (middle and bottom). The PerBUs at the middle and at the bottom are related by a rotation of 180° about *a*.

## 2. Connection mode

Neighboring PerBUs, related along *c* by a rotation of 180° about *a*, are connected along *c* as shown in Figure 2 on next page.

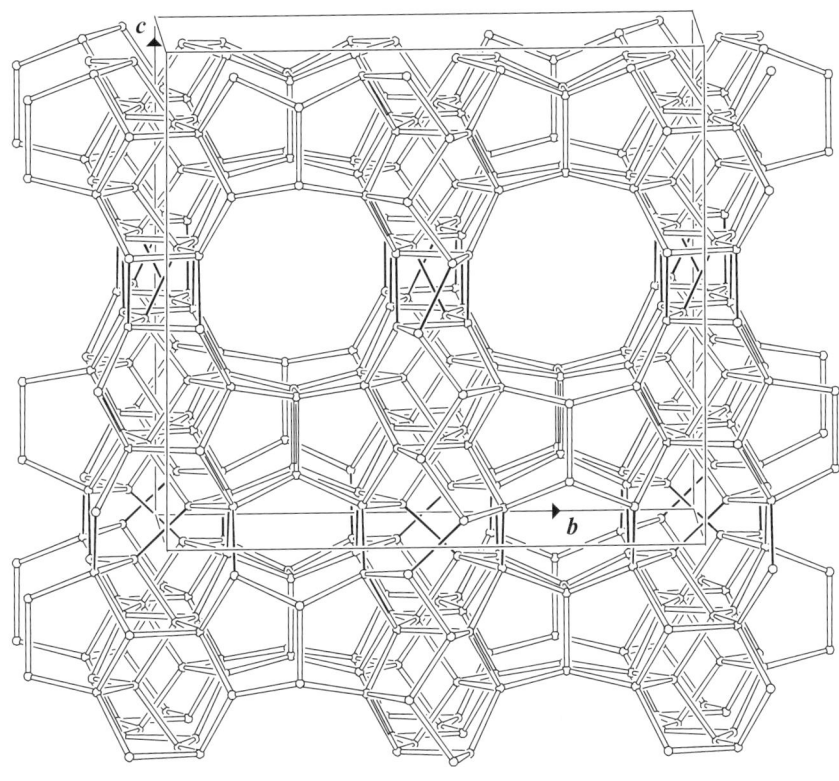

Fig. 2.   Connection mode and unit cell content viewed along *a*.

## 3. Channels and/or cages

Cavities with large side-pockets (Figure 3) are connected through common 10-rings into non-interconnecting 10-ring channels parallel to *a*.

{1 [5⁸6¹⁴10²] [100] (10-ring)}

$\{1\ [5^8 6^{14} 10^2]\ [100]\ (10\text{-ring})\}$

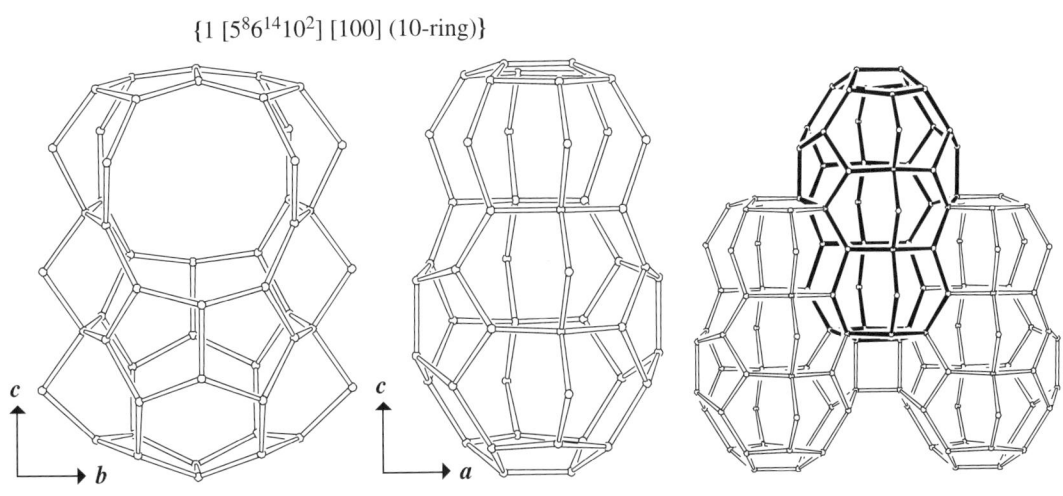

Fig. 3.   Cavity viewed along *a* (left) and along *b* (middle). Fused cavities (right) form 10-ring channels parallel to *a*.

# EZT

# Building Scheme

## 1. Periodic Building Unit

**EZT** can be built using units of 24 T atoms (Figure 1a). The T24-unit consists of a T14-unit and a T16-unit sharing a 6-ring. The T14-unit (in bold) consists of a 3-fold (1,3,5)-connected double 6-ring "capped" on each side by a single T atom (or a $[4^6 6^3]$-cage; compare with **AFS** and **BPH**). The T16-unit is composed of a 4-fold (1,2,4,5)-connected double 6-ring with a two "handles" (or a $[4^2 6^6]$-cage; compare e.g. with **IFR**). The two-dimensional PerBU is obtained by connecting T24-units, related by a shift of $\frac{1}{2}(a+b+c)$, through 4-rings. The PerBU is shown in Figure 1b.

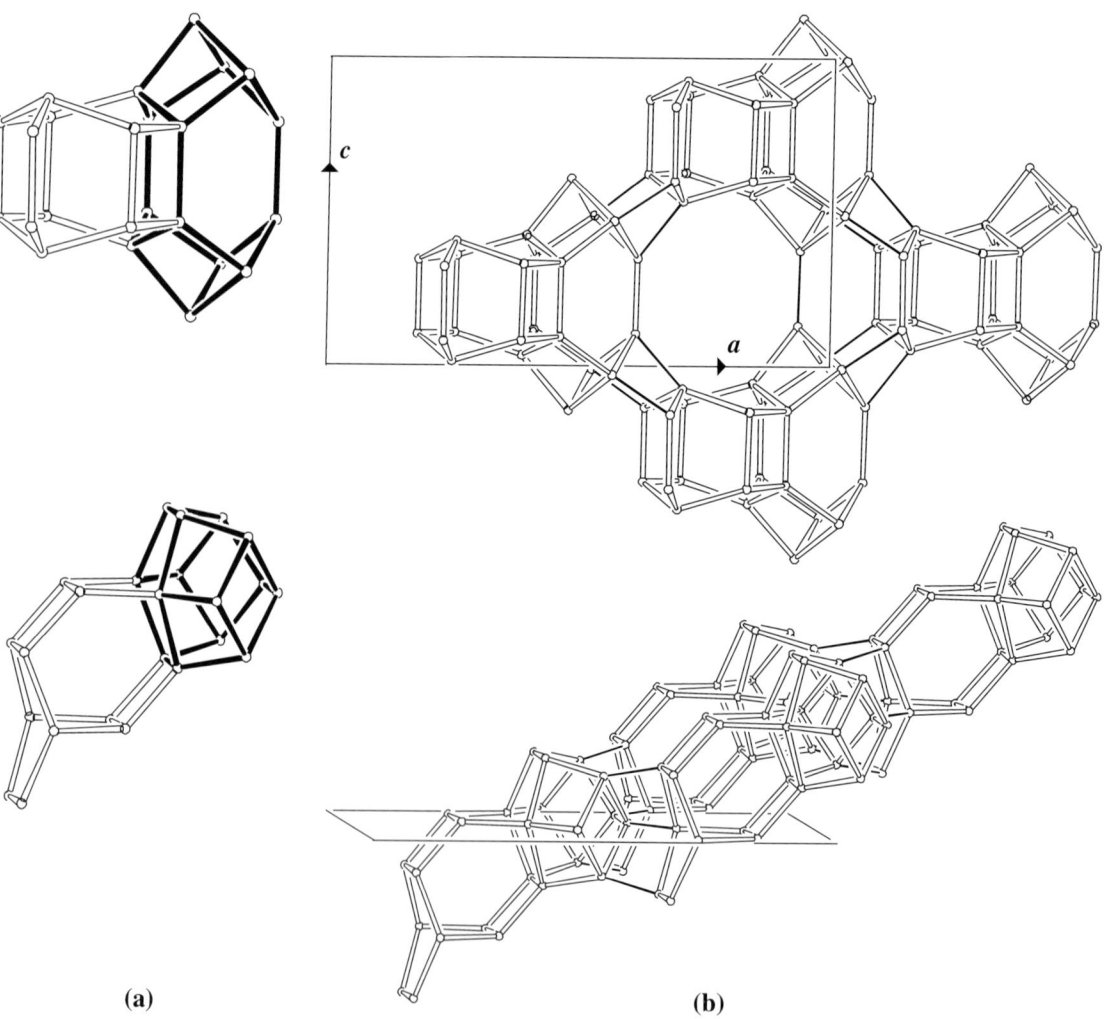

**(a)**                                                                 **(b)**

Fig. 1.   (a) Fusion of T14- and T16-units into the T24 building unit viewed along $b$ (top) and along $c$ (bottom); (b) PerBU viewed along $c$ (top) and along $b$ (bottom).

# Building Scheme

## 2. Connection mode

Neighboring PerBUs, related along $b$ by a pure translation, are connected along $b$ through 4-rings as shown in Figure 2. 8-Ring channels parallel to $c$ are formed.

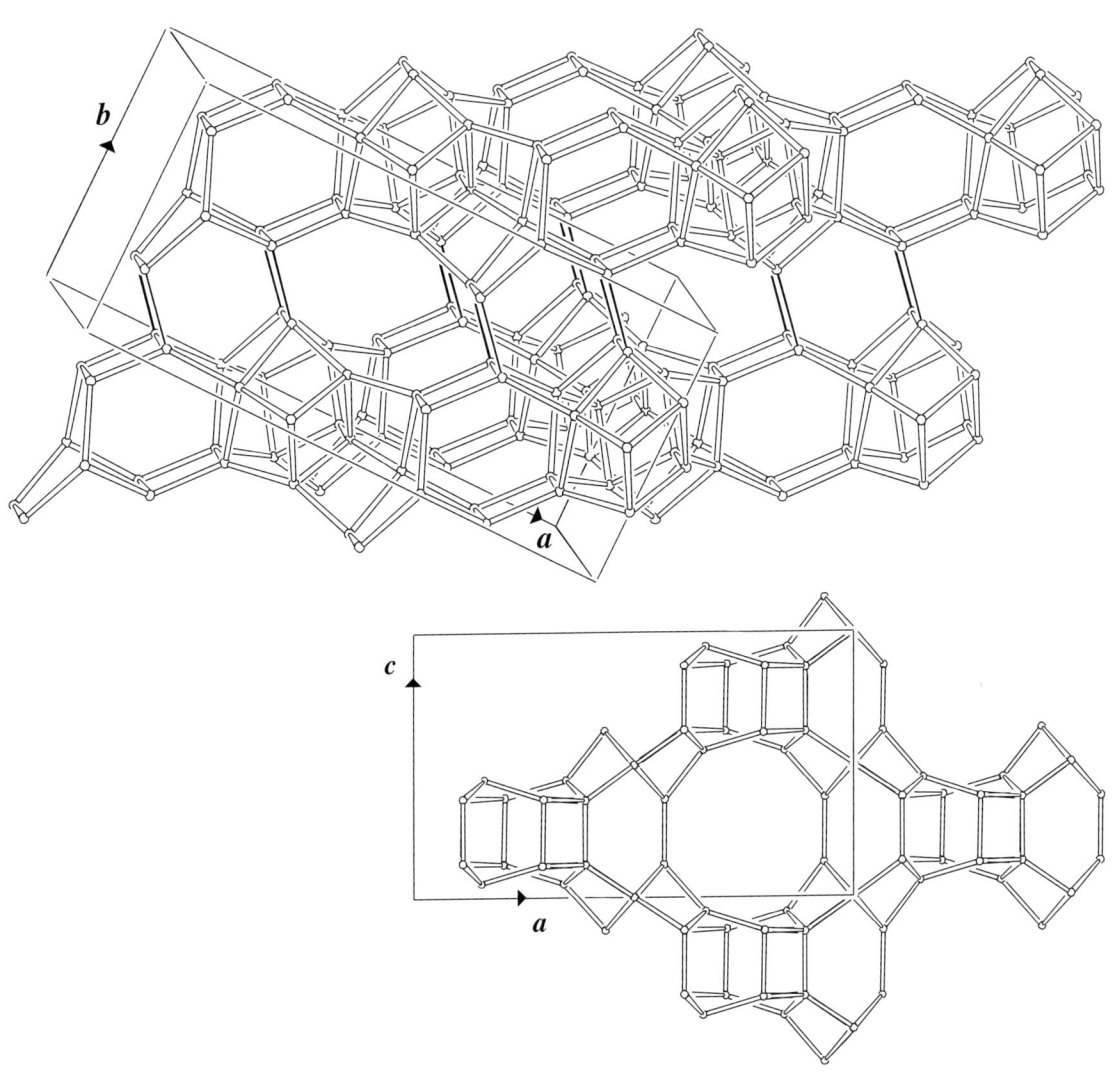

Fig. 2. Connection mode (and unit cell content) viewed along $c$ (top) and unit cell content projected along $b$ (bottom).

# EZT

## Cage/Channel

### 3. Channels and/or cages

Sinusoidal 12-ring channels are parallel to *b* and highly elliptical 8-ring channels are parallel to *c*. The channel intersection is shown in Figure 3 together with the linkage of intersections into 8- and 12-ring channels. Pairs of 8-ring channels are interconnecting along *a* through 12-rings.

$\{2\ [4^8 6^4 8^4 12^2]\ [010]\ (12\text{-ring}),\ [001]\ (8\text{-ring})\}$

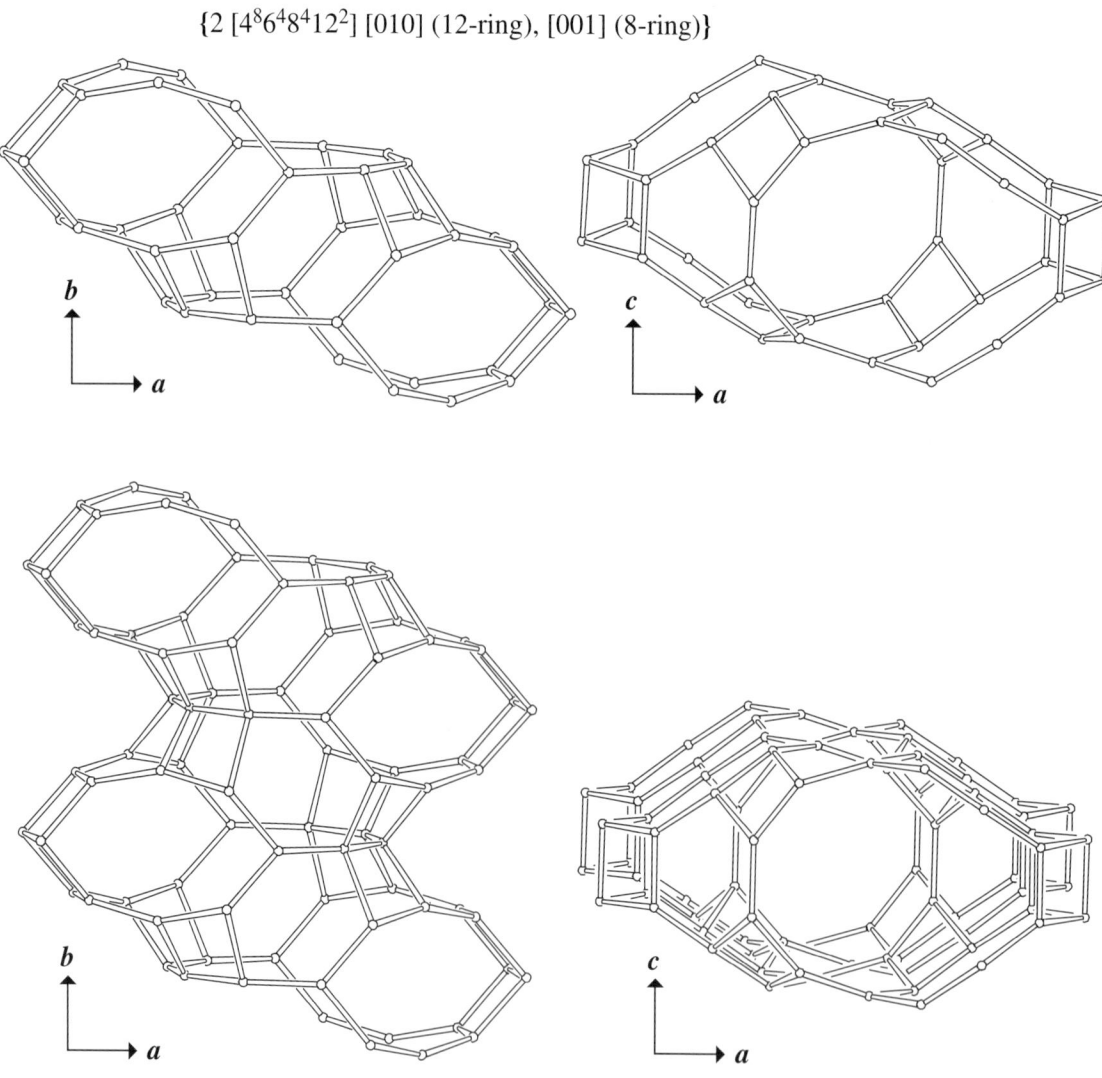

Fig. 3a.　Channel intersection viewed along *c* (left) and along *b* (right). Bottom: connection of intersections into sinusoidal 12-ring channels parallel to *b* viewed perpendicular to the channel axis (left) and along the channel axis (right).

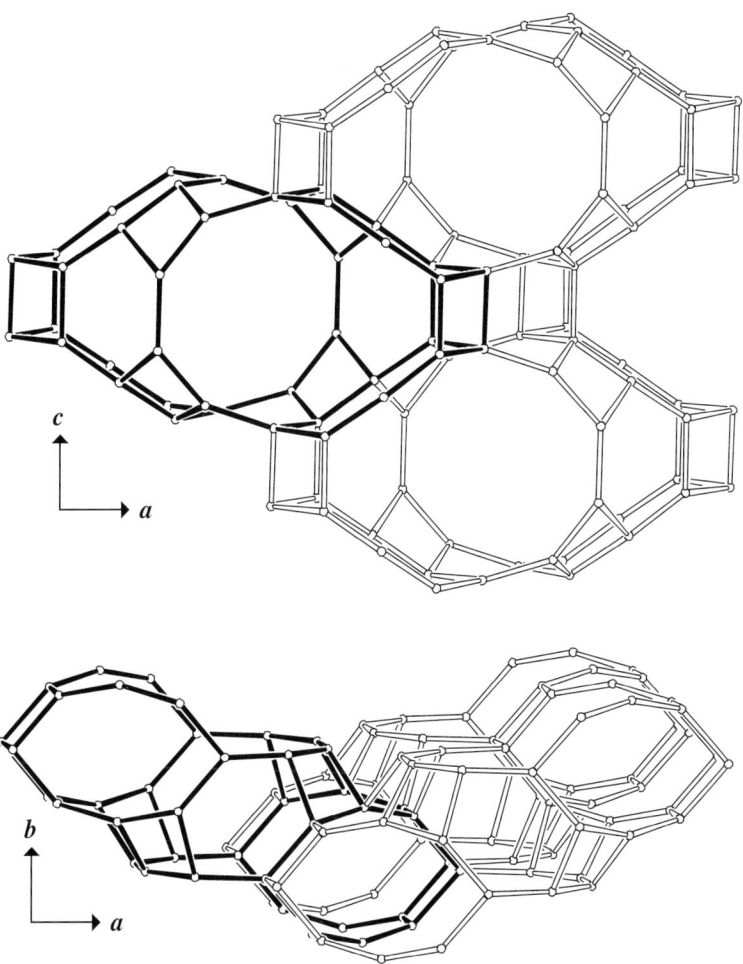

Fig. 3b.  Linkage of intersections into 8-ring channels parallel to **c** viewed perpendicular to the channel axis (top) and along the 8-ring channel axis (bottom).

# FAR                    Building Scheme

## 1. Periodic Building Unit

The two-dimensional PerBU of **FAR** consists of a hexagonal array of non-connected planar 6-rings (bold in Figure 1), which are related along $a$ and $b$ by pure translations. The 6-rings are centered at (0,0) in the $ab$ layer. This position is usually called the **A** position. **FAR** belongs to the ABC-6 family (see Introduction).

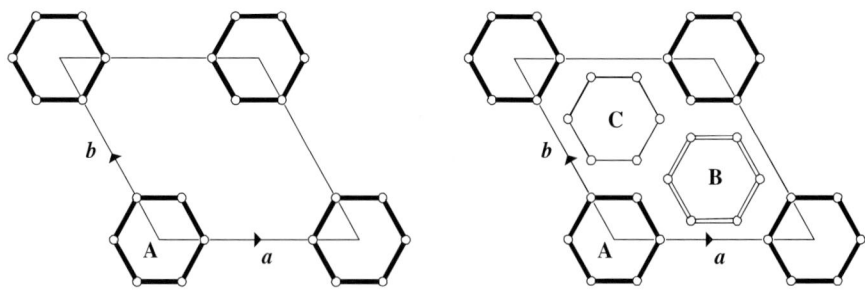

Fig. 1.    PerBU (left) and definition of 6-ring positions used in the stacking modes (right).

## 2. Connection mode

Neighboring PerBUs are connected along $c$ through tilted 4-rings by connection modes **(1)** and **(2)** (see Introduction) as illustrated in Figure 2.

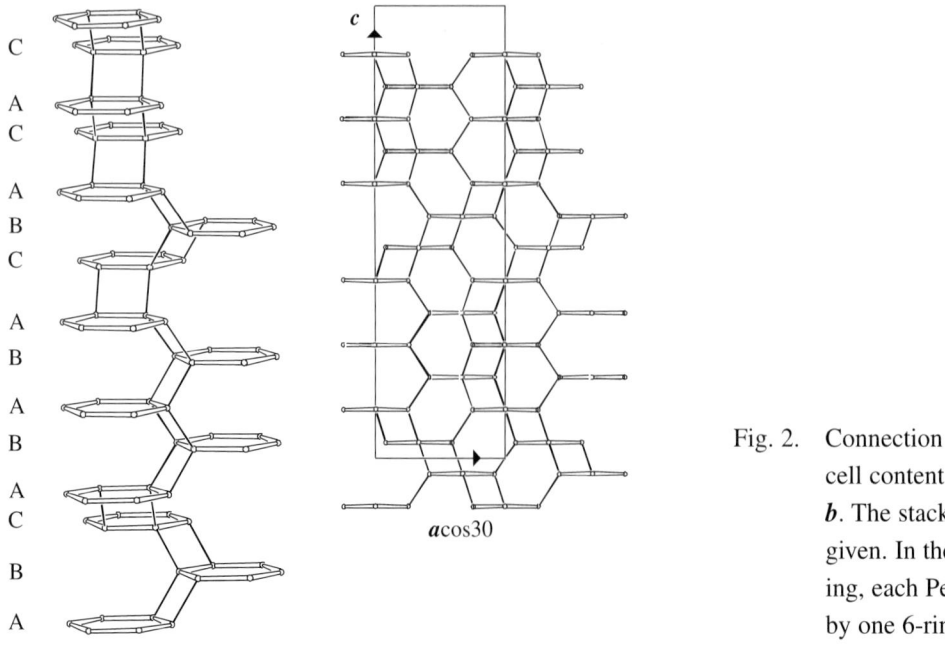

Fig. 2.    Connection mode (left) and unit cell content (right) viewed along $b$. The stacking sequence is given. In the perspective drawing, each PerBU is represented by one 6-ring only.

## 3. Channels and/or cages

The three types of cages in **FAR** are depicted in Figure 3. The *can* cage is also present in **AFG**, **CAN**, **ERI**, **FRA**, **GIU**, **LIO**, **LOS**, **LTL**, **LTN**, **MAR**, **MOZ**, **OFF**, **SAT**, **SBS**, **SBT**, **TOL** and **-WEN**. The *sod* cage is also found in **EMT**, **FAU**, **FRA**, **LTA**, **LTN**, **MAR**, **SOD** and **TSC** and the *lio* cage is also observed in **AFG**, **LIO**, **MAR** and **TOL**. Apertures of "channels" are formed by 6-rings only.

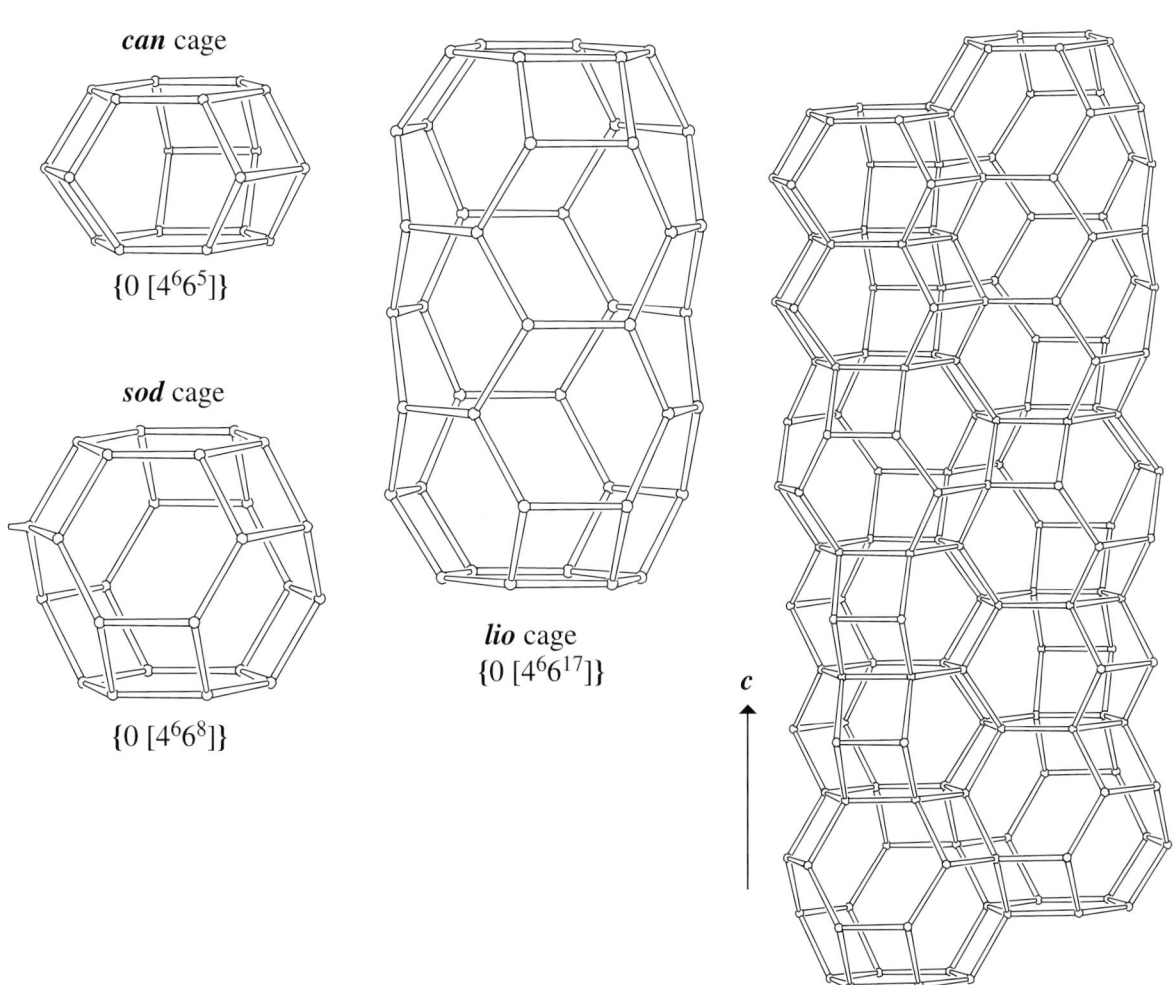

*can* cage
$\{0\,[4^6 6^5]\}$

*sod* cage
$\{0\,[4^6 6^8]\}$

*lio* cage
$\{0\,[4^6 6^{17}]\}$

*c*

Fig. 3. From left to right: *can* cage, *sod* cage, *lio* cage and connection of cages viewed perpendicular to *c*.

# FAU                    **Building Scheme**

## 1. Periodic Building Unit

Cubic **FAU** can be built using the sodalite cage (or *sod* cage) consisting of 24 T atoms (six 4-rings or four 6-rings; Figure 1(a)). The two-dimensional PerBU is obtained when *sod* cages are linked through double 6-rings (D6Rs) into the hexagonal faujasite layer depicted in Figure 1(b). The hexagonal layer corresponds to the (111) layer in cubic **FAU**. (See **EMT**; Compare with **LTA**.)

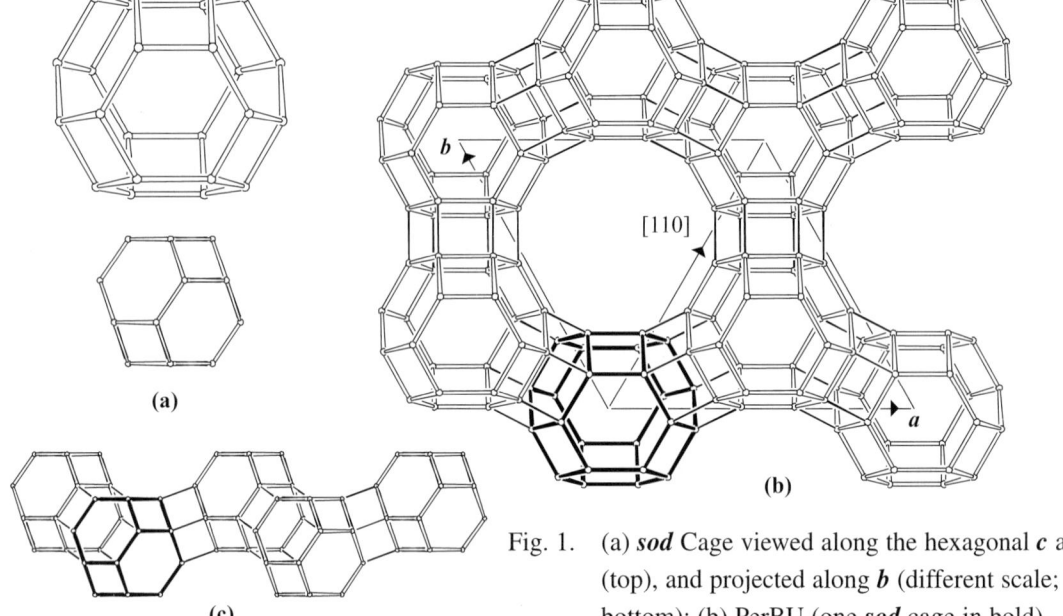

(a)

(b)

(c)

Fig. 1.   (a) *sod* Cage viewed along the hexagonal *c* axis (top), and projected along *b* (different scale; bottom); (b) PerBU (one *sod* cage in bold), viewed along *c*; (c) PerBU projected along *b*.

## 2. Connection mode

Neighboring PerBUs, related by a shift of $\frac{1}{3}(-\boldsymbol{a}+\boldsymbol{b})$, are connected along *c* through D6Rs (Figure 2).

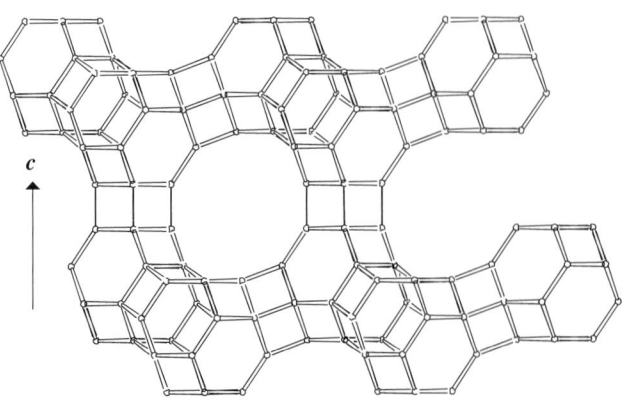

Fig. 2a.   Connection mode viewed along *b* (in hexagonal description).

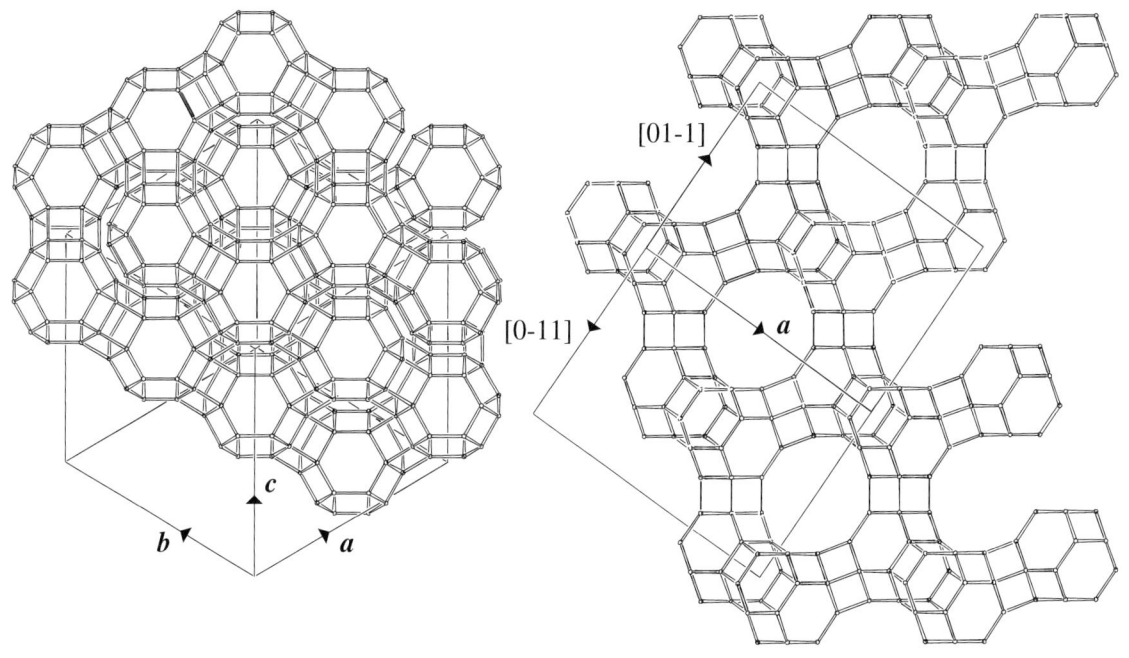

Fig. 2b.     Unit cell content projected along the cubic axes [111] (left) and [011] (right).

## 3. Channels and/or cages

In cubic **FAU** intersecting 12-ring channels are parallel to <110>. The channel intersection is depicted in Figure 3.

{3 [4$^{18}$6$^4$12$^4$] <110> (12-ring)}

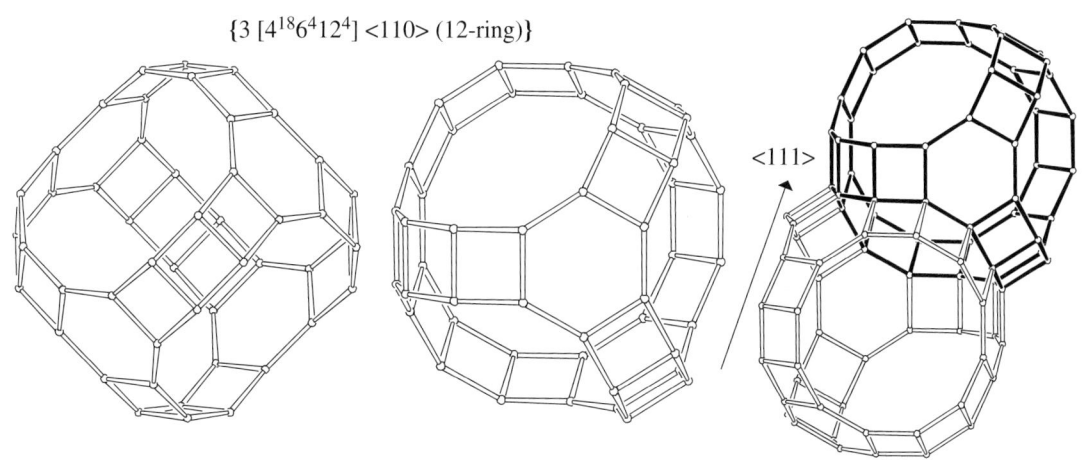

Fig. 3.     Channel intersection viewed along <001> (left) and along <111> (middle). Right: channel intersections are interconnected along <111> through common 12-rings.

# FER

# Building Scheme

## 1. Periodic Building Unit

**FER** can be built using the saw chain (bold in Figure 1) parallel to $a$. Three saw chains are connected into an infinite building unit. A two-dimensional PerBU is obtained when infinite building units, related along $b$ by a rotation of 180° about $b$ and a shift of $\frac{1}{2}c$, are connected along $b$ into the $bc$ layer depicted in Figure 1 (right). (See also **CDO**.)

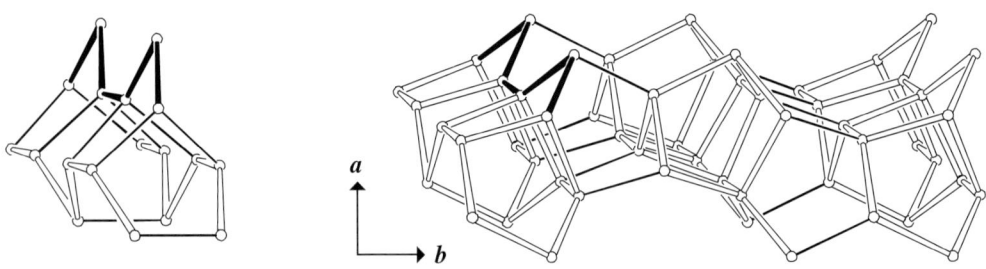

Fig. 1.  Building unit composed of three saw chains (left) and PerBU viewed along $c$ (right).

## 2. Connection mode

Neighboring PerBUs, related along $a$ by a shift of $\frac{1}{2}(a+b+c)$, are connected along $a$ through 6-, 8- and 10-rings as shown in Figure 2.

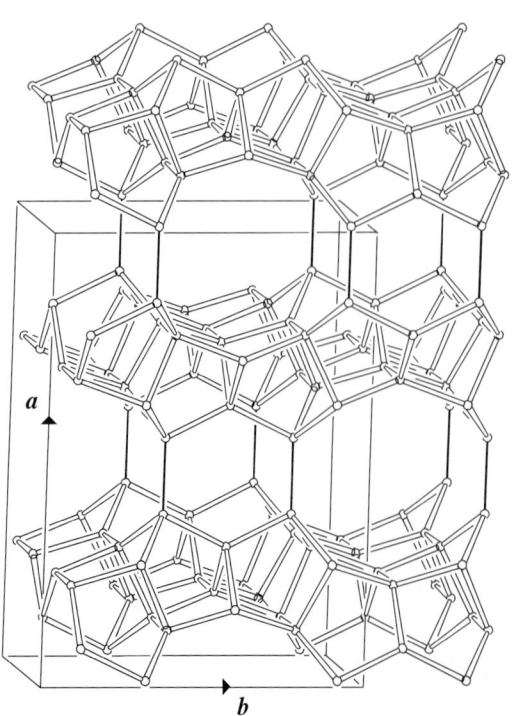

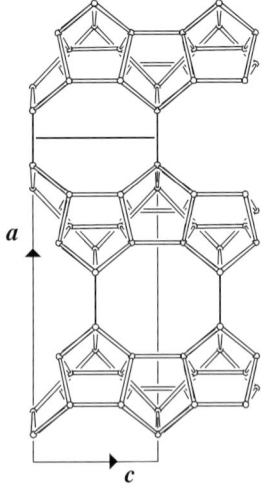

Fig. 2.  Connection mode (and unit cell content) viewed along $c$ (left) and unit cell content projected along $b$ (right).

### 3. Channels and/or cages

Intersecting 8- and 10-ring channels are parallel to **b** and **c**, respectively. The 10-ring channel is topologically equivalent to the 10-ring channel in **DAC**. The channels, with one channel intersection in bold, are shown in Figure 3.

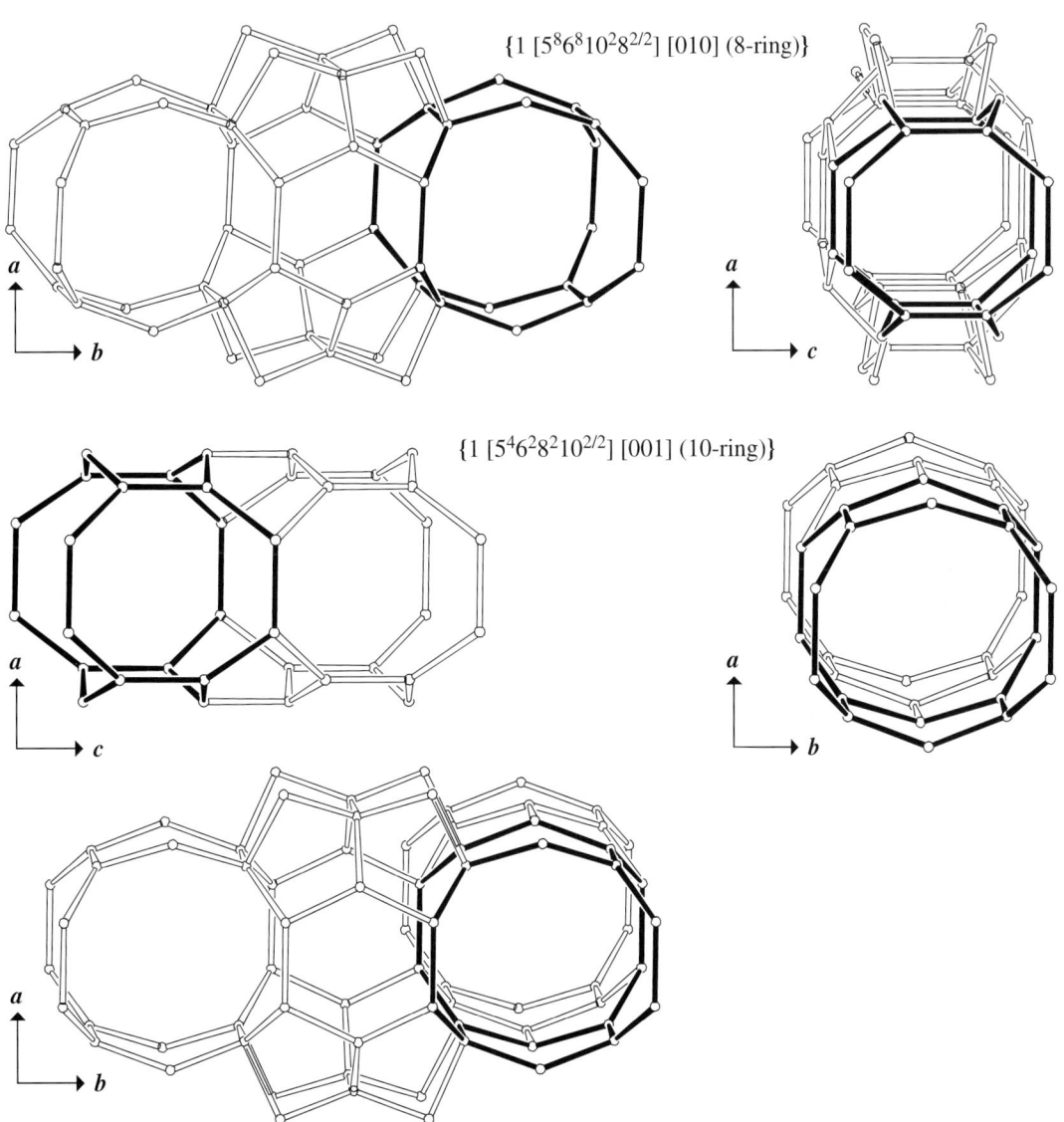

$\{1 \, [5^8 6^8 10^2 8^{2/2}] \, [010] \, (8\text{-ring})\}$

$\{1 \, [5^4 6^2 8^2 10^{2/2}] \, [001] \, (10\text{-ring})\}$

Fig. 3.  8-Ring channel (top) and 10-ring channel (middle) viewed perpendicular to the channel axis (left) and along the channel axis (right). Bottom: interconnection of channels. The channels have $[6^2 8^2 10^2]$-cavities in common (one in bold).

# FRA          Building Scheme

## 1. Periodic Building Unit

The two-dimensional PerBU of **FRA** consists of a hexagonal array of non-connected planar 6-rings (bold in Figure 1), which are related along **a** and **b** by pure translations. The 6-rings are centered at (0,0) in the **ab** layer. This position is usually called the **A** position. **FRA** belongs to the ABC-6 family (see Introduction).

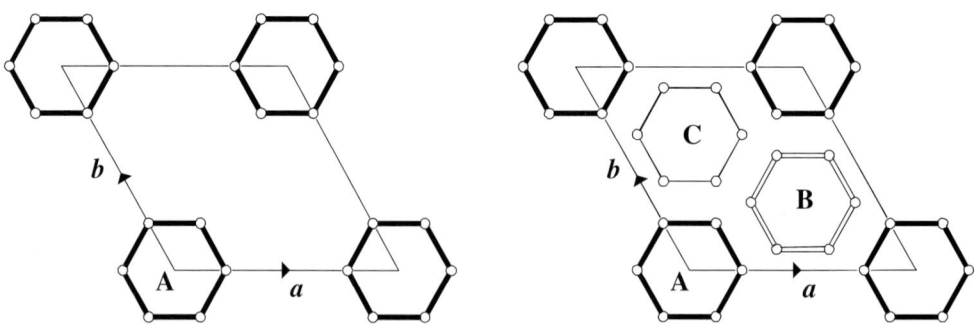

Fig. 1.    PerBU (left) and definition of 6-ring positions used in the stacking modes (right).

## 2. Connection mode

Neighboring PerBUs are connected along **c** through tilted 4-rings by connection modes (**1**) and (**2**) (see Introduction) as illustrated in Figure 2.

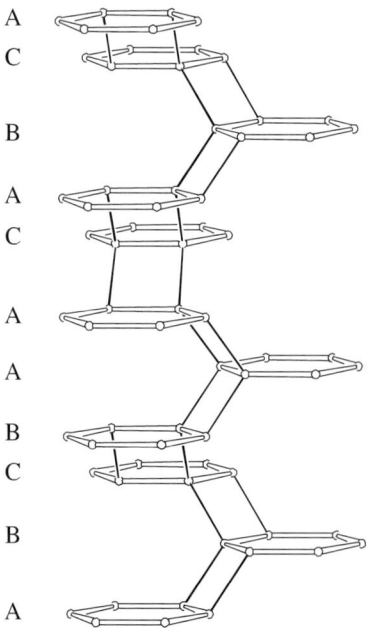

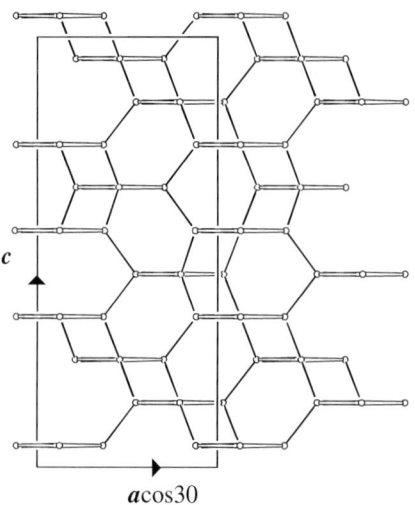

Fig. 2.    Connection mode (left) and unit cell content (right) viewed along **b**. The stacking sequence is given. In the perspective drawing, each PerBU is represented by one 6-ring only.

## 3. Channels and/or cages

The three types of cages in **FRA** are depicted in Figure 3. The *can* cage is also present in **AFG**, **CAN**, **ERI**, **FAR**, **GIU**, **LIO**, **LOS**, **LTL**, **LTN**, **MAR**, **MOZ**, **OFF**, **SAT**, **SBS**, **SBT**, **TOL** and **-WEN**. The *sod* cage is also found in **EMT**, **FAU**, **GIU**, **LTA**, **LTN**, **MAR**, **SOD** and **TSC**. Finally, the *los* cage is also observed in **LIO**, **LOS** and **TOL**. Apertures of "channels" are formed by 6-rings only.

*can* cage          *sod* cage          *los* cage

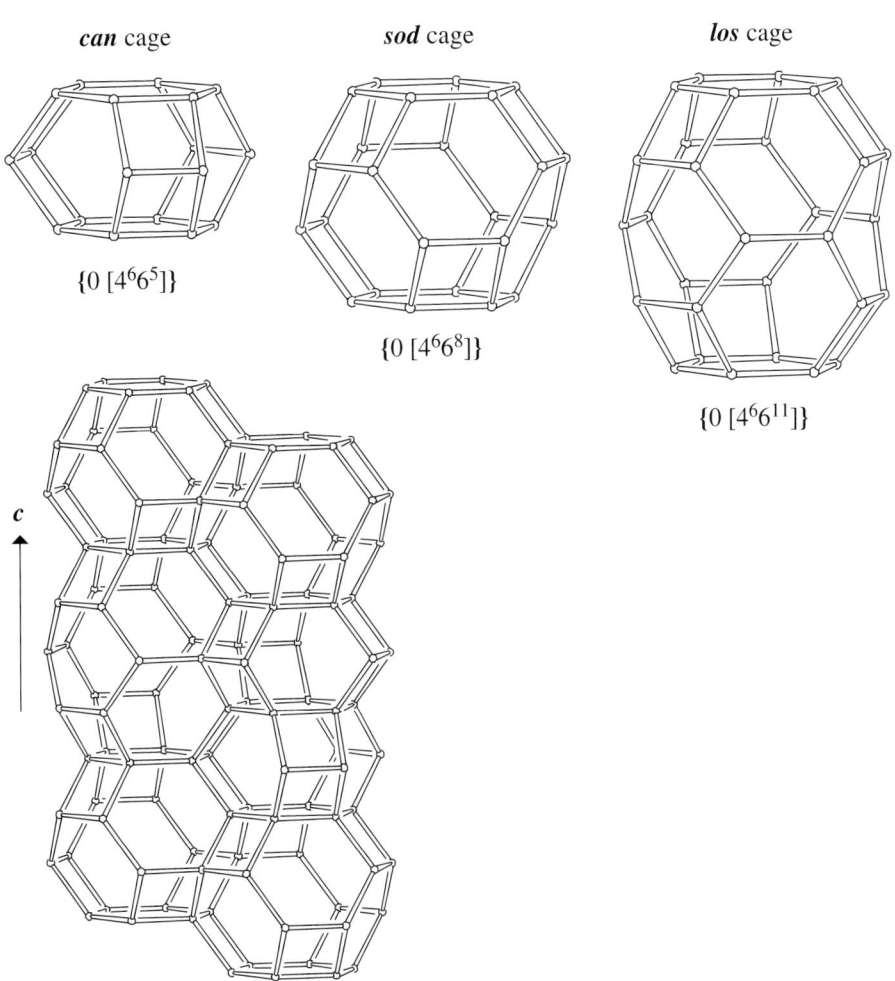

$\{0\ [4^6 6^5]\}$

$\{0\ [4^6 6^8]\}$

$\{0\ [4^6 6^{11}]\}$

Fig. 3.   Top (from left to right): *can* cage, *sod* cage and *los* cage viewed perpendicular to *c*. Bottom: connection of cages.

# GIS          Building Scheme

## 1. Periodic Building Unit

Tetragonal **GIS** can be built using the crankshaft chain (bold in Figure 1, left) running parallel to **a**. A one-dimensional PerBU is obtained when two crankshaft chains and two 4-rings are connected into a channel with an 8-ring aperture. The channel wall consists of 4- and 12-rings. The repeat unit of the PerBU consists of a 4-fold (1,2,3,4)-connected double 8-ring (bold in Figure 1, right). (Compare with **SIV**.)

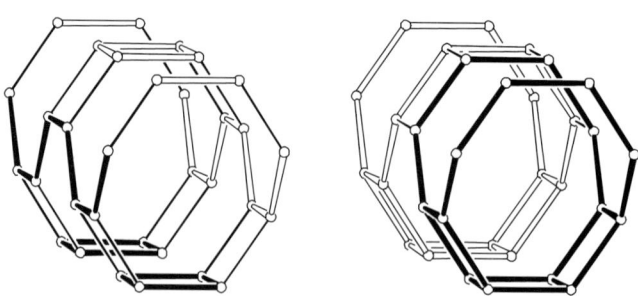

Fig. 1.    PerBU, constructed from crankshaft chains and 4-rings (left) and PerBU constructed from 4-fold connected double 8-rings (right), viewed along **a**.

## 2. Connection mode

Neighboring PerBUs, related along **b** and **c** by pure translations, are connected along **b** and **c** through double-crankshaft chains (Figure 2).

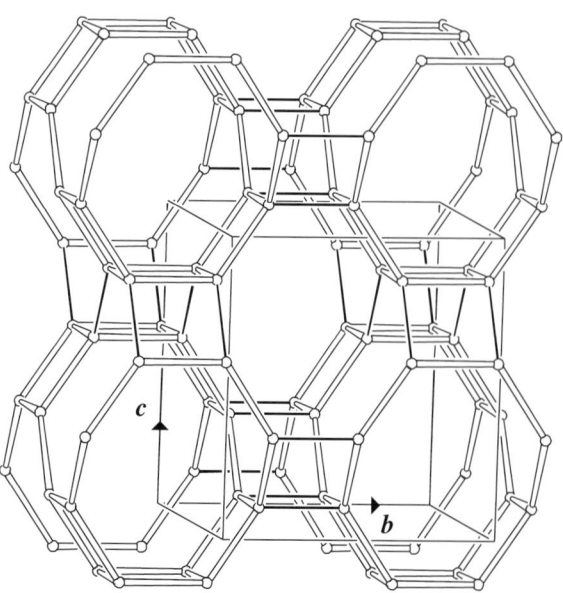

Fig. 2.    Connection mode and unit cell content viewed along **a** (or **b**). For clarity, only $1\frac{1}{2}$ repeat unit along **a** of each PerBU is drawn.

## 3. Channels and/or cages

Equal intersecting 8-ring channels are parallel to <100>. The 8-ring channel is (topologically) equivalent to (one of) the 8-ring channels in **ATT**, **PHI** and **SIV**. Channel and intersecting channels are depicted in Figure 3.

$\{1\ [4^{12}8^48^{2/2}]<100>\ (8\text{-ring})\}$

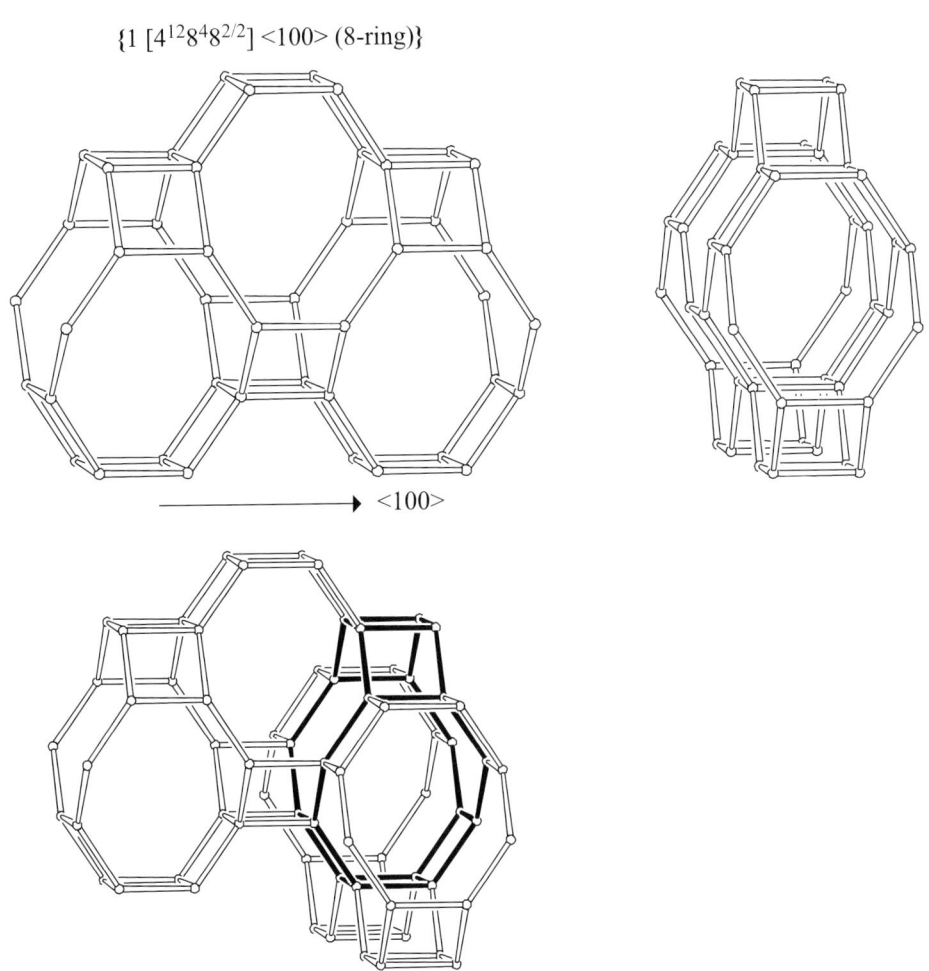

Fig. 3.   Top: 8-ring channel parallel to <100> viewed perpendicular to the channel axis (left) and along the channel axis (right). Bottom: intersecting channels have [$4^68^4$]-cavities in common (one in bold).

# GIU

# Building Scheme

## 1. Periodic Building Unit

The two-dimensional PerBU of **GIU** consists of a hexagonal array of non-connected planar 6-rings (bold in Figure 1), which are related along *a* and *b* by pure translations. The 6-rings are centered at (0,0) in the *ab* layer. This position is usually called the **A** position. **GIU** belongs to the ABC-6 family (see Introduction).

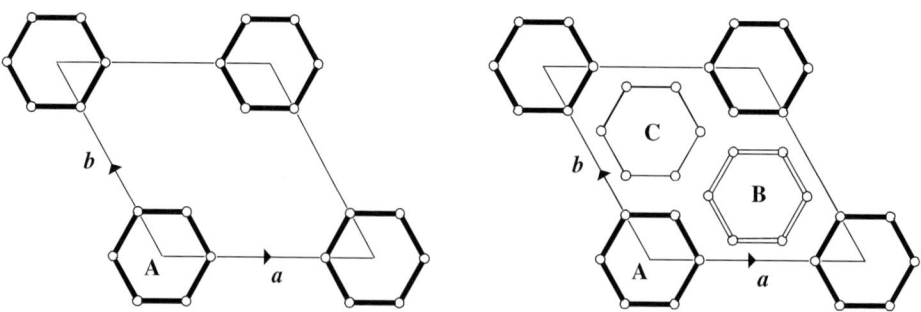

Fig. 1.    PerBU (left) and definition of 6-ring positions used in the stacking modes (right).

## 2. Connection mode

Neighboring PerBUs are connected along *c* through tilted 4-rings by connection modes (**1**) and (**2**) (see Introduction) as illustrated in Figure 2.

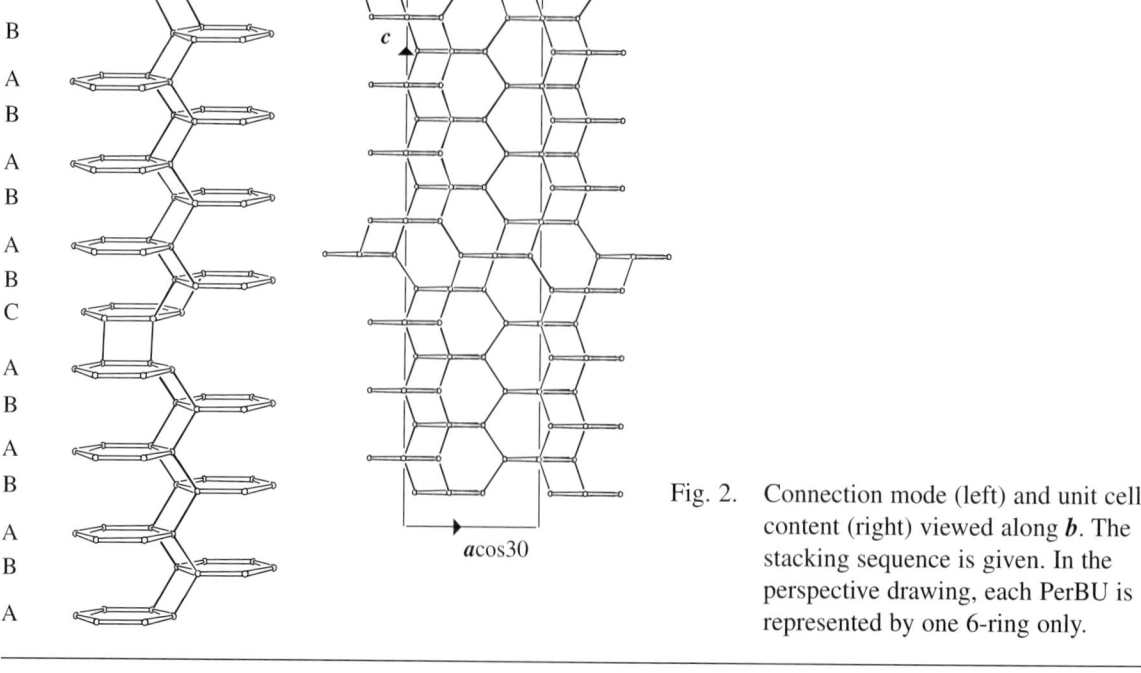

Fig. 2.    Connection mode (left) and unit cell content (right) viewed along *b*. The stacking sequence is given. In the perspective drawing, each PerBU is represented by one 6-ring only.

## 3. Channels and/or cages

The three types of cages in **GIU** are depicted in Figure 3. The *can* cage is also present in **AFG**, **CAN**, **ERI**, **FAR**, **FRA**, **LIO**, **LOS**, **LTL**, **LTN**, **MAR**, **MOZ**, **OFF**, **SAT**, **SBS**, **SBT**, **TOL** and **-WEN**. The *sod* cage is also found in **EMT**, **FAU**, **FRA**, **LTA**, **LTN**, **MAR**, **SOD** and **TSC**. Apertures of "channels" are formed by 6-rings only.

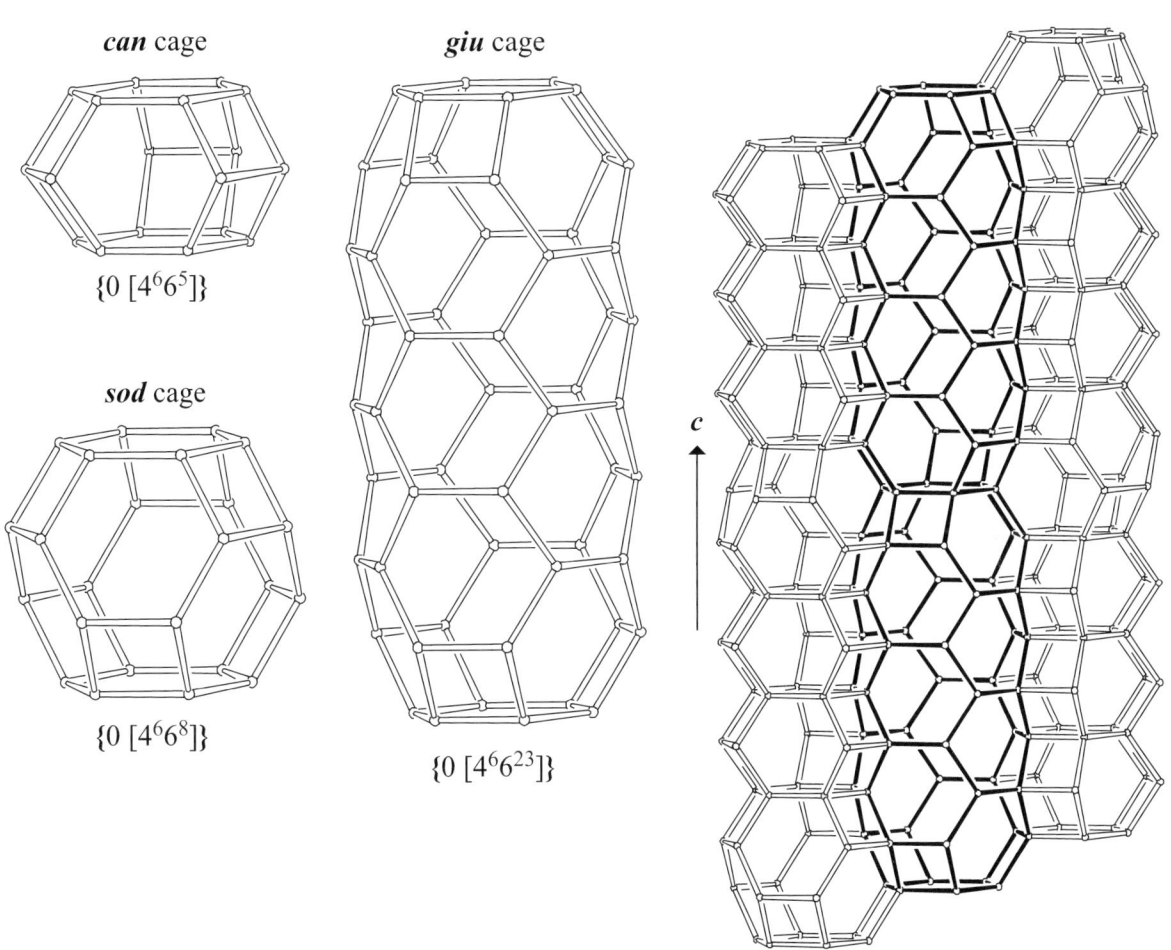

*can* cage

$\{0\,[4^6 6^5]\}$

*sod* cage

$\{0\,[4^6 6^8]\}$

*giu* cage

$\{0\,[4^6 6^{23}]\}$

*c*

Fig. 3.   *can* Cage, *sod* cage, *giu* cage and connection of cages viewed perpendicular to *c*.

# GME                    Building Scheme

## 1. Periodic Building Unit

The two-dimensional PerBU of **GME** consists of a hexagonal array of non-connected planar 6-rings (bold in Figure 1), which are related along *a* and *b* by pure translations. The 6-rings are centered at (0,0) in the *ab* layer. This position is usually called the **A** position. **GME** belongs to the ABC-6 family (see Introduction).

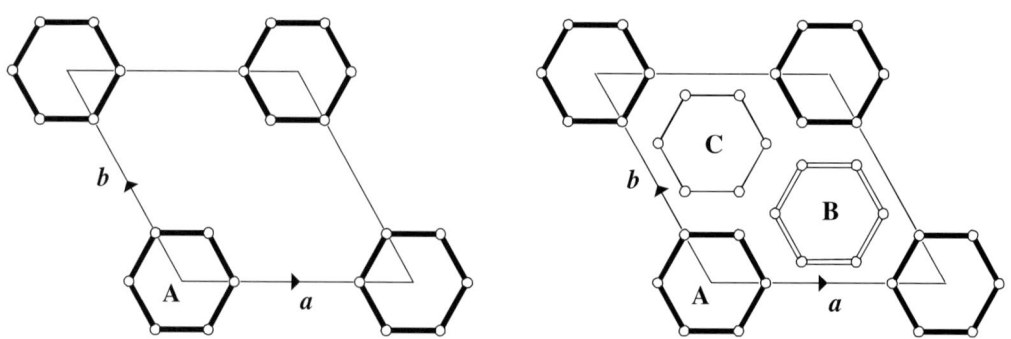

Fig. 1.   PerBU (left) and definition of 6-ring positions used in the stacking modes (right).

## 2. Connection mode

Neighboring PerBUs can be connected along *c* through tilted 4-rings in three different ways (see Introduction). In **GME** all three connection modes between the PerBUs are observed. *gmel* Cavities are formed. (See Figure 2 and **AFX**.)

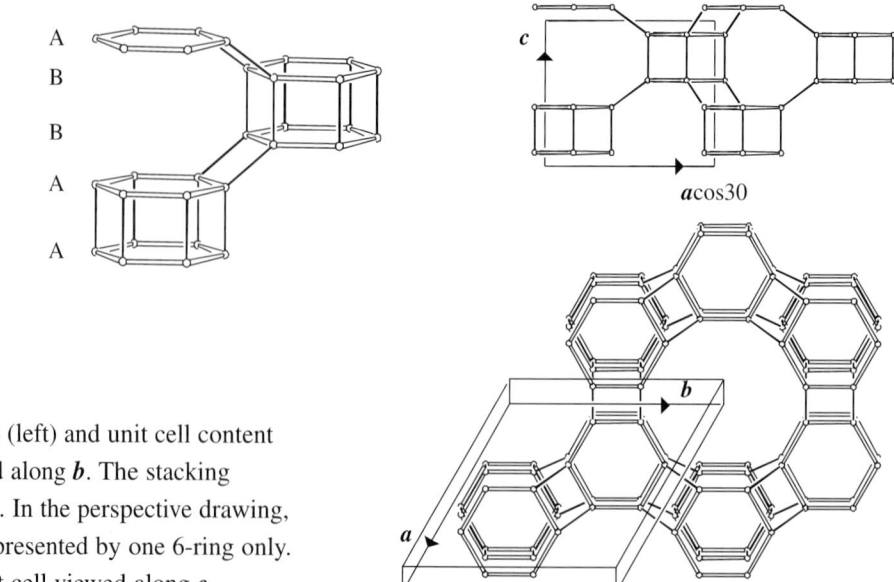

Fig. 2.   Connection mode (left) and unit cell content (top right) viewed along *b*. The stacking sequence is given. In the perspective drawing, each PerBU is represented by one 6-ring only. Bottom right: unit cell viewed along *c*.

## 3. Channels and/or cages

Interconnecting 12-ring channels (Figure 3) are parallel to $c$. Two-dimensional 8-ring channels are perpendicular to $c$. A column of *gmel* cavities (bold in Figure 3) is formed when the 12-ring channels are connected around a 3-fold axis. **GME** can also be built from *gmel* cavities alone (see also: **AFX**). The *gmel* cavity is also present in **AFT**, **AFX**, **EAB**, **EON**, **MAZ** and **OFF**.

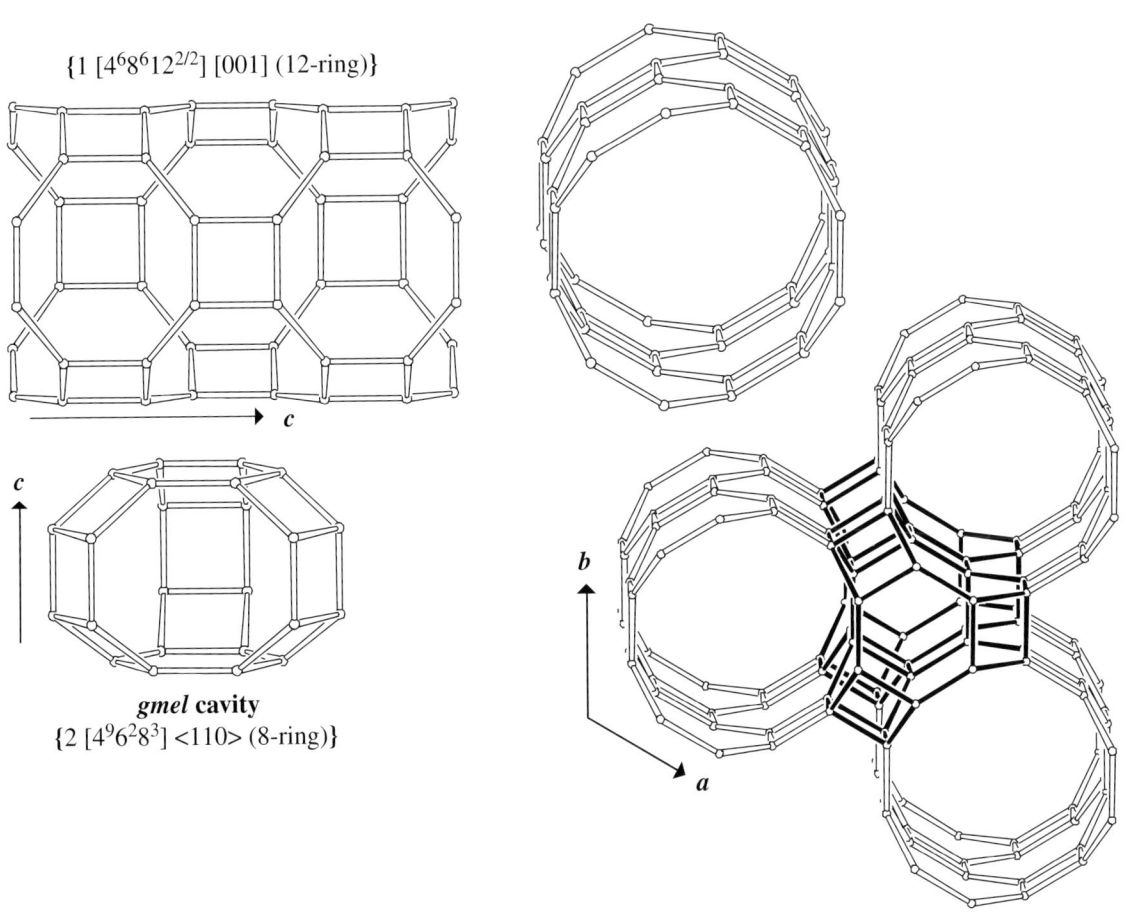

$\{1 \, [4^6 8^6 12^{2/2}] \, [001] \, (12\text{-ring})\}$

*gmel* **cavity**
$\{2 \, [4^9 6^2 8^3] \, <110> \, (8\text{-ring})\}$

Fig. 3.   Top: 12-ring channel viewed perpendicular to the channel axis (left) and along the channel axis (right). Bottom right: 12-Ring channels, connected around a 3-fold axis, form a column of *gmel* cavities (in bold), which are connected through double 6-rings. The two-dimensional 8-ring channel systems are perpendicular to $c$. View along the 3-fold axis parallel to $c$. Bottom left: *gmel* cavity viewed perpendicular to $c$.

# GON

**Building Scheme**

## 1. Periodic Building Unit

**GON** can be built using the zigzag chain (bold in Figure 1, left) running parallel to *c*. The one-dimensional PerBU is obtained when eight zigzag chains are connected into a channel with a 12-ring aperture. The channel wall consists of fused 6-rings.

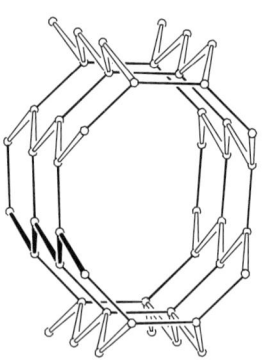 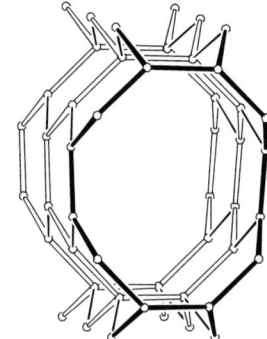

Fig. 1. PerBU, constructed from eight zigzag chains (left) and from (T16) repeat units consisting of a 12-ring and four additional single T atoms (right), viewed along *c*.

## 2. Connection mode

Neighboring PerBUs, related along *a* by a pure translation and along *b* by a shift of $\frac{1}{2}(a+b)$, are connected through (fused) 4-, 5-, 6- and 8-rings as can be seen in Figure 2.

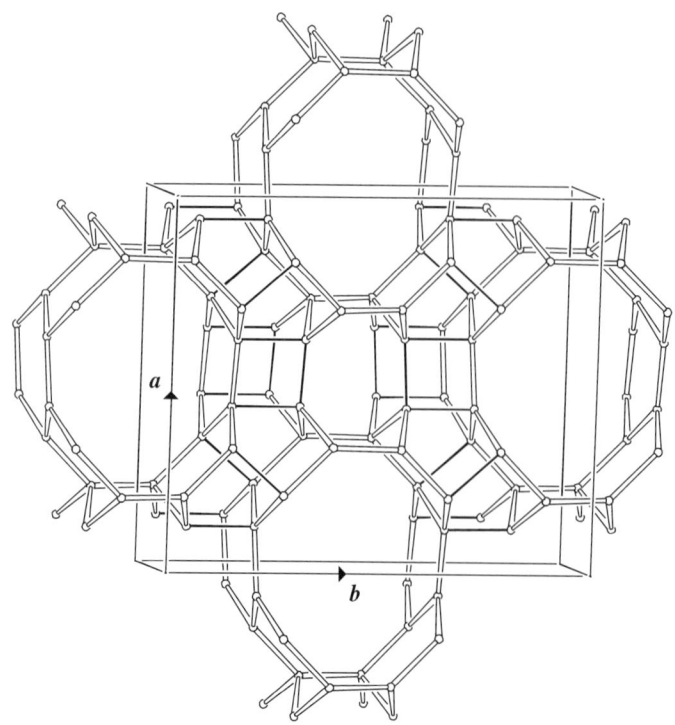

Fig. 2. Connection mode and unit cell content viewed along *c*. Only two repeat units along *c* of each PerBU is drawn for clarity.

# Cage/Channel        **GON**

## 3. Channels and/or cages

Non-interconnecting 8- and 12-ring channels are parallel to $c$. The 8-ring channel is topologically equivalent to the 8-ring channels (Type 1) in **ATN** and **BCT**. The 12-ring channel is topologically equivalent to the 12-ring channel in **MTW**. Both channel walls consist of fused 6-rings as depicted in Figure 3.

{1 $[6^812^{2/2}]$ [001] (12-ring)}

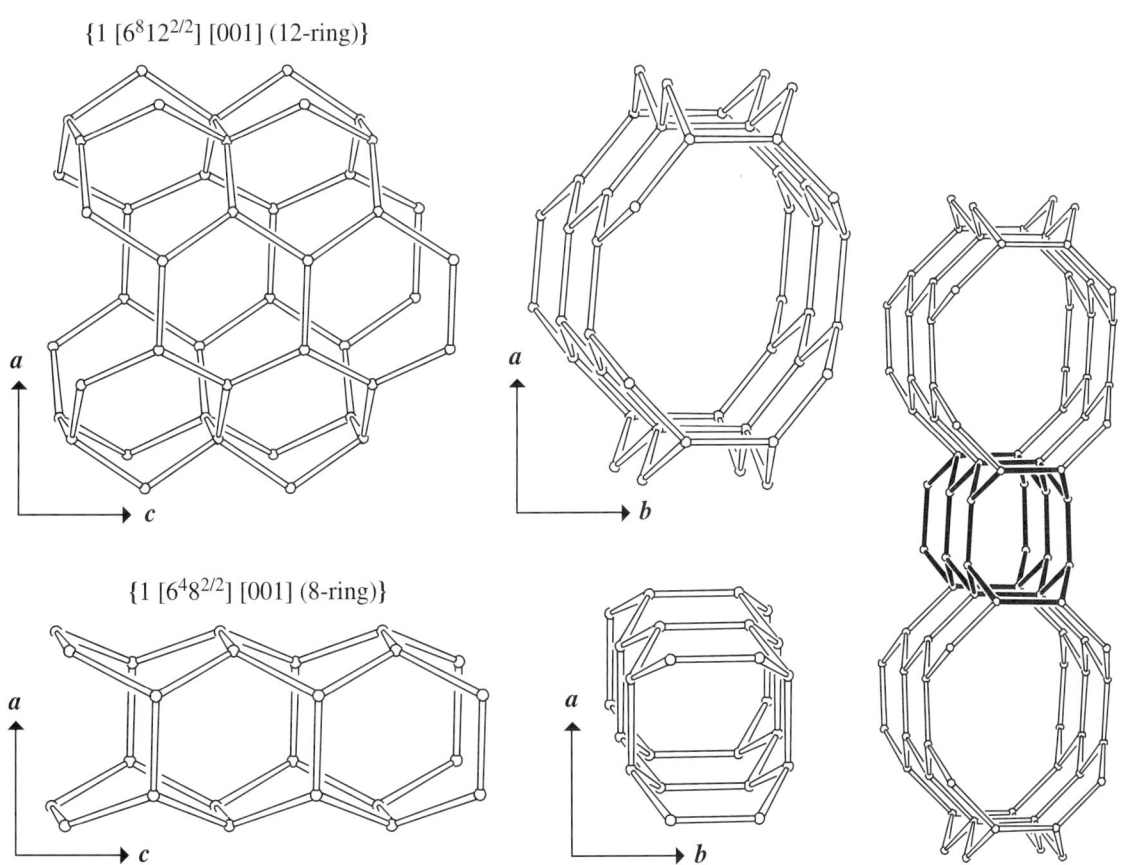

{1 $[6^48^{2/2}]$ [001] (8-ring)}

Fig. 3.    12-ring channel (top) and 8-ring channel (bottom) viewed perpendicular to the channel axis (left) and along the channel axis (middle). Right: linkage between the channels viewed along the channel axis. The linked channels can be described as a double layer of (fused) 6-ring chairs and 6-ring boats.

# GOO                    Building Scheme

## 1. Periodic Building Unit

**GOO** can be built using a building unit of 8 T atoms (bold in Figure 1) consisting of a double 4-ring with two disconnected edges. T8-units, related along **c** by a screw rotation of 180° about **c**, are connected into chains along **c** through (deformed) 4-rings. A two-dimensional PerBU is obtained when neighboring chains, related along [110] by a shift of $\frac{1}{2}(a+b)$, are connected through 4-rings into the (−110) layer shown in Figure 1.

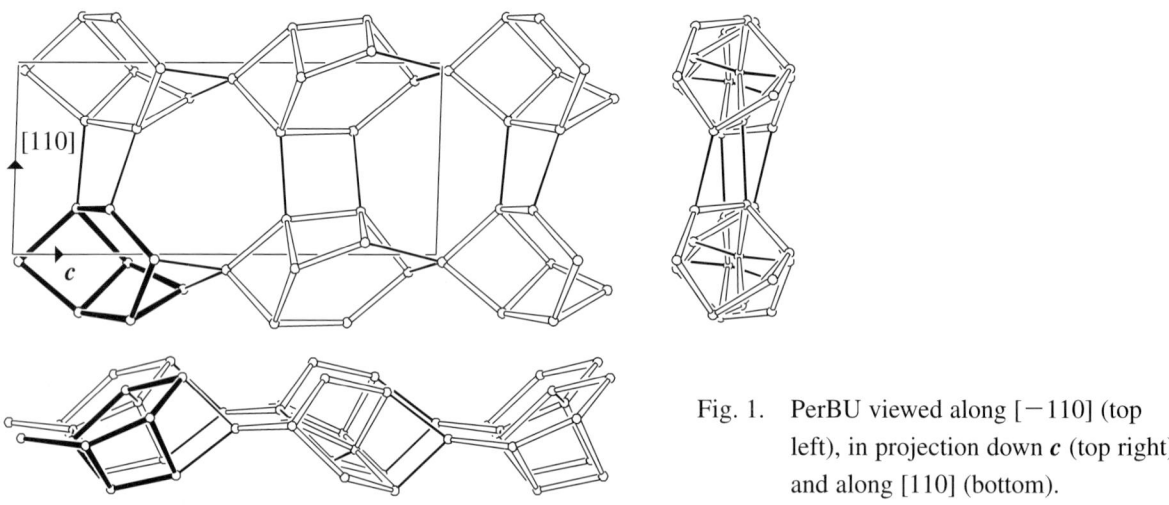

Fig. 1.   PerBU viewed along [−110] (top left), in projection down **c** (top right) and along [110] (bottom).

## 2. Connection mode

Neighboring PerBUs, related along [−110] by a shift of $\frac{1}{2}(-a+b)$, are connected into the (110) layer through 4-rings. The connection mode is equal to that in the (−110) layer (compare Figures 1 and 2).

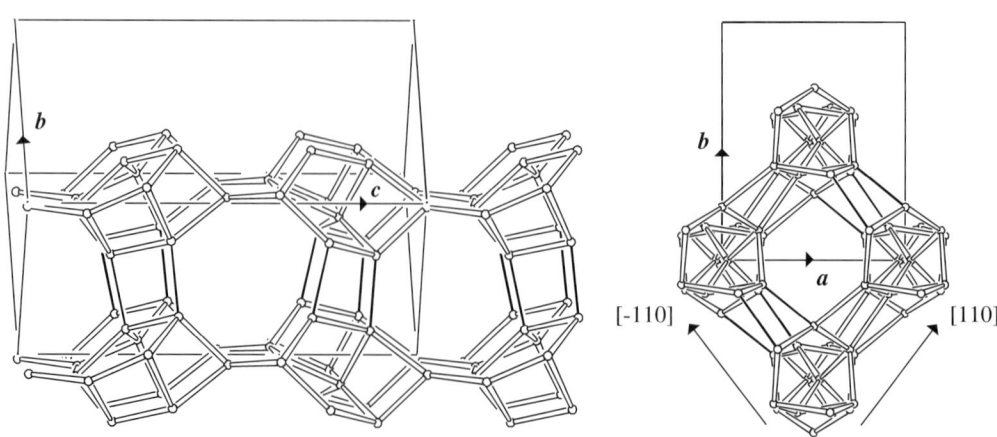

Fig. 2.   Connection mode (and unit cell content) viewed along [110] (left) and unit cell content projected along **c** (right).

## 3. Channels and/or cages

Intersecting 8-ring channels are parallel to [110], [−110] and *c*. The channel intersection and the linkage of channel intersections into channels (one intersection in bold) are depicted in Figure 3.

{3 [4⁴6²8⁶] [110] (8-ring), [-110] (8-ring), [001] (8-ring)}

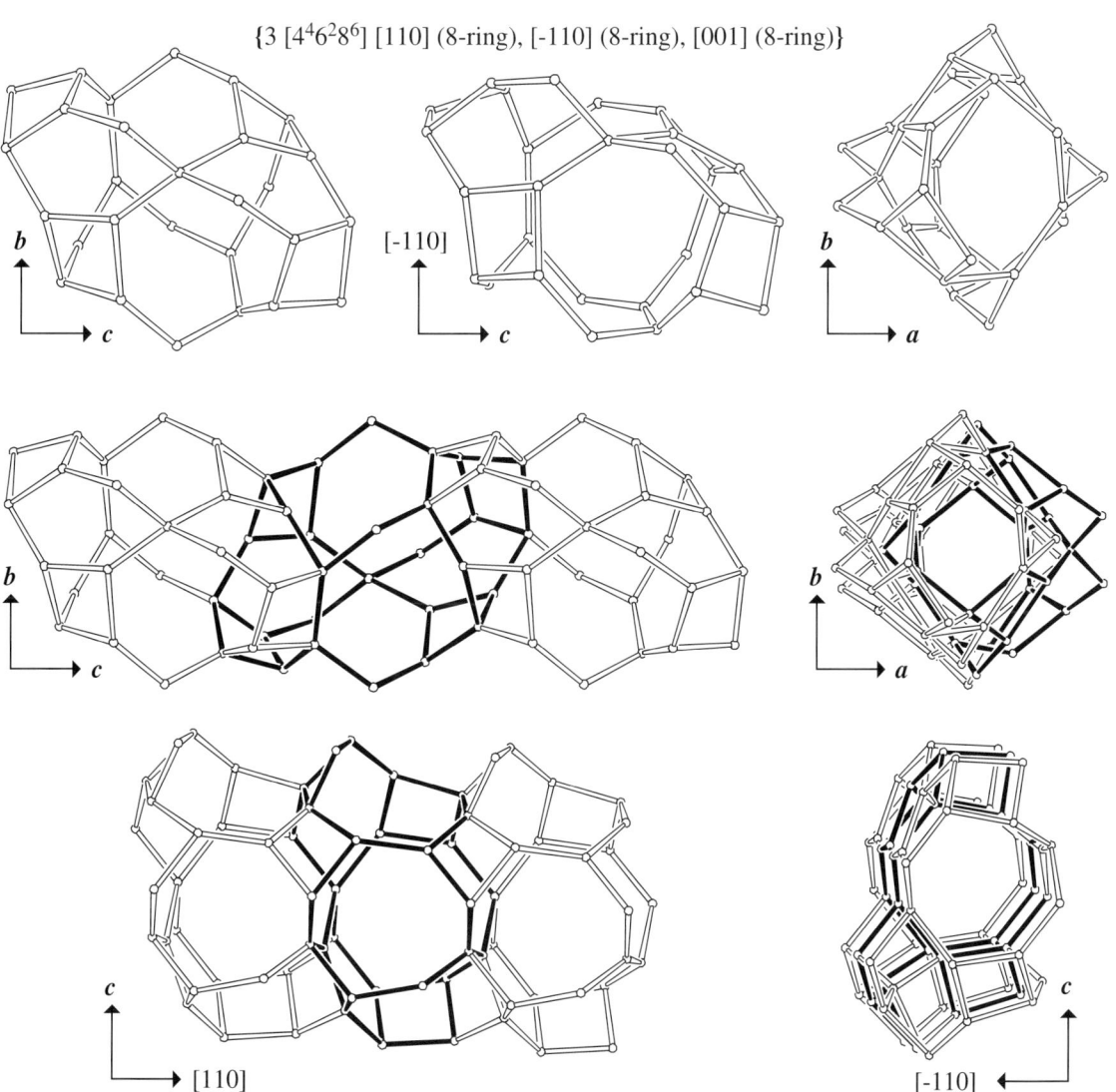

Fig. 3.  Top: intersection of channels viewed along (from left to right) *a*, [110] (the view along [−110] is equivalent) and along *c*. Middle: fusion of channel intersections along *c* viewed along *a* (left) and along *c* (right). Bottom: fusion of intersections along [110] (or [−110]) viewed along [−110] (or [110]; left), and along [110] (or [−110]; right).

# HEU                    Building Scheme

## 1. Periodic Building Unit

**HEU** can be built using T9-units: a double 4-ring with an additional bridging T atom (a $[4^25^26^1]$-cage, or 4-4 = 1 unit; bold in Figure 1). T9-units, related along [102] by a rotation of 180° about an axis parallel to **b** and passing through the bridging T atom, are connected into chains parallel to [102]. The two-dimensional PerBU is obtained when neighboring chains, related along **c** by a pure translation, are connected through (fused) 4- and 5-rings into the **ac** layer. (See also **RRO** and **STI**.)

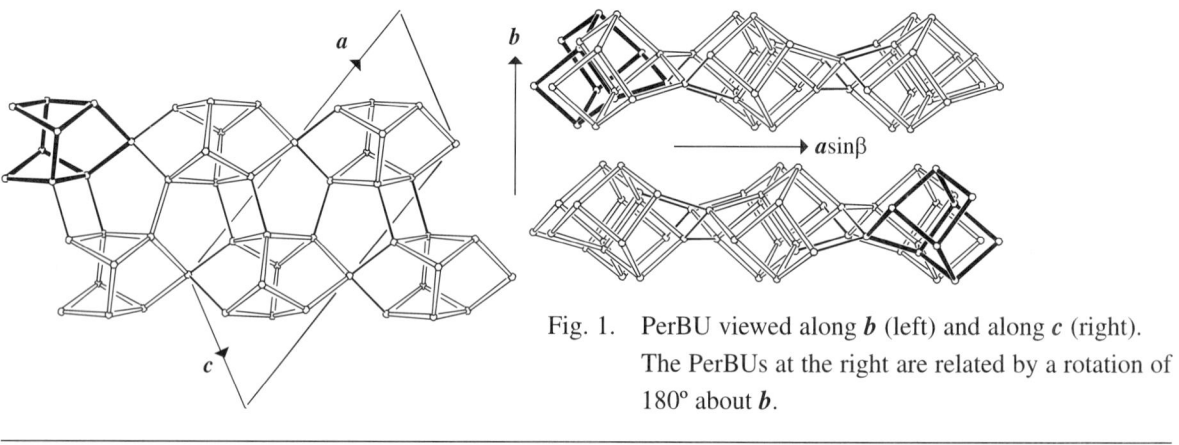

Fig. 1.   PerBU viewed along **b** (left) and along **c** (right). The PerBUs at the right are related by a rotation of 180° about **b**.

## 2. Connection mode

Neighboring PerBUs, related along **b** by a screw rotation of 180° about **b**, are connected along **b** through 8- and 10-rings as depicted in Figure 2.

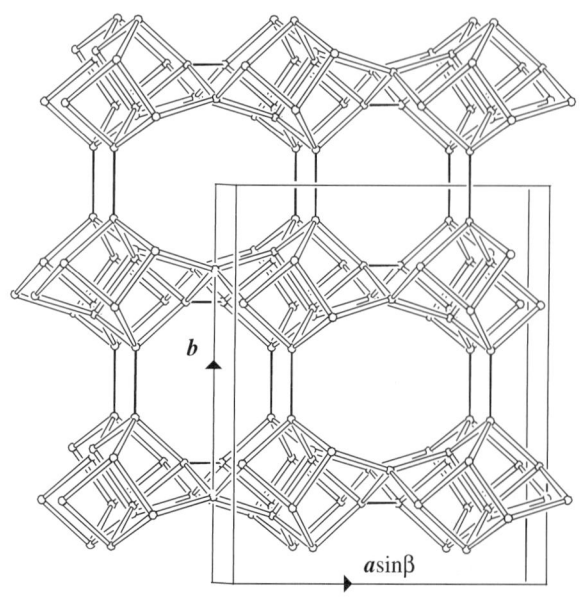

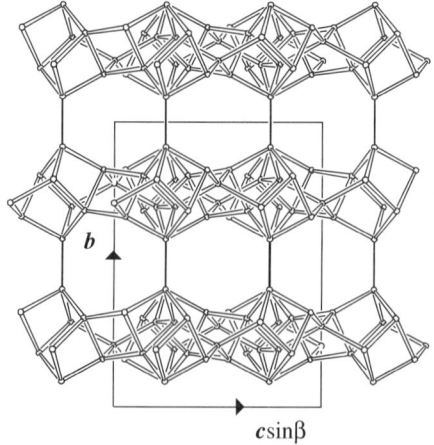

Fig. 2.   Connection mode (and unit cell content) viewed along **c** (left) and unit cell content projected along **a** (right).

## 3. Channels and/or cages

Interconnecting 8- and 10-ring channels parallel to $c$ intersect with another type of 8-ring channels parallel to [102] leading to a two-dimensional channel system with two types of channel intersections. Channel intersections, and their linkage through common 8-rings into the 8-ring channels parallel to [102], are depicted in Figure 3.

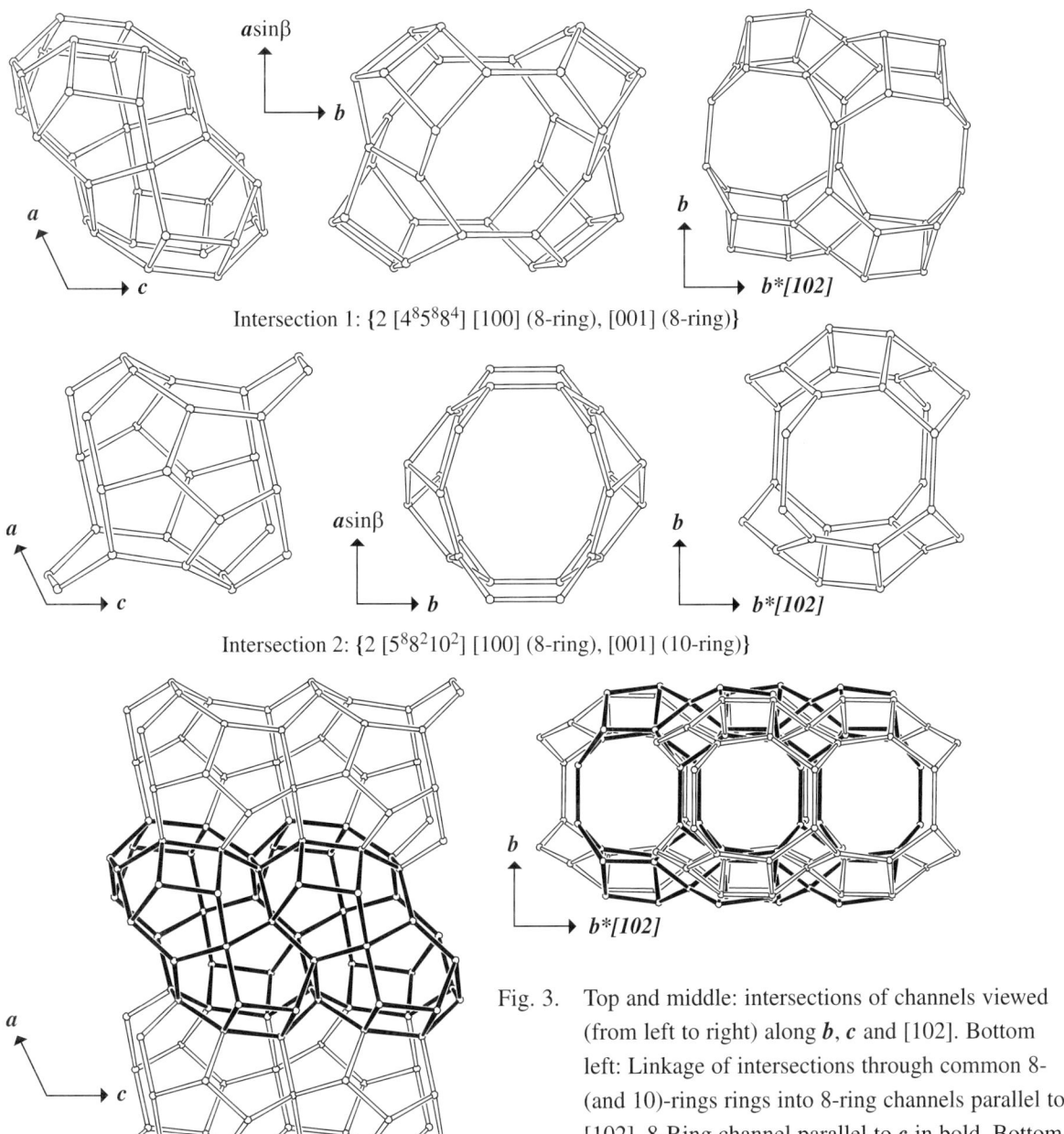

Intersection 1: {2 [$4^8 5^8 8^4$] [100] (8-ring), [001] (8-ring)}

Intersection 2: {2 [$5^8 8^2 10^2$] [100] (8-ring), [001] (10-ring)}

Fig. 3.　Top and middle: intersections of channels viewed (from left to right) along $b$, $c$ and [102]. Bottom left: Linkage of intersections through common 8- (and 10-)rings rings into 8-ring channels parallel to [102]. 8-Ring channel parallel to $c$ in bold. Bottom right: view along the channel axis parallel to [102].

# IFR                    Building Scheme

## 1. Periodic Building Unit

**IFR** can be built using units composed of 16 T atoms: a 4-fold (1,2,4,5)-connected double 6-ring with "handles" (a [$4^2 6^6$]-cage, or two 6-2 units; bold in Figure 1). The one-dimensional PerBU is obtained when T16-units, related along $c$ by pure translations, are connected through 4-rings into a chain along $c$ as shown in Figure 1 (left).

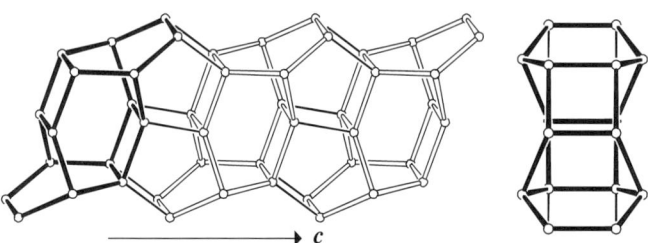

Fig. 1.    PerBU viewed normal to the chain axis $c$ (left) and along the chain axis (right). (The PerBU can as well be constructed using two [$4^3 5^2 6^1$]-cages sharing a 4-ring.)

## 2. Connection mode

Neighboring PerBUs, related along $a$ (and $b$) by a translation of $\frac{1}{2}(a+b)$, are connected along $a$ (and $b$) through 4-rings, as depicted in Figure 2.

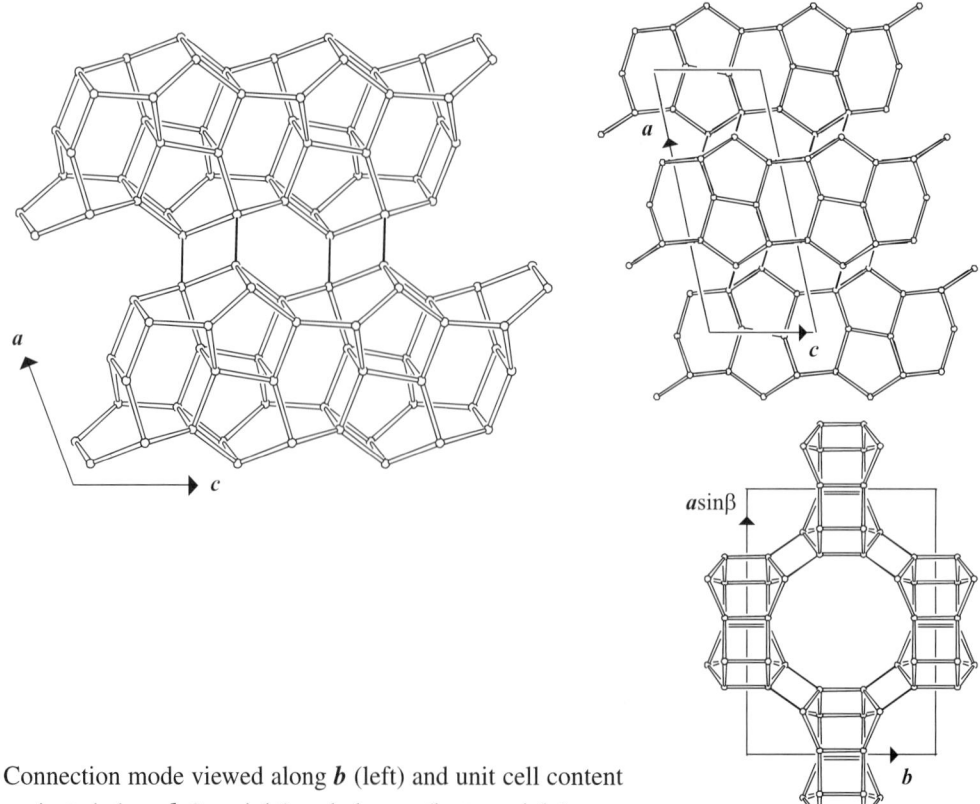

Fig. 2.    Connection mode viewed along $b$ (left) and unit cell content projected along $b$ (top right) and along $c$ (bottom right).

## 3. Channels and/or cages

Non-interconnecting 12-ring channels are parallel to $c$. The repeat unit in the channel is depicted in Figure 3 together with the channel obtained by connecting the cavities through common 12-rings.

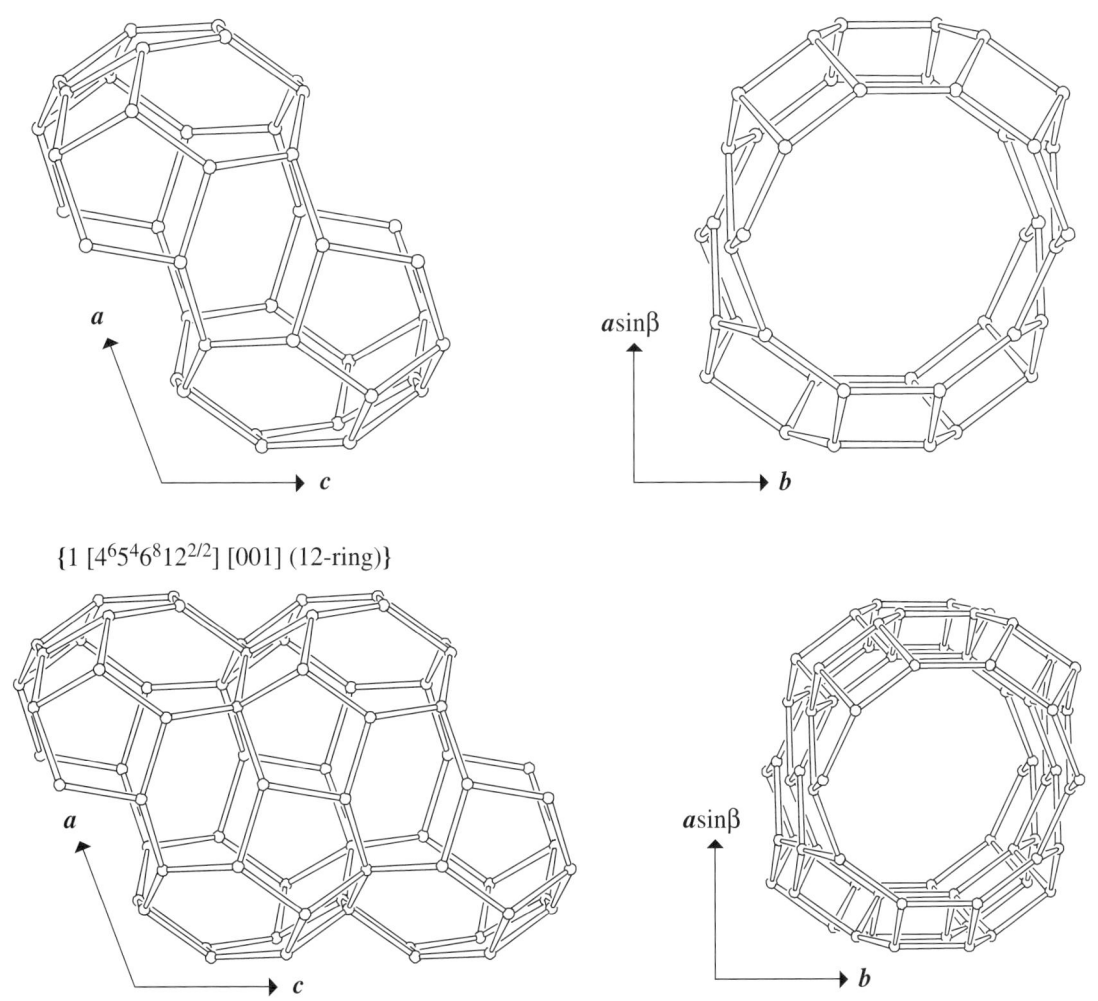

$\{1 \, [4^6 5^4 6^8 12^{2/2}] \, [001] \, (12\text{-ring})\}$

Fig. 3. Top: cavity viewed along $b$ (left) and along $c$ (right). Bottom: fusion of cavities into 12-ring channels parallel to $c$ viewed perpendicular to the channel axis (left) and along the channel axis (right).

# IHW

# Building Scheme

## 1. Periodic Building Unit

**IHW** can be built using centrosymmetrical units composed of 28 T atoms: two 5-1 units and two 5-3 units that form two $[4^1 5^8]$ cages (sharing the 4-ring) with two additional T atoms (Figure 1(a); one 5-1 and one 5-3 unit in bold). A two-dimensional PerBU is obtained when T28-units, related along $a$ (and $b$) by a shift of $\frac{1}{2}(a+b)$, are connected along $a$ (and $b$) through 5-rings (Figure 1(b)).

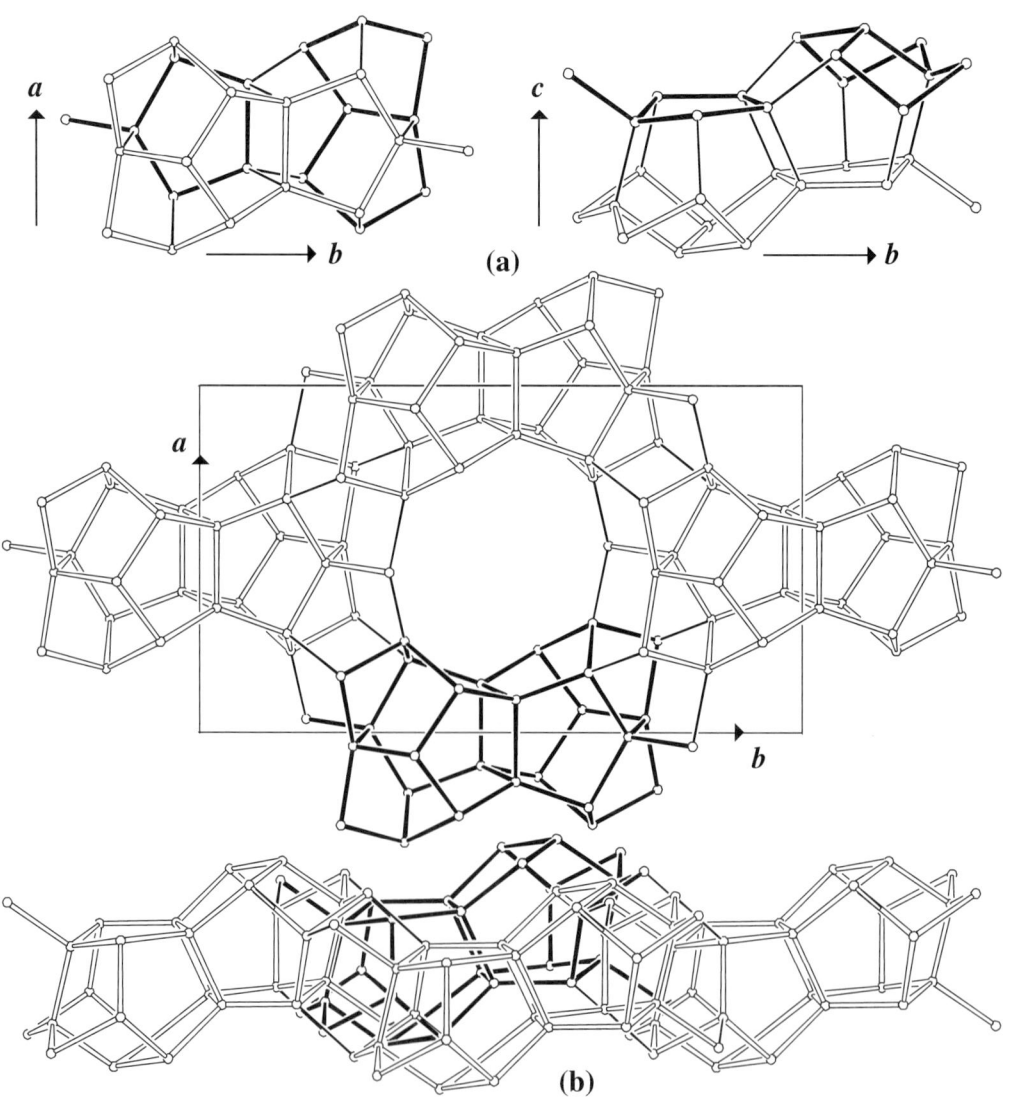

Fig. 1.   (a) T28-unit viewed along $c$ (left) and along $a$ (right); (b) PerBU viewed along $c$ (top) and down $a$ (bottom).

## 2. Connection mode

Neighboring PerBUs, related along $c$ by a screw rotation of 180° about $b$, are connected along $c$ through (fused) 5-, 6- and 8-rings as shown in Figure 2.

# Cage/Channel

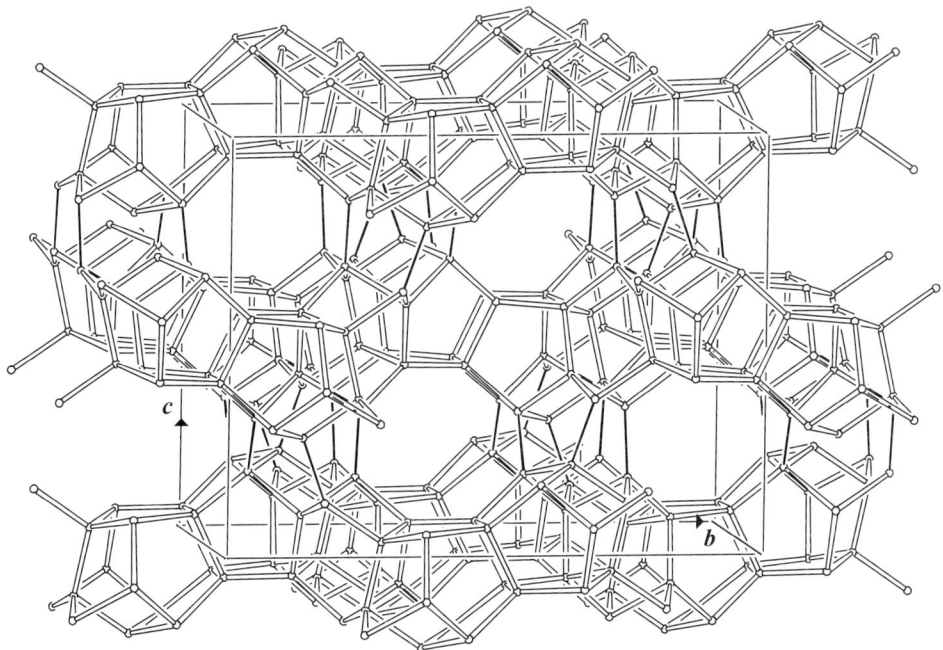

Fig. 2.   Connection mode and unit cell content viewed along *a*.

## 3. Channels and/or cages

Large cavities with side-pockets (Figure 3) are connected through common 8-rings into pairs of interconnecting 8-ring channels parallel to *a*.

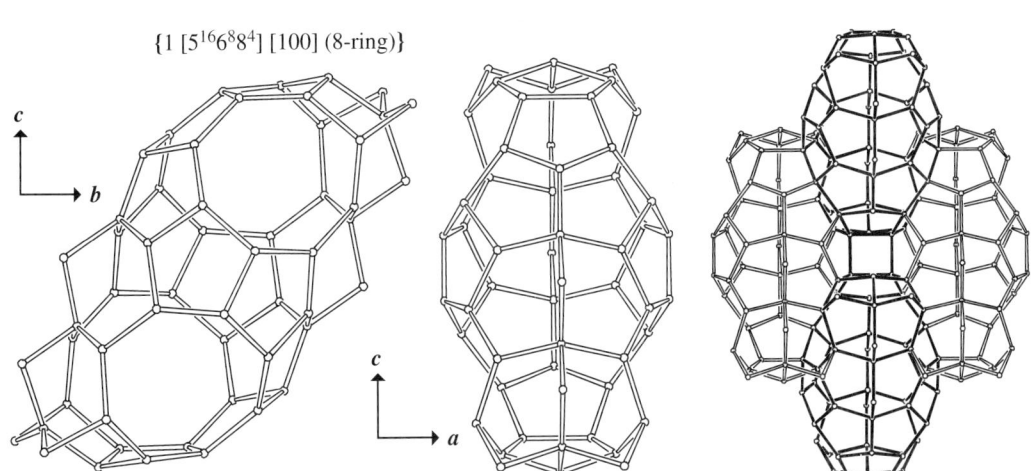

Fig. 3.   Large cavity viewed along *a* (left) and along *b* (middle). Right: linkage of the large cavities. The 8-ring channels are interconnecting along *c* through 12-rings in the cavities.

# IMF           Building Scheme

## 1. Periodic Building Unit

**IMF** can be built using units of 36 T atoms built from three different sub-units of 12 T atoms each (one T36-unit bold Figure 1). All three T12-units consist of two 5-1 units which are connected to a $[4^15^46^1]$-cage attached to a single T atom (T12-unit1), or to $[5^4]$- and $[5^26^1]$-cages (sharing a 5-ring) bearing a dimer (T12-unit2) or two single T atoms (T12-unit3). T36-units, related by a screw rotation of 180° about **b** are connected into chains along **b**. Chains, related by a screw rotation of 180° about **a** are connected along **a** into the Periodic Building Unit (PerBU). The two-dimensional PerBU equals the **ab** layer as shown in Figure 1.

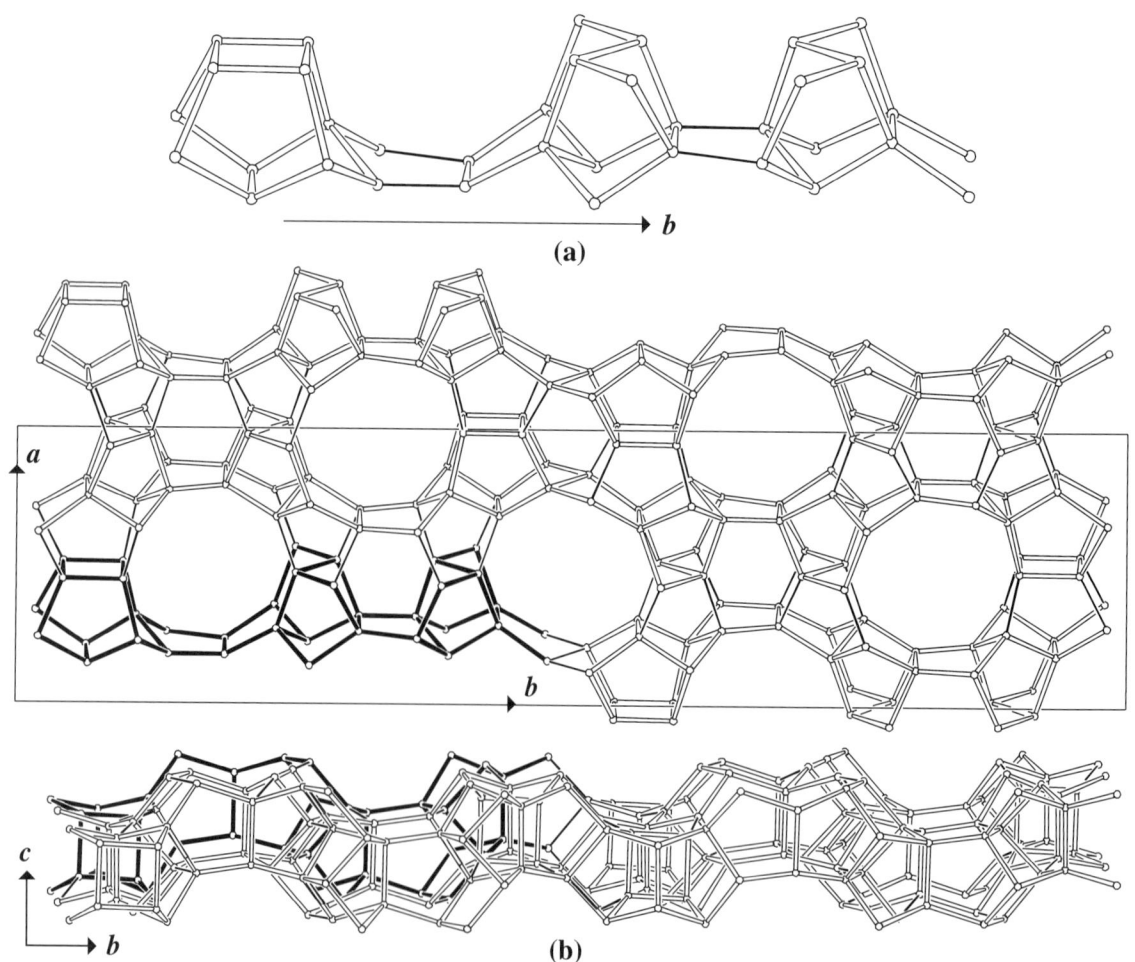

Fig. 1.    (a) T36-unit consisting of three different T12-units; (b) PerBU viewed along **c** (top) and along **a** (bottom).

# Building Scheme

## 2. Connection mode

Neighboring PerBUs, related by a rotation of 180° about **c**, are connected along **c** as depicted in Figure 2.

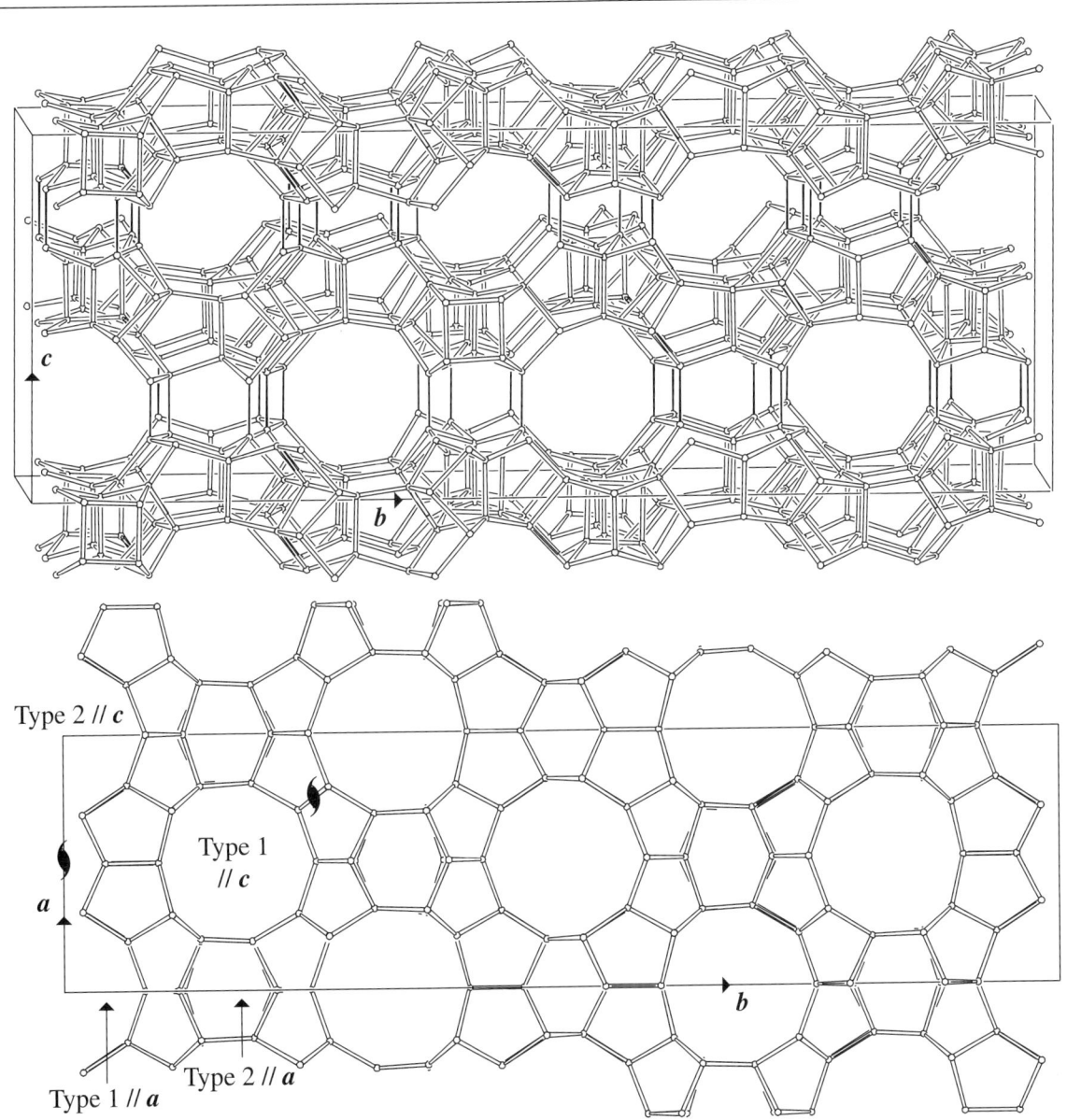

Fig. 2. Connection mode (and unit cell content) viewed along **a** (top). Projection of the unit cell content along **c** (bottom). The position of the 2-fold screw axes and the different types of 10-ring channels parallel to **a** and **c** are indicated (see Figure 3).

# IMF

**Cage/Channel**

## 3. Channels and/or cages

There are two types of intersecting 10-ring channels parallel to *c* and parallel to *a*, as indicated in Figure 2 and as illustrated in Figures 3(a) and 3(b). Pairs of 10-ring channels (of different types) parallel to *c* are interconnecting along *b* through channels of type 1 parallel to *a*, as shown in Figure 3(c). The interconnecting range along *b* is limited to a slab formed by two pairs of channels, related by a 2-fold screw axis parallel to *c*.

*Type 1:*
{1 [$5^8 6^8 10^4 10^{2/2}$] [001] (10-ring)}

*Type 2:*
{1 [$4^2 5^{16} 6^9 10^3 10^{2/2}$] [001] (10-ring)}

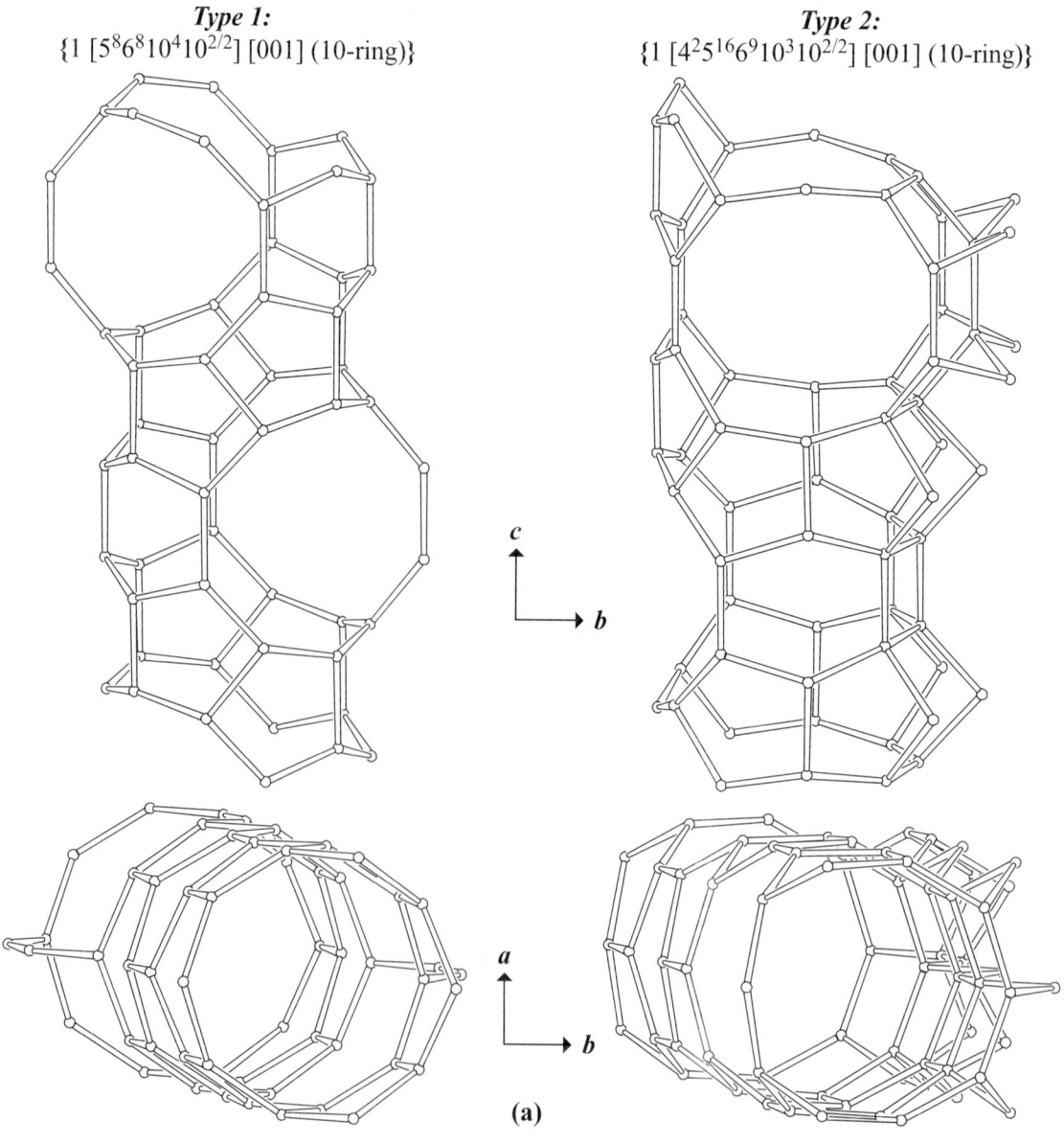

**(a)**

Fig. 3a.    10-ring channels parallel to *c* as viewed along *a* (top) and along the channel axis (bottom).

{1 [$4^2 5^8 6^5 10^3 10^{2/2}$] [100] (10-ring)}   {1 [$4^2 5^{12} 6^5 10^2 10^{2/2}$] [100] (10-ring)}

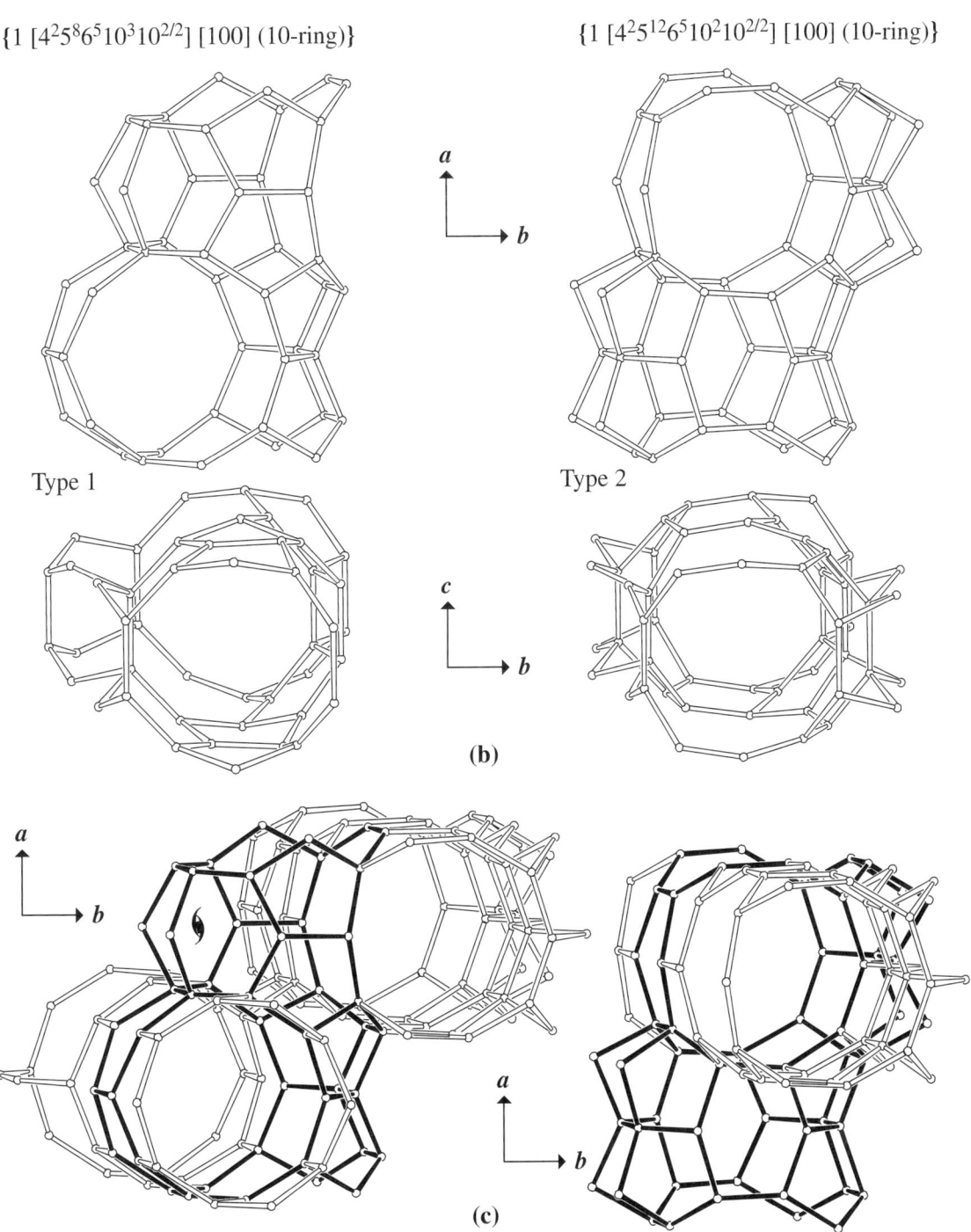

Fig. 3. [Continued].   (b) 10-ring channels parallel to *a* as viewed along *c* (top) and along *a* (bottom); (c) Left: Two types of channels parallel to *c* are interconnecting along *b* through channels of type 1 parallel to *a*. Right: Intersecting channels of type 2.

# ISV

# Building Scheme

## 1. Periodic Building Unit

Tetragonal **ISV** can be built using T16-units: four fused 6-rings or eight fused 5-rings (one unit in bold in Figure 1(a)). T16-units, related along **a** by pure translations, are connected into chains parallel to **a**. Neighboring chains, related along **b** by a rotation of 180° about the chain axis, are connected along **b** through 4-rings into the two-dimensional PerBU depicted in Figure 1(b). (Compare with **\*BEA** and **BEC**.)

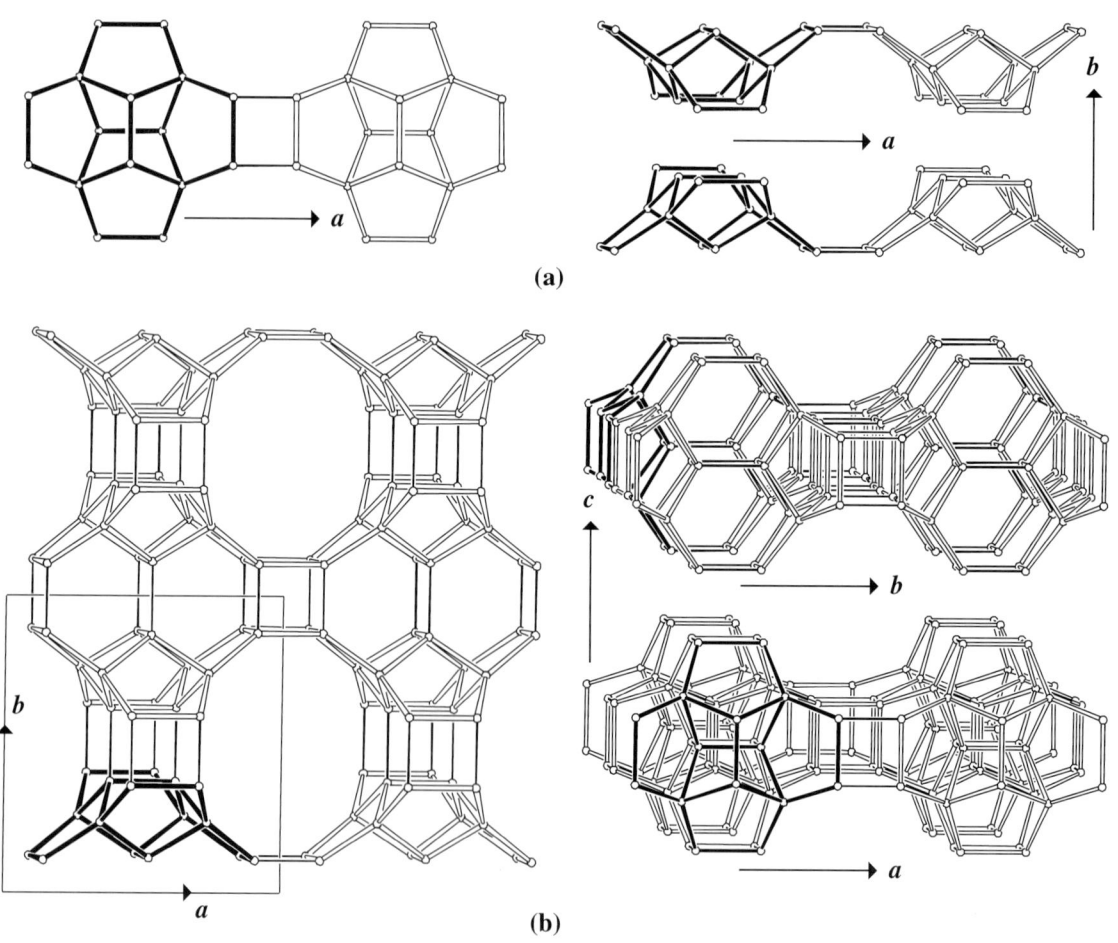

**(a)**

**(b)**

Fig. 1.    (a) Chain of T16 units. Chain viewed along **b** (left) and along **c** (right). The chains on the right differ by a rotation of 180° about **a**; (b) PerBU viewed along **c** (left), down **a** (top right) and along **b** (bottom right). The PerBUs, depicted at the right, are related by a rotation of 90° about **c**.

## 2. Connection mode

Neighboring PerBUs, related by a screw rotation of 90° about **c** followed by a shift of $\frac{1}{2}$**c** (i.e. related by a $4_2$-axis), are connected along **c** through double 4-rings (see Figure 2 on next page).

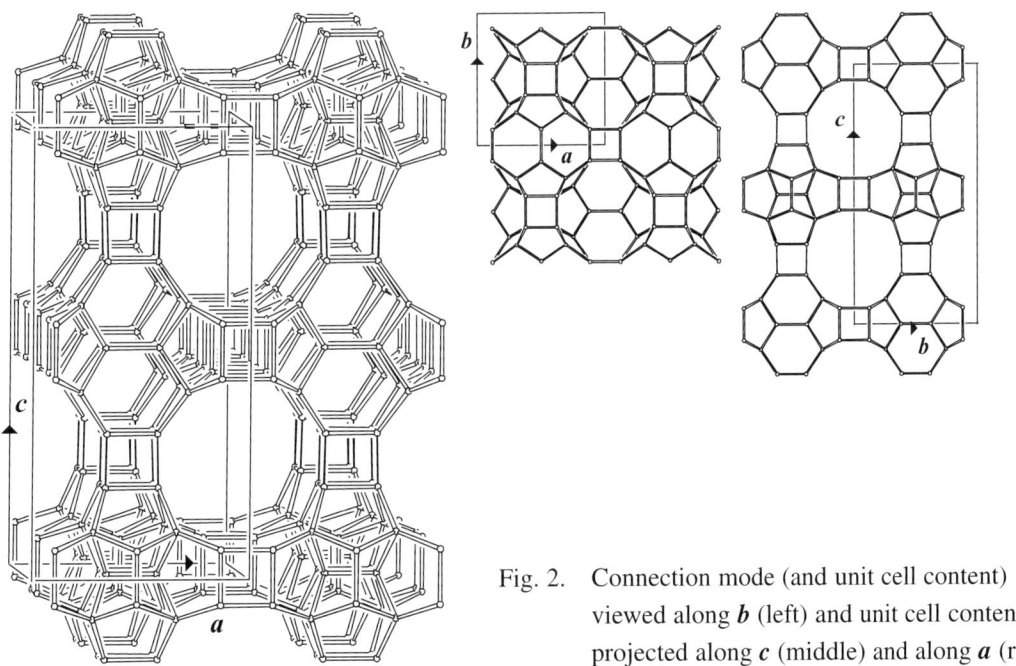

Fig. 2.   Connection mode (and unit cell content) viewed along *b* (left) and unit cell content projected along *c* (middle) and along *a* (right).

## 3.  Channels and/or cages

"Double" 12-ring channels, interconnecting through 12-rings perpendicular to *c*, are parallel to <100> (Figure 3). Diffusion through the 12-ring channel parallel to *c* is seriously obstructed.

{1 [$4^2 5^4 6^4 12^2$] [001] (12-ring) / 1 [$4^4 5^2 6^2 12^3 12^{2/2}$] <100> (12-ring)}

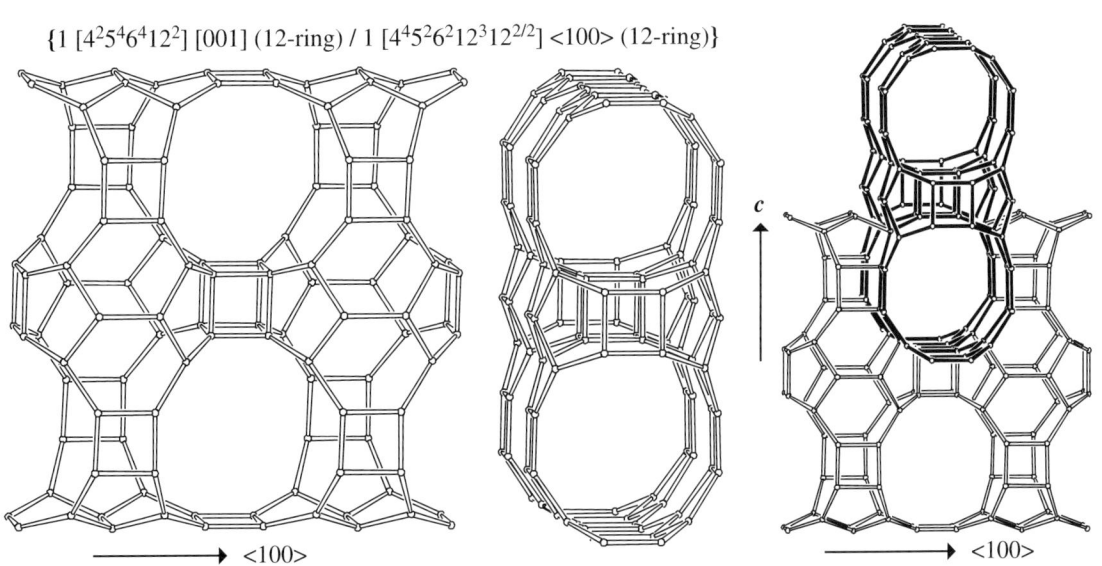

Fig. 3.   "Double" 12-ring channel viewed normal to the channel axis (left) and along the channel axis (middle). Right: intersecting double-channels have [$4^2 12^4$]-cavities in common.

# ITE                                            Building Scheme

## 1. Periodic Building Unit

**ITE** can be built using units of 16 T atoms consisting of two sets of three (fused) 4-rings that are related by a rotation of $\pm 90°$ about **b**. T16-units (one in bold in Figure 1), related along **b** by pure translations, are connected along **b** into left- and right-handed chains of $[4^45^4]$-cages sharing 4-rings. The two-dimensional PerBU is obtained when left- and right-handed chains are linked along **c** through 6-rings into the **bc** layer shown in Figure 1. (See also **RTH**.)

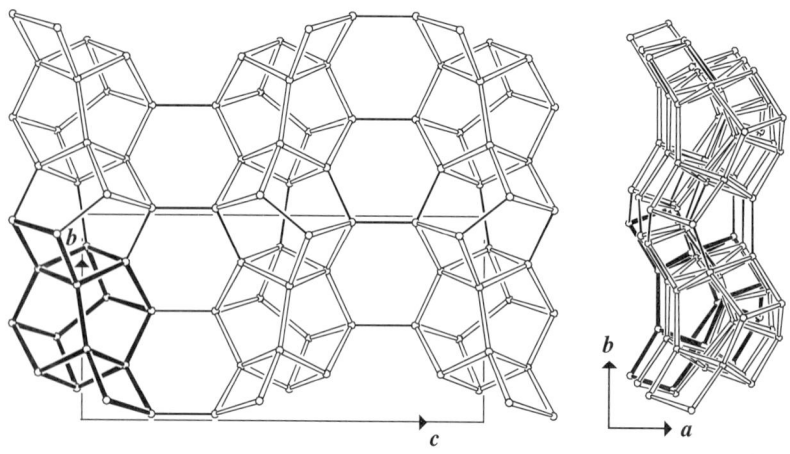

Fig. 1.   PerBU viewed along **a** (left) and projected down **c** (right).

## 2. Connection mode

Neighboring PerBUs, related by a shift of $\frac{1}{2}(\boldsymbol{a} + \boldsymbol{b})$, are connected along **a** through 4-rings (Figure 2).

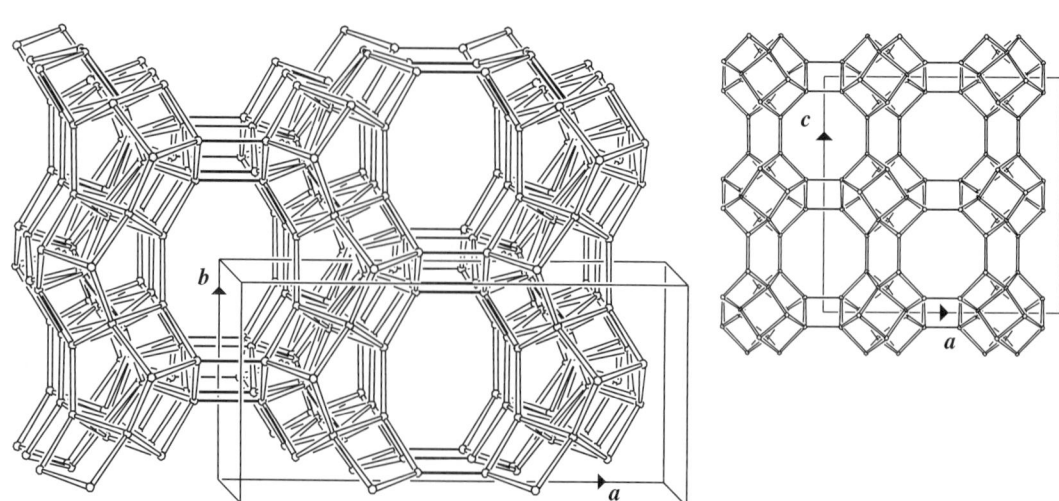

Fig. 2.   Connection mode (and unit cell content) viewed along **c** (left) and unit cell content projected along **b** (right).

# Cage/Channel

## 3. Channels and/or cages

Intersecting (different) 8-ring channels are parallel to **b** and **c**. The intersection, topologically equivalent to the intersection in **RTH**, is depicted in Figure 3 together with the linkage of intersections.

$\{2 [4^6 5^8 6^4 8^4] [010] \text{ (8-ring)}, [001] \text{ (8-ring)}\}$

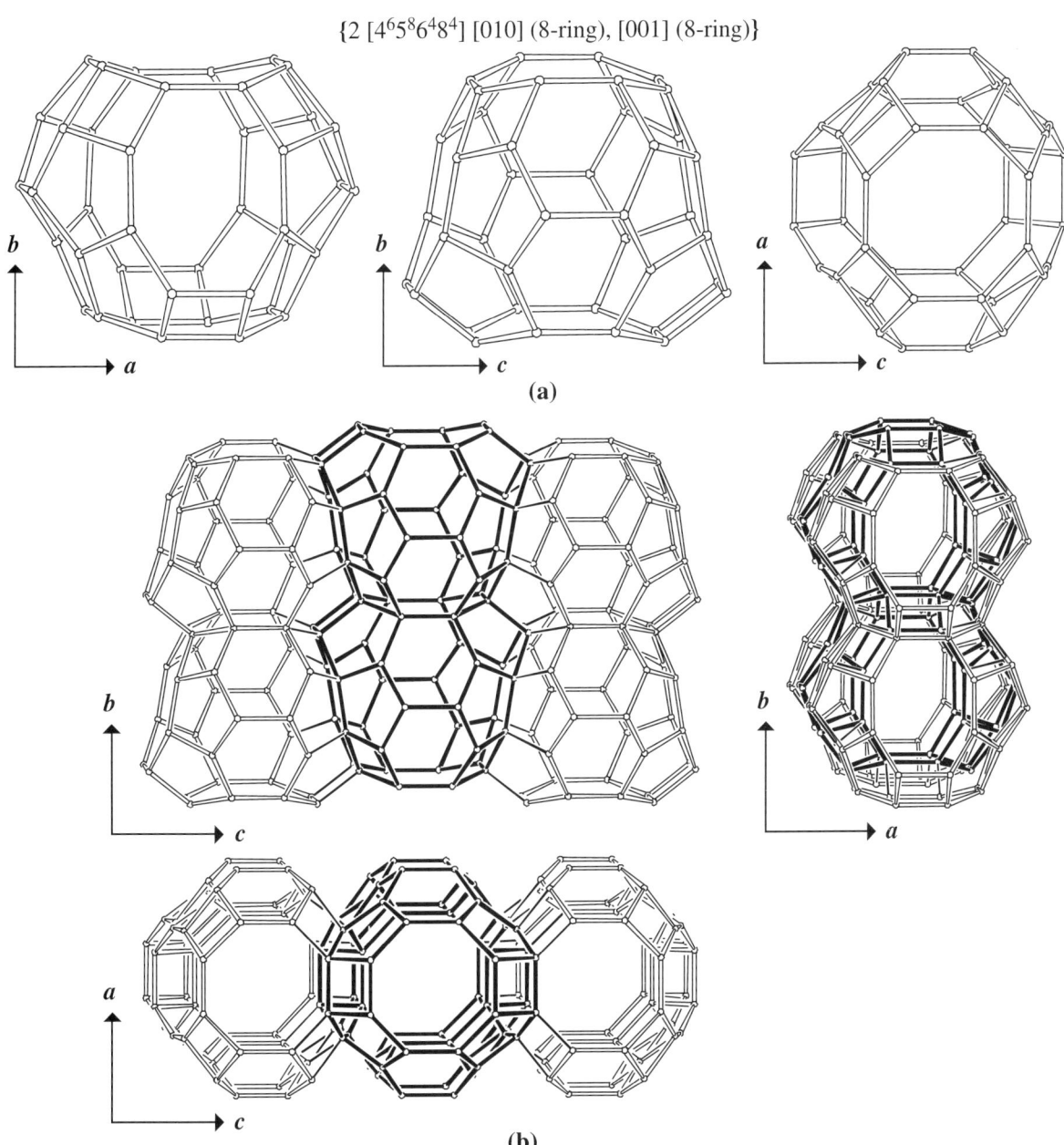

Fig. 3.  (a) Channel intersection viewed (from left to right) along **c**, **a** and **b**; (b) Linkage of intersections into 8-ring channels parallel to **b** and **c** viewed along **a** (top left), along the 8-ring channel axis parallel to **c** (top right) and along the 8-ring channel axis parallel to **b** (bottom).

# ITH

# Building Scheme

## 1. Periodic Building Unit

**ITH** can be built using units of 28 T atoms (see inset Figure 1). T 28-units, related along *a* and *b* by pure translations, are connected through 4-rings (and 6-rings) into the two-dimensional PerBU shown in Figure 1. (Compare with *BEA, BEC, CON, ISV, IWR, IWW and MSE.)

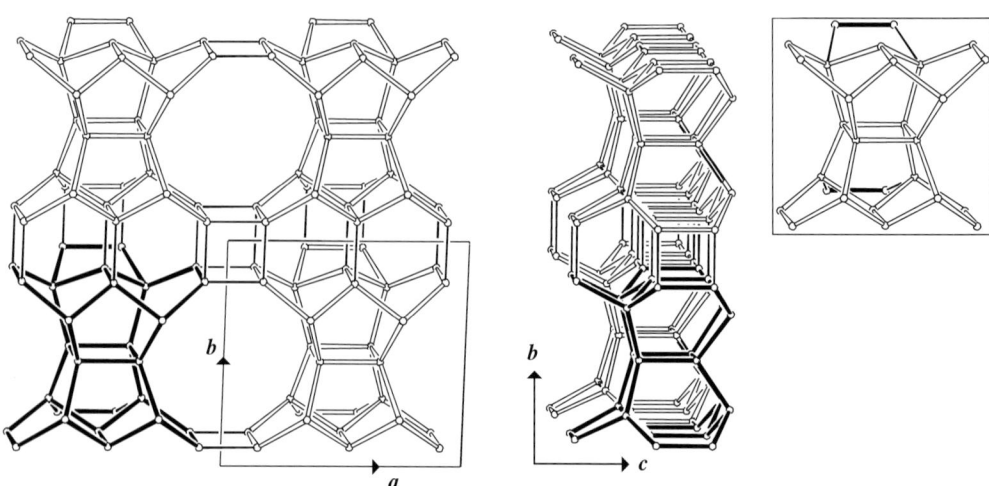

Fig. 1.  PerBU viewed along *c* (left) and along *a* (right). The inset shows the T28-unit: two $[4^15^26^4]$-cages sharing a 4-ring, and two dimers (in bold).

## 2. Connection mode

Neighboring PerBUs, related by a shift of $\frac{1}{2}(\boldsymbol{b}+\boldsymbol{c})$, are connected through 5- and 9-rings (Figure 2).

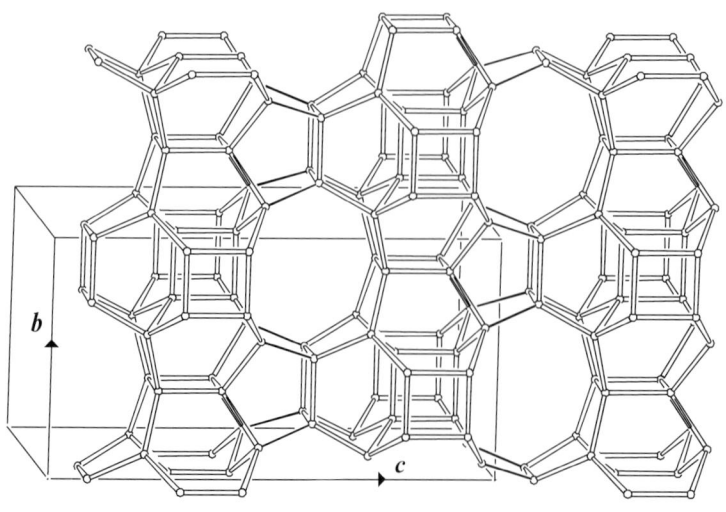

Fig. 2a.  Connection mode and unit cell content viewed along *a*. Only one T28-unit along *a* is drawn for clarity.

# Cage/Channel     **ITH**

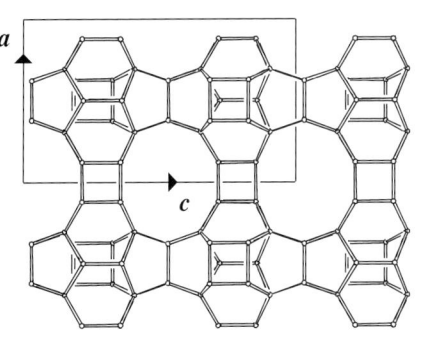 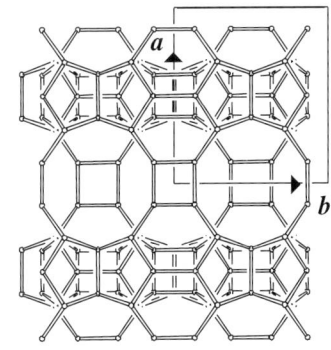

Fig. 2b.   Unit cell content projected along **b** (left) and along **c** (right).

## 3. Channels and/or cages

Intersecting 9- and 10-ring channels are parallel to **a** and **b**, respectively (Figure 3). Diffusion through the 10-ring channel parallel to **c** is seriously obstructed (see Figure 2, right).

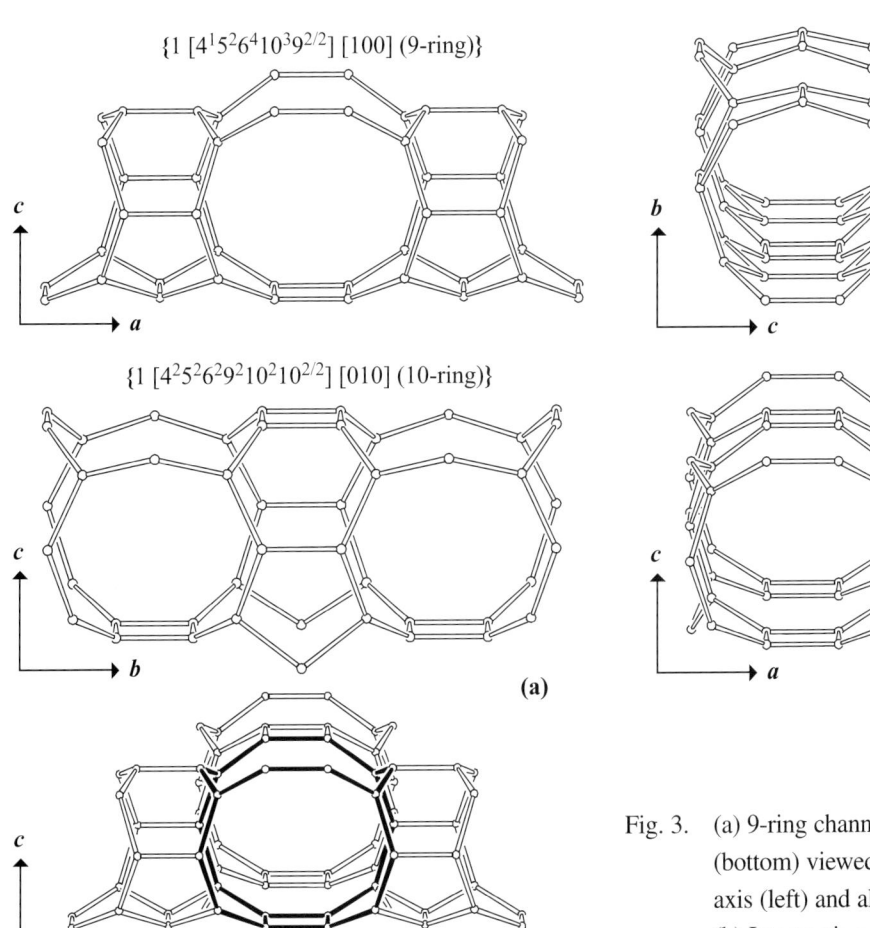

{1 [$4^1 5^2 6^4 10^3 9^{2/2}$] [100] (9-ring)}

{1 [$4^2 5^2 6^2 9^2 10^2 10^{2/2}$] [010] (10-ring)}

**(a)**

**(b)**

Fig. 3.   (a) 9-ring channel (top) and 10-ring channel (bottom) viewed perpendicular to the channel axis (left) and along the channel axis (right); (b) Intersecting channels, with a [$4^1 9^2 10^3$]-cavity (bold) in common, viewed along **b**.

# ITW  Building Scheme

## 1. Periodic Building Unit

Finite building units of 12 T atoms are composed of one (finite) zigzag chain and a double 4-ring (bold in Figure 1). The two-dimensional PerBU is obtained when these T12-units, related along $a$ and $c$ by pure translations, are connected into a layer with an oblique repeat unit. Zigzag chains parallel to $a$ are formed.

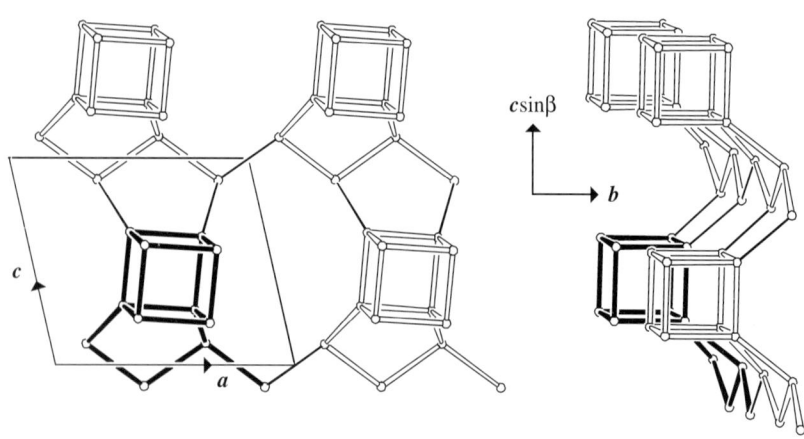

Fig. 1.  PerBU viewed along $b$ (left) and along $a$ (right).

## 2. Connection mode

Neighboring PerBUs, related along $b$ by a shift of $\frac{1}{2}(a+b)$, are connected along $b$ as shown in Figure 2. Arrays of double 4-rings are connected through zigzag chains.

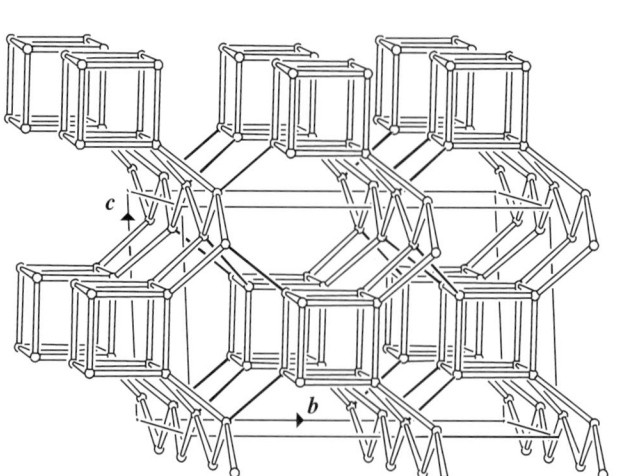

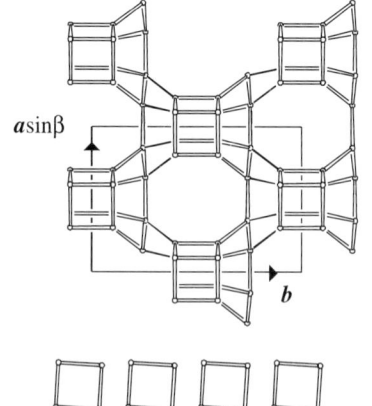

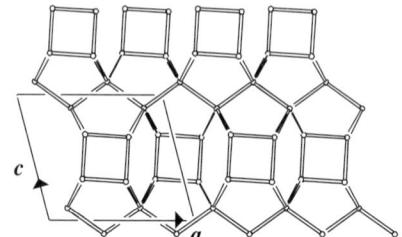

Fig. 2.  Connection mode (and unit cell content) viewed along $a$ (left) and unit cell content projected along $c$ (top right) and along $b$ (bottom right).

# Cage/Channel

## 3. Channels and/or cages

Interconnecting "double" 8-ring channels parallel to **a**, intersect with 8-ring channels parallel to **c**. The channel intersection and the linkage of intersections into channels are depicted in Figure 3.

{2 [$4^6 5^8 6^4 8^4$] [100] (8-ring),[001] (8-ring)}

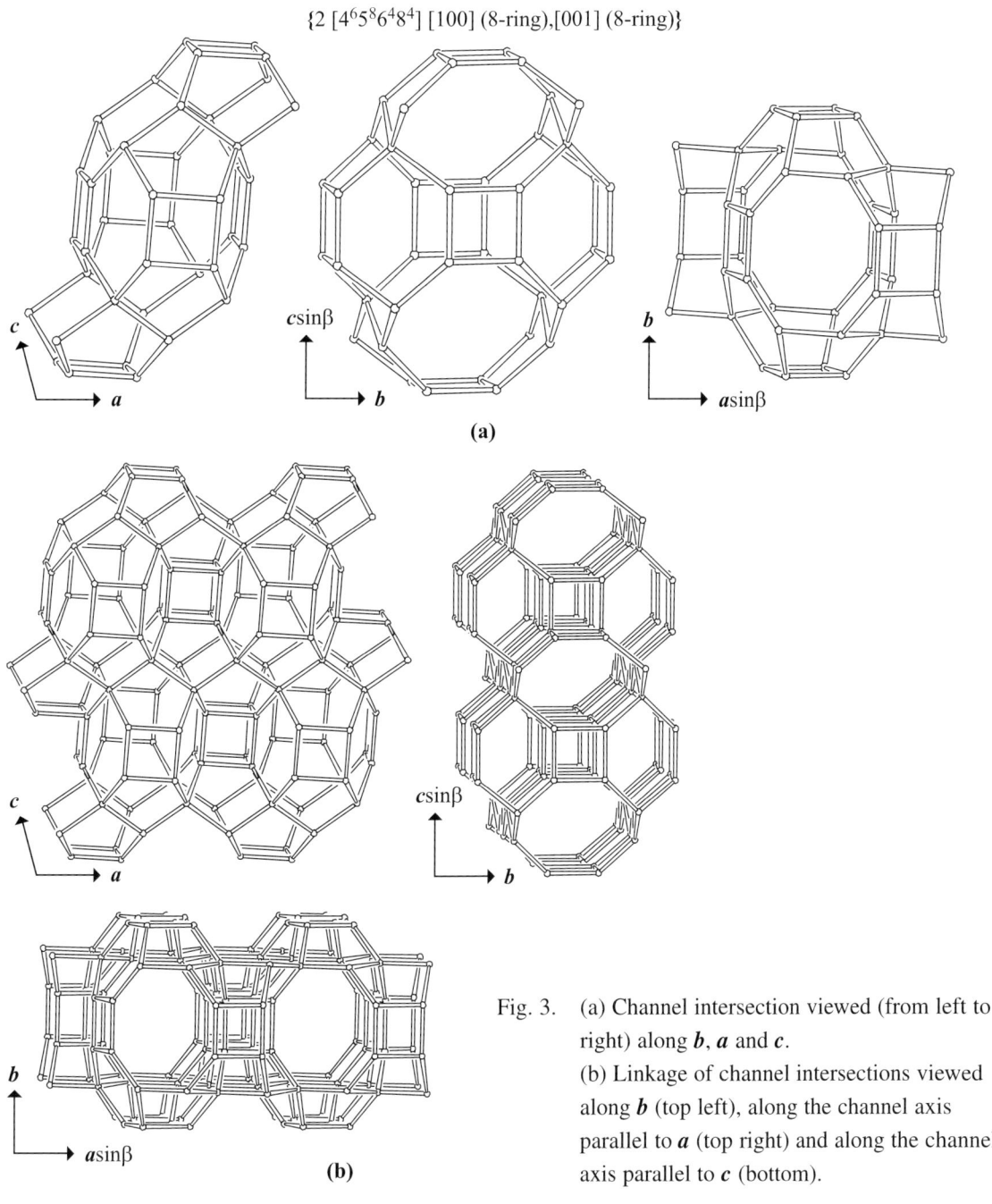

Fig. 3. (a) Channel intersection viewed (from left to right) along **b**, **a** and **c**.
(b) Linkage of channel intersections viewed along **b** (top left), along the channel axis parallel to **a** (top right) and along the channel axis parallel to **c** (bottom).

# IWR                                    Building Scheme

## 1. Periodic Building Unit

**IWR** can be built using chains parallel to *c* constructed from units of 14 T atoms (a $[4^15^26^4]$-cage, or two 1-5-1 units; bold in Figure 1) that are related along *c* by pure translations. A two-dimensional PerBU is obtained when parallel chains, related along *b* by a screw rotation of 180° about *b*, are connected into the *bc* layer through 4- and 5-rings (Figure 1). (See also **CON**; compare with **ITH**.)

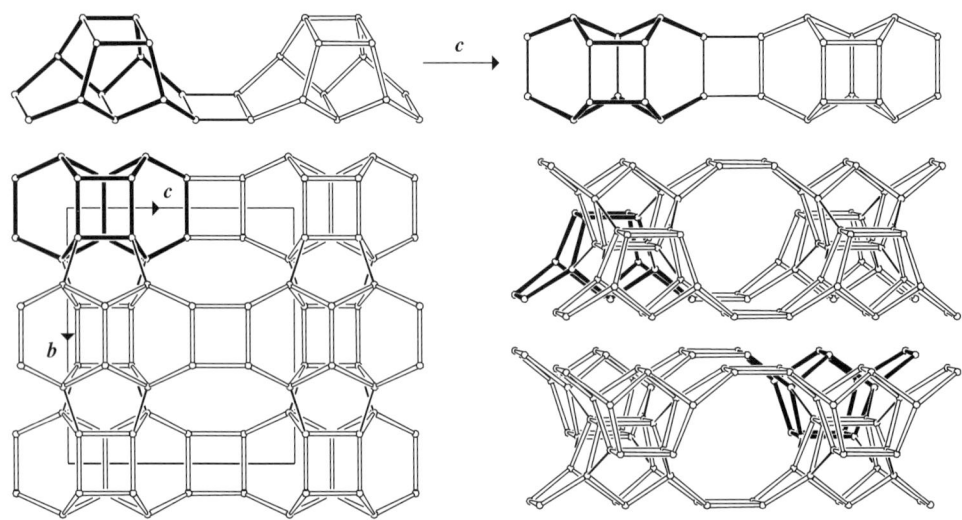

Fig. 1.   Top: chains of T14-units. The chain at the right is rotated over 90° about *c* with respect to the left one. Bottom: PerBU viewed along *a* (left), and along *b* (right). The PerBUs, depicted at the right, are related by a rotation of 180° about *b*.

## 2. Connection mode

Neighboring PerBUs, related by a rotation of 180° about *b* and no lateral shift along *c*, are connected along *a* through 4-rings, as depicted in Figure 2. Fused 4-6-6 ring sequences are formed.

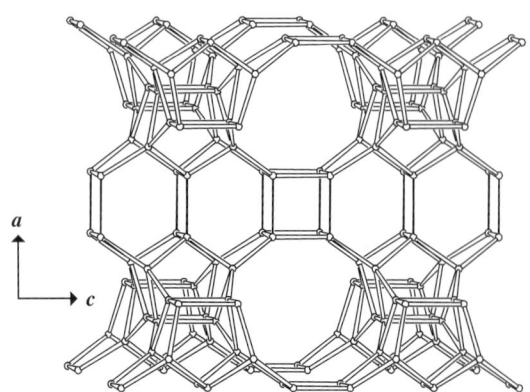

Fig. 2a.   Connection mode viewed along *b*.

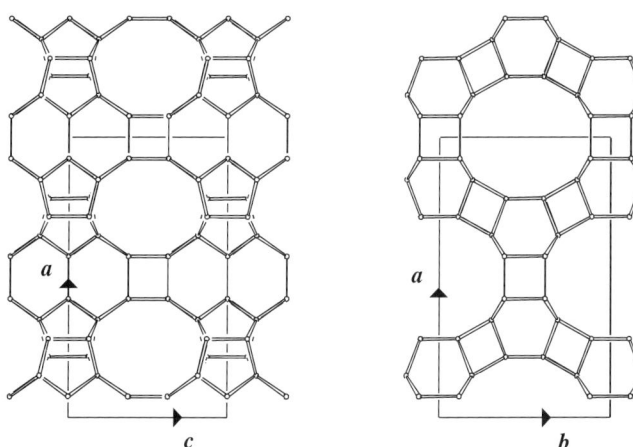

Fig. 2b.   Unit cell content projected along **b** (left), and along **c** (right).

## 3. Channels and/or cages

Interconnecting straight 10- and 12-ring channels are parallel to **b** and **c**, respectively. The 12-ring channel and the 10-ring channel are topologically equivalent to the 12-ring channel in **BEC** and the 10-ring channel in **CON**. An additional sinusoidal 12-ring channel, intersecting with the straight channels, is parallel to **a**. The channels are shown in Figure 3.

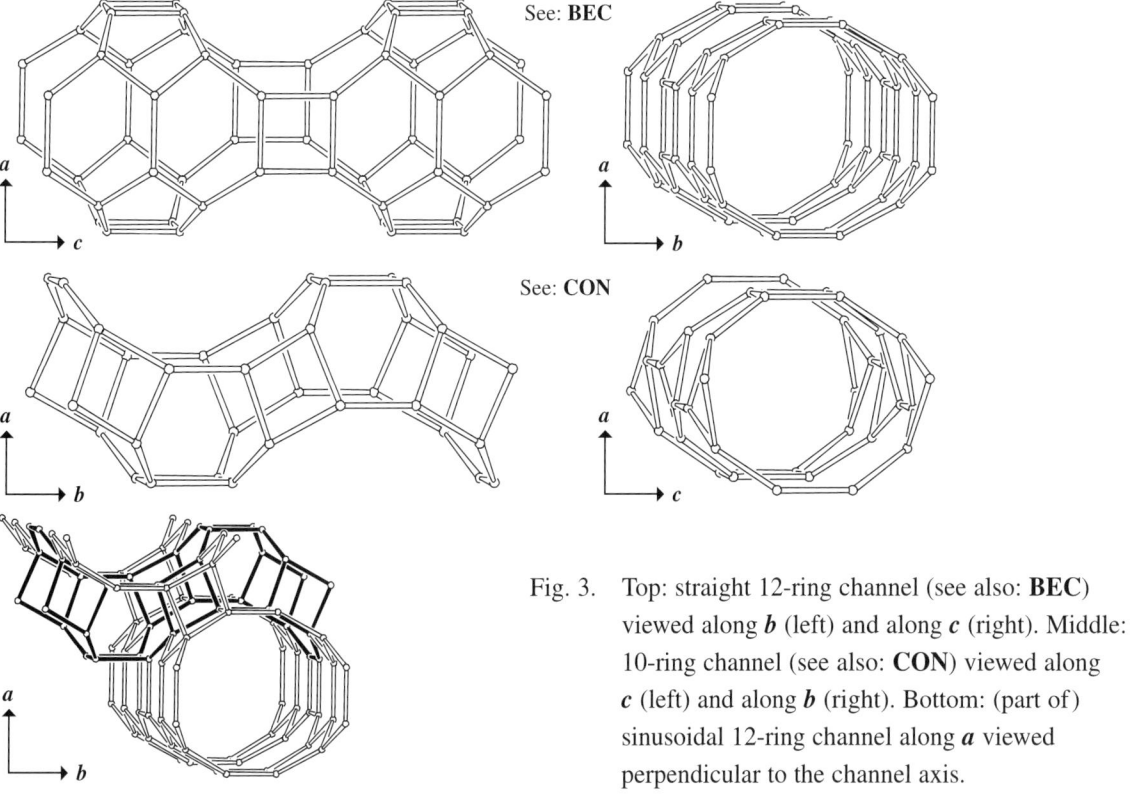

See: **BEC**

See: **CON**

Fig. 3.   Top: straight 12-ring channel (see also: **BEC**) viewed along **b** (left) and along **c** (right). Middle: 10-ring channel (see also: **CON**) viewed along **c** (left) and along **b** (right). Bottom: (part of) sinusoidal 12-ring channel along **a** viewed perpendicular to the channel axis.

# IWV                    Building Scheme

## 1. Periodic Building Unit

**IWV** can be built using building units composed of 19 T atoms: two 5-1 units and a 5-2 unit (Figure 1). The two-dimensional PerBU is obtained when T19-units, related along **c** by a screw rotation of 180° about **c** and along **a** by a screw rotation of 180° about **a**, are connected into the **ac** layer shown in Figure 1. (Compare with PerBUs in **NES**, **NON** and **EUO**.)

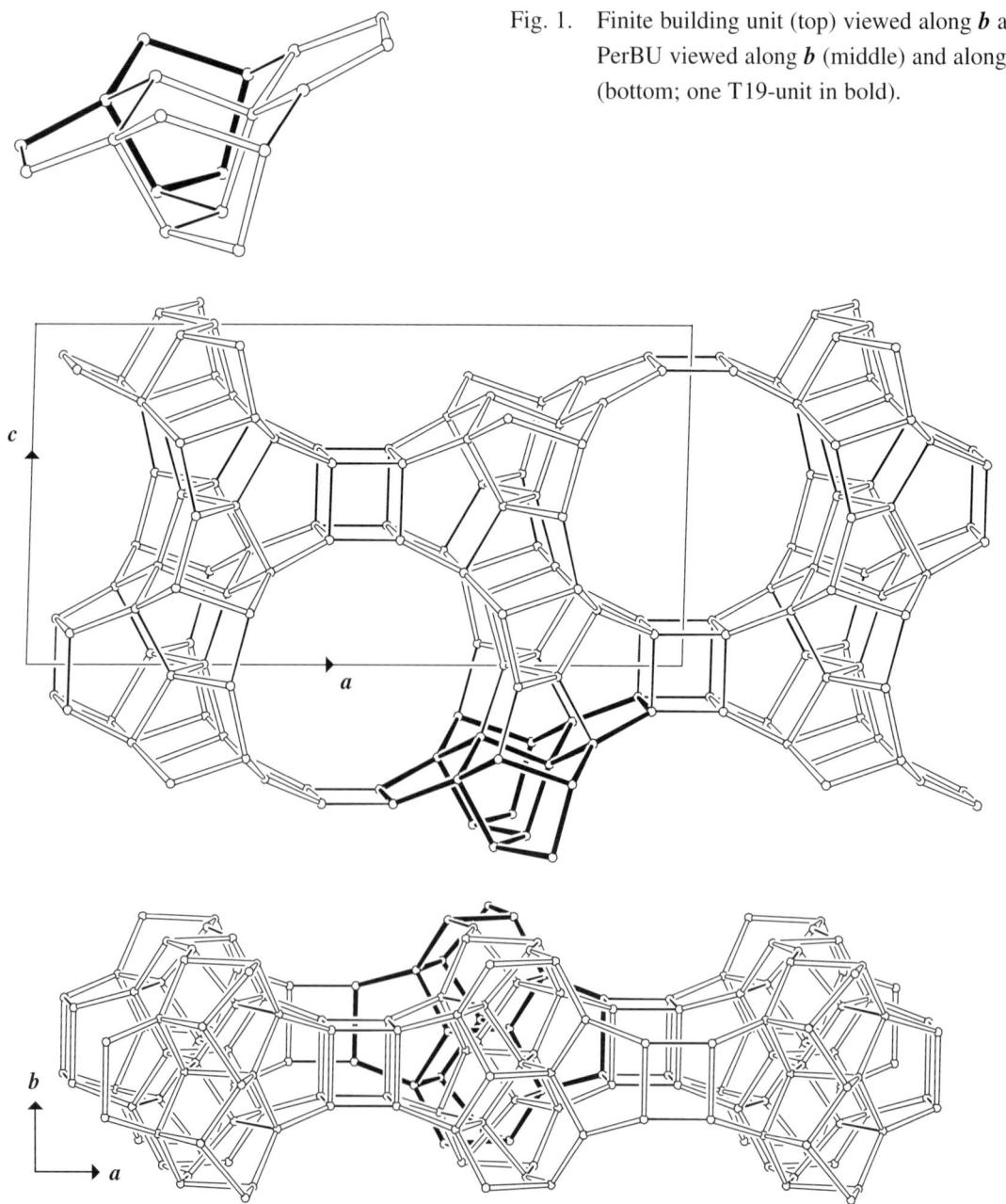

Fig. 1.  Finite building unit (top) viewed along **b** and PerBU viewed along **b** (middle) and along **c** (bottom; one T19-unit in bold).

# Building Scheme

## 2. Connection mode

Neighboring PerBUs, related by a shift of $\frac{1}{2}\boldsymbol{a}$ (or $\frac{1}{2}\boldsymbol{c}$), are connected along $\boldsymbol{b}$ as shown in Figure 2.

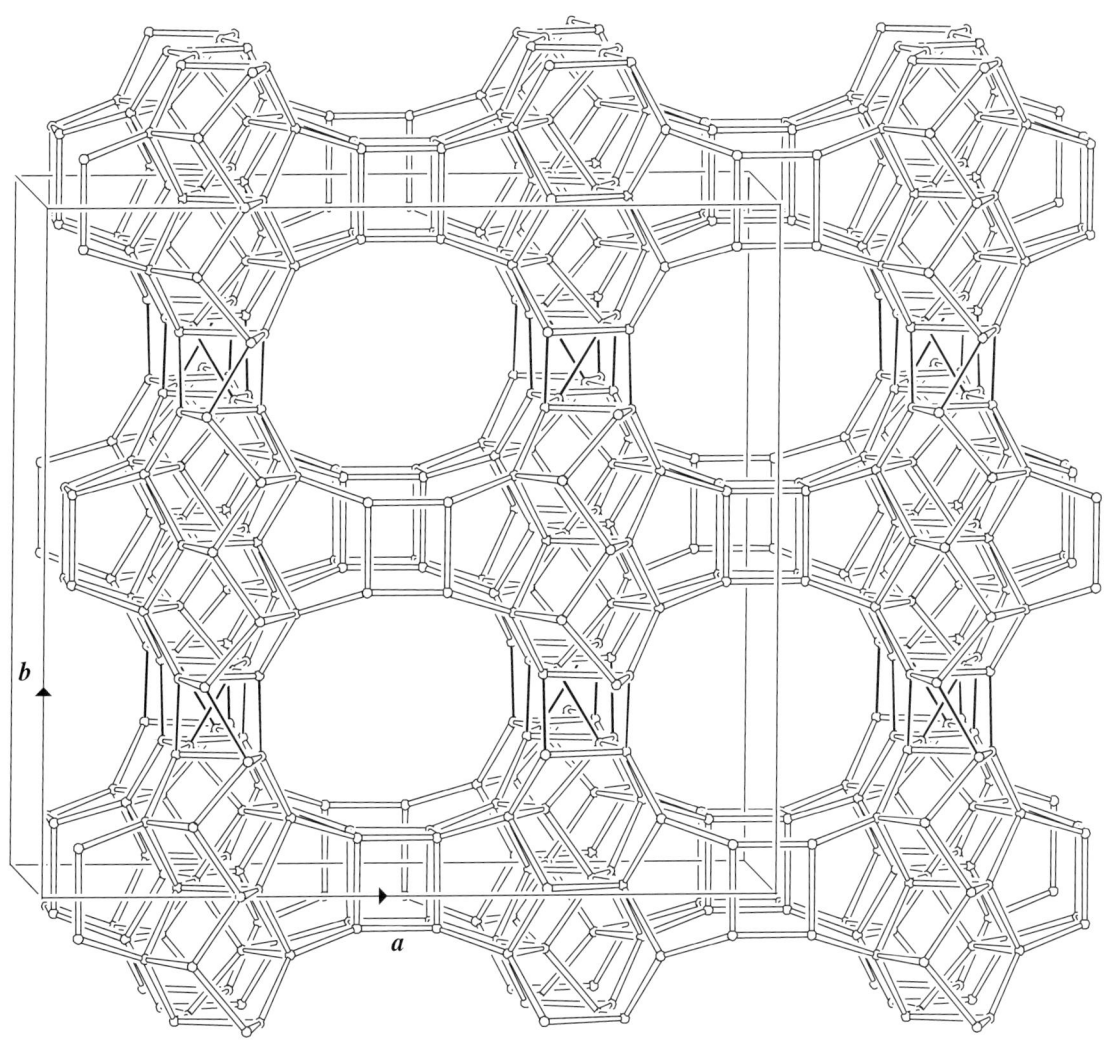

Fig. 2a.   Connection mode (and unit cell content) in **IWV** viewed along $\boldsymbol{c}$.

# IWV

## Cage/Channel

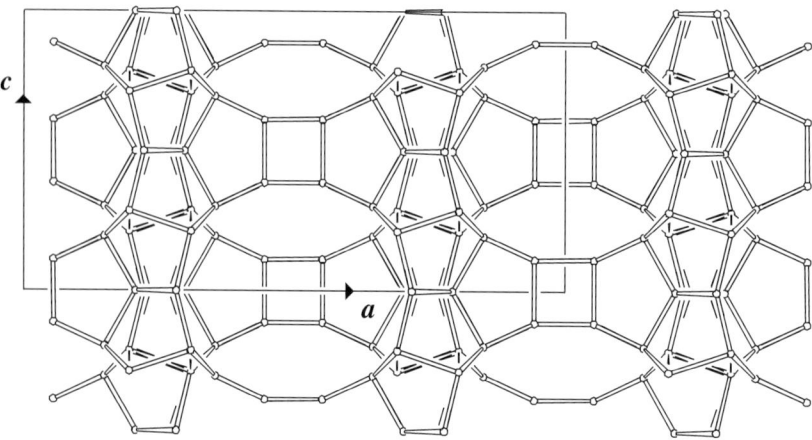

Fig. 2b.   Unit cell content projected along **b**.

### 3.  Channels and/or cages

Large cavities are connected along **c** through common 12-rings into 12-ring channels parallel to **c** as illustrated in Figure 3 on next page. The 12-ring channels are interconnecting through 14-rings in the large cavity.

$\{1\ [4^2 5^8 6^{16} 12^4]\ [001]\ (12\text{-ring})\}$

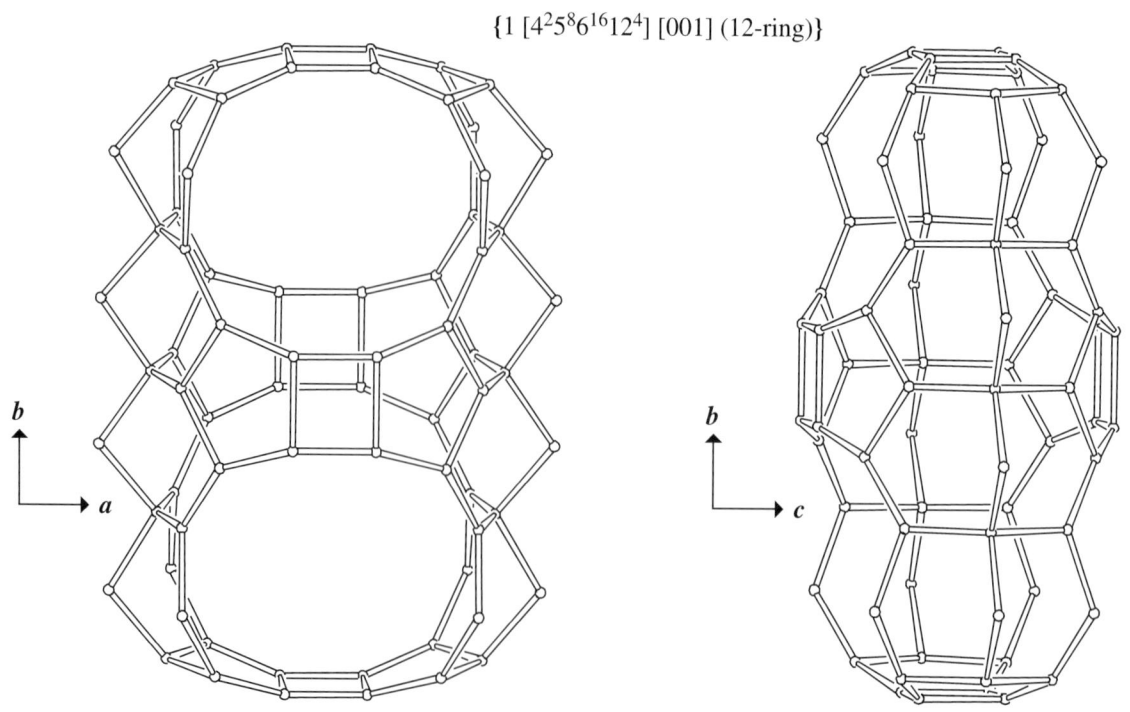

Fig. 3a.   Large cavity viewed along **c** (left) and along **a** (right).

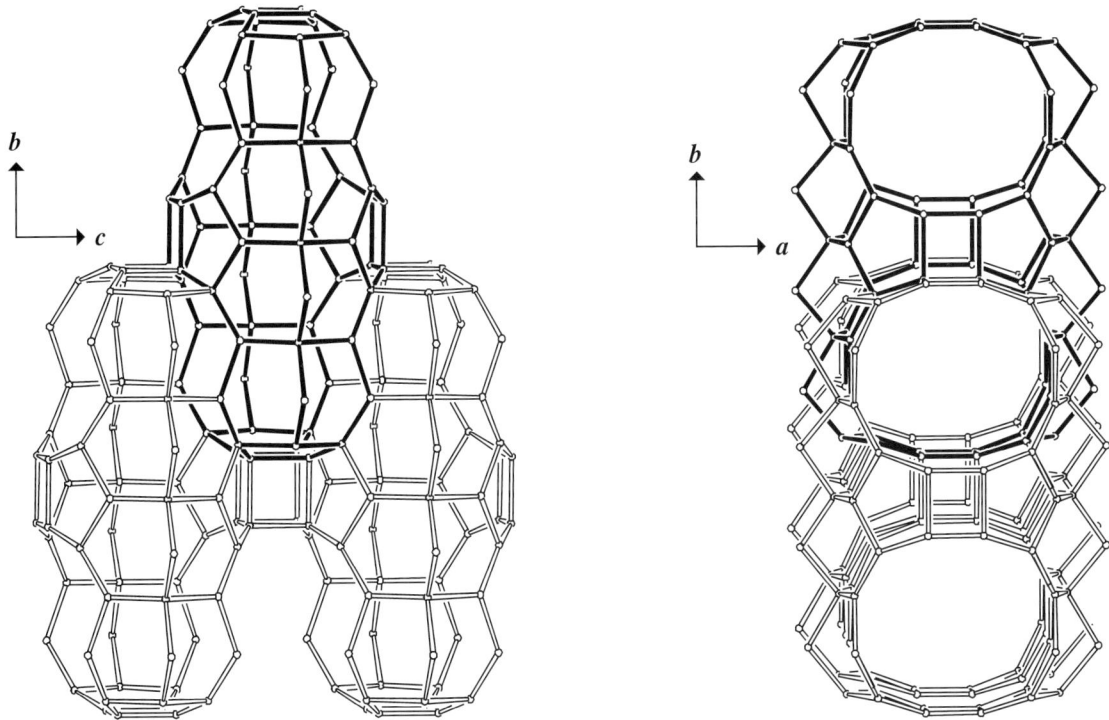

Fig. 3b.    Cavities are linked into interconnecting 12-ring channels parallel to *c*. Pairs of 12-ring channels are
interconnecting along *b* through 14-rings in the cavity. View along *a* (left) and along *c* (right).

# IWW
**Building Scheme**

## 1. Periodic Building Unit

**IWW** can be built using units of 28 T atoms (see inset in Figure 1). The two-dimensional PerBU is obtained when T28-units, related along **b** and **c** by pure translations, are connected into the **bc** layer depicted in Figure 1. (Compare with **\*BEA**, **BEC**, **CON**, **ISV**, **ITH**, **IWR** and **MSE**.)

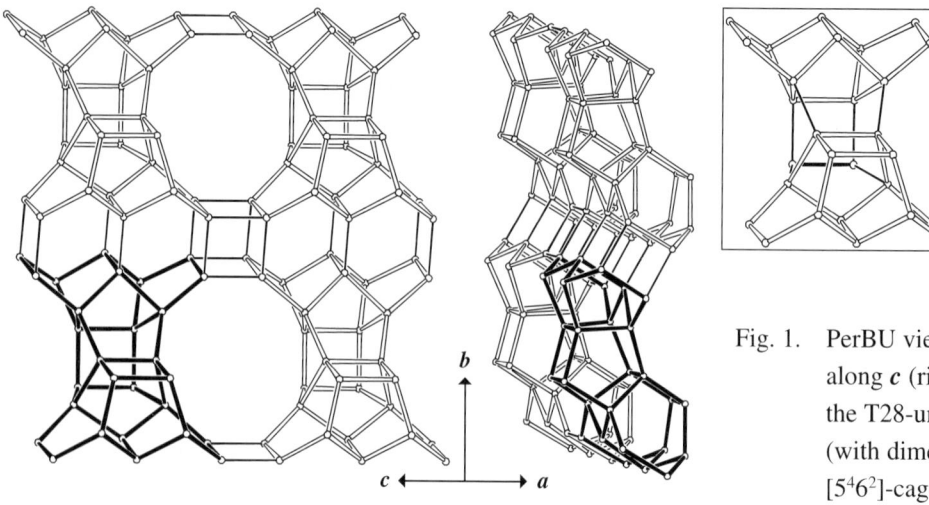

Fig. 1.   PerBU viewed along **a** (left) and along **c** (right). The inset shows the T28-unit: a $[4^1 5^2 6^4]$-cage (with dimer; bold) attached to a $[5^4 6^2]$-cage.

## 2. Connection mode

Neighboring PerBUs are alternately related along **a** by a rotation of 180° about **c** (and a shift of $\frac{1}{2}$**b**) and by a rotation of 180° about **b** (and a shift of $\frac{1}{2}$**b**) (Figure 2). 8- and 12-rings are formed.

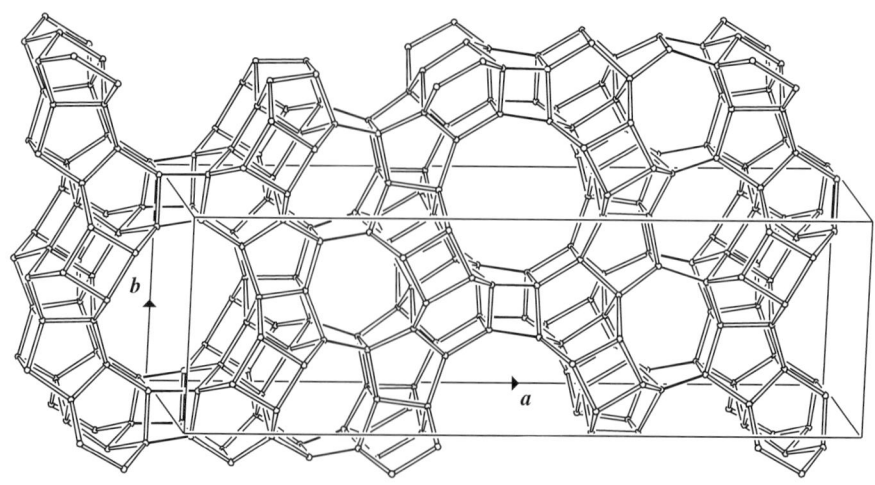

Fig. 2a.   Connection mode and unit cell content viewed along **c**. Only one T28-unit along **c** has been drawn for clarity.

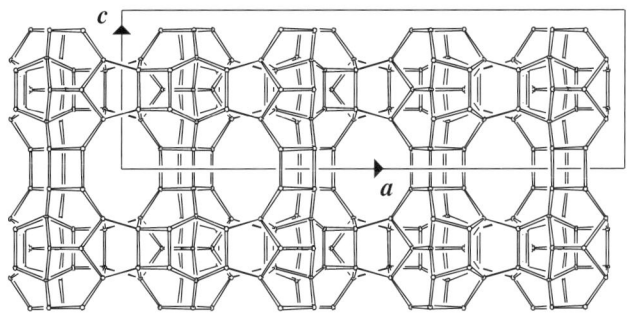

Fig. 2b.    Unit cell content projected along **b**.

### 3. Channels and/or cages

Straight 8- and 12-ring channels are parallel to **c** and (straight and sinusoidal) 10-ring channels are parallel to **b**. There are two types of channel intersections depicted in Figure 3(a): Intersection 1 between the straight 10-ring channel parallel to **b** and the 12-ring channel, and Intersection 2 between the sinusoidal 10-ring channel parallel to **b** and the 8-ring channel. Intersections 1 and 2 are interconnecting through common 10-rings as illustrated in Figure 3(b).

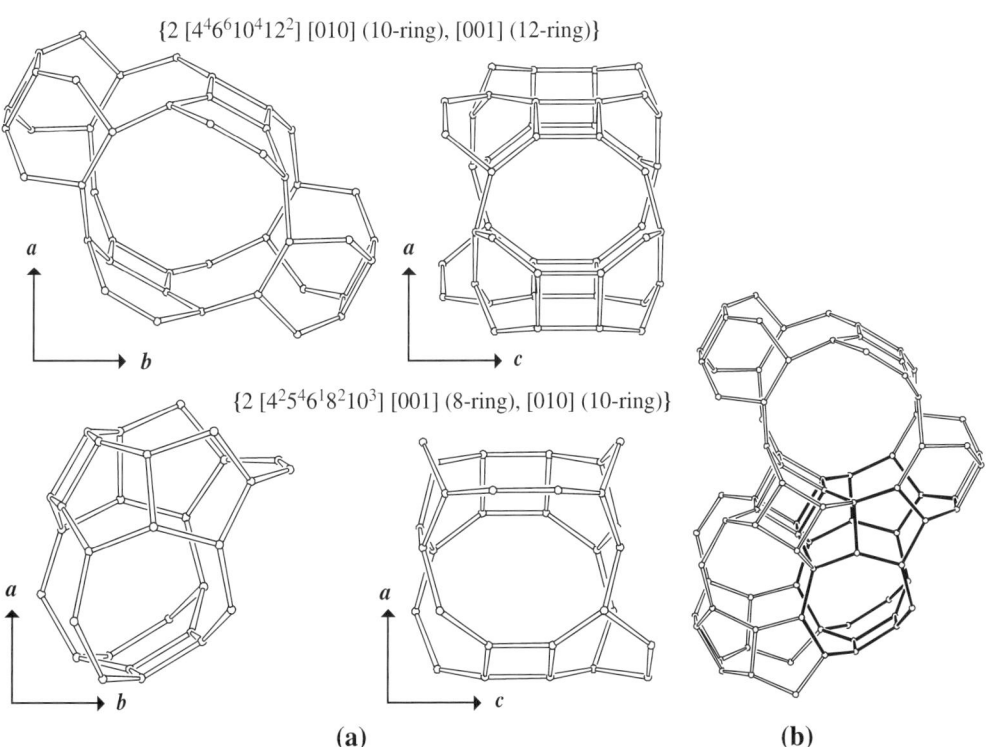

Fig. 3.    (a) Intersection 1 (top) and Intersection 2 (bottom) viewed along **c** (left) and **b** (right); (b) linkage of channel intersections, viewed along **c**, illustrating the different (interconnecting) channel systems.

# JBW                          Building Scheme

## 1. Periodic Building Unit

**JBW** can be built using the zigzag chain (bold in Figure 1) parallel to **a**. The two-dimensional PerBU is obtained when zigzag chains are connected along **c** into a layer of (fused) 6-ring chairs capped with additional zigzag chains as shown in Figure 1.

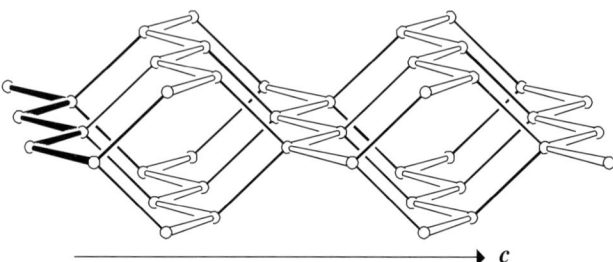

Fig. 1.   PerBU, constructed from three independent zigzag chains, viewed along **a**.

## 2. Connection mode

Neighboring PerBUs, related along **b** by pure translations, are connected along **b** through double zigzag chains (Figure 2).

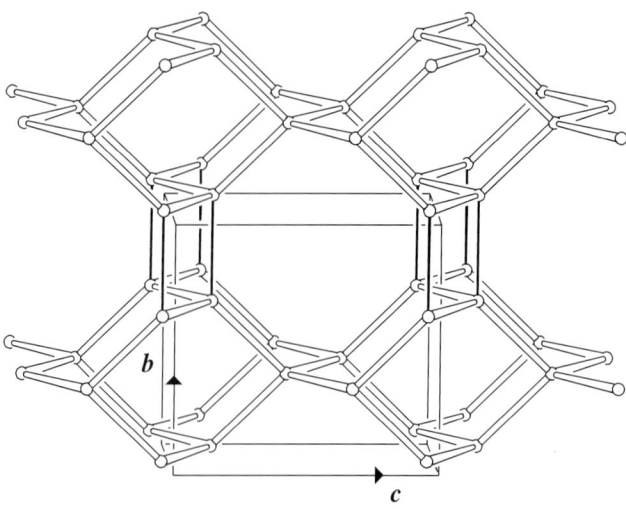

Fig. 2.   Connection mode and unit cell content viewed along **a**. Only two repeat units along the zigzag chain are drawn for clarity.

### 3. Channels and/or cages

Non-interconnecting 8-ring channels (Figure 3) are parallel to *a*. The channel is topologically equivalent to the channel in **ABW**.

$$\{1\ [4^4 6^4 8^{2/2}]\ [100]\ (8\text{-ring})\}$$

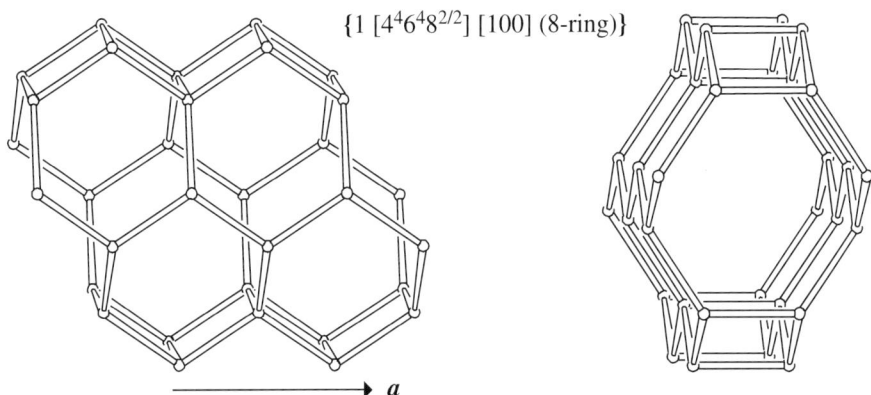

*a*

Fig. 3.    Channel (with side-pockets) viewed perpendicular to the channel axis (left) and along the channel axis (right).

# KFI                    Building Scheme

## 1. Periodic Building Unit

The two-dimensional PerBU in cubic **KFI** is the double 6-ring layer depicted in Figure 1. Double 6-rings (one in bold), related along **a** and **b** by screw rotations of 180° about **a** and **b**, are connected into the **ab** layer through 4-rings. (See also **SAV**; compare with **RHO**.)

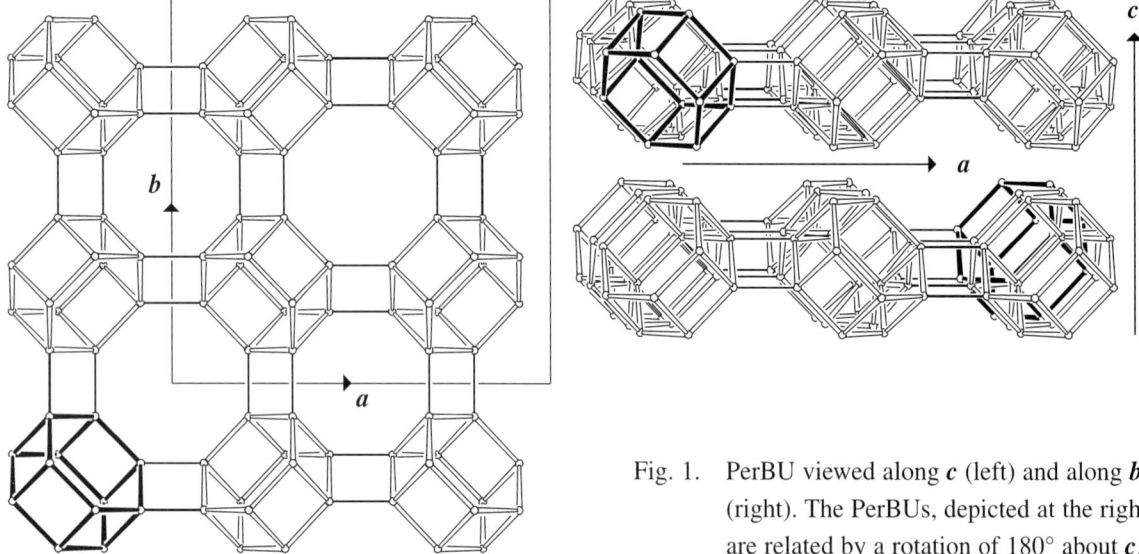

Fig. 1.   PerBU viewed along **c** (left) and along **b** (right). The PerBUs, depicted at the right, are related by a rotation of 180° about **c**.

## 2. Connection mode

Neighboring PerBUs, related along **c** by a screw rotation of 180° about **c**, are connected along **c** through 4-rings as depicted in Figure 2.

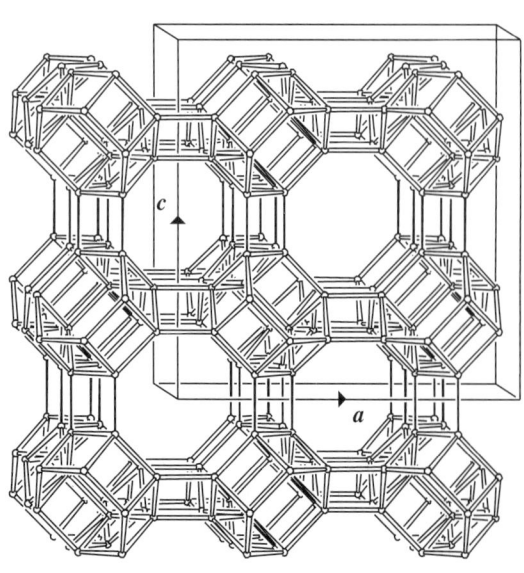

Fig. 2.   Connection mode and unit cell content viewed along **b**. **rho** Cavities are formed.

## 3. Channels and/or cages

Intersecting 8-ring channels are parallel to the cell axes. The channel intersections are equal to the *mer* and *rho* cavities (Figure 3). The *mer* cavity is also observed in **MER**, **MOZ** and **PAU**. **KFI** can also be built from *rho* cavities (see also **RHO**). The *rho* cavity is also present in **-CLO**, **LTA**, **LTN**, **PAU**, **RHO**, **TSC** and **UFI**.

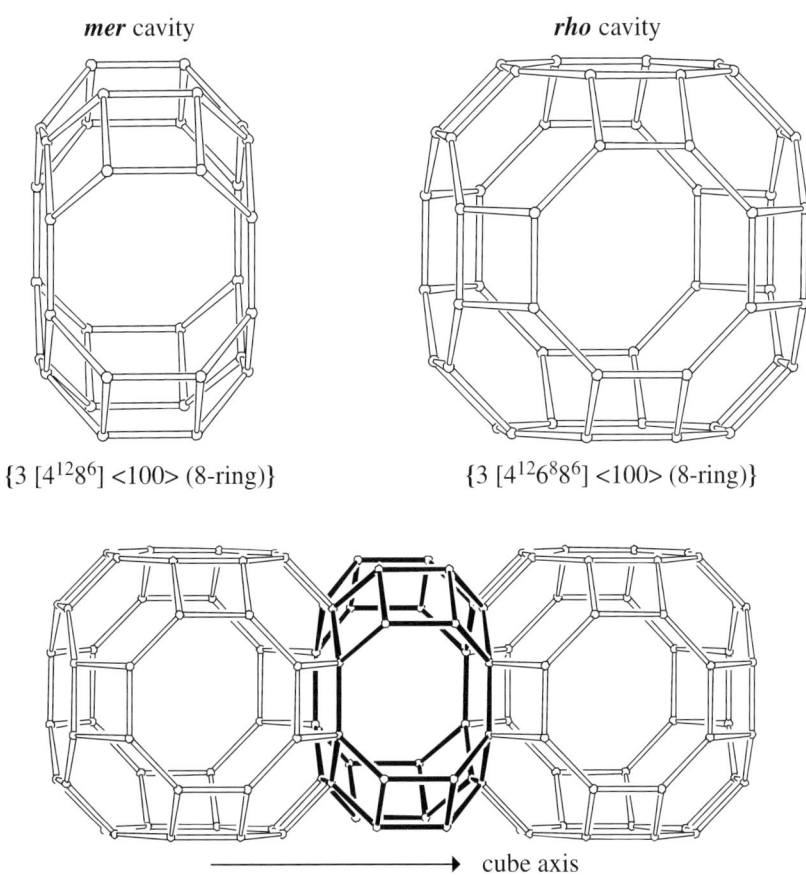

*mer* cavity          *rho* cavity

{3 [$4^{12}8^6$] <100> (8-ring)}          {3 [$4^{12}6^88^6$] <100> (8-ring)}

cube axis

Fig. 3.    *mer* Cavity (top left) and *rho* cavity (top right) viewed along a cube axis. The cavities are linked into 8-ring channels by sharing 8-rings (bottom; *mer* cavity in bold).

# LAU                    Building Scheme

## 1. Periodic Building Unit

**LAU** can be built using the 4-fold (1,2,4,5)-connected double 6-ring of 12 T atoms (a $[4^2 6^4]$-cage; bold in Figure 1). The one-dimensional PerBU is obtained when T12-units, related along $c$ by pure translations, are connected through 4-rings into a chain along $c$ as shown in Figure 1. (The PerBU can as well be built using a different $[4^2 6^4]$-cage, as can be seen from Figure 1.)

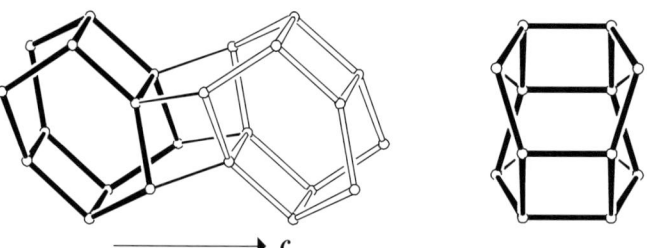

Fig. 1.   PerBU constructed from 4-fold connected double 6-rings viewed normal to the chain axis $c$ (left), and along the chain axis (right).

## 2. Connection mode

Neighboring PerBUs, related along $a$ (and $b$) by a shift of $\frac{1}{2}(a+b)$, are connected along $a$ (and $b$) by 4-rings as illustrated in Figure 2.

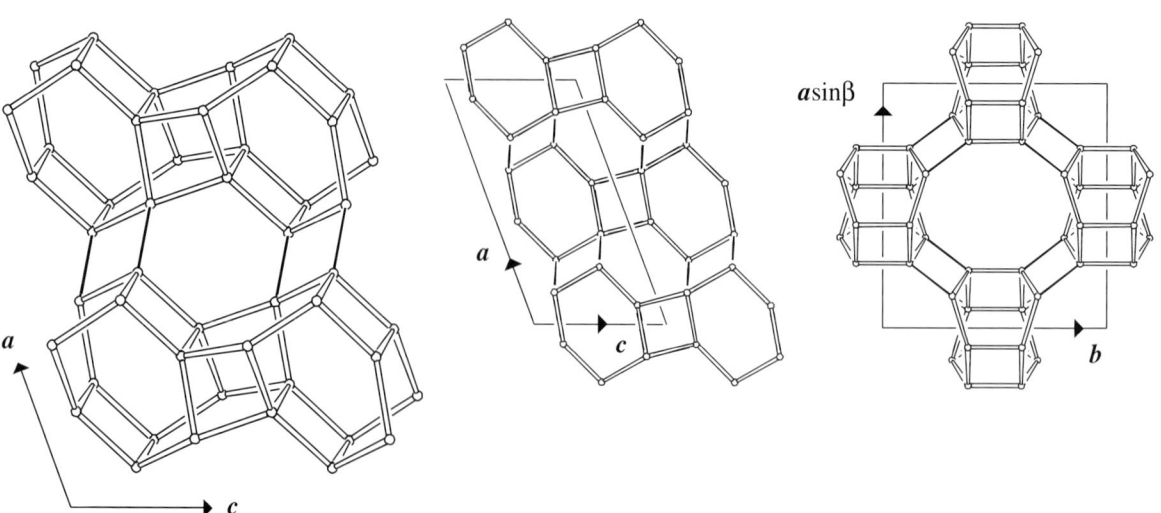

Fig. 2.   Connection mode viewed along $b$ (left) and unit cell content projected along $b$ (middle) and along $c$ (right).

# Cage/Channel                                                 **LAU**

## 3. Channels and/or cages

Non-interconnecting 10-ring channels are parallel to $c$. The repeat unit of the channel is equal to the cavity shown in Figure 3.

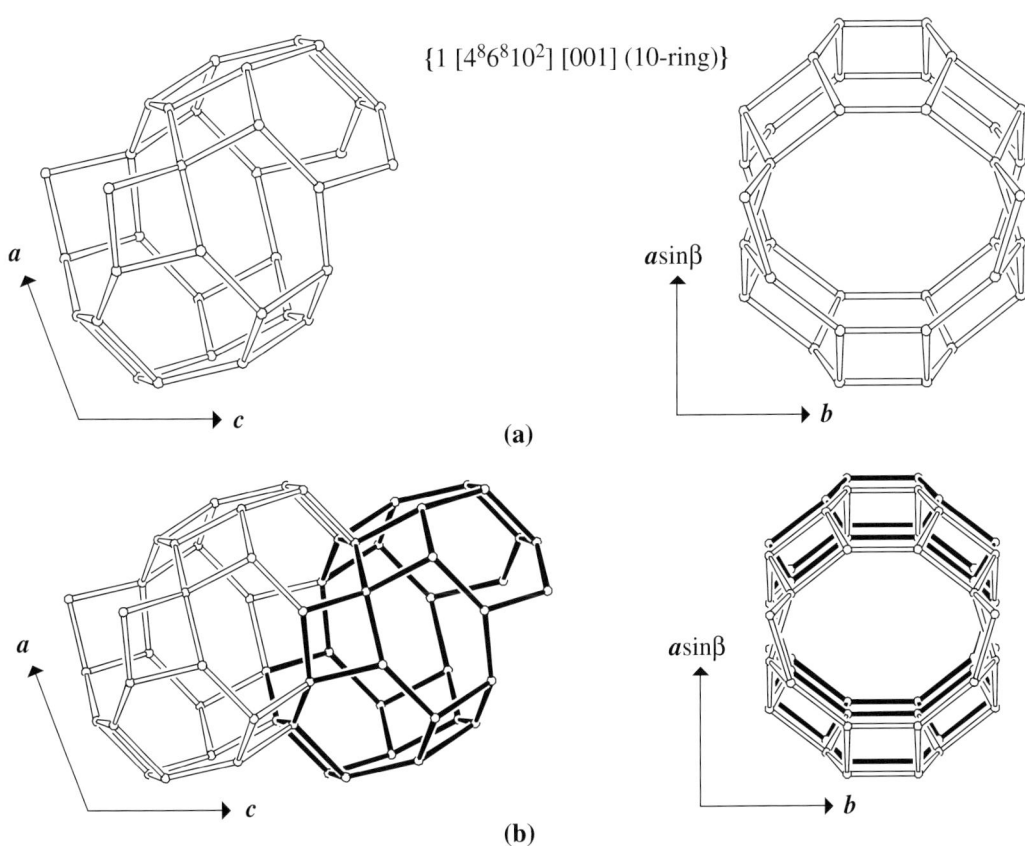

$\{1 \; [4^8 6^8 10^2] \; [001] \; (10\text{-ring})\}$

**(a)**

**(b)**

Fig. 3.  (a) Cavity viewed along $b$ (left) and along $c$ (right); (b) channel viewed perpendicular to the channel axis (left) and along the channel axis (right). The channel consists of cavities that are connected through common 10-rings.

# LEV                    Building Scheme

## 1. Periodic Building Unit

The two-dimensional PerBU of **LEV** consists of a hexagonal array of non-connected planar 6-rings (bold in Figure 1), which are related along **a** and **b** by pure translations. The 6-rings are centered at (0,0) in the **ab** layer. This position is usually called the **A** position. **LEV** belongs to the ABC-6 family (see Introduction).

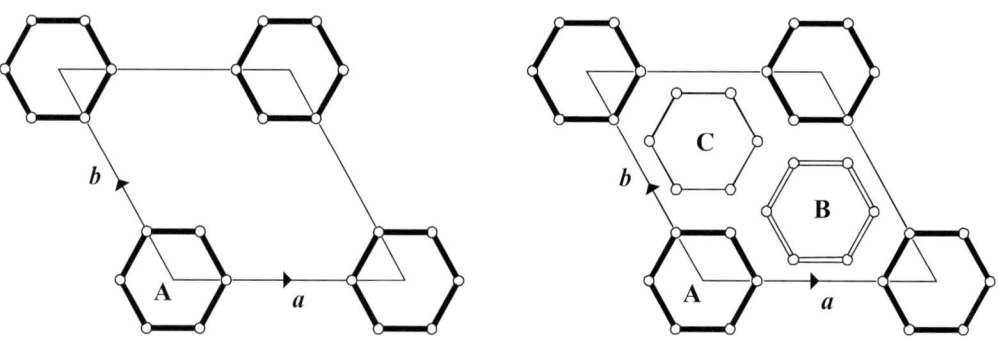

Fig. 1.    PerBU (left) and definition of 6-ring positions used in the stacking modes (right).

## 2. Connection mode

Neighboring PerBUs can be connected along **c** through tilted 4-rings in three different ways (see Introduction). In **LEV** all three connection modes between the PerBUs are observed (Figure 2).

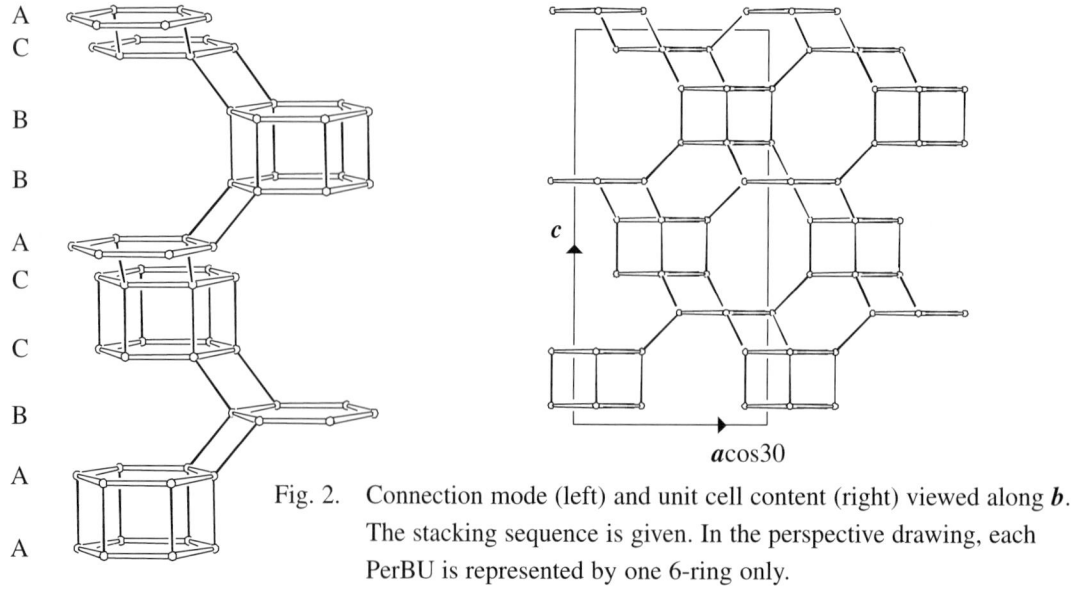

Fig. 2.    Connection mode (left) and unit cell content (right) viewed along **b**. The stacking sequence is given. In the perspective drawing, each PerBU is represented by one 6-ring only.

### 3. Channels and/or cages

The *lev* cavity is depicted in Figure 3. Non-interconnecting two-dimensional 8-ring channel systems, consisting of fused *lev* cavities, are perpendicular to *c*.

*lev* **cavity**
$\{2\ [4^9 6^5 8^3]\ <100>\ (8\text{-ring})\}$

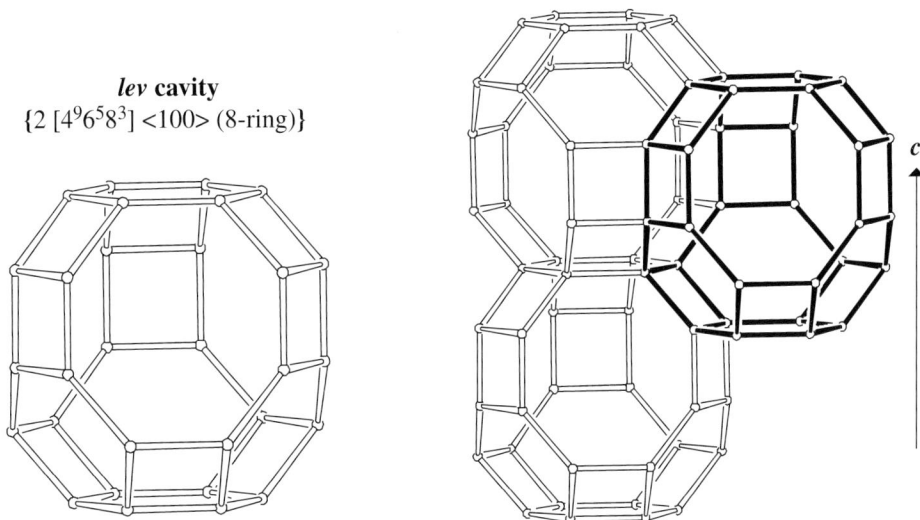

Fig. 3.  *lev* Cavity (left) and fused cavities (right) viewed along <120>. Non-interconnecting two-dimensional 8-ring channel systems are perpendicular to *c*.

# LIO                    Building Scheme

## 1. Periodic Building Unit

The two-dimensional PerBU of **LIO** consists of a hexagonal array of non-connected planar 6-rings (bold in Figure 1), which are related along *a* and *b* by pure translations. The 6-rings are centered at (0,0) in the *ab* layer. This position is usually called the **A** position. **LIO** belongs to the ABC-6 family (see Introduction).

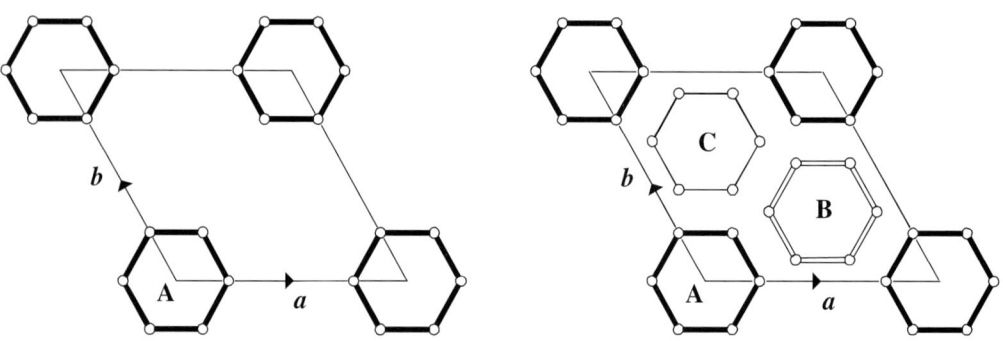

Fig. 1.   PerBU (left) and definition of 6-ring positions used in the stacking modes (right).

## 2. Connection mode

Neighboring PerBUs are connected along *c* through tilted 4-rings by connection modes (**1**) and (**2**) (see Introduction) as illustrated in Figure 2.

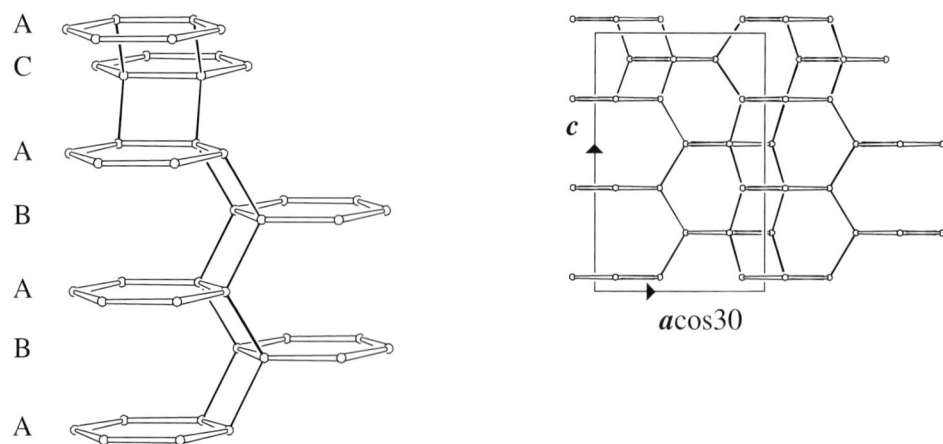

Fig. 2.   Connection mode (left) and the unit cell content (right) viewed along *b*. The stacking sequence is given. In the perspective drawing, each PerBU is represented by one 6-ring only.

### 3. Channels and/or cages

The three types of cages in **LIO** are depicted in Figure 3. The ***can*** cage is also present in **AFG**, **CAN**, **ERI**, **FAR**, **FRA**, **GIU**, **LOS**, **LTL**, **LTN**, **MAR**, **MOZ**, **OFF**, **SAT**, **SBS**, **SBT**, **TOL** and **-WEN**. The *lio* cage is also found in **AFG**, **FAR**, **MAR** and **TOL**. Finally, the *los* cage is also observed in **FRA**, **LOS** and **TOL**. Apertures of "channels" are formed by 6-rings only.

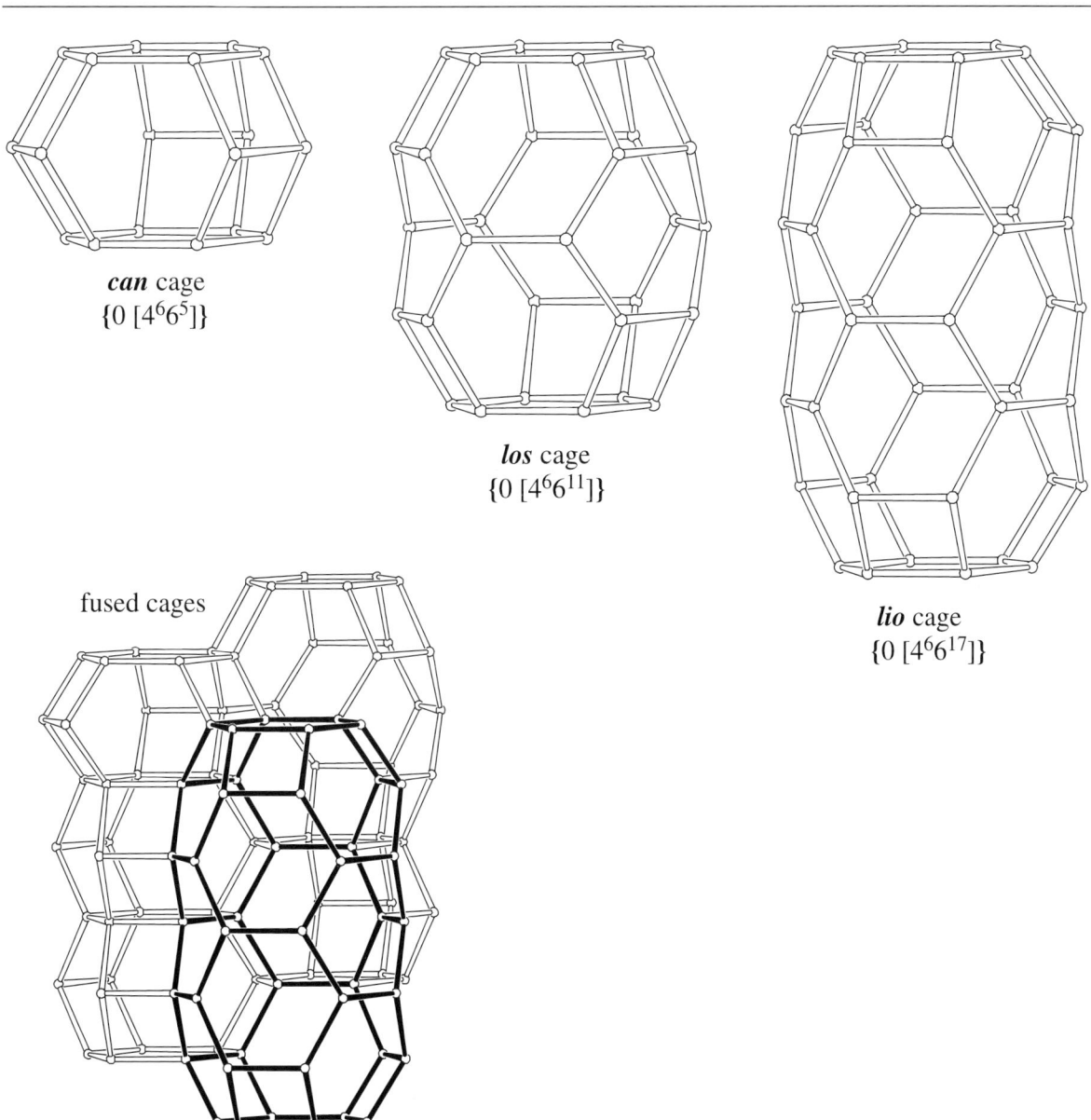

*can* cage
$\{0\,[4^{6}6^{5}]\}$

*los* cage
$\{0\,[4^{6}6^{11}]\}$

*lio* cage
$\{0\,[4^{6}6^{17}]\}$

fused cages

Fig. 3.   ***can*** Cage (top left), ***los*** cage (top middle), ***lio*** cage (top right) and connection of cages (bottom; *lio* cage in bold) viewed perpendicular to *c*.

# -LIT                    Building Scheme

## 1. Periodic Building Unit

**-LIT** can be built using units of 12 T atoms. The T12-unit consists of a 3-fold (1,2,3)-connected double 6-ring (bold in Figure 1, left). A one-dimensional PerBU is obtained when T12-units, related along **b** by pure translations, are connected in such a way that two crankshafts chains and (fused) 4-rings are formed (bold in Figure 1, right). Each crankshaft chain bears two independent terminal oxygen atoms.

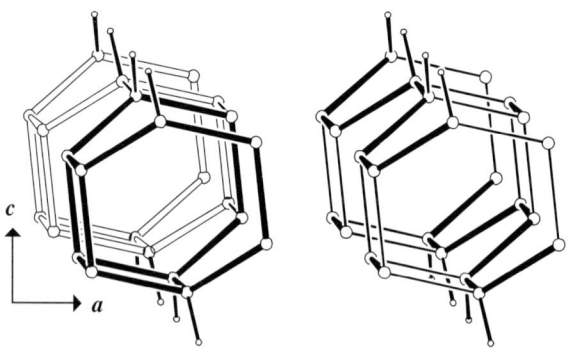

Fig. 1.   PerBU, constructed using 3-fold connected double 6-rings (left) or from two crankshaft chains and dimers (right), viewed along **b**. Terminal oxygen atoms are indicated by bold bonded small circles.

## 2. Connection mode

Neighboring PerBUs, related along **a** by a screw rotation of 180° about **a** followed by a shift of $\frac{1}{2}$**c**, are connected through crankshaft chains as shown in Figure 2.

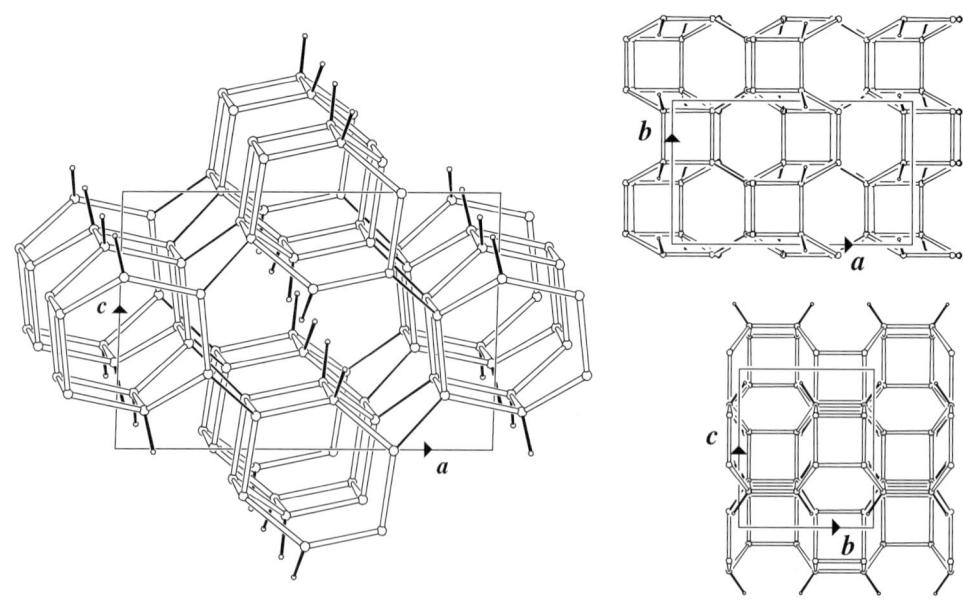

Fig. 2.   Connection mode (and unit cell content) viewed along **b** (left) and unit cell content projected along **c** (top right) and along **a** (bottom right).

# Cage/Channel **-LIT**

## 3. Channels and/or cages

Non-interconnecting 10-ring channels are parallel to **b** (Figure 3).

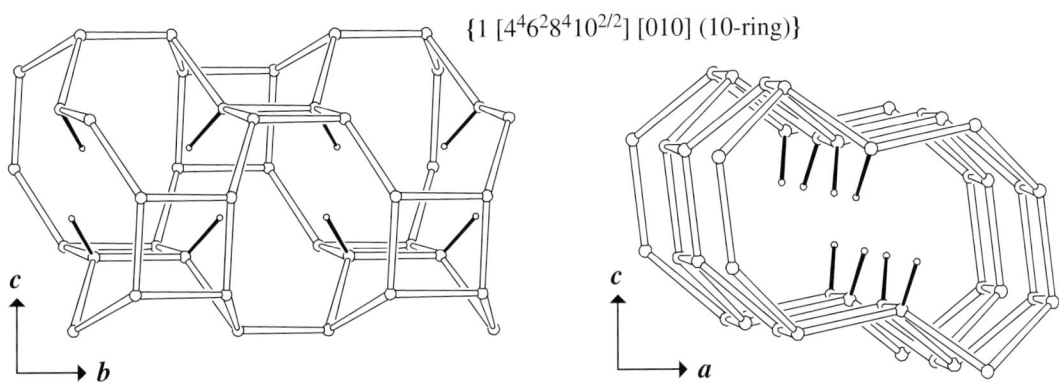

$\{1\ [4^46^28^410^{2/2}]\ [010]\ (10\text{-ring})\}$

Fig. 3.    10-Ring channel viewed perpendicular to the channel axis (left) and along the channel axis (right). The terminal oxygen atoms (bold bonded small circles) seriously hamper the diffusion through the channel.

# LOS                                    Building Scheme

## 1. Periodic Building Unit

The two-dimensional PerBU of **LOS** consists of a hexagonal array of non-connected planar 6-rings (bold in Figure 1), which are related along *a* and *b* by pure translations. The 6-rings are centered at (0,0) in the *ab* layer. This position is usually called the **A** position. **LOS** belongs to the ABC-6 family (see Introduction).

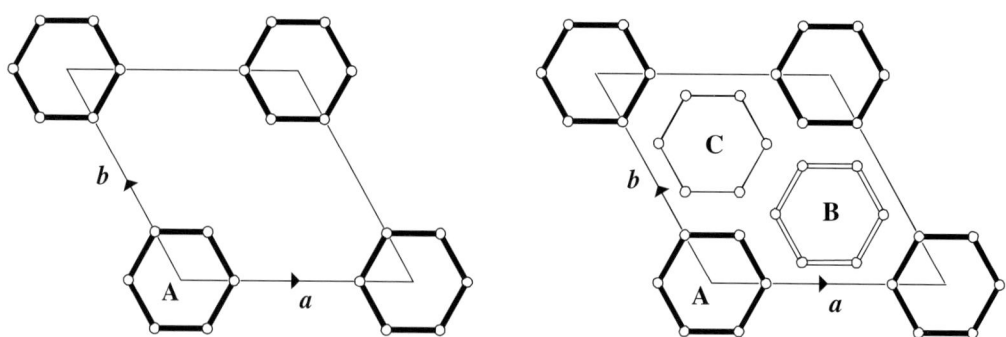

Fig. 1.    PerBU (left) and definition of 6-ring positions used in the stacking modes (right).

## 2. Connection mode

Neighboring PerBUs are connected along *c* through tilted 4-rings by connection modes (**1**) and (**2**) (see Introduction) as illustrated in Figure 2.

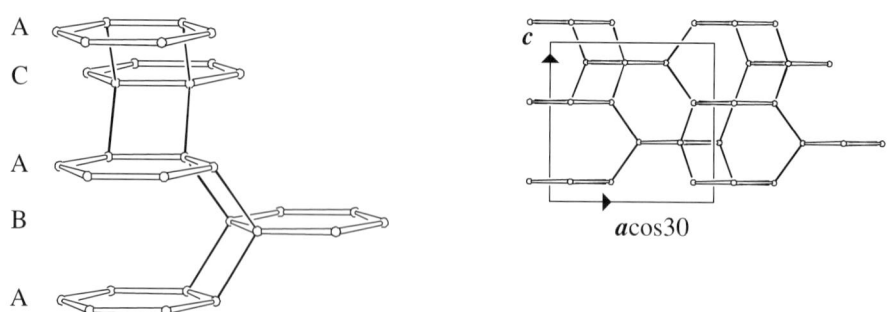

Fig. 2.    Connection mode (left) and unit cell content (right) viewed along *b*. The stacking sequence is given. In the perspective drawing, each PerBU is represented by one 6-ring only.

### 3. Channels and/or cages

The two types of cages in **LOS** are depicted in Figure 3. The *can* cage is also present in **AFG**, **CAN**, **ERI**, **FAR**, **FRA**, **GIU**, **LIO**, **LTL**, **LTN**, **MAR**, **MOZ**, **OFF**, **SAT**, **SBS**, **SBT**, **TOL** and **-WEN**. The *los* cage is also observed in **FRA**, **LIO** and **TOL**. Apertures of "channels" are formed by 6-rings only.

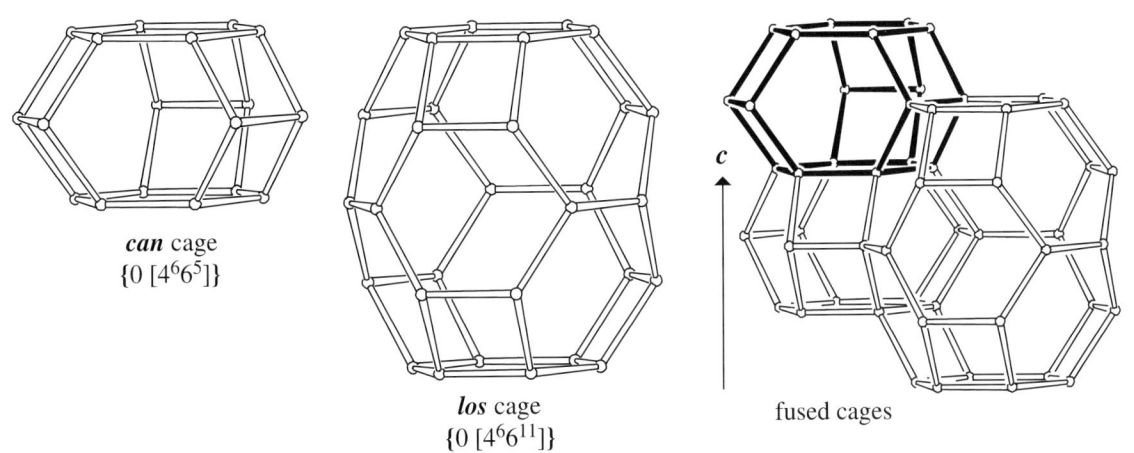

*can* cage
$\{0\ [4^6 6^5]\}$

*los* cage
$\{0\ [4^6 6^{11}]\}$

fused cages

Fig. 3.    *can* Cage (left), *los* cage (middle) and connection of cages (right) viewed perpendicular to *c*.

# LOV                    Building Scheme

## 1. Periodic Building Unit

Tetragonal **LOV** can be built using units of 9 T atoms: two 4-rings connected through a single T atom (bold in Figure 1). The two-dimensional PerBU, composed of T9-units related along *a* and *b* by pure translations, is equal to the layer depicted in Figure 1. (See also **RSN** and **VSV**.)

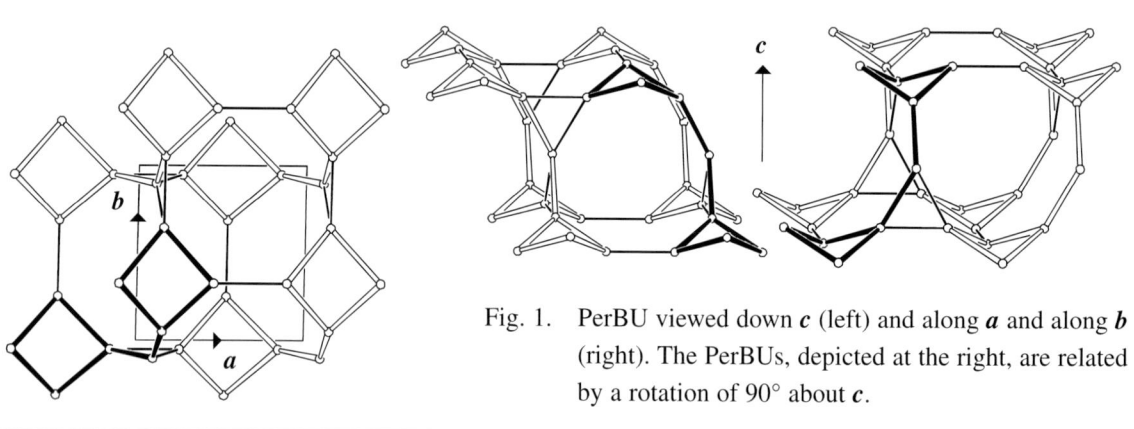

Fig. 1.   PerBU viewed down *c* (left) and along *a* and along *b* (right). The PerBUs, depicted at the right, are related by a rotation of 90° about *c*.

## 2. Connection mode

Neighboring PerBUs, related along *c* by a rotation of 90° about *c* and a shift of $\frac{1}{2}c$ followed by a lateral shift of zero along both *a* and *b*, are connected along *c* through 4- and 6-rings (Figure 2).

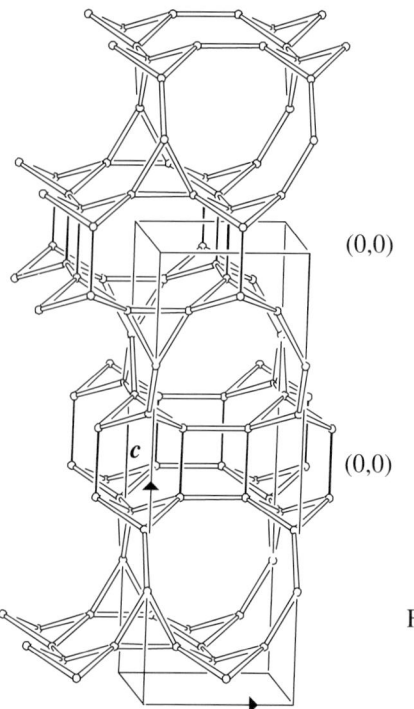

(0,0)

(0,0)

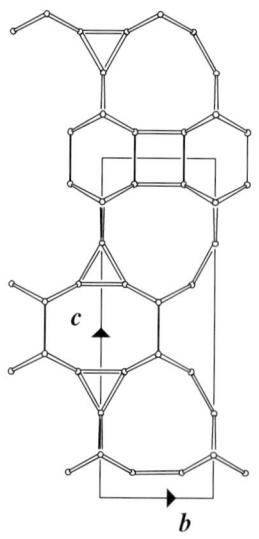

Fig. 2.   Connection mode (and unit cell content) viewed along *b* and projection of the unit cell content along *a* (right). The lateral shifts between neighboring PerBUs are given in fractions of *a* and *b*. Successive PerBUs are related by a 4₂ axis.

### 3. Channels and/or cages

Interconnecting 8-ring channels and intersecting 9-ring channels are parallel to <100>. The 8-ring channels are topologically equivalent to (one of) those in **DFT**, **LTL**, **MOZ** and **RSN**, and the 9-ring channels to those in **NAB**, **RSN** and **VSV**. The 9-ring channels are interconnecting along $c$ through 8-rings perpendicular to $c$ (Figure 3).

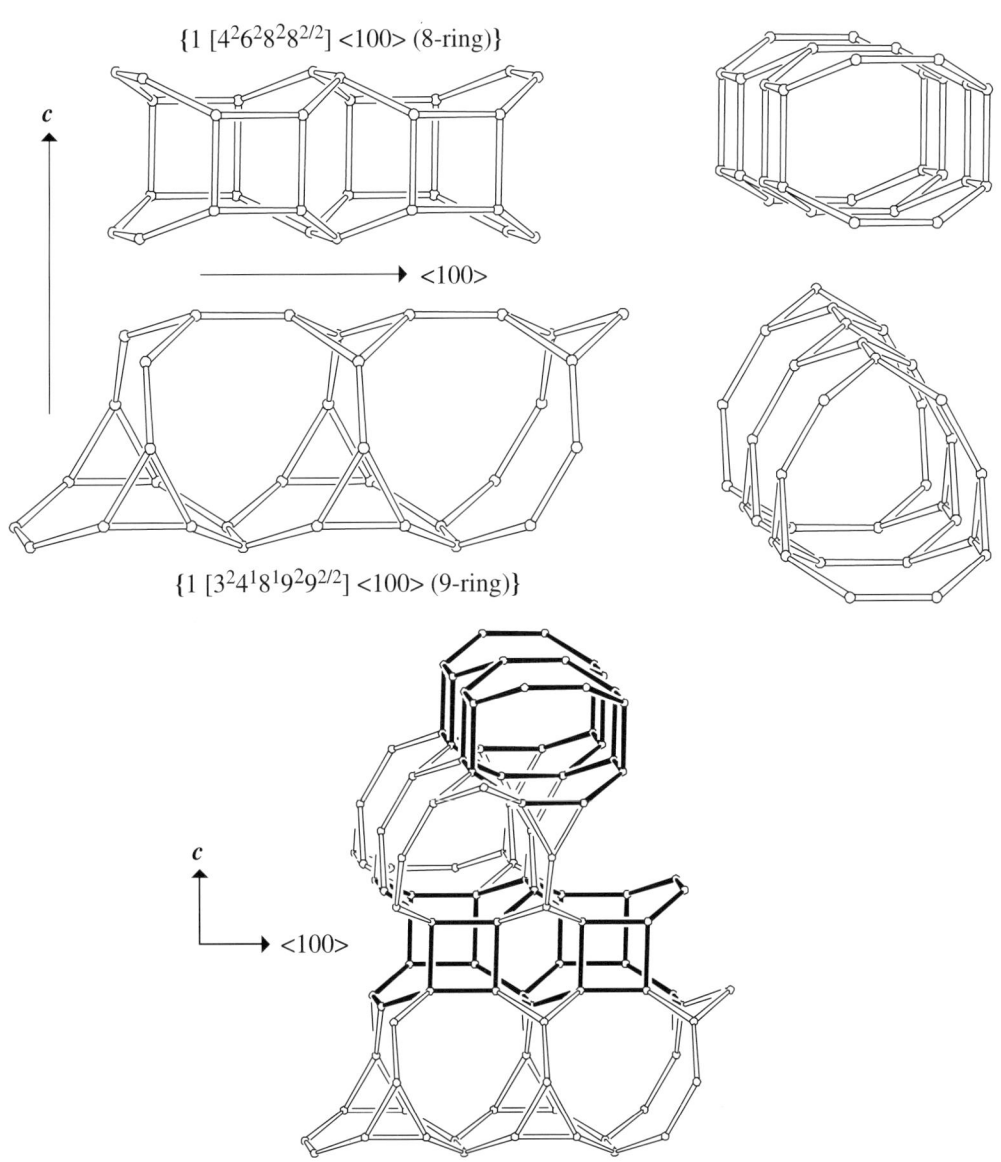

Fig. 3.    8-Ring channel (top) and 9-ring channel (middle) viewed perpendicular to the channel axis (left) and along the channel axis (right). Bottom: interconnection of the channels along $c$ (8-ring channels in bold). The intersecting 9-ring channels have [$9^4$]-cavities in common. (See **NAB**.)

# LTA

# Building Scheme

### 1. Periodic Building Unit

Cubic **LTA** can be built using the sodalite cage (or *sod* cage) consisting of 24 T atoms (six 4-rings or four 6-rings; Figure 1) as zero-dimensional PerBU. (Compare with **EMT** and **FAU**.)

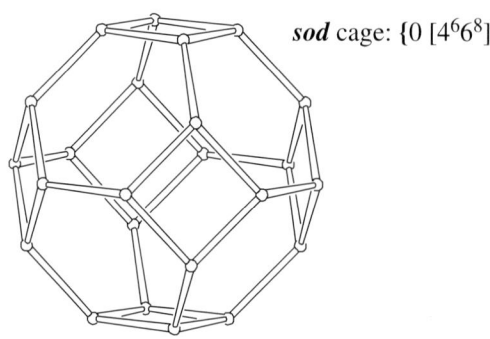

*sod* cage: $\{0\ [4^6 6^8]\}$

Fig. 1.   PerBU (or *sod* cage) viewed down one of the cube axes.

### 2. Connection mode

The three-dimensional **LTA** framework is obtained when PerBUs, related along the cube axes by pure translations, are linked through double 4-rings as shown in Figure 2 for one cubic face.

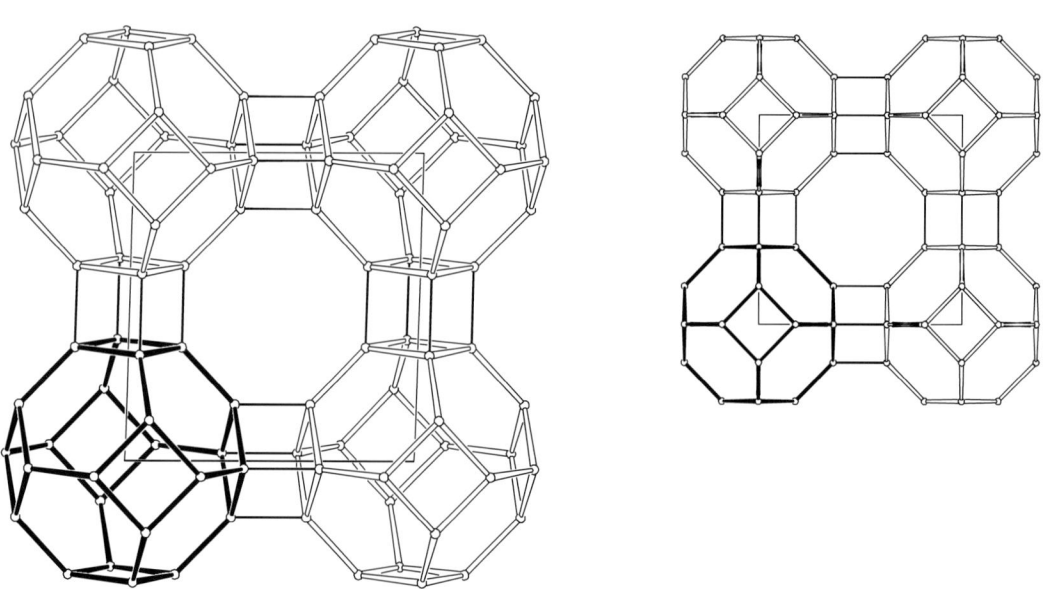

Fig. 2.   Connection mode in one cubic face of **LTA** (left) and projection of the unit cell content (right) viewed along a cube axis.

### 3. Channels and/or cages

Intersecting 8-ring channels are parallel to <100>. The channel intersection is equal to the ***rho*** cavity depicted in Figure 3. The ***rho*** cavity is also observed in **-CLO**, **KFI**, **LTN**, **PAU**, **RHO**, **TSC** and **UFI**. The ***sod*** cage is also found in **EMT**, **FAU**, **FRA**, **GIU**, **LTN**, **MAR**, **SOD** and **TSC**. The ***rho*** cavities are linked through common 8-rings.

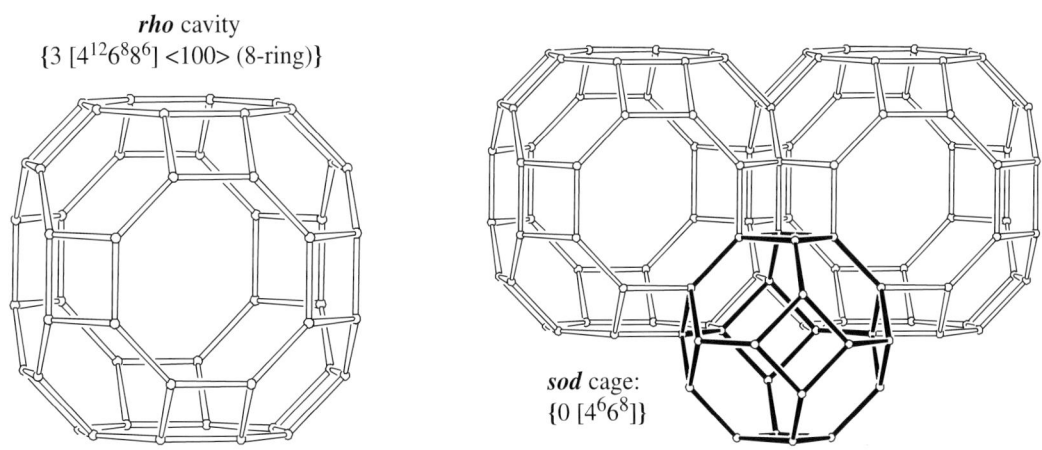

***rho*** cavity
$\{3\ [4^{12}6^{8}8^{6}]\ <100>\ (8\text{-ring})\}$

***sod*** cage:
$\{0\ [4^{6}6^{8}]\}$

Fig. 3.   ***rho*** Cavity (left) and linkage of cavities (right) viewed along a cube axis.

# LTL                    Building Scheme

## 1. Periodic Building Unit

Hexagonal **LTL** can be built using the saw chain (bold in Figure 1) running parallel to *c*. Six saw chains are connected into an one-dimensional PerBU consisting of a column of *can* cages. The *can* cages are connected through double 6-rings (Figure 1). (See also **ERI**, **MOZ** and **OFF**.)

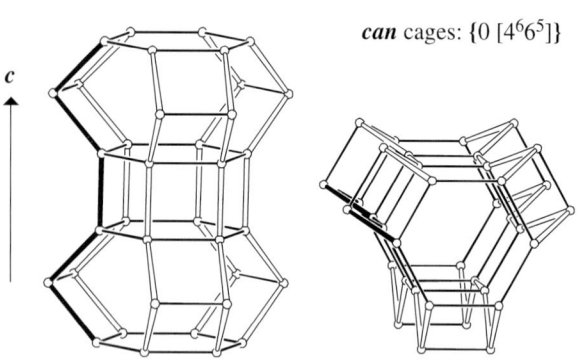

*can* cages: {0 [$4^6 6^5$]}

Fig. 1. PerBU, composed of a column of *can* cages, viewed perpendicular to *c* (left) and along *c* (right).

## 2. Connection mode

Neighboring PerBUs, related by a rotation of 60° about the hexagonal *c* axis, are connected into the three-dimensional framework through 8-rings (Figure 2).

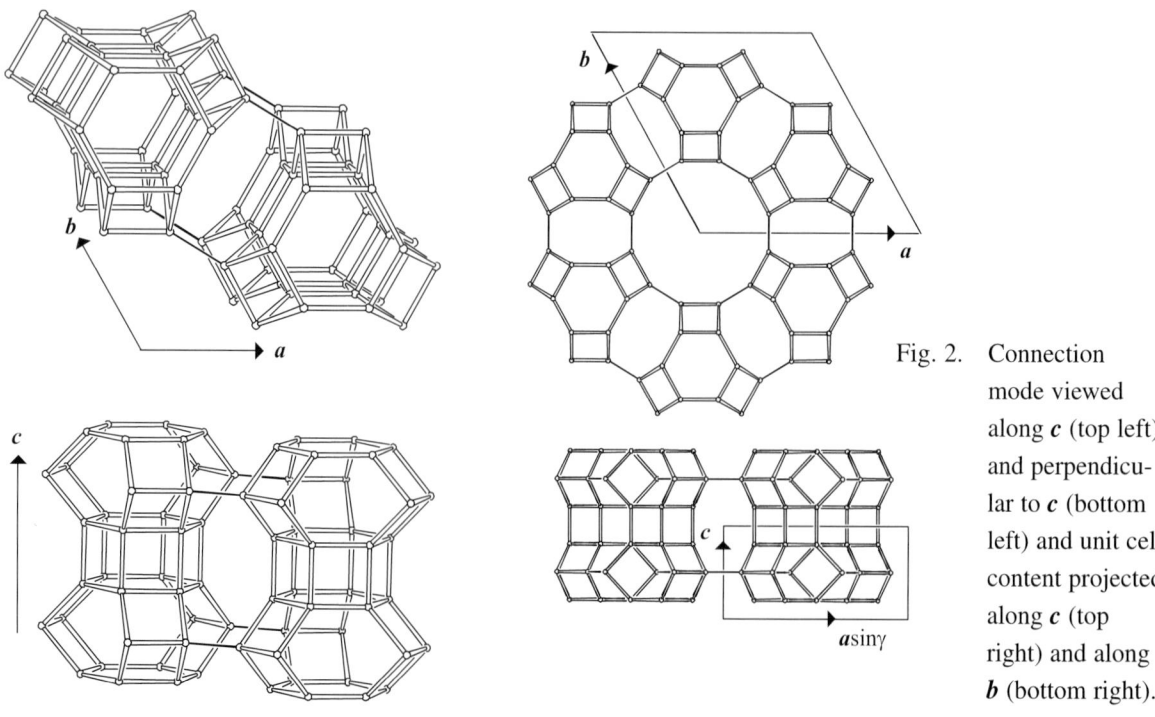

Fig. 2. Connection mode viewed along *c* (top left) and perpendicular to *c* (bottom left) and unit cell content projected along *c* (top right) and along *b* (bottom right).

## 3. Channels and/or cages

Interconnecting 8-ring and 12-ring channels are parallel to $c$ (Figure 3). The 8-ring channel is topologically equivalent to 8-ring channels in **DFT**, **LOV**, **MOZ** and **RSN**. The 12-ring channel, equivalent to the 12-ring channel in **MOZ**, is composed of *ltl* cavities. The 8-ring and 12-ring channels are interconnected through common 8-rings perpendicular to <100>. The **can** cage is also present in **AFG**, **CAN**, **ERI**, **FAR**, **FRA**, **GIU**, **LIO**, **LOS**, **LTN**, **MAR**, **MOZ**, **OFF**, **SAT**, **SBS**, **SBT**, **TOL** and **-WEN**.

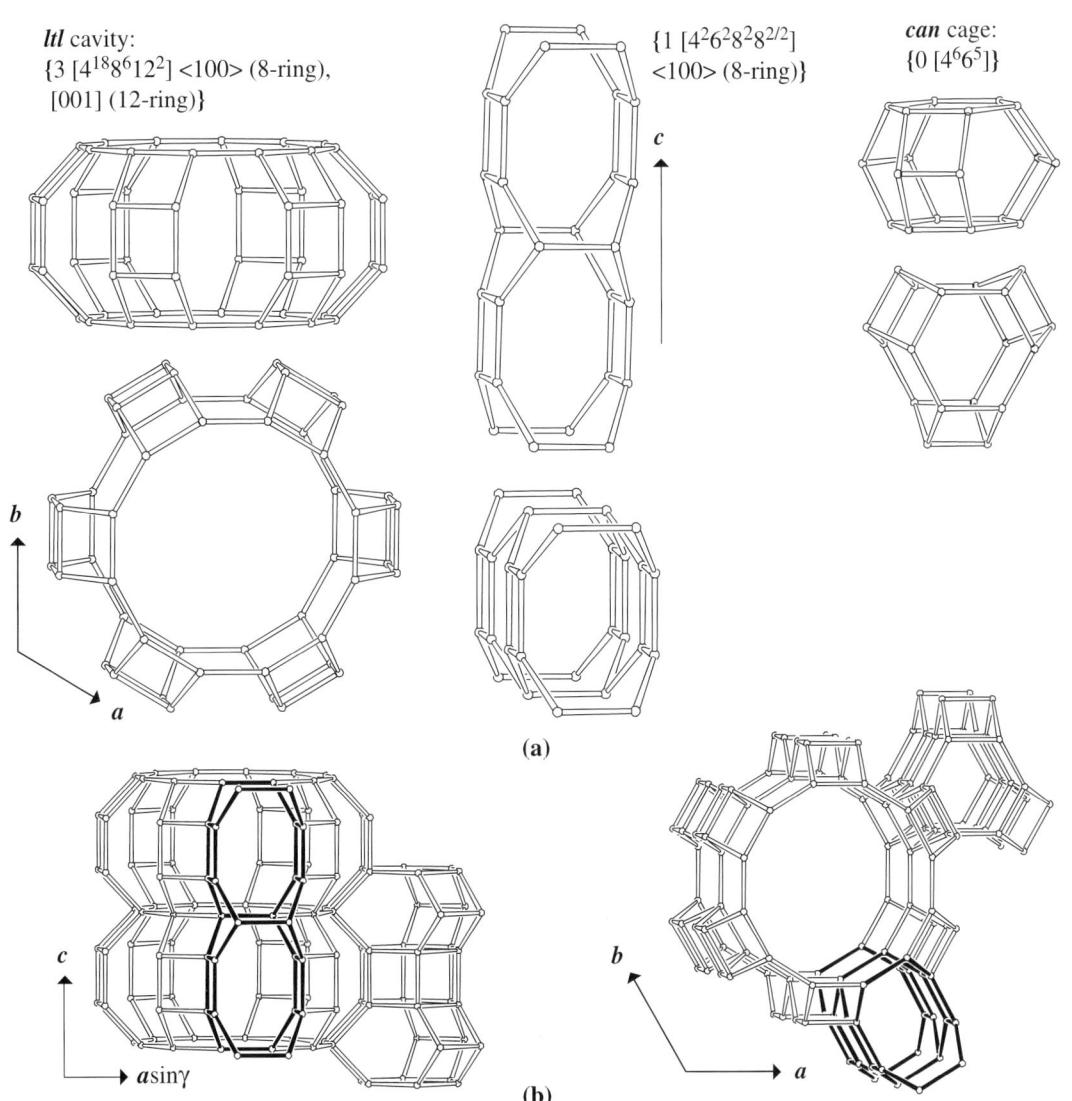

Fig. 3. (a) From left to right: *ltl* cavity, 8-ring channel and **can** cage viewed perpendicular to $c$ (top) and along $c$ (bottom). (b) Linkage of *ltl* cavities, 8-ring channel (in bold) and **can** cages viewed along <010> (left) and along $c$ (right).

# LTN                    Building Scheme

## 1. Periodic Building Unit

In cubic **LTN** two types of building units can be distinguished. The first building unit (BU1) consists of 120 T atoms. BU1 (Figure 1a) is obtained when tetrahedral-related 6-rings in the **rho** cavity are replaced by **can** cages (Figure 1c) and double 6-rings, respectively. The second building unit (BU2) consists of 72 T atoms. BU2 (Figure 1b) is obtained when four **can** cages replace four (tetrahedral related) 6-rings in the **sod** cage. Both building units exhibit 4-fold inversion symmetry. Neighboring BU1s, related by a shift of $\frac{1}{2}(a+b)$, form cube faces (see Figure 1d on next page). The two-dimensional PerBU is obtained when the empty spaces in the BU1 cubic face are filled with BU2 units. BU1 and BU2, related by a shift of $\frac{1}{2}a$ (or $\frac{1}{2}b$), are connected through 6- and 4-rings as shown in Figure 1(e1) on next page.

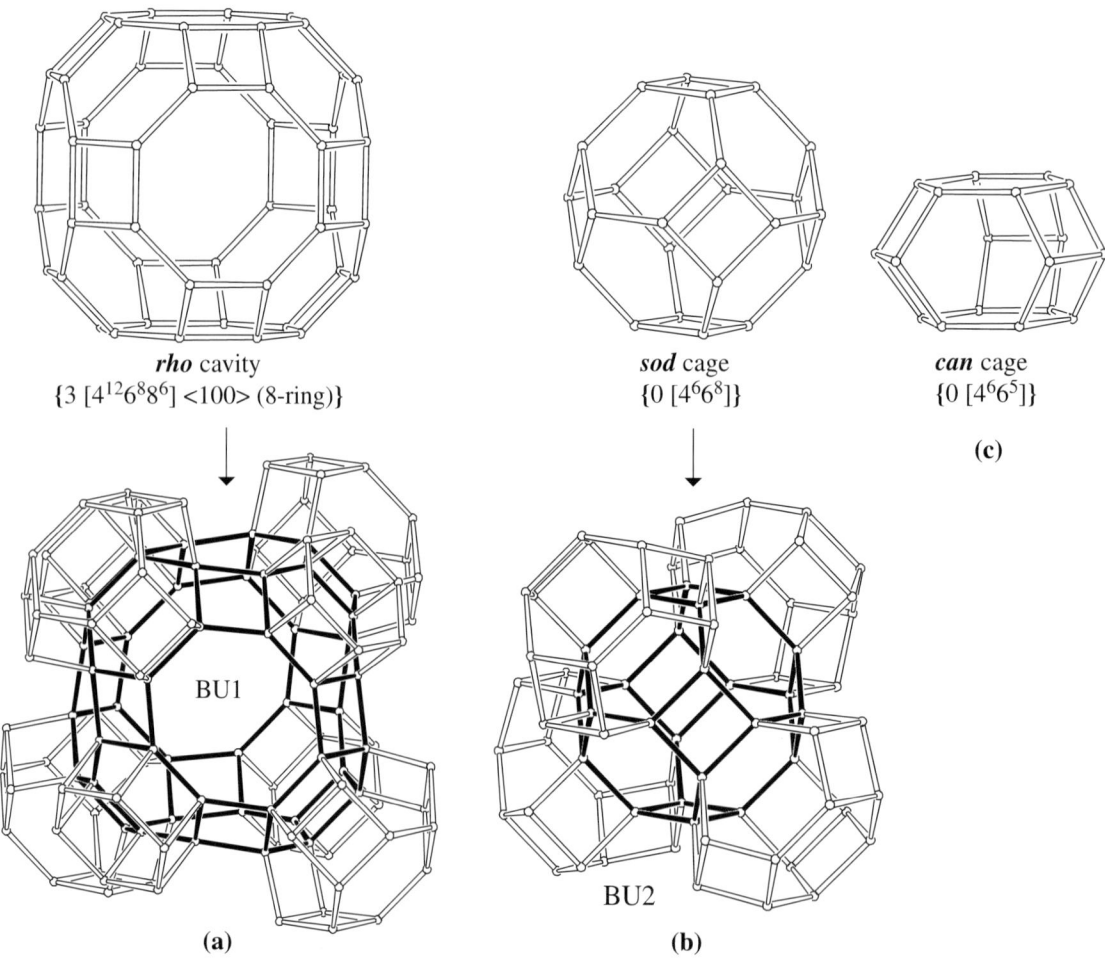

**rho** cavity
{3 [$4^{12}6^86^6$] <100> (8-ring)}

**sod** cage
{0 [$4^66^8$]}

**can** cage
{0 [$4^66^5$]}

(c)

BU1

BU2

(a)

(b)

Fig. 1(a–c).   (a) **rho** Cavity (top) and BU1 constructed from four **can** cages and four D6Rs (bottom; **rho** cavity in bold); (b) **sod** cage (top) and BU2 built from four **can** cages (bottom; **sod** cage in bold); and (c) **can** cage.

# Building Scheme

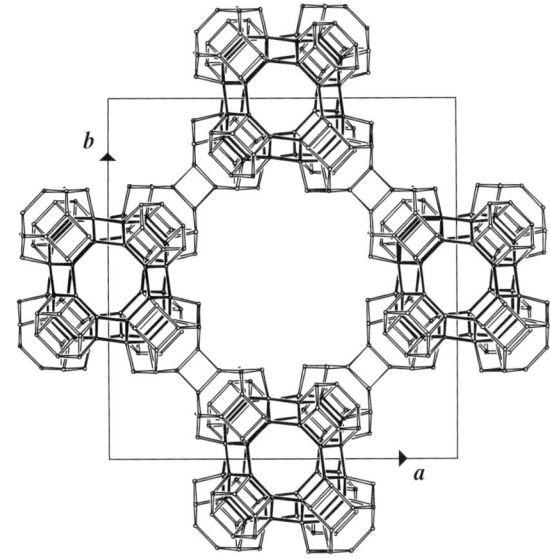

Fig. 1d.   Cube face built from BU1s. ***can*** Cages from different BU1s are connected through 4-rings into a cube face with empty spaces.

Fig. 1(e1).   PerBU viewed along ***c***. BU2s (in bold) fill the empty spaces in the ***ab*** plane.

**(d)**

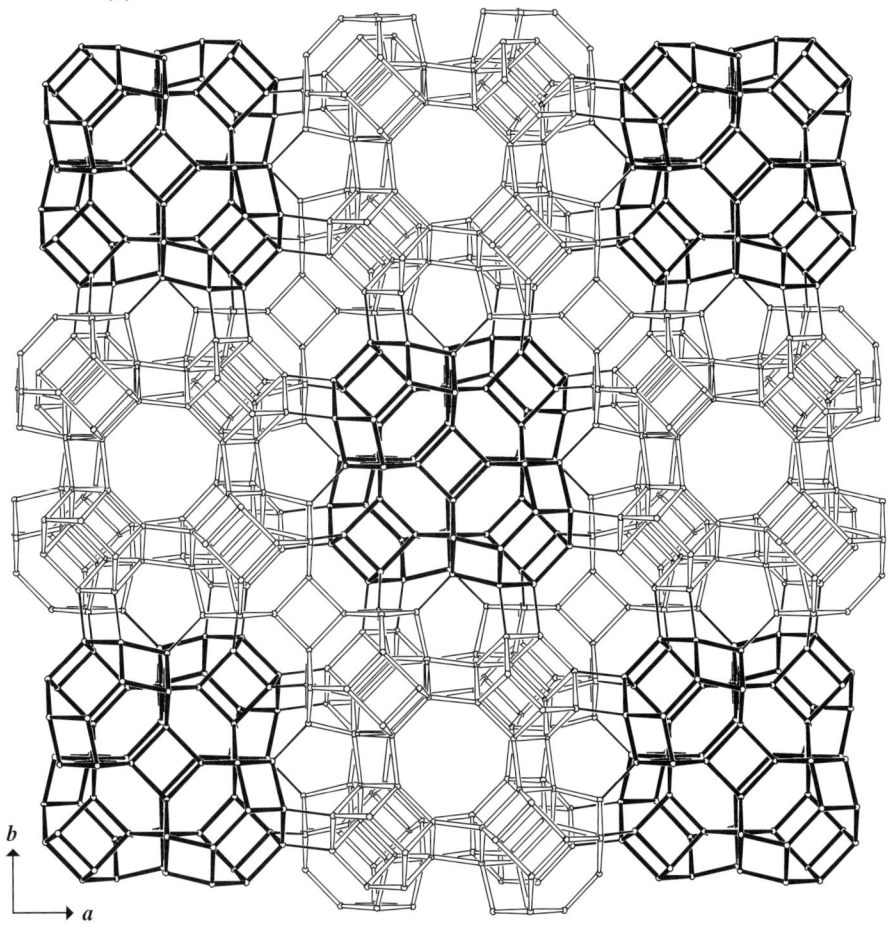

**(e1)**

# LTN

# Building Scheme

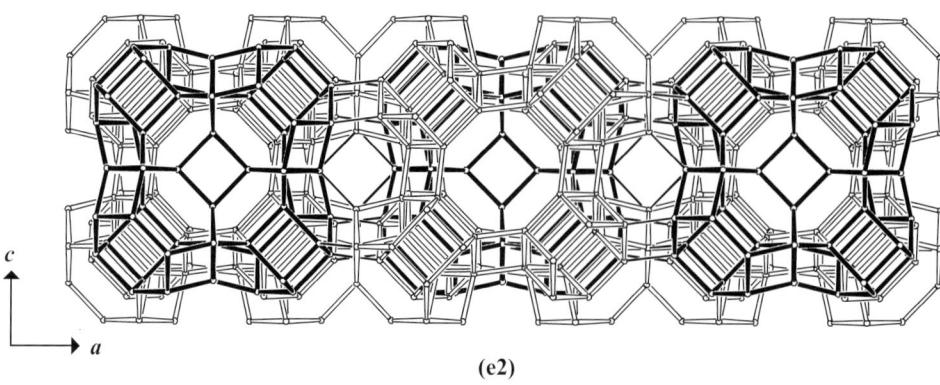

(e2)

Fig. 1(e2).   PerBU projected along **b**.

## 2. Connection mode

Neighboring **ab** planes, related by a shift of $\frac{1}{2}\boldsymbol{b}$ (or $\frac{1}{2}\boldsymbol{a}$), are connected along **c** to form the three-dimensional structure shown in Figure 2. The connection modes in the **ac** and **bc** planes are equivalent to those shown in Figure 1 for the **ab** plane.

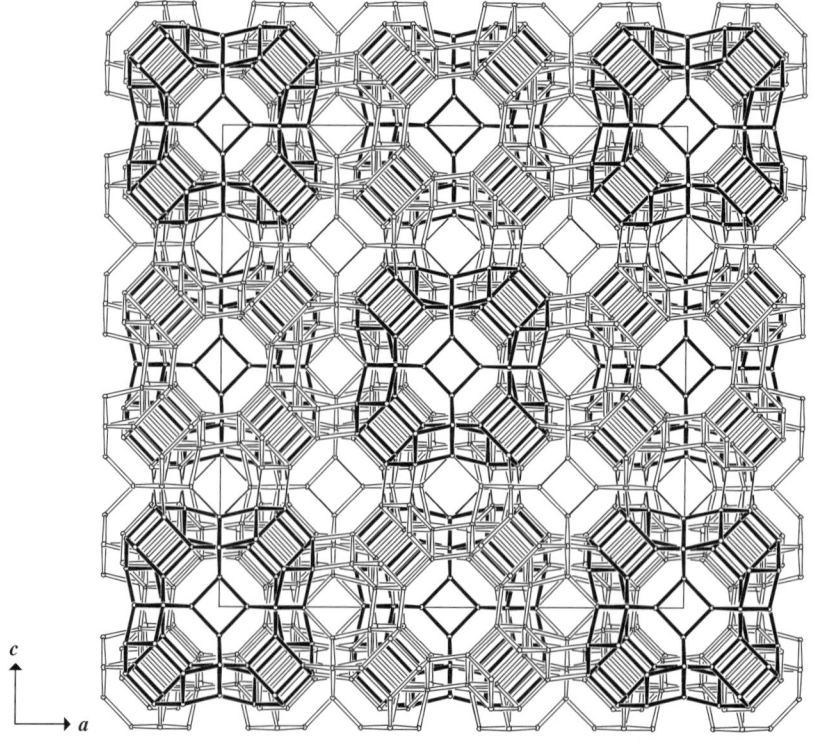

Fig. 2.   Connection mode and unit cell content projected along **b**. BU2s are in bold.

## 3. Channels and/or cages

There are three types of cages. The *rho* cavity is also observed in **-CLO, KFI, LTA, PAU, RHO, TSC** and **UFI**. The *sod* cage is also found in **EMT, FAU, FRA, GIU, LTA, MAR, SOD** and **TSC**. The *can* cage is also present in **AFG, CAN, ERI, FAR, FRA, GIU, LIO, LOS, LTL, MAR, MOZ, OFF, SAT, SBS, SBT, TOL** and **-WEN**. Apertures of "channels" are formed by 6-rings only. Cavities and cages are shown in Figure 1a.

# MAR

# Building Scheme

## 1. Periodic Building Unit

The two-dimensional PerBU of **MAR** consists of a hexagonal array of non-connected planar 6-rings (bold in Figure 1), which are related along *a* and *b* by pure translations. The 6-rings are centered at (0,0) in the *ab* layer. This position is usually called the **A** position. **MAR** belongs to the ABC-6 family. (See Introduction.)

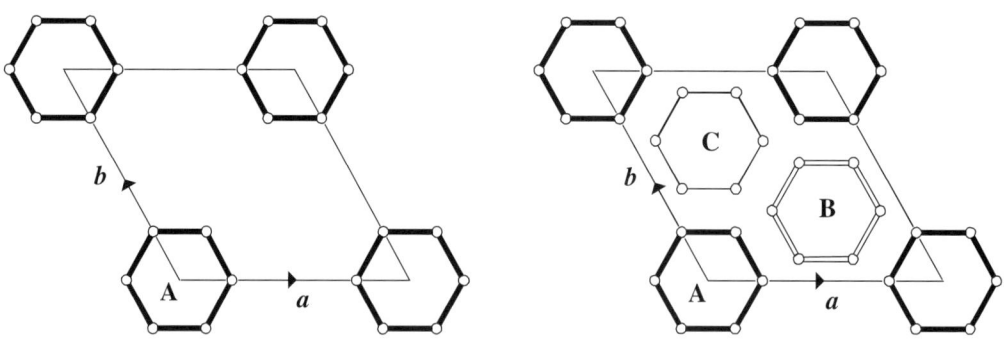

Fig. 1.    PerBU (left) and definition of 6-ring positions used in the stacking modes (right).

## 2. Connection mode

Neighboring PerBUs are connected along *c* through tilted 4-rings by connection modes (**1**) and (**2**) (see Introduction) as illustrated in Figure 2.

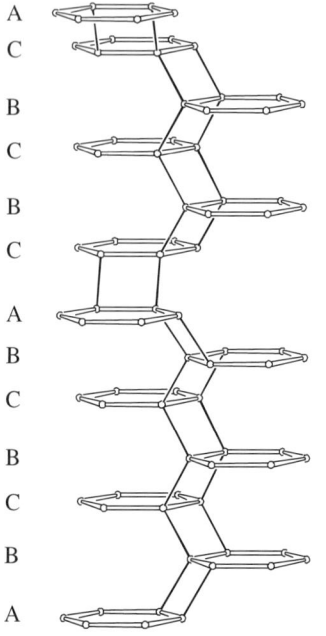

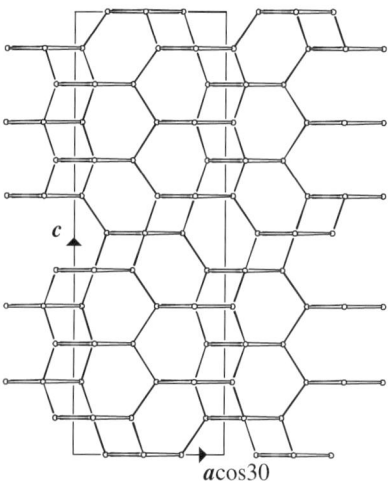

Fig. 2.    Connection mode (left) and unit cell content (right) viewed along *b*. The stacking sequence is given. In the perspective drawing, each PerBU is represented by one 6-ring only.

## 3. Channels and/or cages

The three types of cages in **MAR** are depicted in Figure 3. The *can* cage is also present in **AFG**, **CAN**, **ERI**, **FAR**, **FRA**, **GIU**, **LIO**, **LOS**, **LTL**, **LTN**, **MOZ**, **OFF**, **SAT**, **SBS**, **SBT**, **TOL** and **-WEN**. The *sod* cage is also found in **EMT**, **FAU**, **FAR**, **FRA**, **GIU**, **LTA**, **LTN**, **SOD** and **TSC**. Finally, the *lio* cage is also found in **AFG**, **FAR**, **LIO** and **TOL**. Apertures of "channels" are formed by 6-rings only.

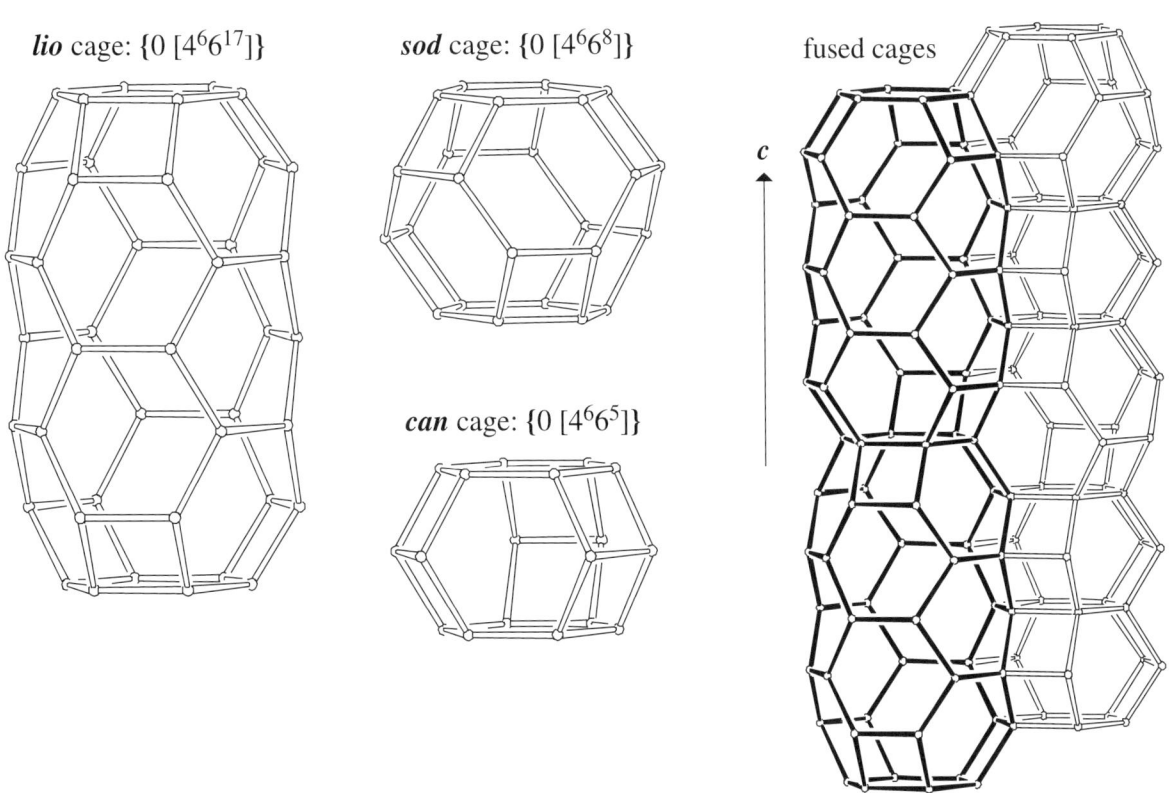

*lio* cage: $\{0\,[4^6 6^{17}]\}$    *sod* cage: $\{0\,[4^6 6^8]\}$    fused cages

*can* cage: $\{0\,[4^6 6^5]\}$

Fig. 3.   *lio* Cage (left), *sod* cage and *can* cage (middle) and connection of cages (right; *lio* cages in bold) viewed perpendicular to *c*.

# MAZ

## Building Scheme

### 1. Periodic Building Unit

Hexagonal **MAZ** can be built using the saw chain (bold in Figure 1) running parallel to $c$. Six saw chains are connected into a one-dimensional PerBU consisting of a column of *gmel* cavities sharing 6-rings (Figure 1).

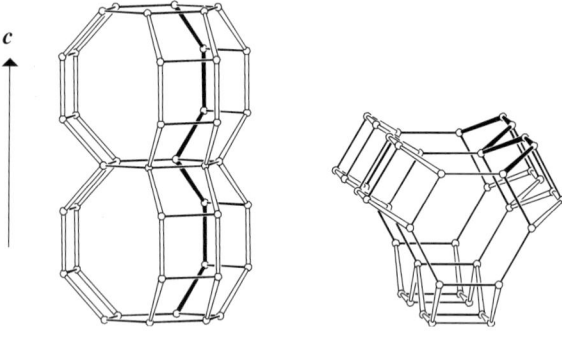

Fig. 1.  PerBU composed of (fused) *gmel* cavities viewed perpendicular to $c$ (left) and along $c$ (right).

### 2. Connection mode

Neighboring PerBUs, related by a rotation of 60° about $c$ and a shift of $\frac{1}{2}c$ (i.e. related by a $6_3$-axis), are connected into the *ab* plane through 5- and 8-rings (Figure 2).

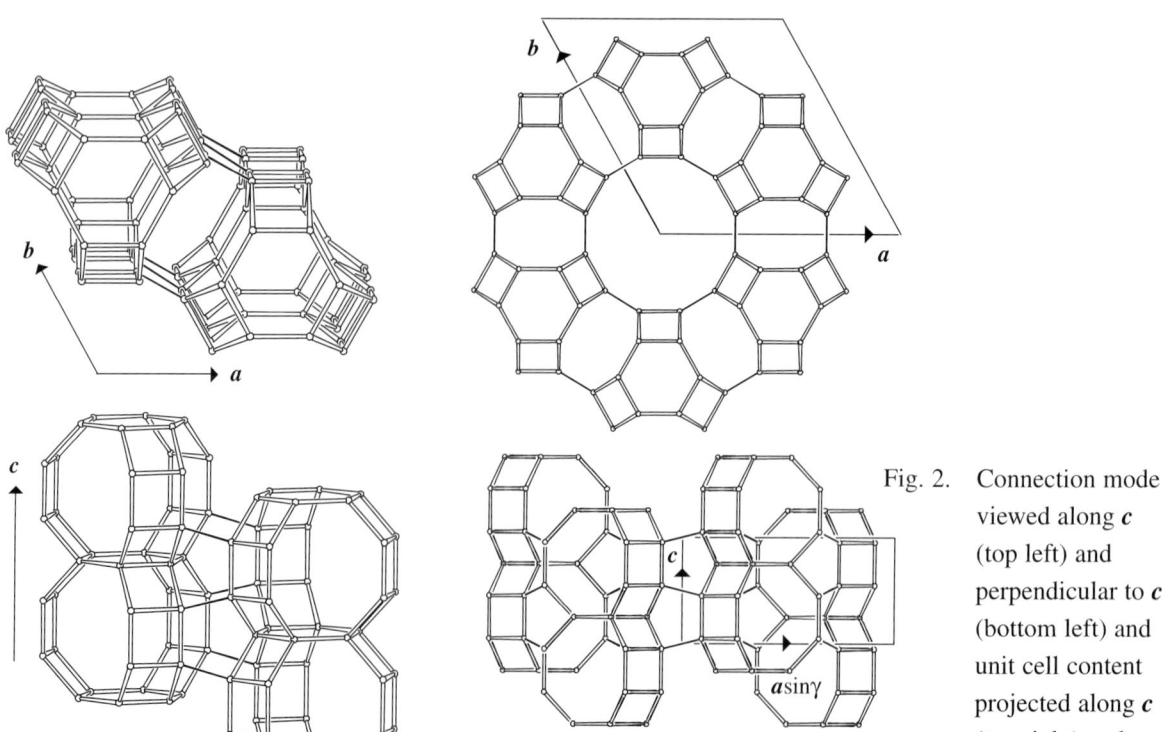

Fig. 2.  Connection mode viewed along $c$ (top left) and perpendicular to $c$ (bottom left) and unit cell content projected along $c$ (top right) and along $b$.

# Cage/Channel      **MAZ**

## 3. Channels and/or cages

Non-interconnecting 8-ring and 12-ring channels are parallel to $c$ (Figure 3). The 8-ring channel is topologically equivalent to (one of) the 8-ring channels in **EON**, **MON**, **MOR**, **RSN** and **VSV**. Columns of **gmel** cavities are interconnected through 8-ring channels parallel to $c$ leading to a rather complicated three-dimensional channel system with 8-ring windows. The **gmel** cavity is also present in **AFT**, **AFX**, **EAB**, **EON**, **GME** and **OFF**.

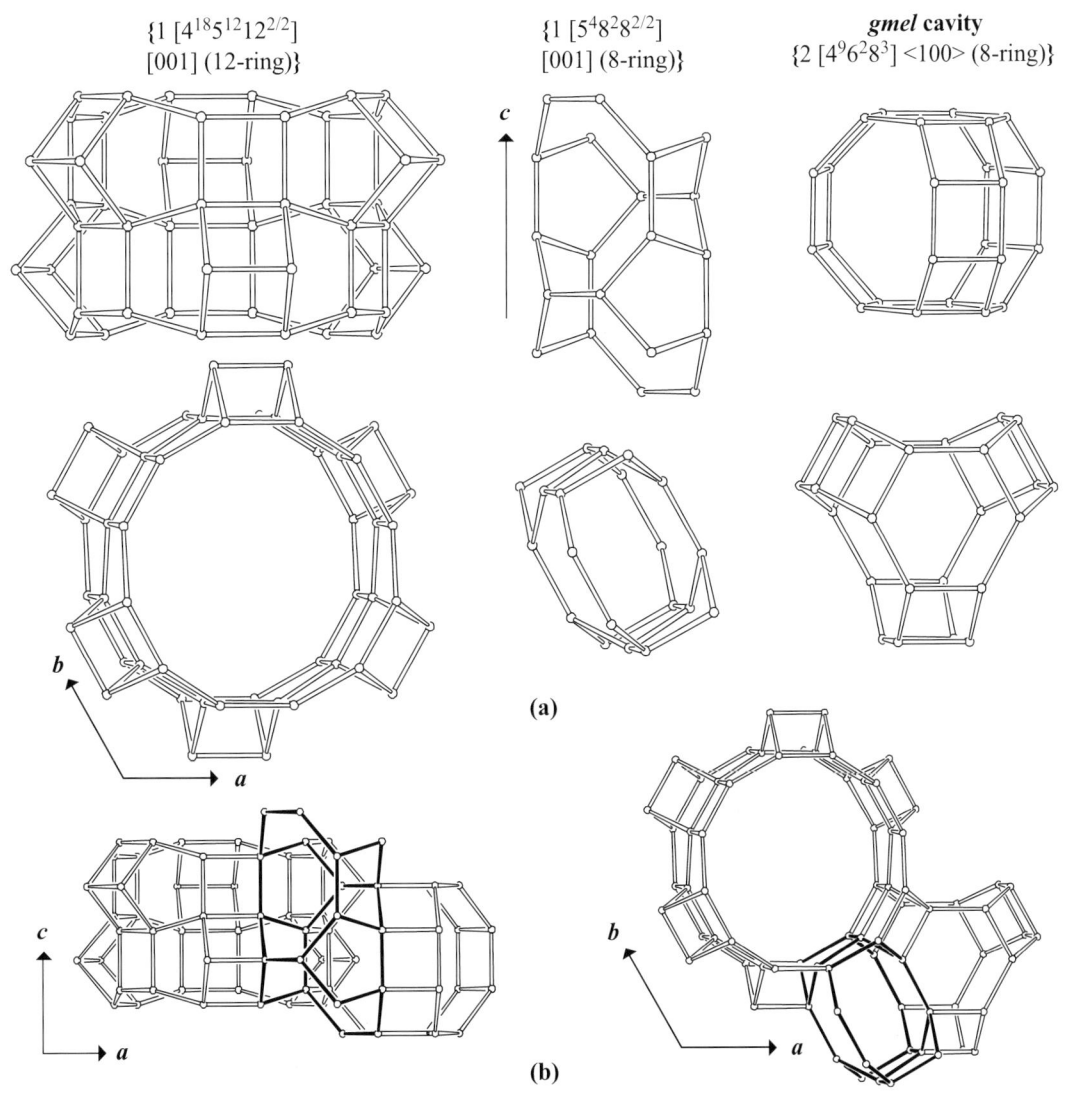

$\{1 [4^{18}5^{12}12^{2/2}]$
$[001] (12\text{-ring})\}$

$\{1 [5^{4}8^{2}8^{2/2}]$
$[001] (8\text{-ring})\}$

**gmel cavity**
$\{2 [4^{9}6^{2}8^{3}] <100> (8\text{-ring})\}$

(a)

(b)

Fig. 3.  (a) From left to right: 12-ring channel, 8-ring channel and **gmel** cavity, viewed along <120> (top) and along $c$ (bottom). (b) Linkage of 12-ring channel, 8-ring channel (in bold) and **gmel** cavity viewed along <120> (left) and along $c$ (right).

# MEI

## Building Scheme

### 1. Periodic Building Unit

Hexagonal **MEI** can be built using units of 17 T atoms (bold in Figure 1). The T 17-unit consists of two 6*1 units connected through a 3-ring (or of two $[3^1 4^3 5^3]$-cages sharing a 3-ring). The one-dimensional PerBU is obtained when T 17-units, related along $c$ by pure translations, are connected into a chain along $c$.

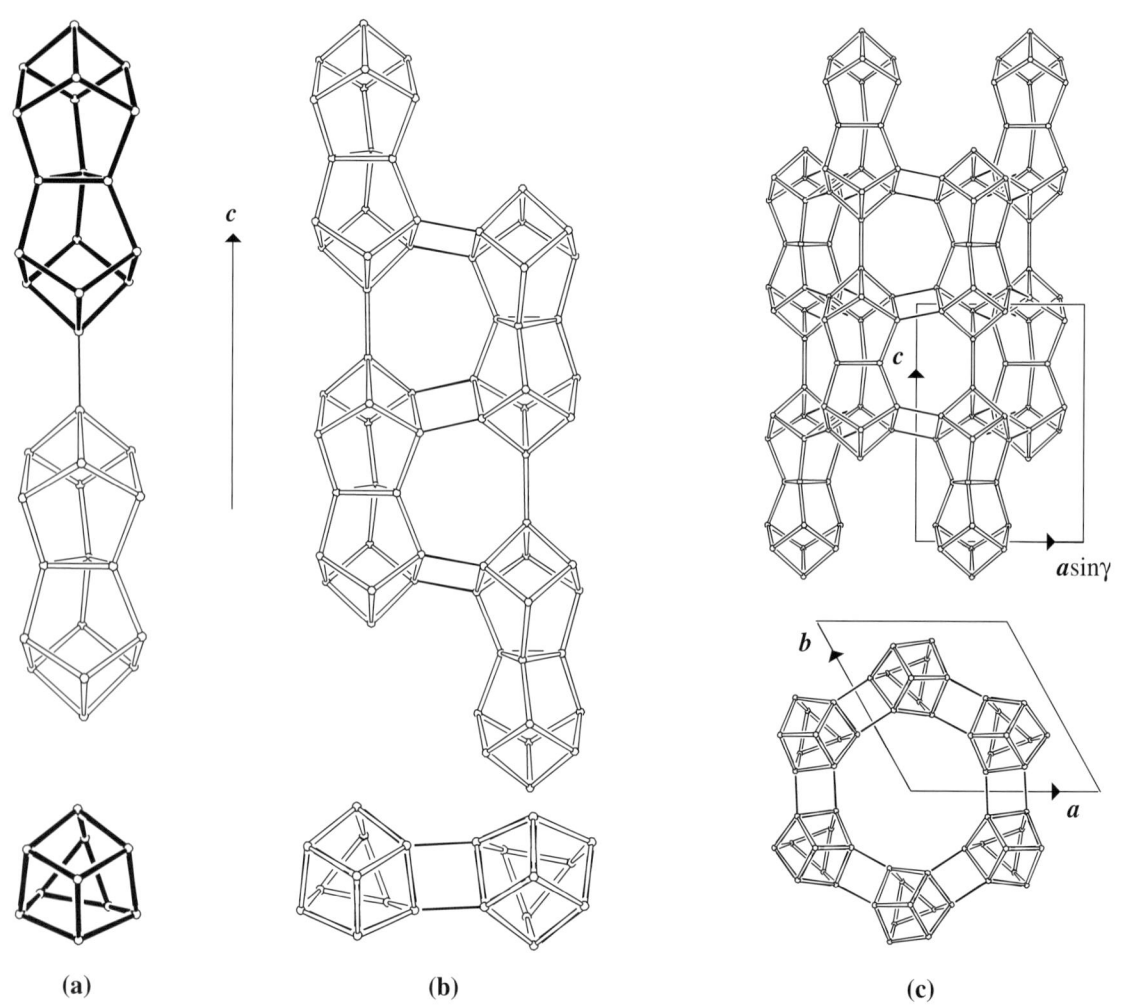

(a)   (b)   (c)

Fig. 1.   (a) PerBU viewed along <010> (top) and along $c$ (bottom); (b) connection mode viewed along <010> (top) and along $c$ (bottom); and (c) projection of the unit cell content along <010> (top) and along $c$ (bottom).

### 2. Connection mode

Neighboring PerBUs, related by a rotation of 60° about $c$ and a shift of $\frac{1}{2} c$ (i.e. related by a $6_3$-axis), are connected into the **ab** plane through 4- and 7-rings as illustrated in Figure 1b.

## 3. Channels and/or cages

Two-dimensional 7-ring channel systems, perpendicular to $c$, and straight 12-ring channels, parallel to $c$, intersect. The channel intersection is depicted in Figure 2 together with their linkage into 12-ring channels parallel to $c$.

$\{3\ [4^{12}5^{6}7^{6}12^{2}]\ <100>\ (7\text{-ring}),\ [001]\ (12\text{-ring})\}$

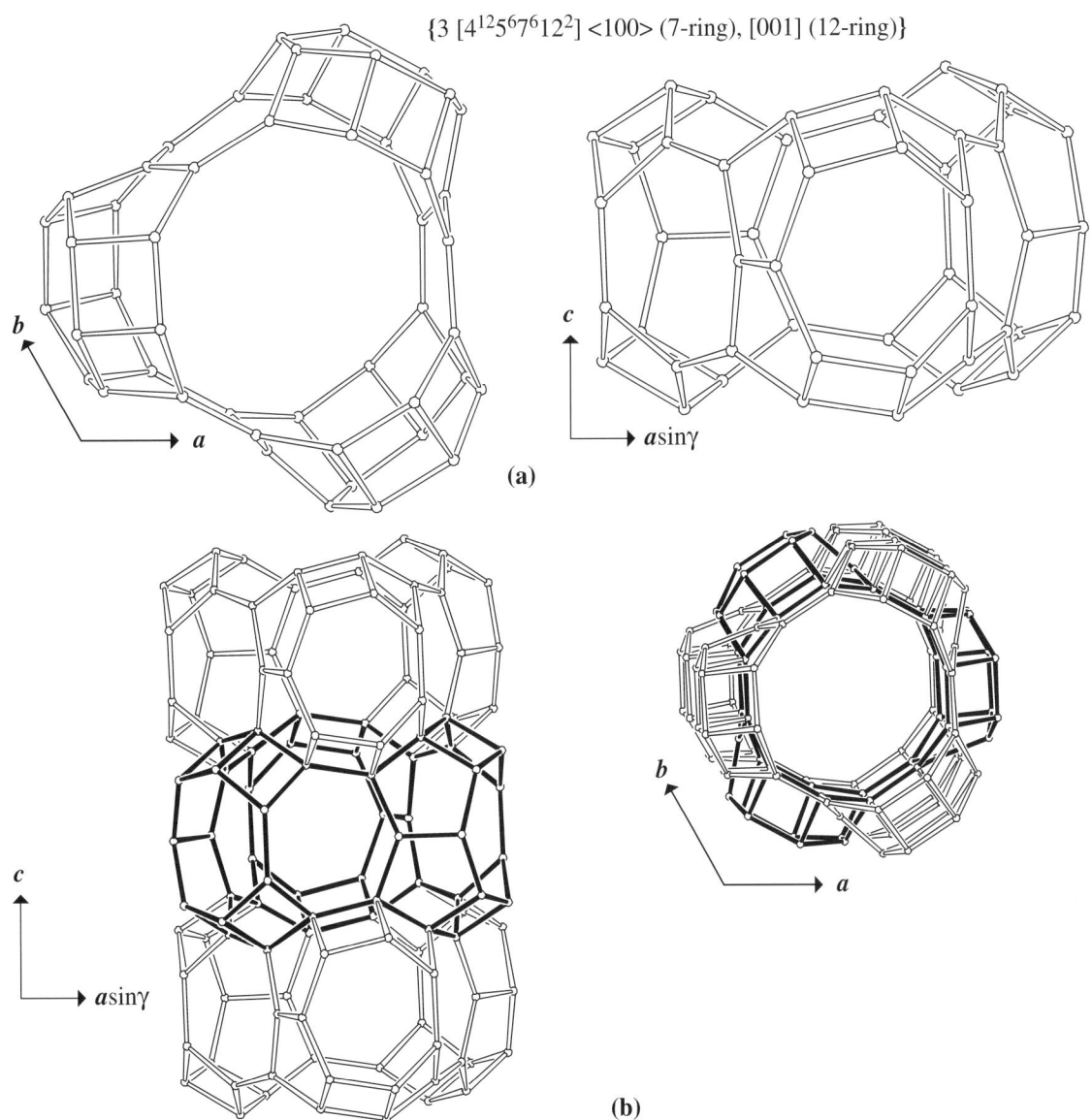

Fig. 2.   (a) Channel intersection viewed along $c$ (left) and <010> (right); (b) linkage of the channel intersections along $c$ viewed along <010> (left; one intersection in bold) and along the 12-ring channel axis parallel to $c$ (right).

# MEL

## Building Scheme

### 1. Periodic Building Unit

Tetragonal **MEL** can be built using units of 12 T atoms (one in bold in Figure 1). The T12-unit consists of two 5-1 units. T12-units, related by a screw rotation of 180° about **c**, form left- and right-handed chains along **c**. Left- and right-handed chains are connected along **b** into a two-dimensional PerBU. The PerBU equals the **bc** layer (or pentasil layer) shown in Figure 1. (See also **MFI**.)

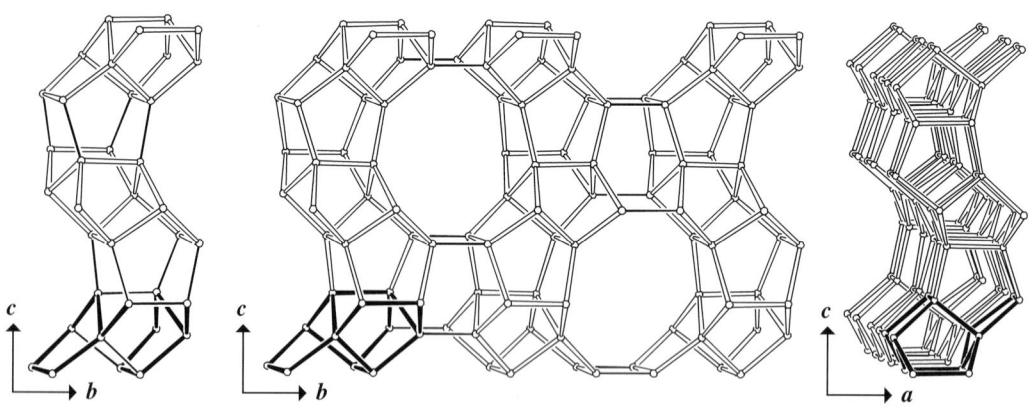

Fig. 1. Polar chain (left) viewed along **a** and PerBU viewed along **a** (middle) and along **b** (right).

### 2. Connection mode

Neighboring PerBUs, related along **a** by a rotation of 180° about **c** and a shift of $\frac{1}{2}$ **b**, are connected along **a** through 4- and 6-rings as depicted in Figure 2.

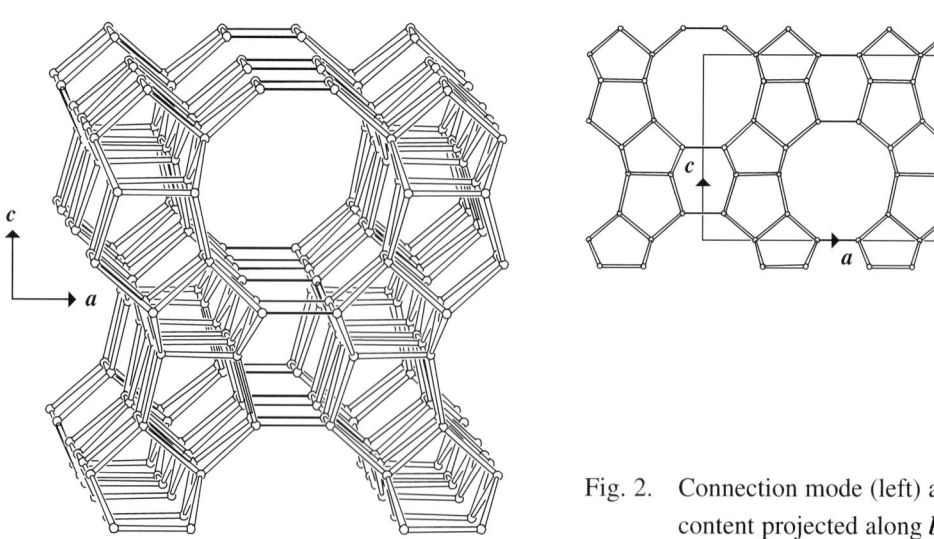

Fig. 2. Connection mode (left) and unit cell content projected along **b** (or **a**).

## 3. Channels and/or cages

Intersecting 10-ring channels are parallel to <100>. There are two types of channel intersections. Intersections, channel and the interconnection of channels are depicted in Figure 3.

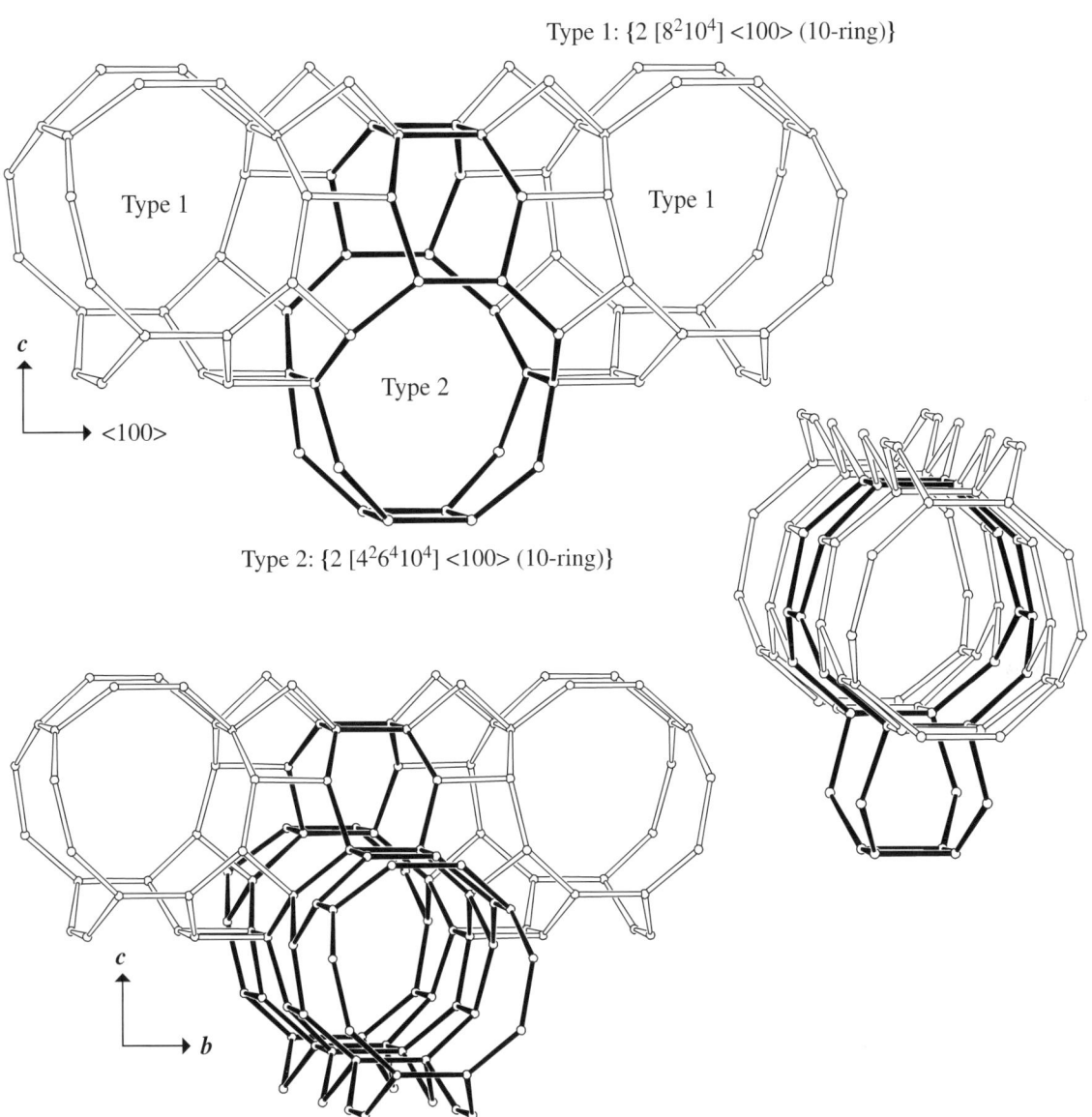

Type 1: {2 [$8^2 10^4$] <100> (10-ring)}

Type 2: {2 [$4^2 6^4 10^4$] <100> (10-ring)}

Fig. 3. Top: interconnection of channel intersections (Type 2 in bold) into straight 10-ring channels parallel to <100> viewed perpendicular to the channel axis (left) and along the channel axis (middle right). Bottom: example of intersecting 10-ring channels (one in bold). The other linkage is obtained when the bold channel is shifted over $\frac{1}{2}(\mathbf{a} + \mathbf{b} + \mathbf{c})$.

# MEP                    Building Scheme

## 1. Periodic Building Unit

Cubic **MEP** can be built using units of 46 T atoms. The T46-unit consists of a $[5^{12}6^2]$-cage attached to a "zigzag" 12-ring and to a tetra-(1,1,4,4)-substituted 6-ring (bold in Figure 1). A one-dimensional PerBU is obtained when T46-units, related along $b$ by pure translations, are connected along $b$ into a column of $[5^{12}6^2]$-cages sharing 6-rings and into a chain of 6-rings as shown in Figure 1a. (See also **DDR**, **DOH** and **MTN**.)

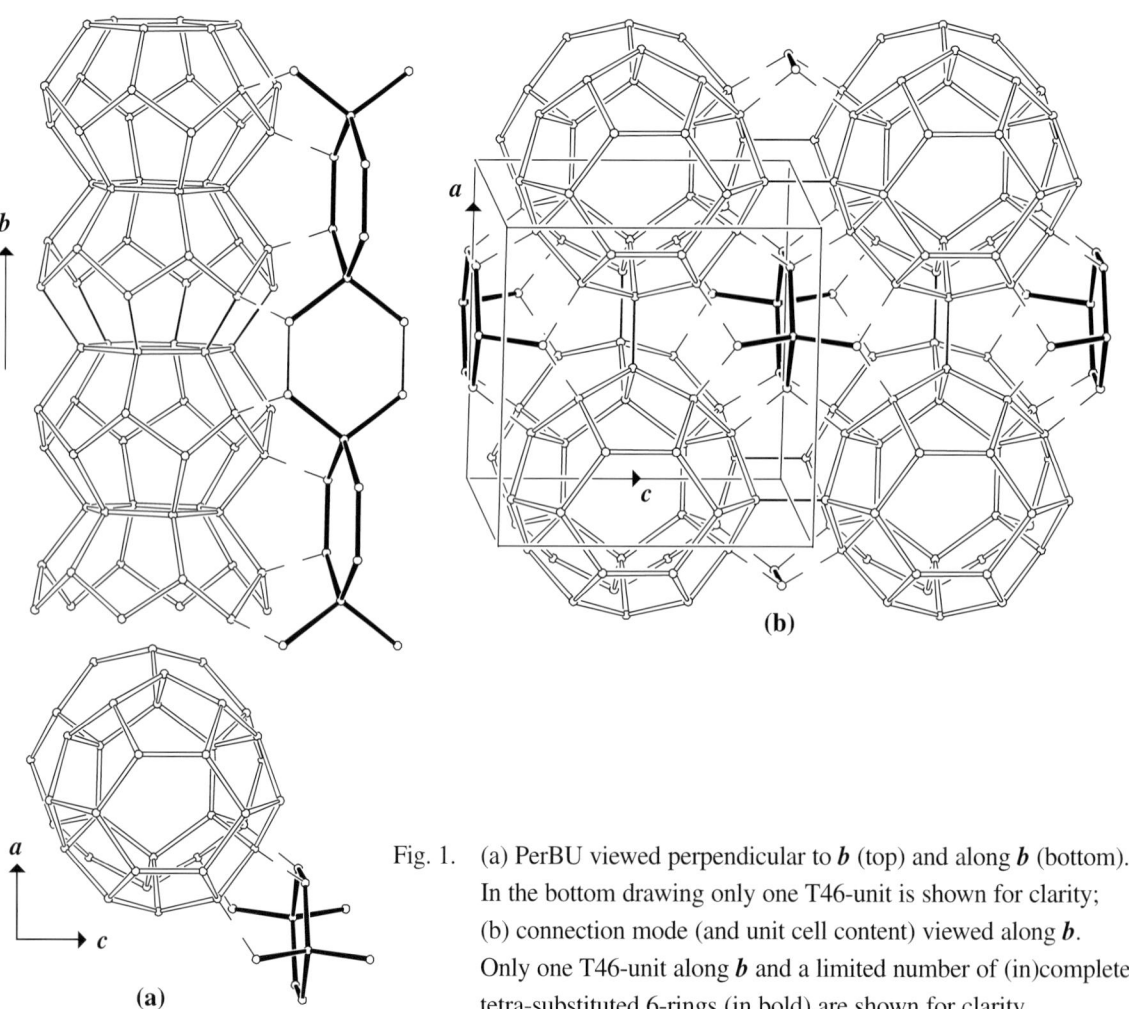

Fig. 1.  (a) PerBU viewed perpendicular to $b$ (top) and along $b$ (bottom). In the bottom drawing only one T46-unit is shown for clarity; (b) connection mode (and unit cell content) viewed along $b$. Only one T46-unit along $b$ and a limited number of (in)complete tetra-substituted 6-rings (in bold) are shown for clarity.

## 2. Connection mode

Neighboring PerBUs, related along $a$ and $c$ by pure translations, are connected through 5-rings into the three-dimensional framework. Columns of $[5^{12}6^2]$-cages parallel to $c$ sharing 5-rings with $[5^{12}]$-cages are generated as depicted in Figure 1b.

## 3. Channels and/or cages

The two types of cages are depicted in Figure 2. The $[5^{12}]$-cages are also found in **DDR**, **DOH** and **MTN**. Apertures are formed by 6-rings only.

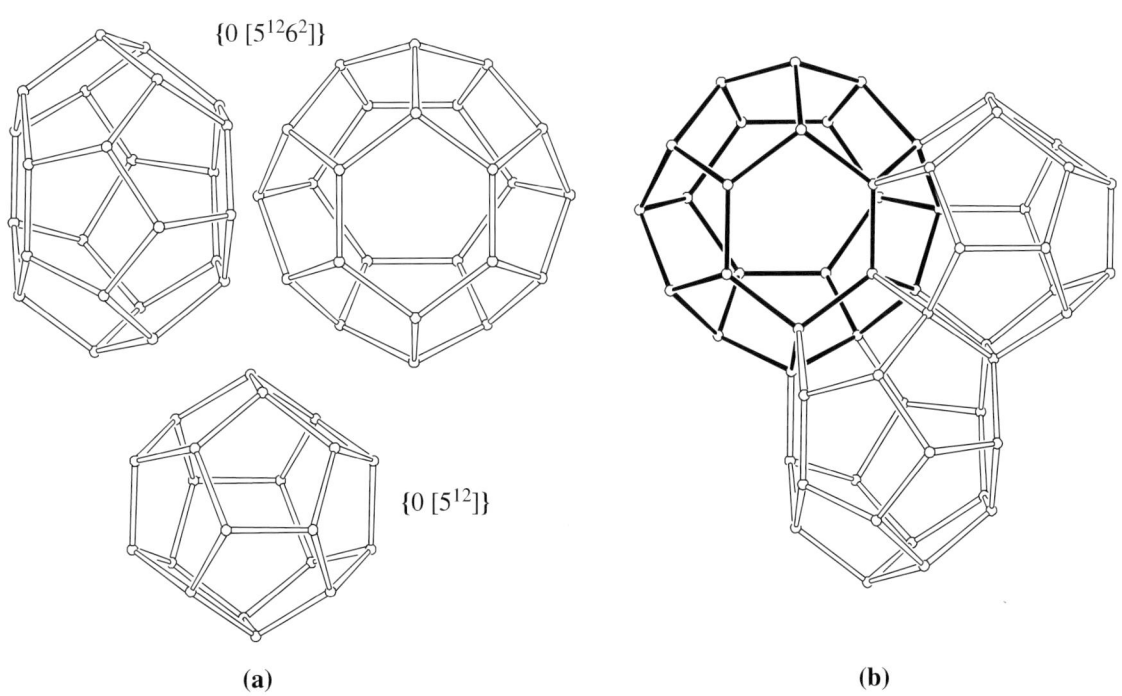

$\{0\ [5^{12}6^2]\}$

$\{0\ [5^{12}]\}$

**(a)**            **(b)**

Fig. 2.   (a) $[5^{12}6^2]$-Cage (top) viewed perpendicular to a cube axis (left) and along that cube axis (right) and $[5^{12}]$-cage (bottom); (b) linkage of the cages viewed along a cube axis.

# MER

# Building Scheme

## 1. Periodic Building Unit

Tetragonal **MER** can be built using the crankshaft chain (bold in Figure 1, left) running parallel to *c*. A one-dimensional PerBU is obtained when eight crankshaft chains are connected into a column of *mer* cavities. The *mer* cavities are connected through double 8-rings.

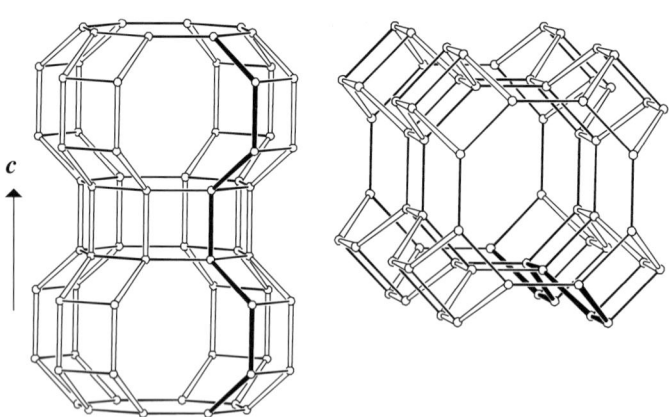

Fig. 1. PerBU constructed from eight crankshaft chains that form an 8-ring channel parallel to *c*. The PerBU consists of a column of *mer* cavities. PerBU viewed perpendicular to *c* (left) and along *c* (right).

## 2. Connection mode

Neighboring PerBUs, related along *a* and *b* by pure translations, are connected along *a* and *b* through 4-rings as shown in Figure 2.

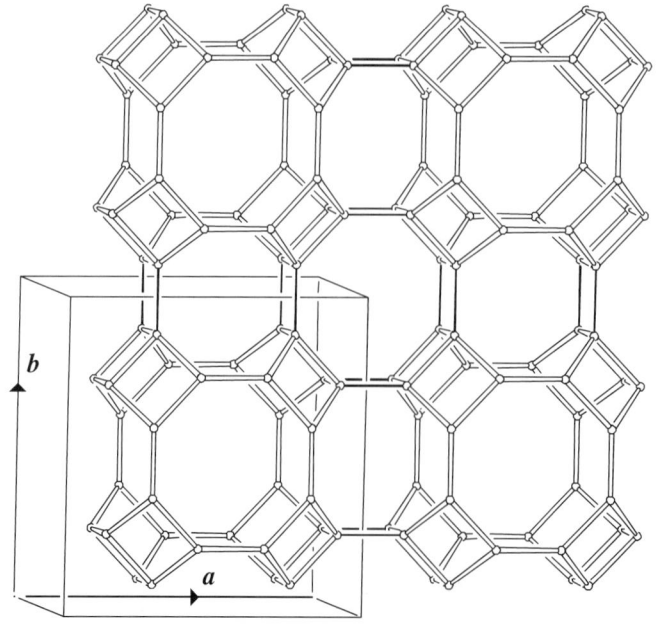

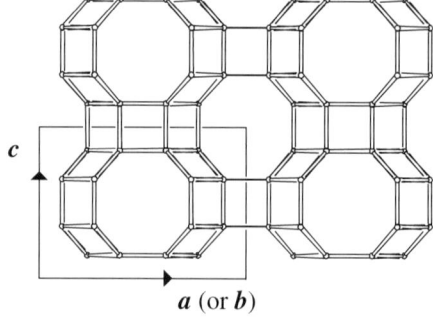

Fig. 2. Connection mode (and unit cell content) viewed along *c* and unit cell content projected along *a* (or *b*; right). In the perspective drawing, only one repeat unit along *c* of each PerBU is drawn for clarity.

## 3. Channels and/or cages

*mer* Cavities, also present in **KFI**, **MOZ** and **PAU**, are connected into two different interconnecting 8-ring channels parallel to *c* and into a third type of 8-ring channel parallel to <100> (Figure 3). The first type of 8-ring channels parallel to *c* equals the PerBU and consists of *mer* cavities which are connected through double 8-rings. The second type is obtained when *mer* cavities are connected along *b* (or *a*) through 4-rings. The 8-ring channel of Type 2 is topologically equivalent to the channel in **ACO** and one of the channels in **DFT**.

Type 1: {1 [$4^{20}8^48^{2/2}$] [001] (8-ring)}

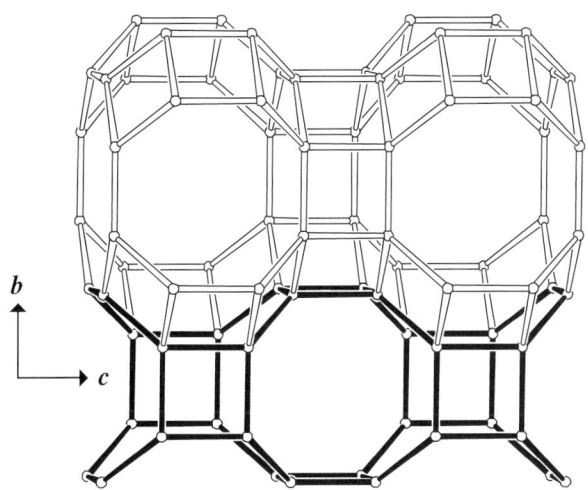

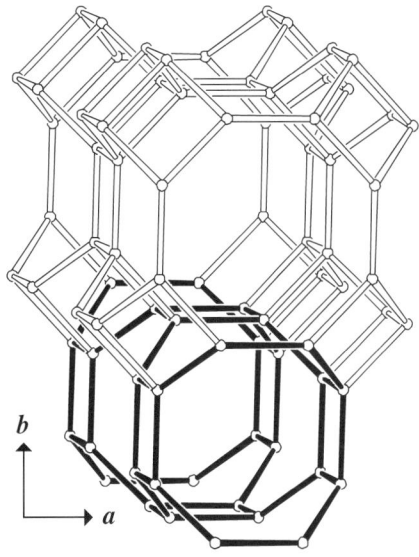

Type 2: {1 [$4^48^48^{2/2}$] [001] (8-ring)}

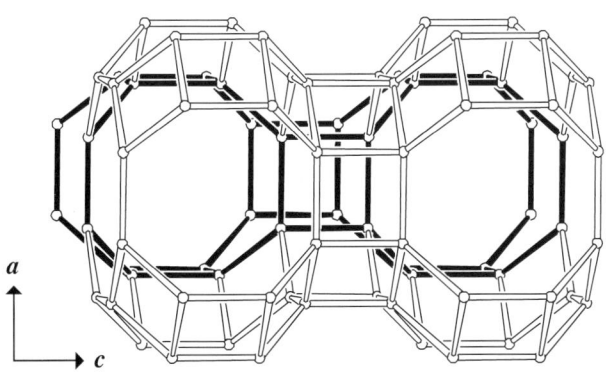

Fig. 3.　Channels of Type 1 and Type 2 (Type 2 in bold) are connected through common 8-rings into 8-ring channels parallel to <010> and into interconnecting 8-ring channels parallel to *c*. View perpendicular to <010> (left), along *c* (right) and along the 8-ring channel axis (Type 3) parallel to <010> (bottom).

# MFI                     Building Scheme

## 1. Periodic Building Unit

**MFI** can be built using units of 12 T atoms (one in bold in Figure 1). The T12-unit consists of two 5-1 units. T12-units, related along $c$ by a screw rotation of 180° about $c$, form left- and right-handed chains along $c$. Left- and right-handed chains are connected along $b$ into a two-dimensional PerBU. The PerBU equals the $bc$ layer (or pentasil layer) shown in Figure 1. (See also **MEL**.)

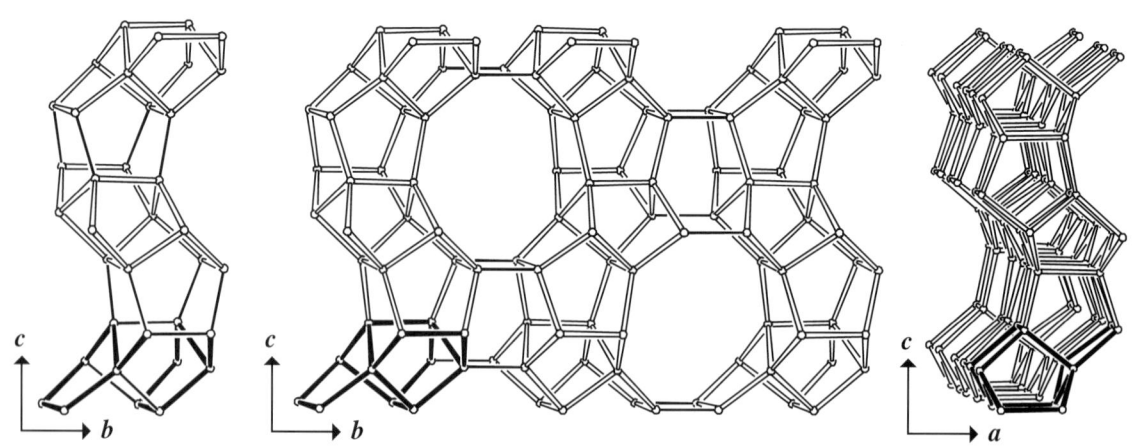

Fig. 1.    Polar chain (left) viewed along $a$ and PerBU viewed along $a$ (middle) and along $b$ (right).

## 2. Connection mode

Neighboring PerBUs, related along $a$ by a screw rotation of 180° about $a$ and a shift of $\frac{1}{2} b$, are connected along $a$ through 6-rings (and 5-rings) as depicted in Figure 2.

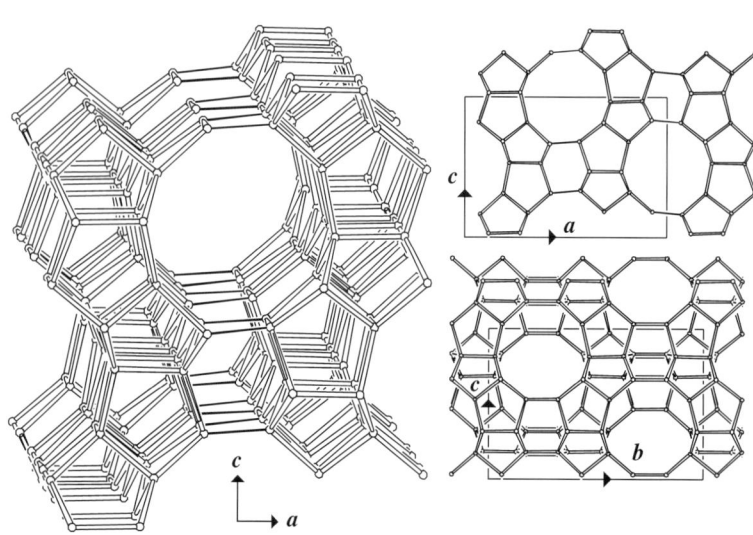

Fig. 2.    Connection mode viewed along $b$ (left) and unit cell content projected along $b$ (top right) and along $a$ (bottom right).

### 3. Channels and/or cages

Straight 10-ring channels parallel to **b** and sinusoidal 10-ring channels parallel to **a** do intersect. The channel intersection is depicted in Figure 3 together with the linkage of intersections into channels.

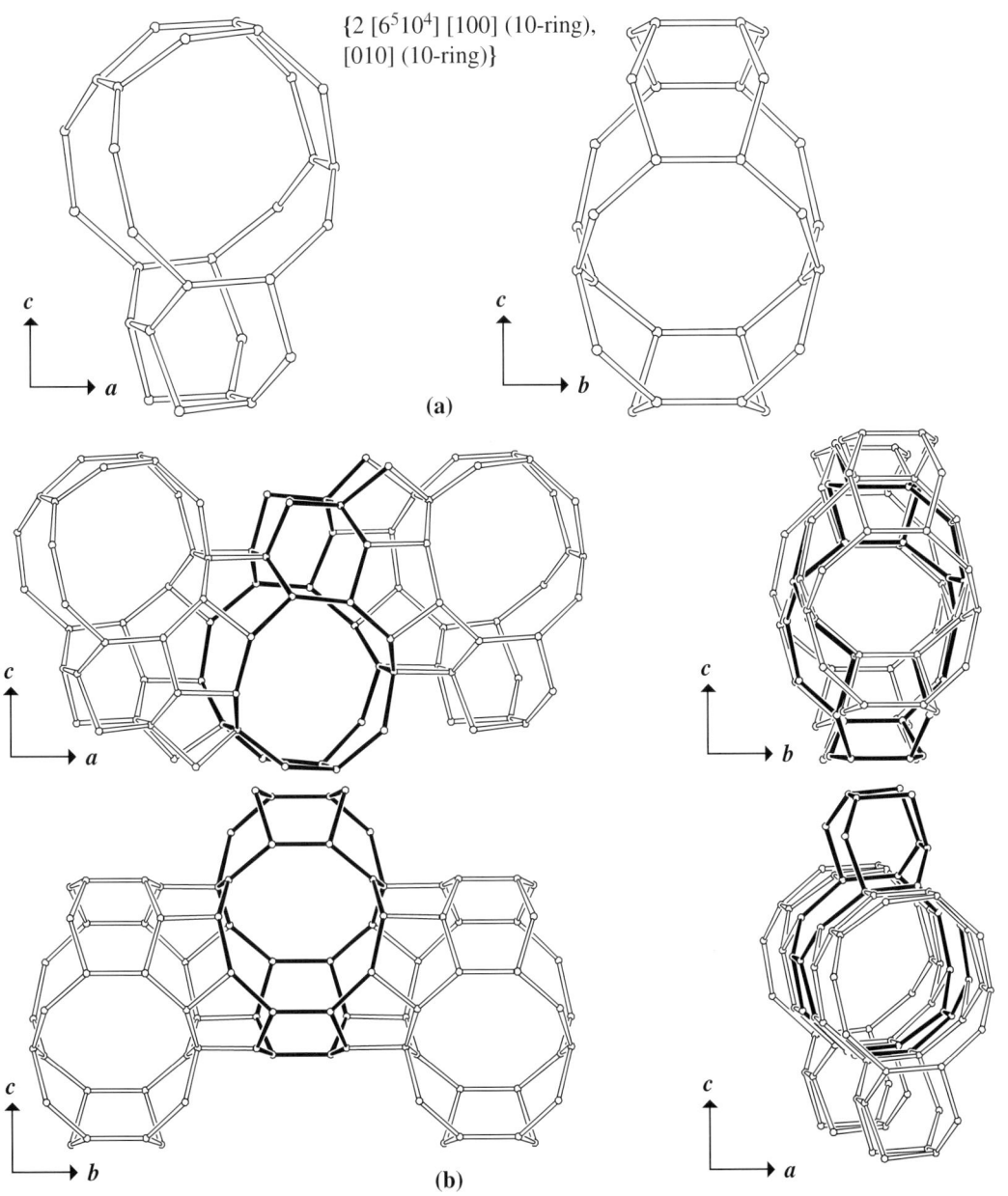

{2 $[6^5 10^4]$ [100] (10-ring), [010] (10-ring)}

Fig. 3. (a) Channel intersection viewed along **b** (left) and along **a** (right); (b) channel intersections (one in bold) are connected into sinusoidal 10-ring channels parallel to **a** (top left) and into straight channels parallel to **b** (bottom left). Right: view along the channel axes.

# MFS                                       Building Scheme

## 1. Periodic Building Unit

**MFS** can be built using the saw chain (bold in Figure 1) running parallel to **a**. Six saw chains (or two 5-1 units and one 6-ring) are connected into a one-dimensional PerBU depicted in Figure 1.

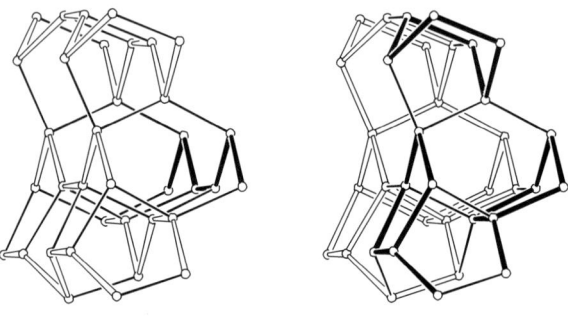

Fig. 1.   PerBU, constructed from six saw chains (left) and from 5-1 units and 6-rings (right), viewed along **a**.

## 2. Connection mode

Neighboring PerBUs, related along **b** (and **c**) by a shift of $\frac{1}{2}(a+b+c)$, are connected along **b** (and **c**) through fused 4-, 5- and 6-rings as shown in Figure 2.

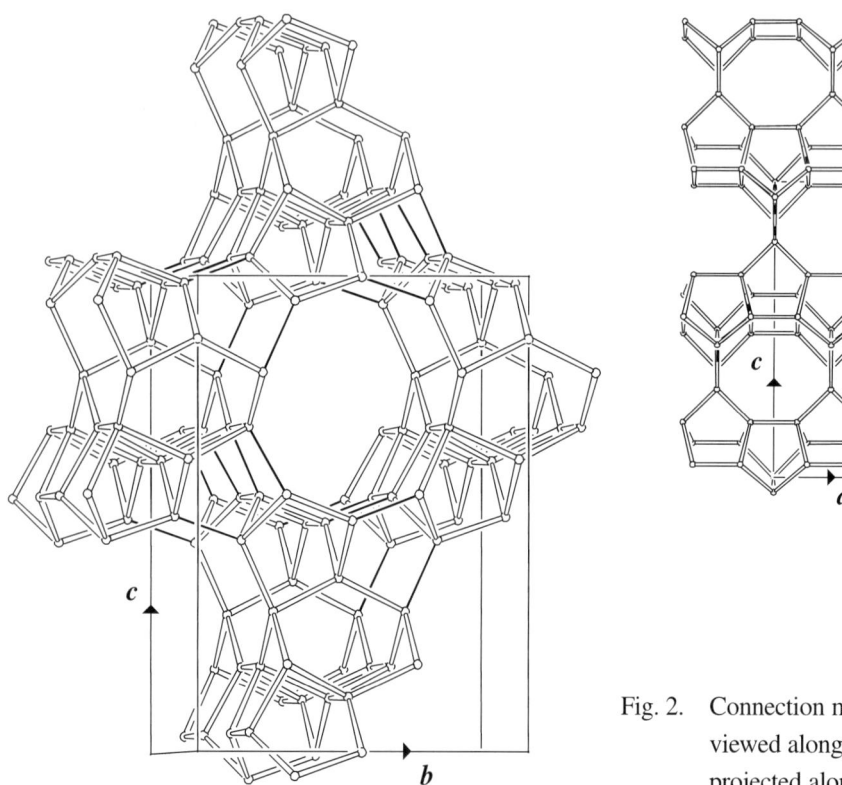

Fig. 2.   Connection mode (and unit cell content) viewed along **a** (left) and unit cell content projected along **b** (right).

### 3. Channels and/or cages

Intersecting 8-ring and 10-ring channels are parallel to **b** and **a**, respectively. The channels and their intersection are illustrated in Figure 3.

{1 [$4^2 5^{10} 6^7 10^2 8^{2/2}$][010] (8-ring)}

{1 [$4^2 5^8 6^4 8^2 10^{2/2}$] [100] (10-ring)}

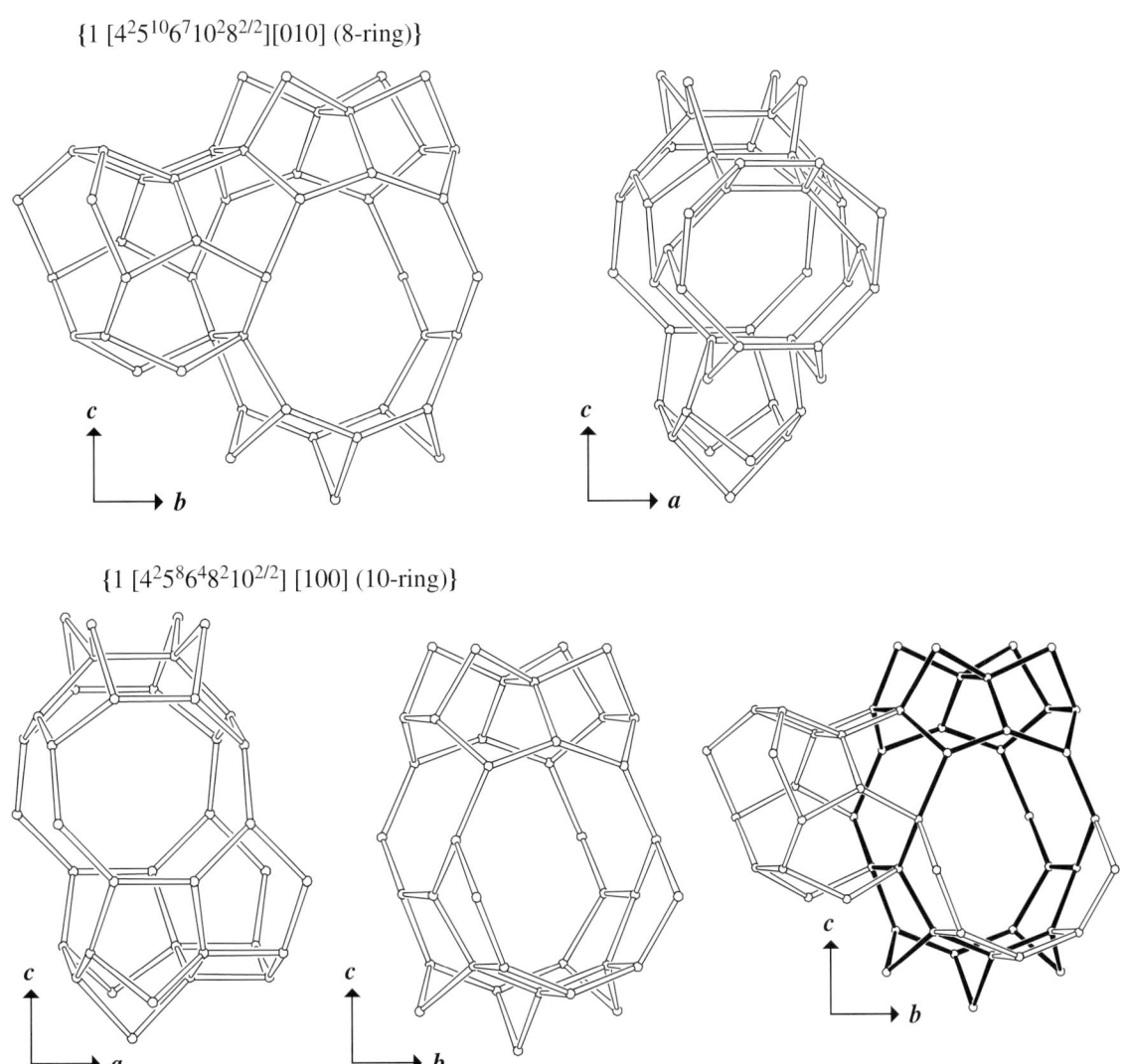

Fig. 3. 8-Ring channel (top) and 10-ring channel (bottom) viewed perpendicular to the channel axis (left) and along the channel axis (right). Only one repeat unit along the channel direction is drawn. Bottom at the far right: intersecting channels, viewed along **a**, have a [$5^6 6^4 8^2 10^2$]-cavity in common (in bold).

# MON
# Building Scheme

## 1. Periodic Building Unit

Tetragonal **MON** can be built using 4-rings. The PerBU equals the 4-ring layer depicted in Figure 1. The 4 rings (one in bold) are related along **a** and **b** by pure translations.

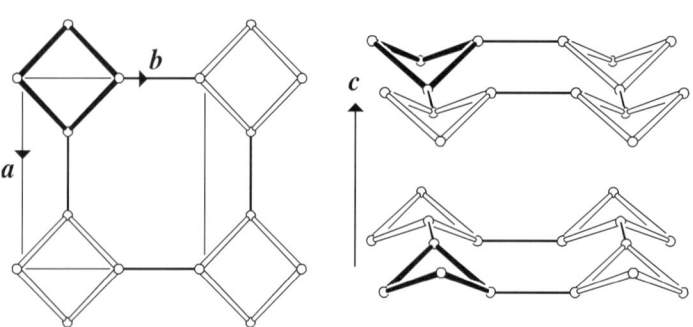

Fig. 1. PerBU viewed down **c** (left), down **a** (top right) and along **b** (bottom right). The PerBUs, depicted at the right, differ by a rotation of 90° about **c** (or by a mirror plane perpendicular to **c**).

## 2. Connection mode

Neighboring PerBUs, related along **c** by a rotation of 90° about **c** accompanied by a lateral shift of $\frac{1}{2}$ **a** or $\frac{1}{2}$ **b**, are connected along **c** through 5-rings as shown in Figure 2. The connectivity codes are denoted as (1/2, 0) or (0, 1/2) depending on whether the lateral shift is along **a** or **b**.

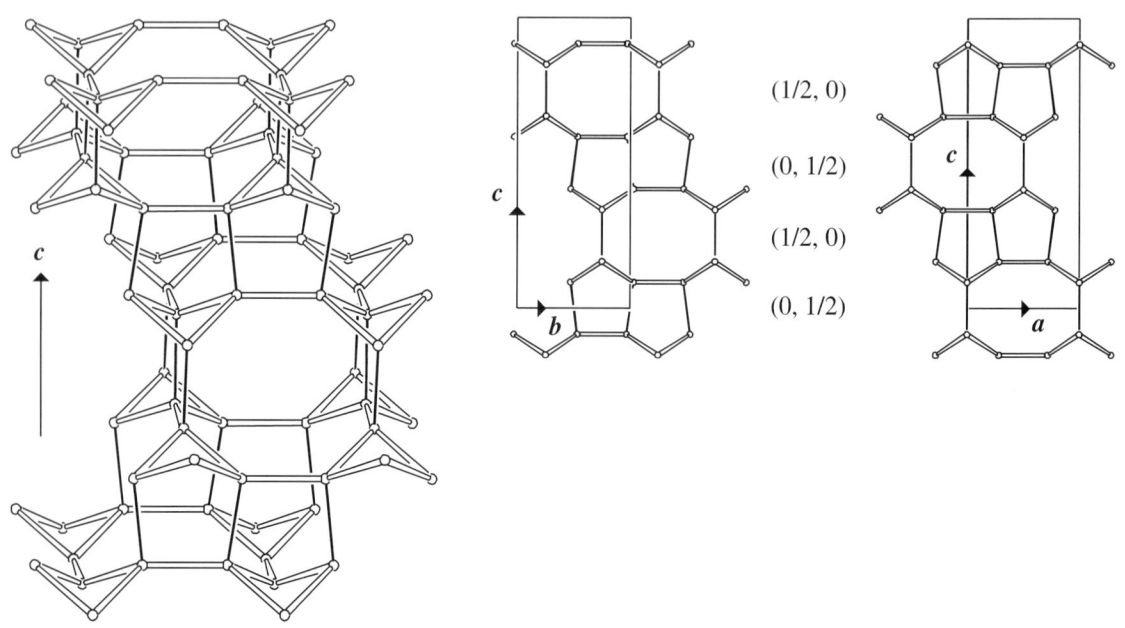

(1/2, 0)

(0, 1/2)

(1/2, 0)

(0, 1/2)

Fig. 2. Connection mode viewed along **a** (left) and unit cell content projected along **b** (top left) and along **a** (top right). The lateral shifts between neighboring PerBUs, related by $4_1$-axis parallel to **c**, are given in fractions of **a** and **b**.

## 3. Channels and/or cages

Interconnecting 8-ring channels are parallel to <100>. The channel, depicted in Figure 3, is topologically equivalent to (one of) the 8-ring channels in **EON**, **MAZ**, **MOR**, **RSN** and **VSV**. The interconnection of channels generate wavy 8-ring channels parallel to $c$.

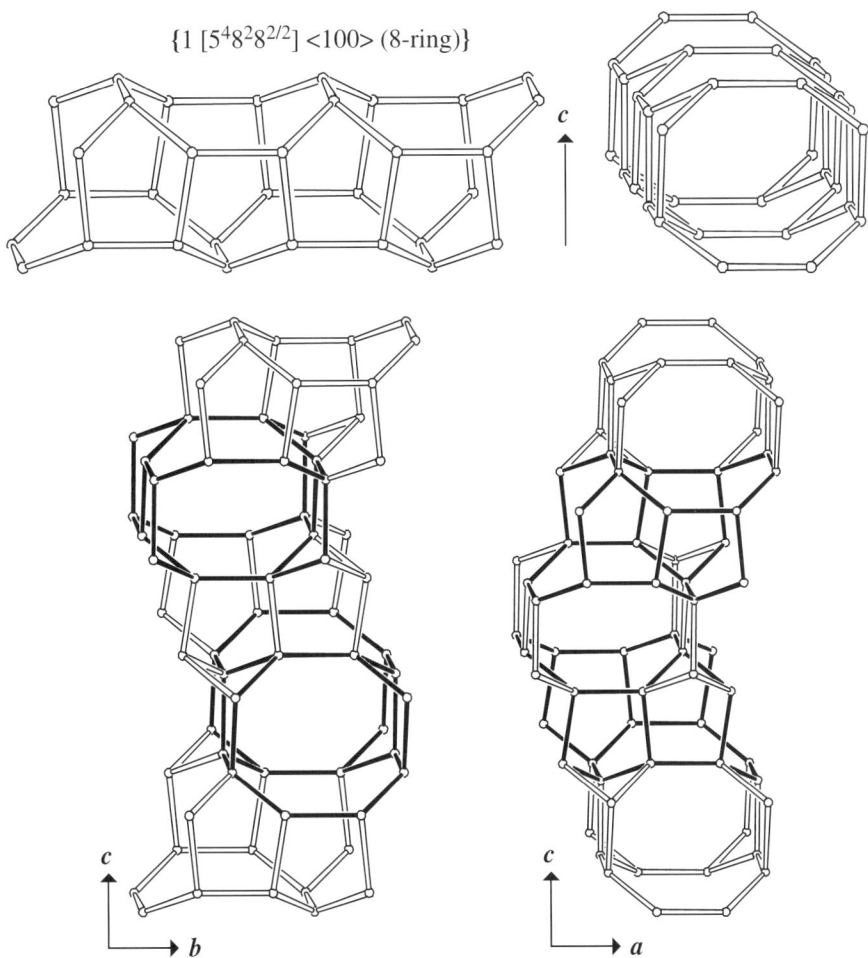

$\{1\ [5^4 8^2 8^{2/2}]\ <100>\ (8\text{-ring})\}$

Fig. 3.    Top: 8-ring channel viewed perpendicular to the channel axis (left) and along the channel axis (right). Bottom: channels are connected along $c$ into a wavy channel along $c$. View along $a$ (left) and along $b$ (right). Only one repeat unit in the channels parallel to $a$ and $b$ is drawn for clarity.

# MOR

## Building Scheme

### 1. Periodic Building Unit

Finite building units of 12 T atoms are composed of two 5-1 units (bold in Figure 1a). The two-dimensional PerBU is obtained when these T12-units, related along $c$ by pure translations and along $b$ by a rotation of 180° about $a$, are connected into the $bc$ layer (Figure 1b). Saw chains along $c$ are formed. (See also **DAC** and **EPI**.)

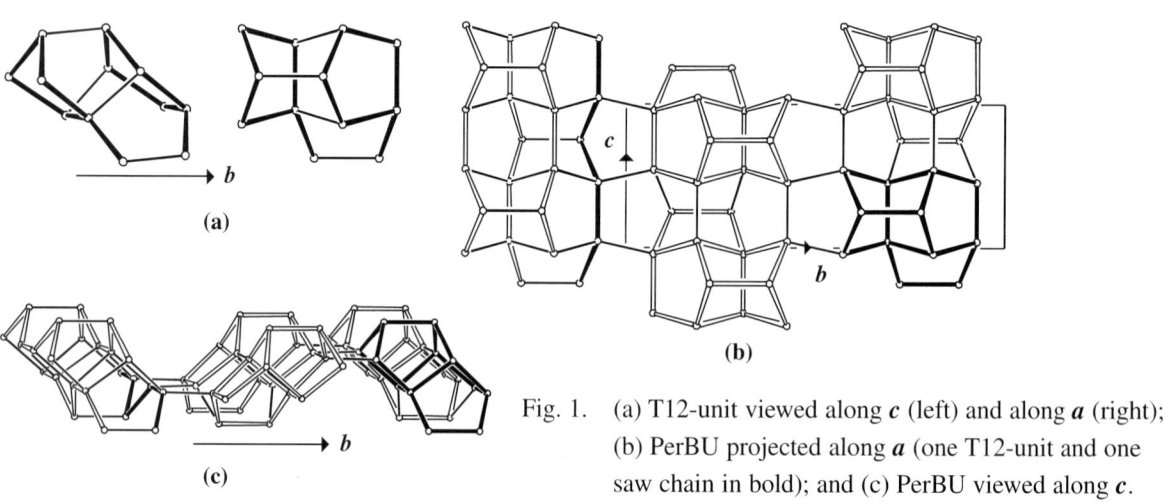

**(a)**

**(b)**

**(c)**

Fig. 1. (a) T12-unit viewed along $c$ (left) and along $a$ (right); (b) PerBU projected along $a$ (one T12-unit and one saw chain in bold); and (c) PerBU viewed along $c$.

### 2. Connection mode

Neighboring PerBUs, related along $a$ by a shift of $\frac{1}{2}(a+b)$, are connected along $a$ through 4-rings as depicted in Figure 2.

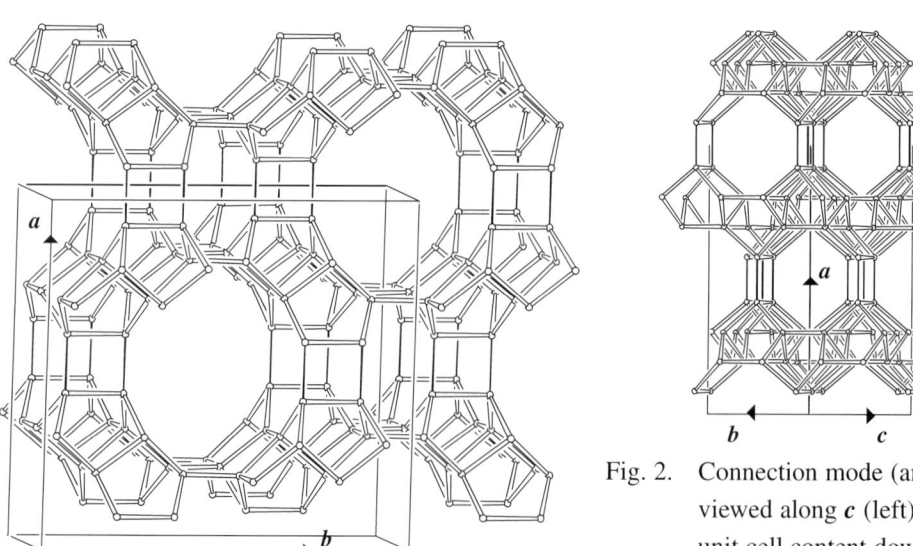

Fig. 2. Connection mode (and unit cell content) viewed along $c$ (left) and projection of the unit cell content down [011] (right).

## 3. Channels and/or cages

Interconnecting 8-ring and 12-ring channels are parallel to *c* (Figure 3). The 8-ring channel is topologically equivalent to (one of) the 8-ring channels in **EON**, **MAZ**, **MON**, **RSN** and **VSV**. The channels are linked into sinusoidal channels (with limiting 8-ring windows) parallel to *b*.

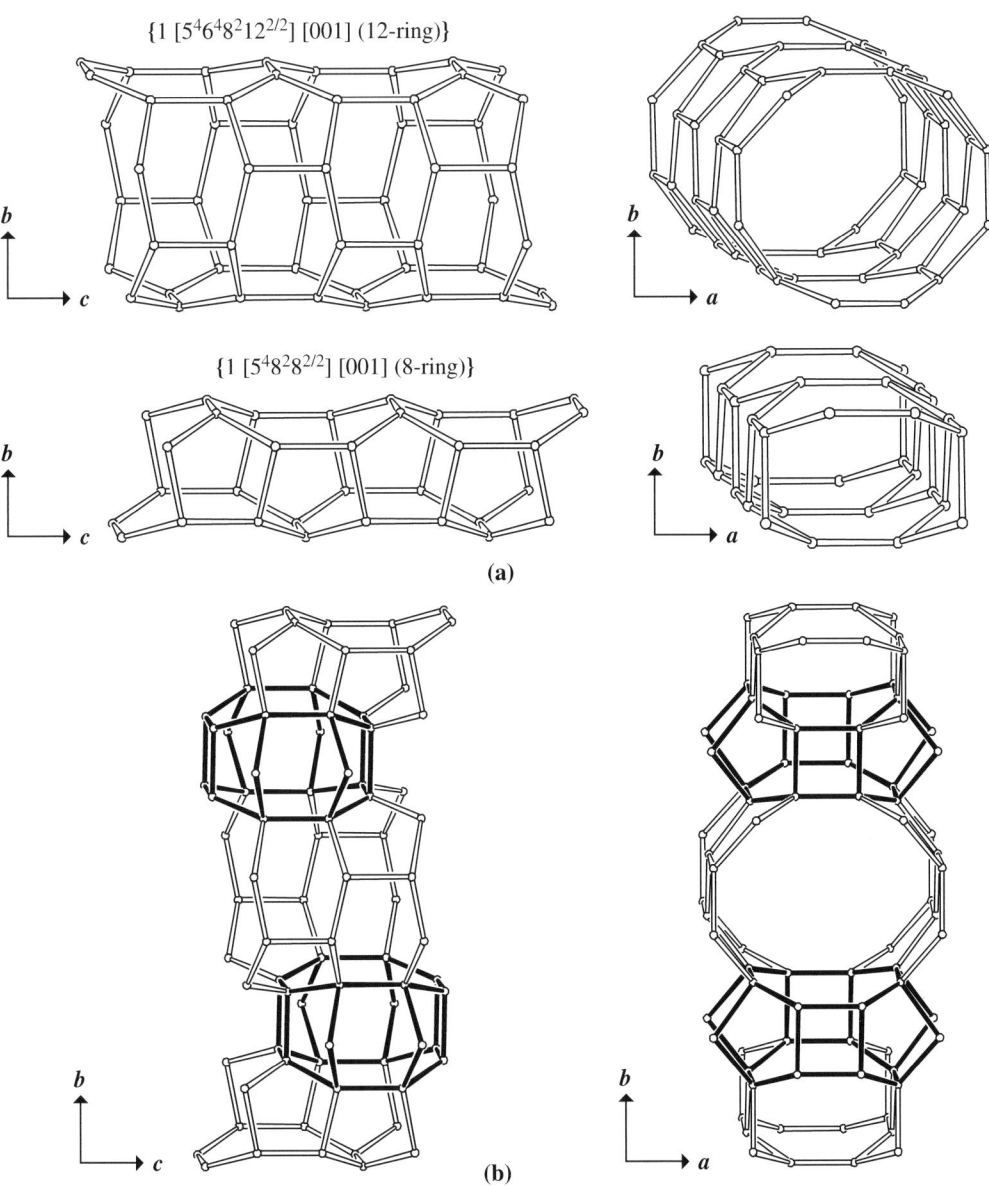

$\{1\ [5^4 6^4 8^2 12^{2/2}]\ [001]\ (12\text{-ring})\}$

$\{1\ [5^4 8^2 8^{2/2}]\ [001]\ (8\text{-ring})\}$

**(a)**

**(b)**

Fig. 3.  (a) 12-Ring and 8-ring channels viewed perpendicular to the channel axis (left) and along the channel axis (right); (b) the interconnected channels form a sinusoidal channel (with limiting 8-ring windows) parallel to *b* (interconnecting cavity in bold).

# MOZ        Building Scheme

## 1. Periodic Building Unit

Hexagonal **MOZ** can be built using the saw chain (bold in Figure 1) running parallel to *c*. Six saw chains are connected into an one-dimensional PerBU consisting of a column of *can* cages that are connected through double 6-rings. (See also **ERI**, **LTL** and **OFF**.)

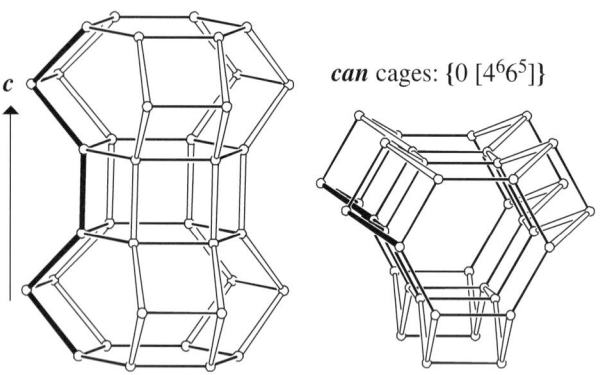

*can* cages: {0 [$4^6 6^5$]}

Fig. 1.   PerBU, a column of *can* cages, viewed perpendicular to *c* (left) and along *c* (right).

## 2. Connection mode

Neighboring PerBUs, related by a 2-fold axis at (1/2,1/2) parallel to *c* and 3-fold axes at <2/3,1/3> parallel to *c*, are connected into the *ab* plane through 8- and 12-rings as shown in Figure 2.

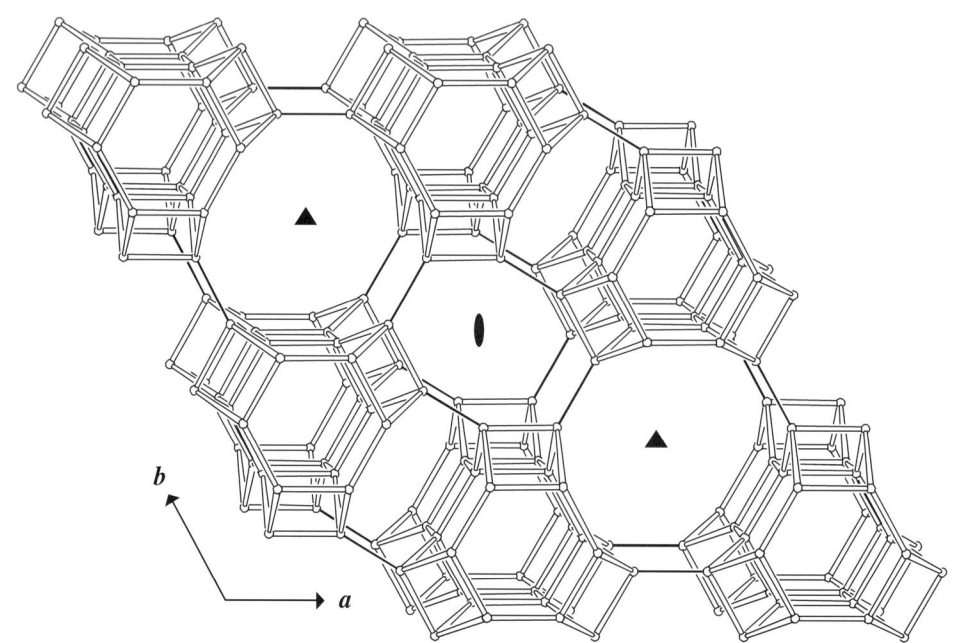

Fig. 2a.   Connection mode viewed along *c*.

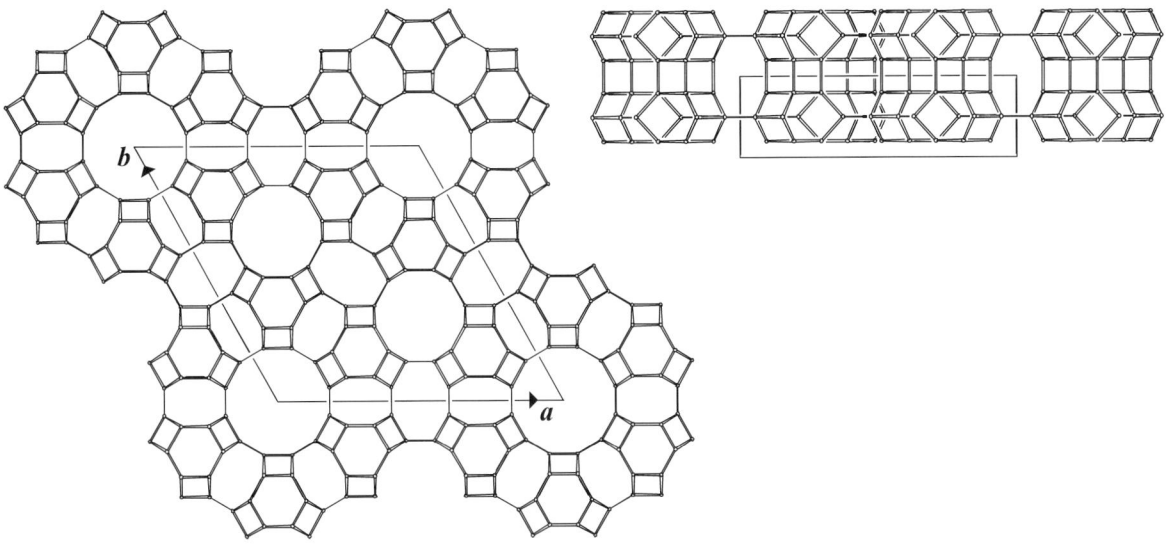

Fig. 2b.  Unit cell content projected along *c* (left) and along *b* (right).

## 3. Channels and/or cages

Two types of 8-ring and 12-ring channels are parallel to *c*. The first type of 8-ring channel is topologically equivalent to (one of) the 8-ring channels in **DFT**, **LOV**, **LTL** and **RSN**. The second type of 8-ring channels consists of *mer* cavities connected along *c* through common 8-rings. The *mer* cavity is also present in **MER**, **KFI** and **PAU**. The two types of 12-ring channels are topologically equivalent to those in **OFF** and **LTL**, respectively. The **LTL**-type 12-ring channel consists of a column of fused *ltl* cavities (see **LTL**). Additional two-dimensional 8-ring channels, intersecting the 12-ring channels, are perpendicular to *c*. Channels and the interconnection of channels are depicted in Figure 3.

*can* cage: $\{0\,[4^6 6^5]\}$

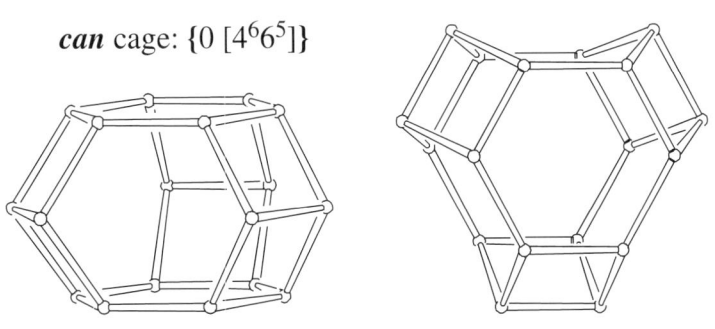

The *can*-cage is also present in:
**AFG, CAN, ERI, FAR, FRA, GIU, LIO, LOS, LTL, LTN, MAR, OFF, SAT, SBS, SBT, TOL** and **-WEN**

Fig. 3a.  *can* Cage.

# MOZ

## Cage/Channel

{1 [$4^2 6^2 8^2 8^{2/2}$] <100> (8-ring)}

{1 [$4^3 6^3 8^3 12^{2/2}$] [001] (12-ring)}

*mer* cavity
{3 [$4^{12} 8^6$] <100> (8-ring),[001] (8-ring)}

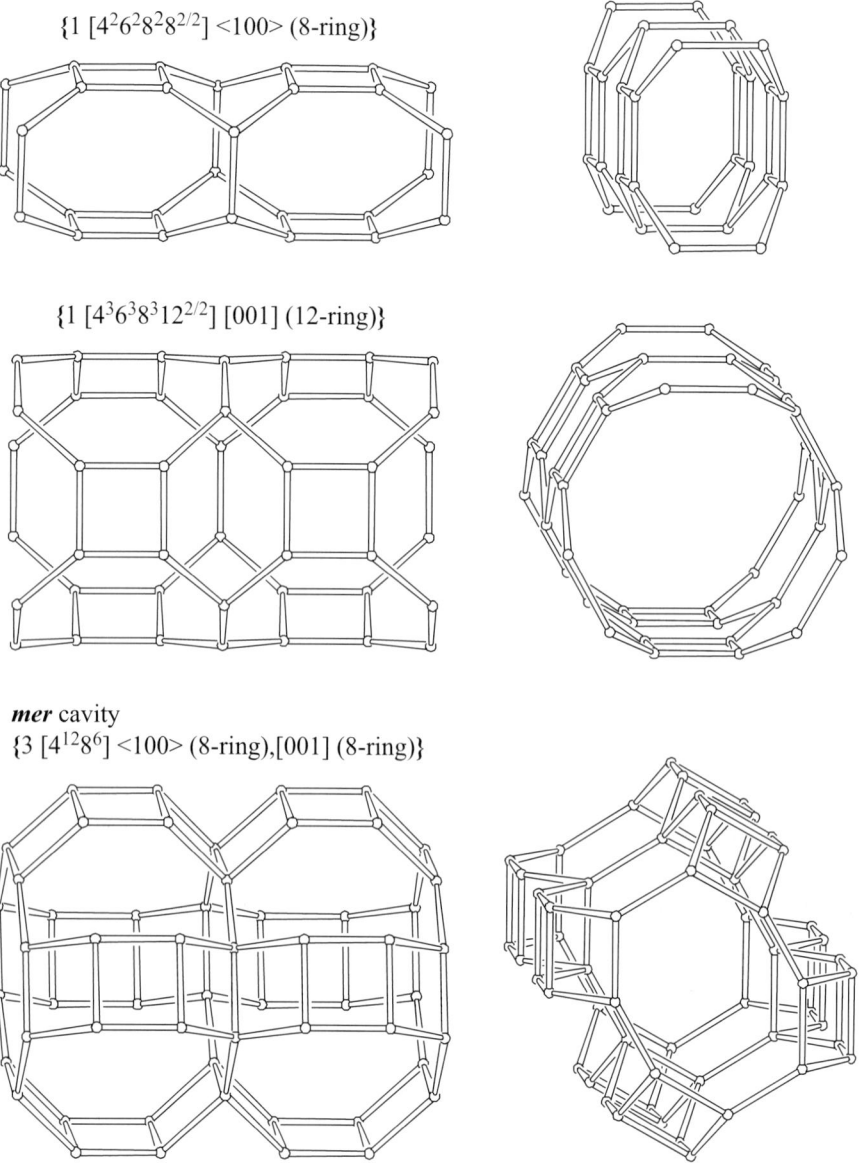

Fig. 3b.    8-Ring channel of Type 1 (top), 12-ring channel of Type 1 (middle) and 8-ring channel of Type 2 (bottom) viewed perpendicular to the channel axis parallel to *c* (left) and along the channel axis (right).

*ltl* cavity: {3 [$4^{18}8^612^2$] <100> (8-ring), [001] (12-ring)}

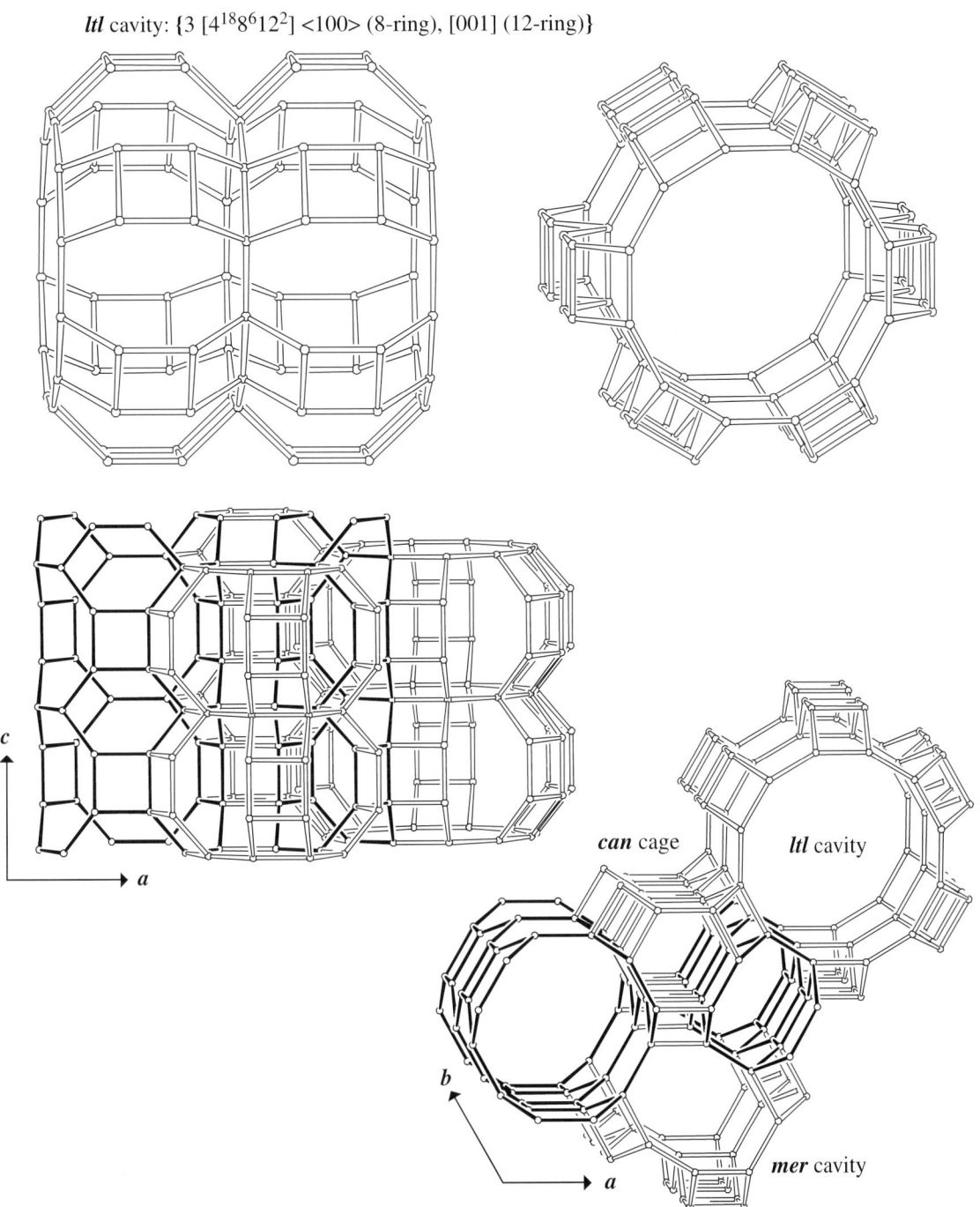

Fig. 3c.   Top: 12-ring channel of Type 2 viewed perpendicular to the channel axis (left) and along the channel axis (right). Bottom: interconnection of **can** cage, **mer** and **ltl** cavities and channels viewed along <010> (left) and along **c** (right). One 8-ring channel (see **LOV**) and one 12-ring channel (see **OFF**) parallel to **c** are in bold. Additional 8-ring channels, intersecting the 12-ring channels, are perpendicular to **c**.

# MSE        Building Scheme

## 1. Periodic Building Unit

Tetragonal **MSE** can be built using units of 28 T atoms. The T28-unit consists of two mirror-related T14-units: a [5$^4$]-cage with three "handles", or two 1-5-1 units (Figure 1, left). T28-units, related by a screw rotation of 180° about **a** (or about **b**), are connected through 4-rings into the two-dimensional PerBU depicted in Figure 1 (right). (Compare the building unit and PerBU with those in *BEA, BEC, CON, ISV, ITH, IWR** and **IWW**.)

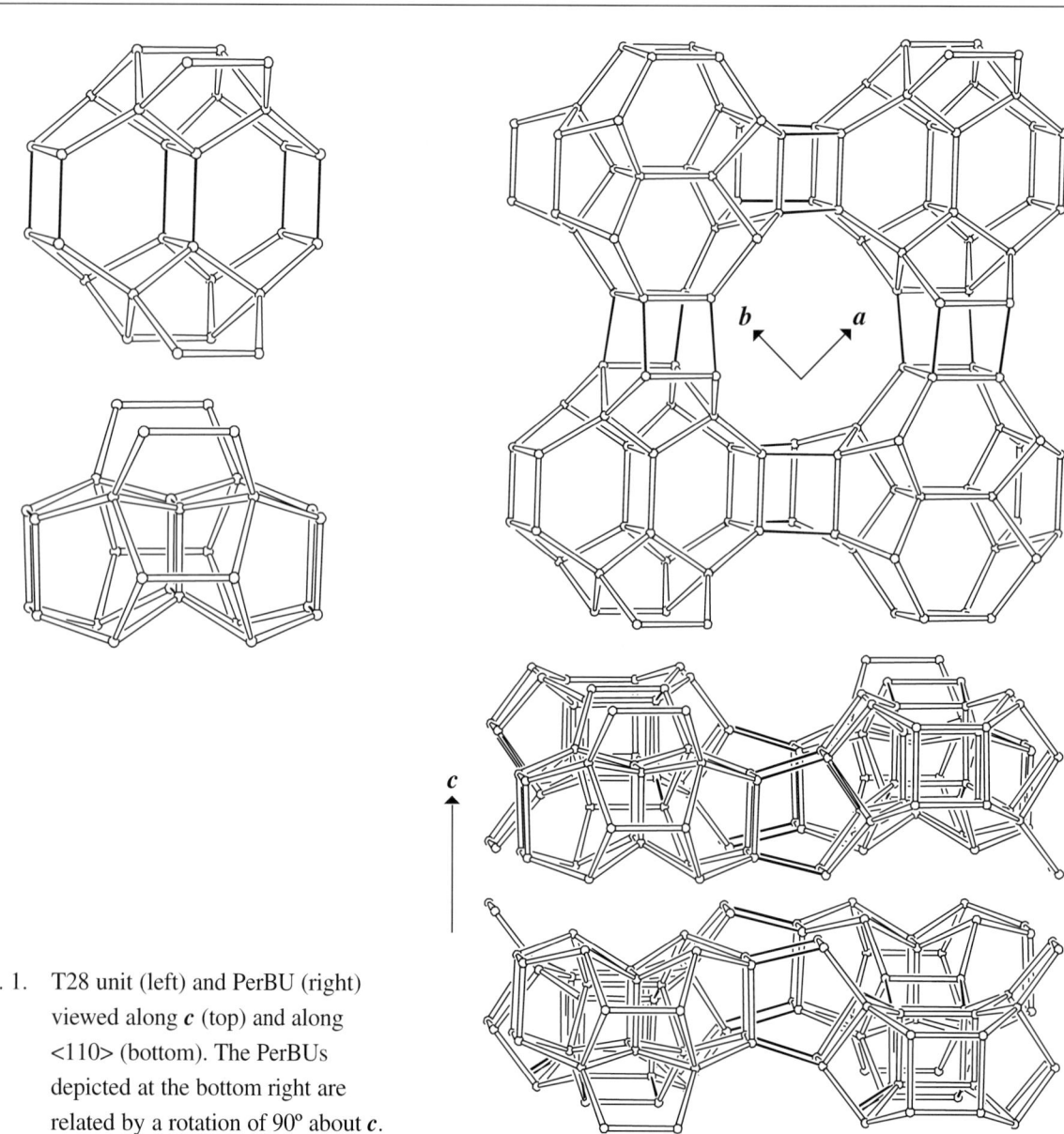

Fig. 1.    T28 unit (left) and PerBU (right)
viewed along **c** (top) and along
<110> (bottom). The PerBUs
depicted at the bottom right are
related by a rotation of 90° about **c**.

## Building Scheme **MSE**

### 2. Connection mode

Neighboring PerBUs, related along *c* by a rotation of 90° about *c* followed by a shift of $\frac{1}{2}$ *c*, are connected through 4-rings as shown in Figure 2. The successive PerBUs are related by a $4_2$-axis.

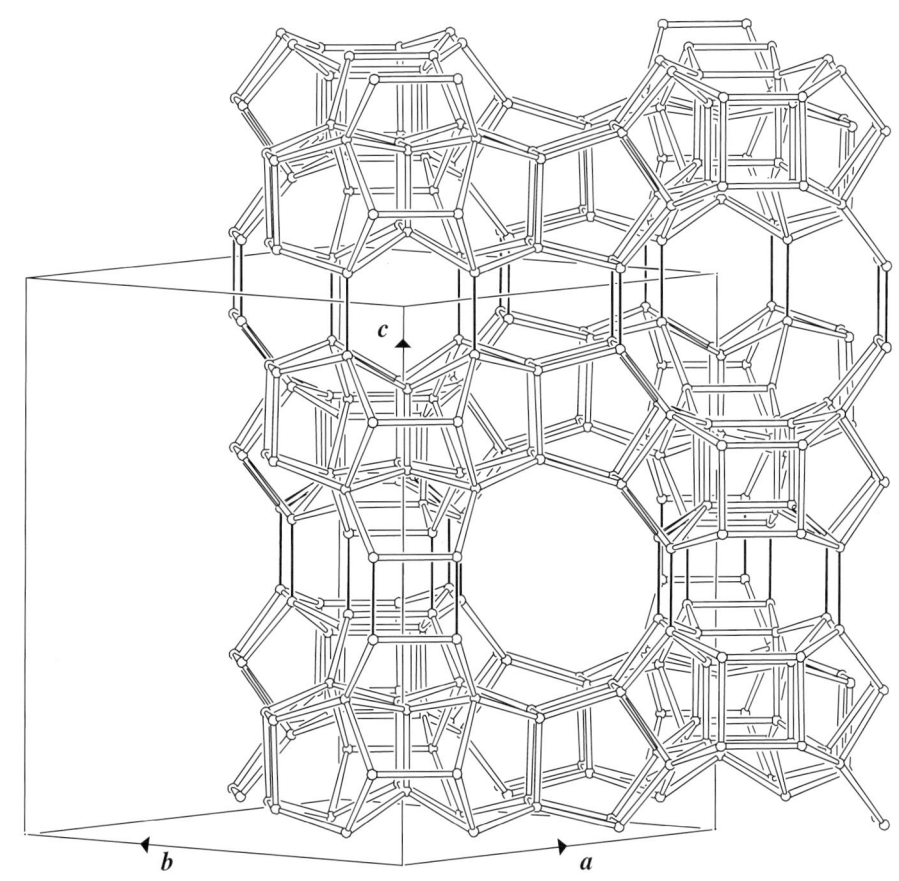

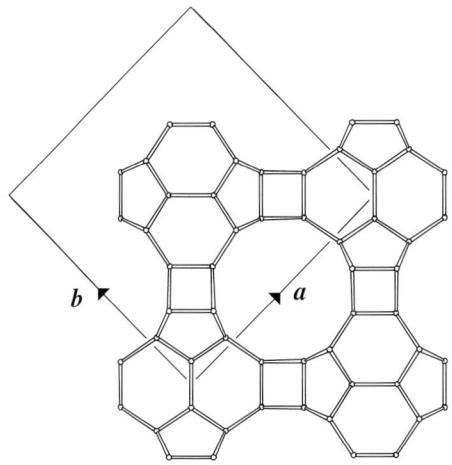

Fig. 2.   Connection mode (and unit cell content) viewed along [110] (top) and unit cell content projected along *c* (bottom).

## 3. Channels and/or cages

12-Ring channels are parallel to $c$. The 12-ring channels intersect with interconnecting undulating 10-ring channels parallel to <110>. The channel intersection, a $[6^2 10^2 12^2]$-cavity, is depicted in Figure 3 together with the linkage of intersections into channels.

{2 $[6^2 10^2 12^2]$ [001] (12-ring), <110> (10-ring)}

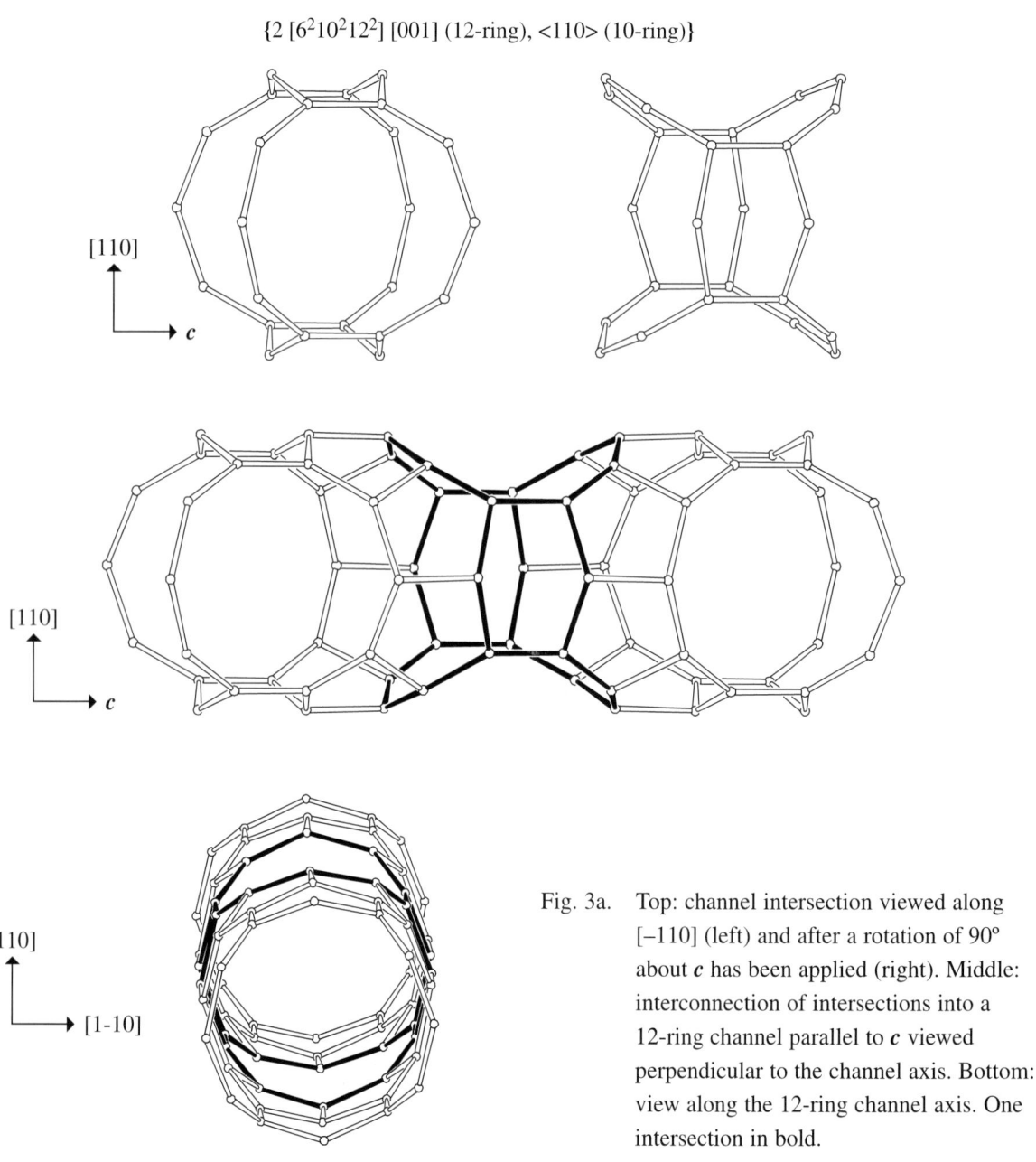

Fig. 3a.   Top: channel intersection viewed along [−110] (left) and after a rotation of 90° about $c$ has been applied (right). Middle: interconnection of intersections into a 12-ring channel parallel to $c$ viewed perpendicular to the channel axis. Bottom: view along the 12-ring channel axis. One intersection in bold.

# Cage/Channel

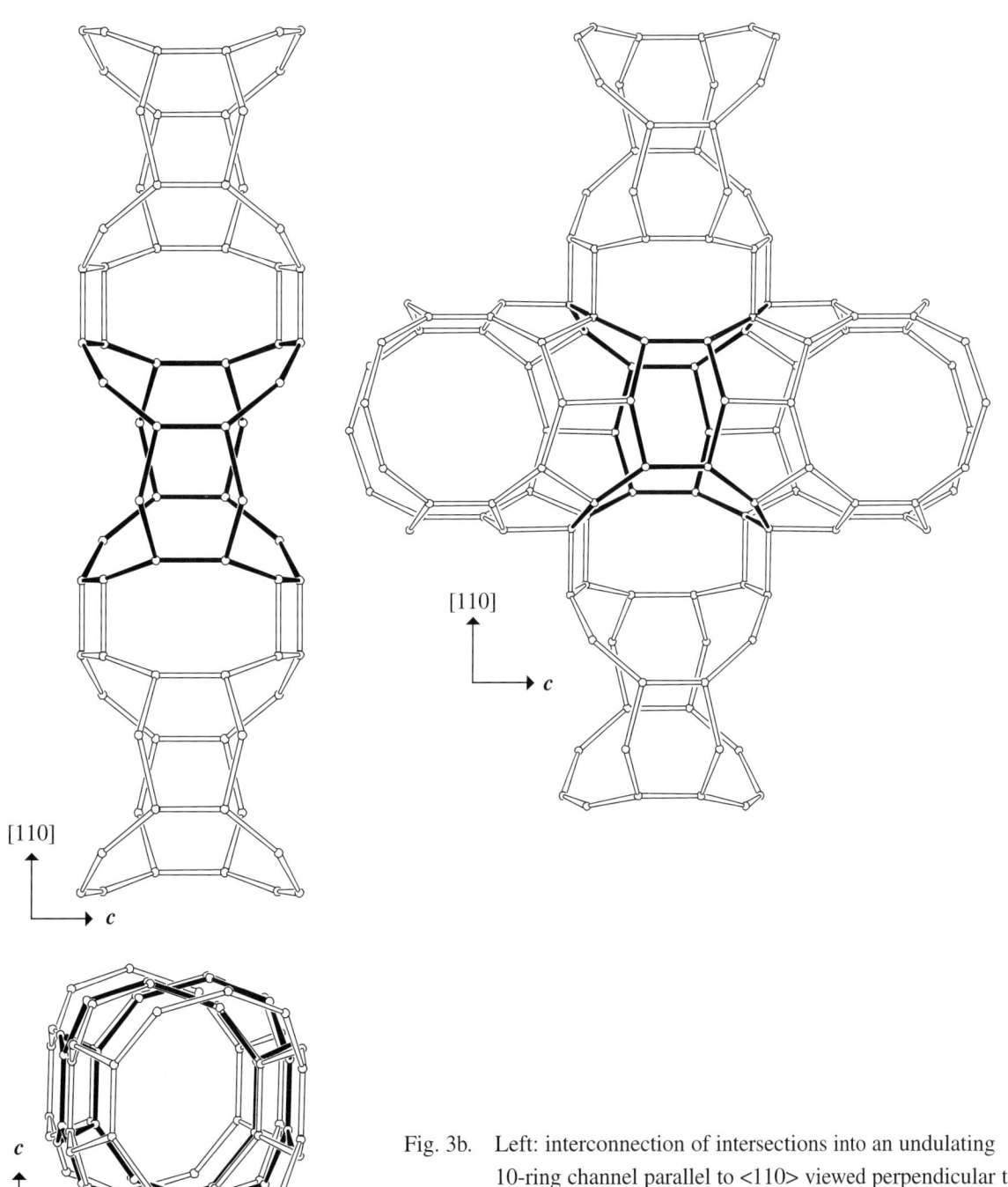

Fig. 3b.   Left: interconnection of intersections into an undulating 10-ring channel parallel to <110> viewed perpendicular to the channel axis (top) and along the channel axis (bottom). Right: intersecting channels viewed along [–110].

# MSO                    Building Scheme

## 1. Periodic Building Unit

Trigonal **MSO** can be built using units of 30 T atoms (Figure 1). The T30-unit consists of two double 6-rings sharing 6-rings with a [$6^8$]-cage. The two-dimensional PerBU equals the hexagonal array of non-connected T30-units (Figure 1). The T30-units, related along *a* and *b* by pure translations, are centered at (0,0) in the *ab* layer. This position is usually called the **A** position.

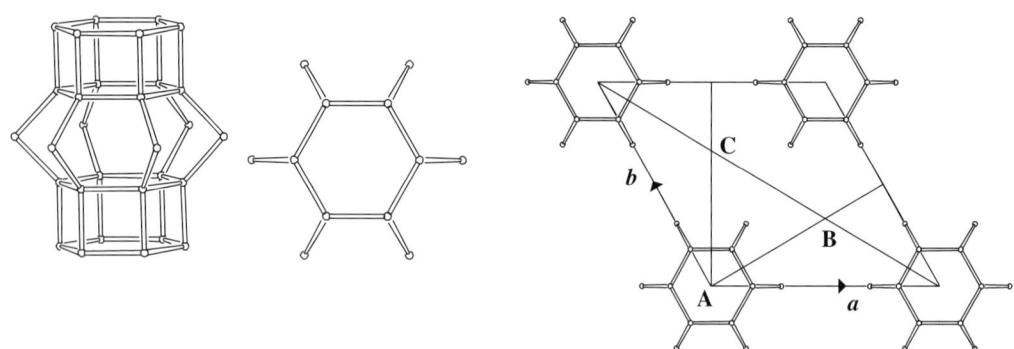

Fig. 1.    Building unit viewed perpendicular to *c* (left) and down *c* (middle) and PerBU of non-connected T30-units viewed along *c* (right).

## 2. Connection mode

Neighboring PerBUs, related along *c* by a shift of $+(\frac{2}{3}\boldsymbol{a}+\frac{1}{3}\boldsymbol{b})$, are connected along *c* through 6-rings (see Figure 2). The definition of the positions of the T30-units in neighboring PerBUs is indicated (compare this connection mode with mode (**1**) in the ABC-6 family; see Introduction).

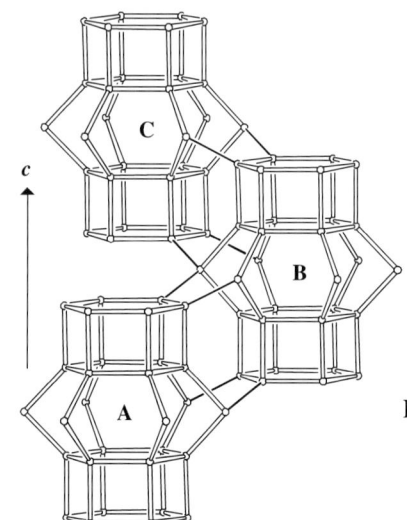

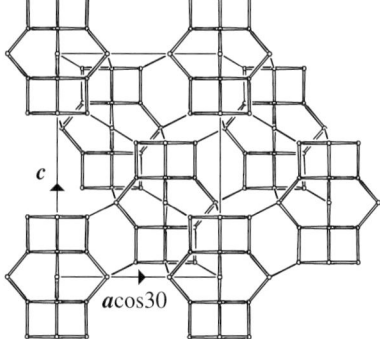

Fig. 2.    Connection mode (left) and unit cell content (right) viewed along *b*. The stacking sequence is given. In the perspective drawing, each PerBU is represented by one T30-unit only.

### 3. Channels and/or cages

The two types of cages are depicted in Figure 3. Cages are connected through (common) 4- and 6-rings. Apertures are formed by 6-rings only.

$\{0\ [4^{12}6^{8}]\}$

$\{0\ [4^{6}6^{20}]\}$

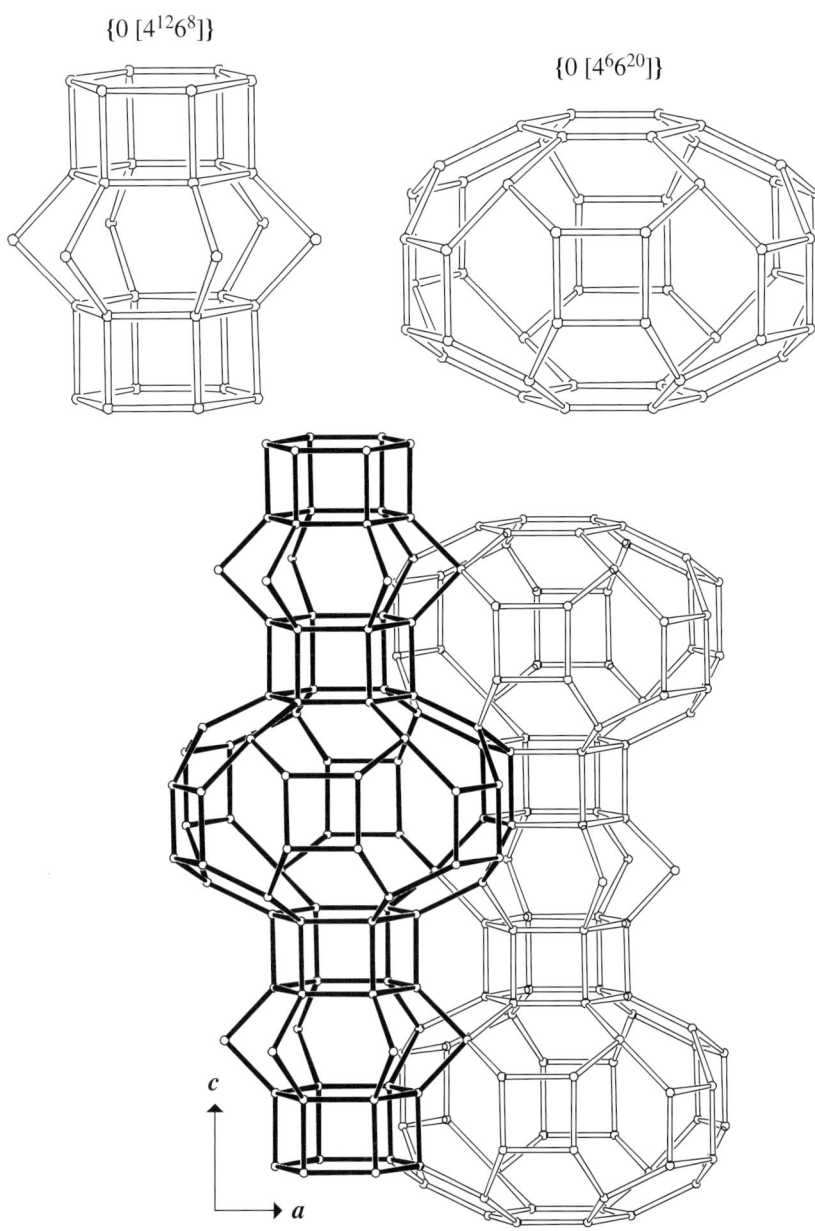

Fig. 3.   Top: cages viewed along <120>. One of the cages equals the T30 building unit.

Bottom: connected cages viewed along <120>.

# MTF                    Building Scheme

## 1. Periodic Building Unit

**MTF** can be built using units of 11 T atoms (5-5 = 1 units; one in bold in Figure 1). T11-units, related along $c$ by a pure translation, are connected into chains along $c$. Two chains, related by a rotation of about 90° about the chain axis, are connected through 4-rings into a "double-chain". Double-chains, related along $b$ by a shift of $\frac{1}{2}(a+b)$, are connected through 6-rings into the undulating $bc$ layer depicted in Figure 1. The two-dimensional PerBU is equal to the $bc$ layer.

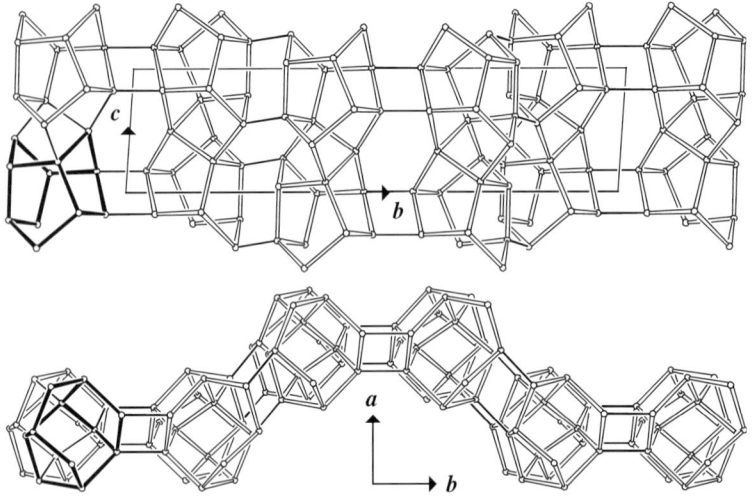

Fig. 1.  PerBU viewed around an axis about 15° inclined with respect to $a$ and $b$ (top) and along $c$ (bottom).

## 2. Connection mode

Neighboring PerBUs, related along $a$ by a pure translation, are connected along $a$ as illustrated in Figure 2. 8-Rings are formed.

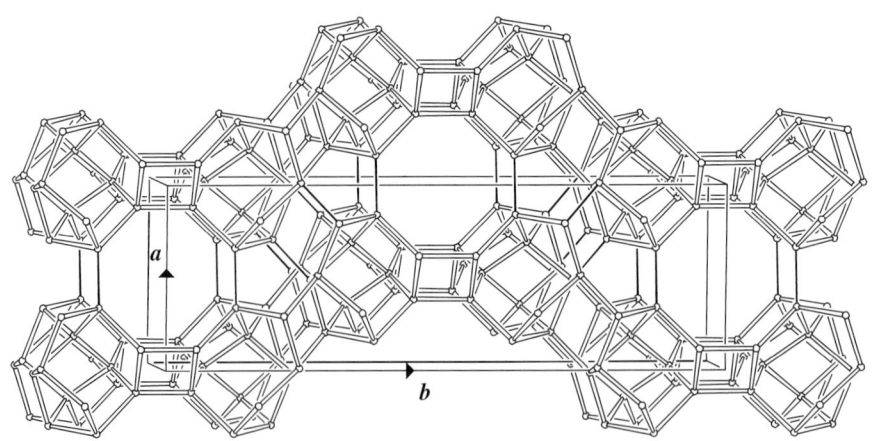

Fig. 2.  Connection mode and unit cell content viewed along $c$.

## 3. Channels and/or cages

The cavity (equal to the repeat unit in the channel) is shown in Figure 3 together with the non-interconnecting 8-ring channels, parallel to *c*, obtained by connecting the cavities through common 8-rings.

$$\{1 \ [4^2 5^8 6^2 8^4] \ [001] \ (\text{8-ring})\}$$

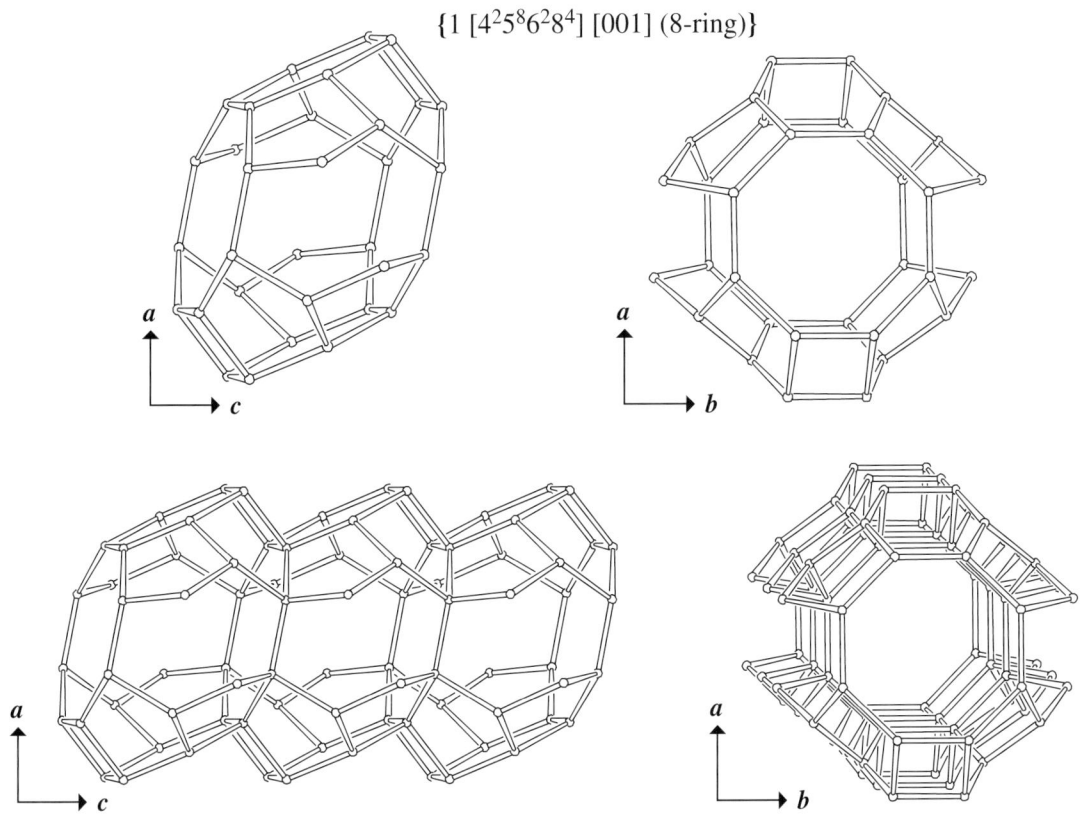

Fig. 3.    Top: cavity viewed along *b* (left) and along *c* (right). Bottom: connection of cavities along *c* into 8-ring channels viewed perpendicular to the channel axis (left) and along the channel axis (right).

# MTN                           Building Scheme

## 1. Periodic Building Unit

In cubic **MTN**, units of 30 T atoms (two 12-rings and a 6-ring) are connected into hexagonal layers perpendicular to [111]. The two-dimensional PerBU is obtained when the "empty" spaces between the T30-units in a hexagonal layer are filled with T2-dimers. [$5^{12}$]-Cages are generated (Figure 1). The T30-units are centered at (0,0) in the hexagonal **ab** layer. This position is usually called the **A** position. At sites **B** and **C**, [$5^{12}$]-cages share faces. The repeat unit of the PerBU consists of 34 T atoms: the T30-unit and two "space-filling" dimers. (See **DDR**, **DOH** and **MEP**.)

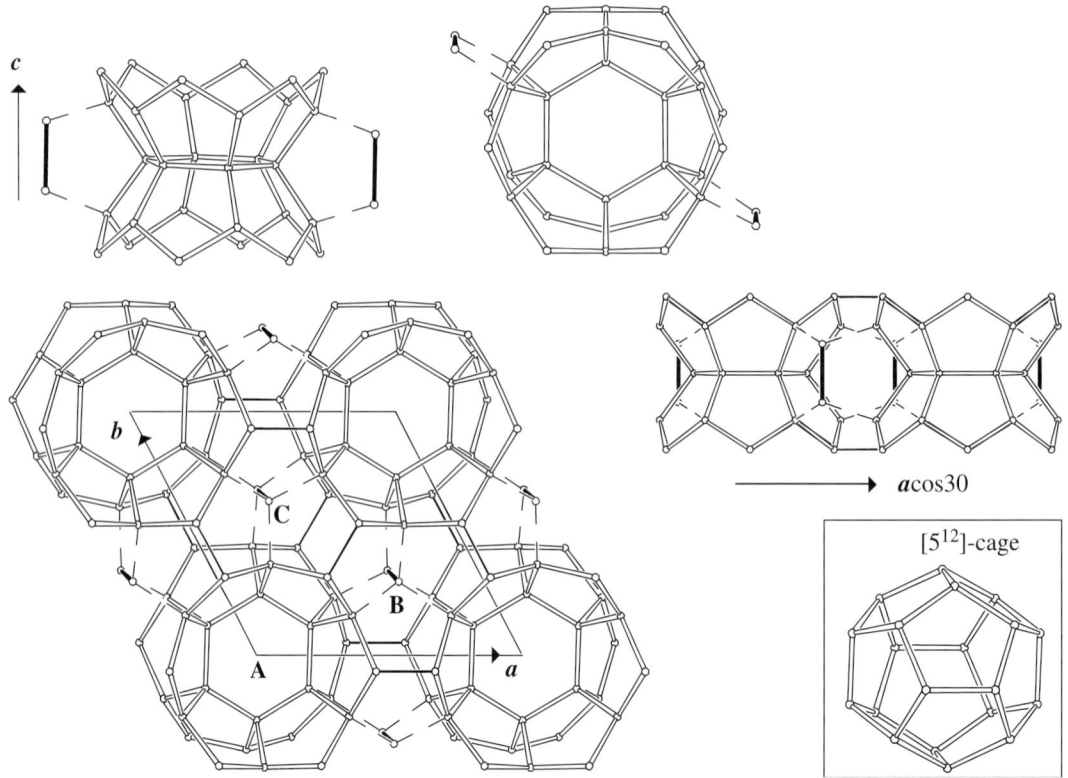

Fig. 1.    T34-unit (top) viewed along [110] (left) and along **c** (right). Hexagonal PerBU viewed along **c** (bottom left) and along **b** (top right). Connections to the space-filling dimers (in bold) are dashed. The inset shows the [$5^{12}$]-cage.

## 2. Connection mode

Neighboring PerBUs, related along **c** by a shift of $+(\frac{2}{3}\boldsymbol{a} + \frac{1}{3}\boldsymbol{b})$, are connected along **c** through 6-rings (see Figure 2 on next page) (compare this connection mode with mode (**1**) in the ABC-6 family, described in the Introduction).

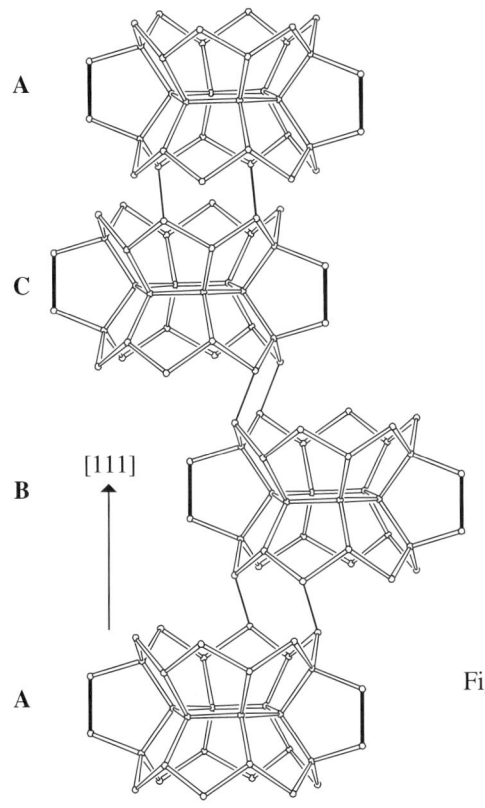

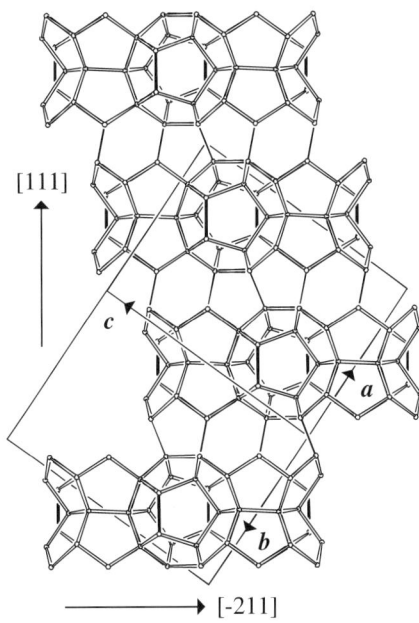

Fig. 2. Connection mode viewed perpendicular to the cubic [111] axis (left) and unit cell content projected along the hexagonal **b** axis (right). The stacking sequence is given. In the perspective drawing, each PerBU is represented by one T34-unit only.

## 3. Channels and/or cages

Two types of inter-layer cages are formed. The $[5^{12}]$-cage is also present in **DDR**, **DOH** and **MEP**. Apertures of "channels" are formed by 6-rings only. Cages and their linkage are shown in Figure 3.

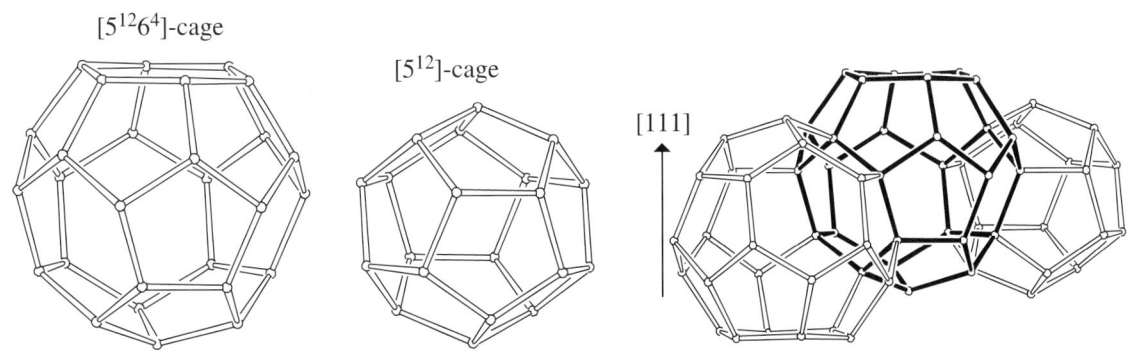

Fig. 3. Inter-layer cages (left and middle) and their linkage (right) viewed perpendicular to [111].

# MTT                              Building Scheme

## 1. Periodic Building Unit

**MTT** can be built using the zigzag chain (one in bold in Figure 1, left). Six zigzag chains form an infinite building unit (Figure 1, left). The repeat unit of this infinite building unit consists of two 5-1 units (bold in Figure 1, middle). The two-dimensional PerBU is obtained when infinite building units, related along $c$ by a pure translation, are connected into the layer shown in Figure 1 (right). (See also **TON**.)

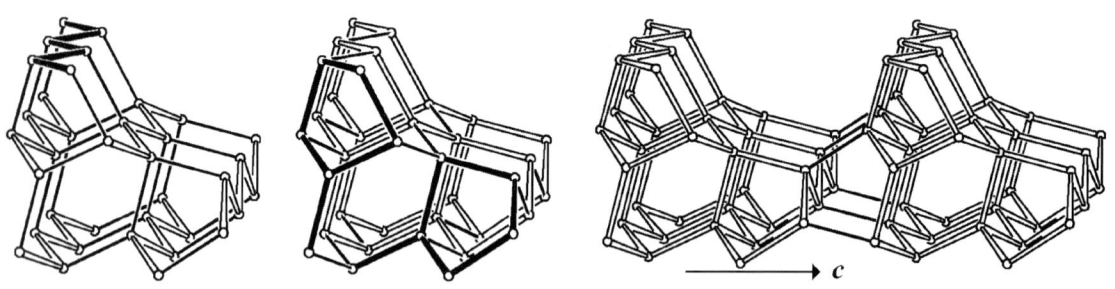

Fig. 1.    Infinite building unit constructed from six zigzag chains (left) and from 5-1 units (middle) and PerBU obtained when infinite building units are connected along $c$ (right).

## 2. Connection mode

Neighboring PerBUs, related along $b$ by a screw rotation of 180° about $b$, are connected along $b$ through (fused) 5- and 6-rings as shown in Figure 2.

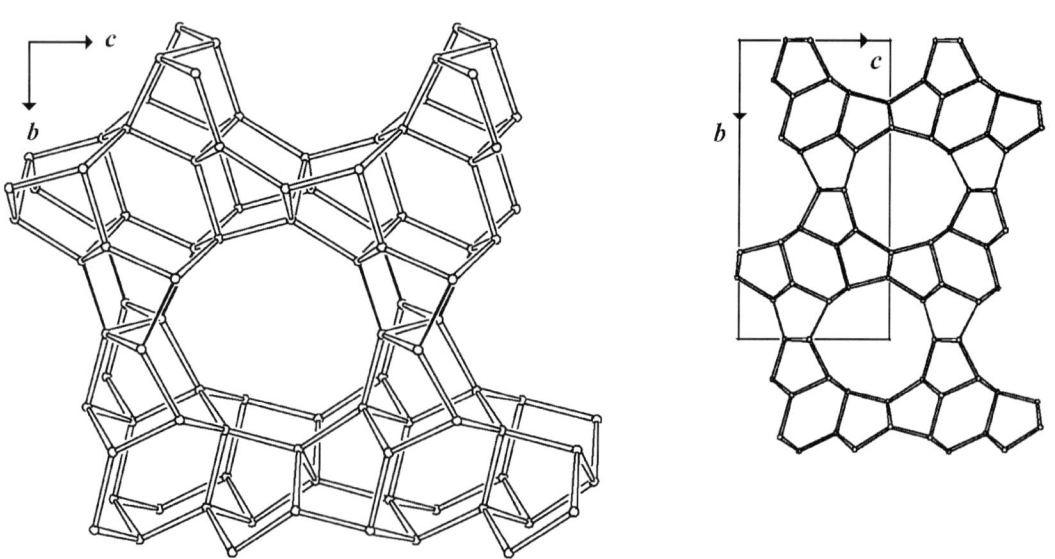

Fig. 2.    Connection mode (left) and unit cell content (right) viewed along $a$. Only two repeat units along $a$ of each PerBU are drawn for clarity.

## 3. Channels and/or cages

One-dimensional non-interconnecting 10-ring channels are parallel to *a*. The channel, depicted in Figure 3, is topologically equivalent to the channel in **TON**.

$\{1\ [6^8 10^{2/2}]\ [001]\ (10\text{-ring})\}$

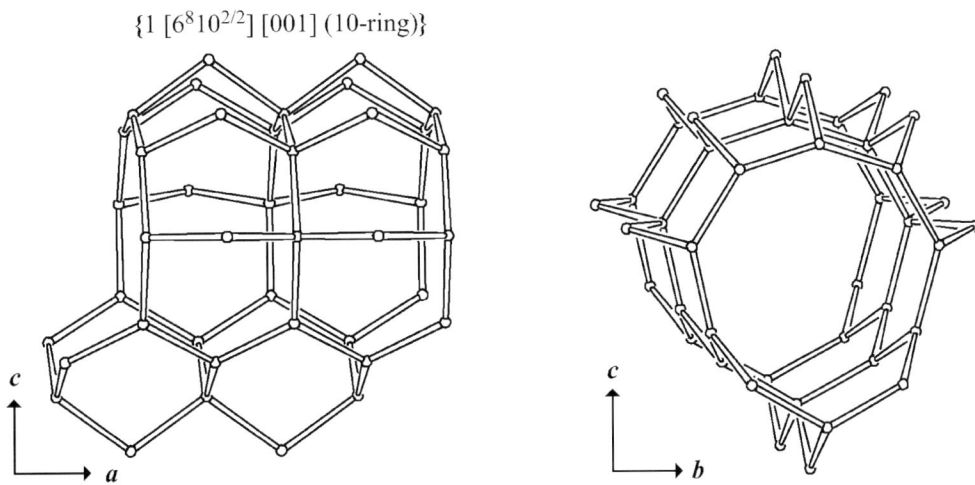

Fig. 3.    Channel viewed perpendicular to the channel axis (left) and along the channel axis (right).

# MTW Building Scheme

## 1. Periodic Building Unit

**MTW** can be built using the zigzag chain (one in bold in Figure 1, left). Seven zigzag chains form an infinite building unit (Figure 1, left). The repeat unit of this infinite building unit consists of 5-1 and 5-3 units (bold in Figure 1, middle). The two-dimensional PerBU is obtained when infinite building units, related by a pure translation along $c$, are connected into the layer shown in Figure 1.

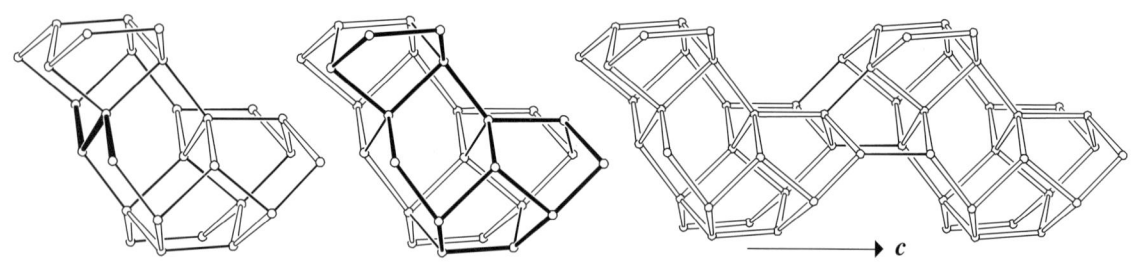

Fig. 1. Infinite building unit constructed from seven zigzag chains (left) and from 5-1 and 5-2 units (middle) and PerBU obtained when infinite building units are connected along $c$ (right).

## 2. Connection mode

Neighboring PerBUs, related along $a$ by a shift of $\frac{1}{2}(a+b)$, are connected along $a$ through 4-rings as depicted in Figure 2.

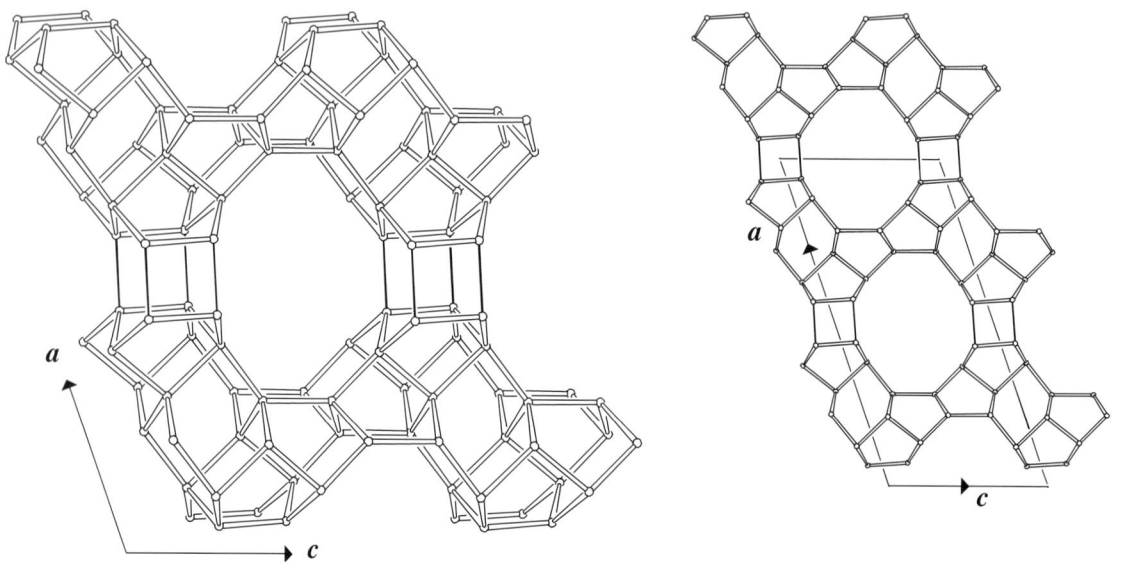

Fig. 2. Connection mode (left) viewed along $b$ and parallel projection of the unit cell content along $b$ (right).

# Cage/Channel

## 3. Channels and/or cages

Non-interconnecting 12-ring channels are parallel to **b**. The channel, depicted in Figure 3, is topologically equivalent to the channel in **GON**.

$\{1\ [6^8 12^{2/2}]\ [010]\ (12\text{-ring})\}$

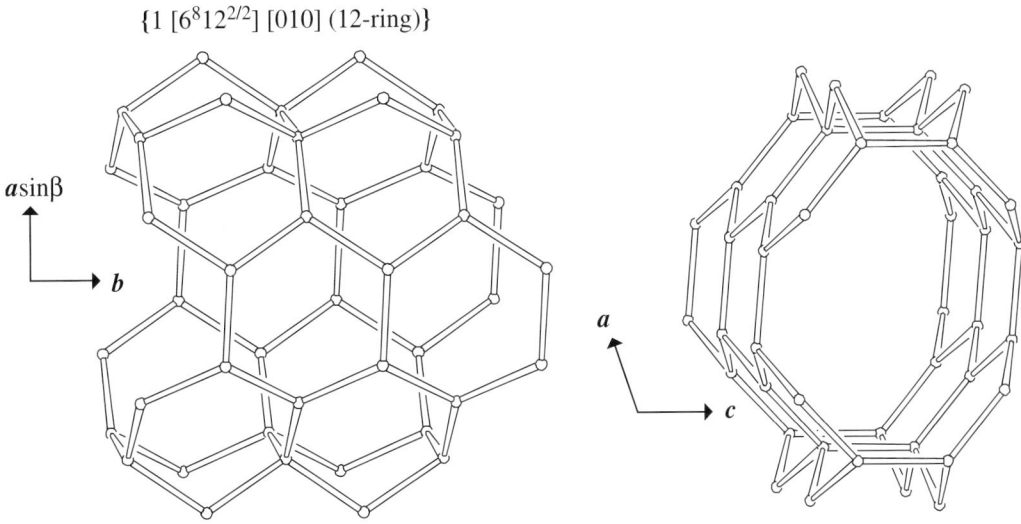

Fig. 3.   Channel viewed perpendicular to the channel axis (left) and along the channel axis (right).

# MWW                    Building Scheme

## 1. Periodic Building Unit

Hexagonal **MWW** can be built using units of 36 T atoms. The T36-unit resembles a "half-cage" (or 12-ring cup) consisting of cylindrical 6-ring band (to which two T2-dimers and two T atoms are bonded) closed at one side by a 6-ring. The one-dimensional PerBU is obtained when T36-units, related by a mirror plane perpendicular to *c*, are connected into columns along *c* through double 6-rings and single T–T bonds (Figure 1).

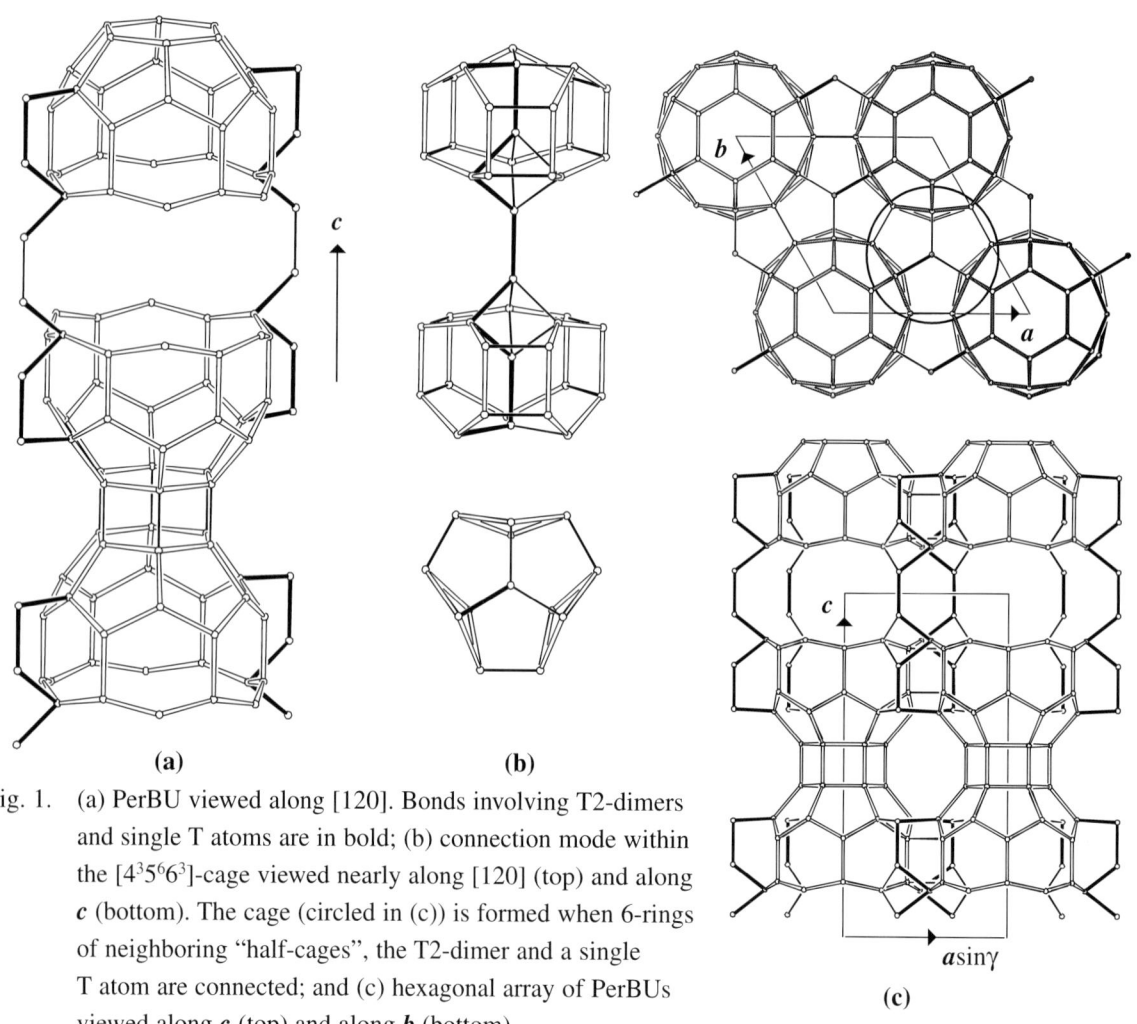

**(a)**                    **(b)**

Fig. 1.   (a) PerBU viewed along [120]. Bonds involving T2-dimers and single T atoms are in bold; (b) connection mode within the $[4^3 5^6 6^3]$-cage viewed nearly along [120] (top) and along *c* (bottom). The cage (circled in (c)) is formed when 6-rings of neighboring "half-cages", the T2-dimer and a single T atom are connected; and (c) hexagonal array of PerBUs viewed along *c* (top) and along *b* (bottom).

**(c)**

## 2. Connection mode

PerBUs, related along *a* and *b* by pure translations, are connected into the hexagonal packing shown in Figure 1c. The hexagonal packing of the PerBUs generates a double layer that contains a new cage, the $[4^3 5^6 6^3]$-cage, consisting of three fused $[4^1 5^2 6^2]$-cages.

## 3. Channels and/or cages

Two types of non-interconnecting 10-ring channels are parallel to <100>. The channel intersection of Type 1 channels (the 10-ring channels within the double layer) is equal to cavity **1** and the channel intersection of Type 2 channels (the two-dimensional 10-ring channel systems between the double layers) is equal to the cavity 2. Both cavities are depicted in Figure 2 together with the linkage of the cavities and cage 1 into channels. The $[4^35^66^3]$-cage is shown in Figure 1b.

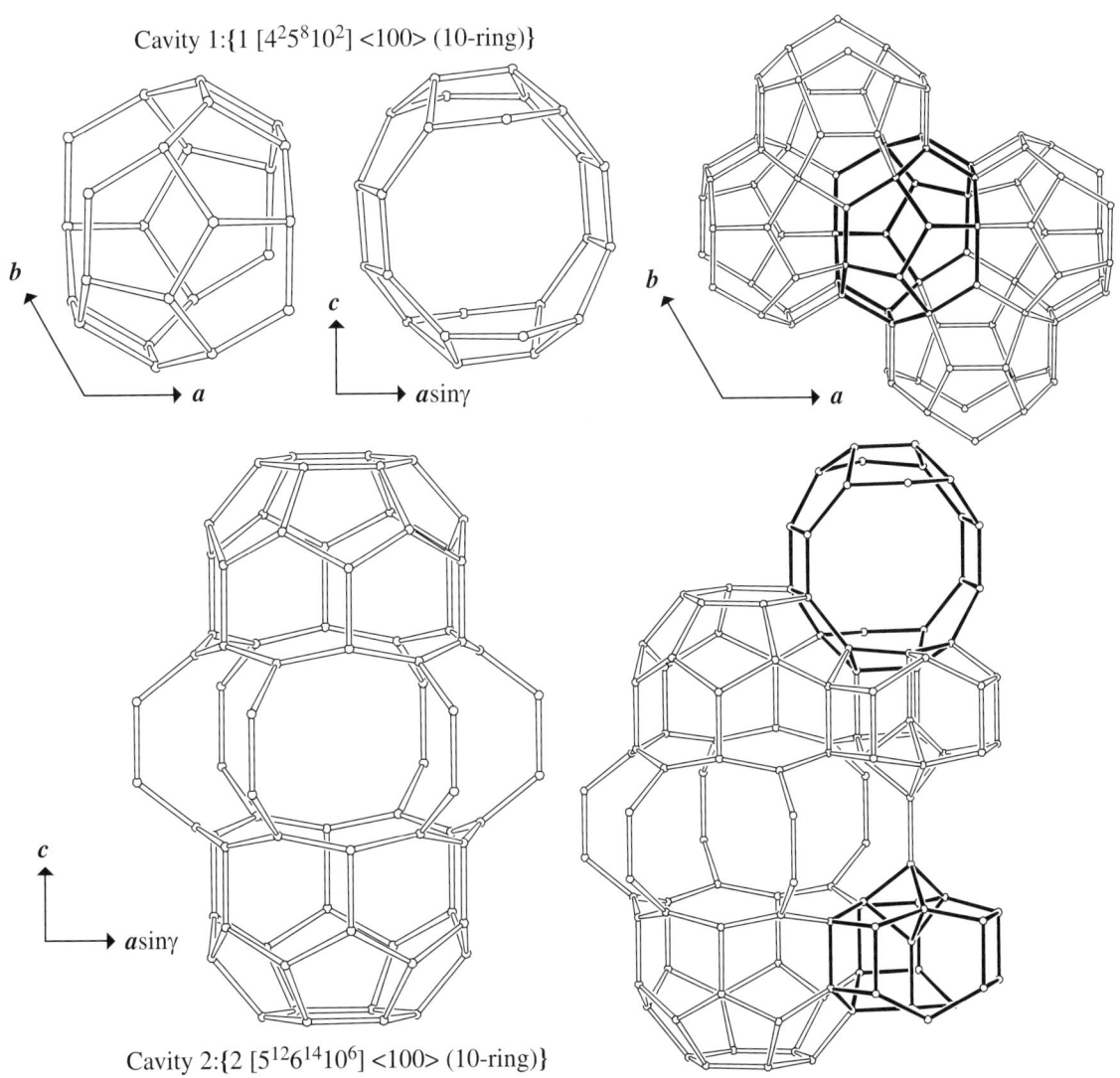

Cavity 1:{1 $[4^25^810^2]$ <100> (10-ring)}

Cavity 2:{2 $[5^{12}6^{14}10^6]$ <100> (10-ring)}

Fig. 2.    Top left: cavity 1 within the double layer viewed along **c** (left) and along <010> (right). Top right: 10-ring channels parallel to <100> viewed along **c**. Bottom left: cavity 2 between double layers viewed along <010>. Bottom right: fusion of cavities 1 (in bold) and 2 and cage 1 (one in bold).

# NAB                      Building Scheme

## 1. Periodic Building Unit

Tetragonal **NAB** can be built using units of 5 T atoms: a 4-ring connected to a single T atom (bold in Figure 1). The two-dimensional PerBU is obtained when T5-units, related along **a** and **b** by pure translations, are connected into the **ab** layer depicted in Figure 1.

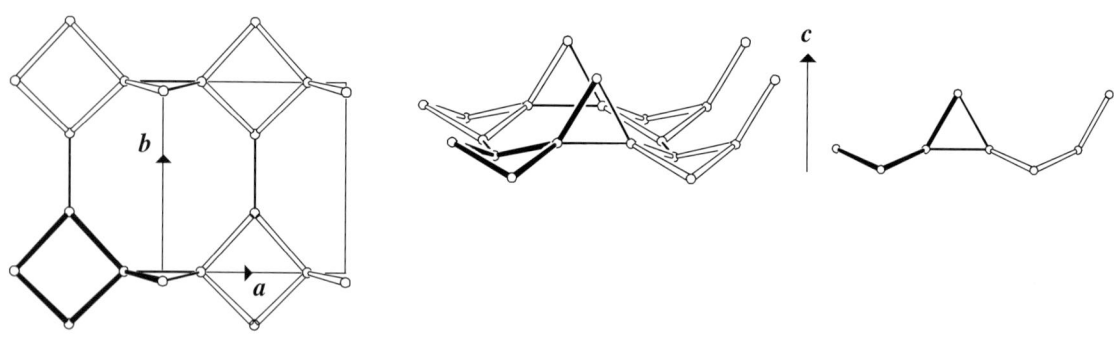

Fig. 1.   PerBU viewed along **c** (left) and in perspective view and in projection along **b** (middle and right).

## 2. Connection mode

Neighboring PerBUs, related along **c** by a screw rotation of 180° about **c** followed by a shift of $\frac{1}{2}(a+b)$ are connected along **c** through 3-rings as illustrated in Figure 2.

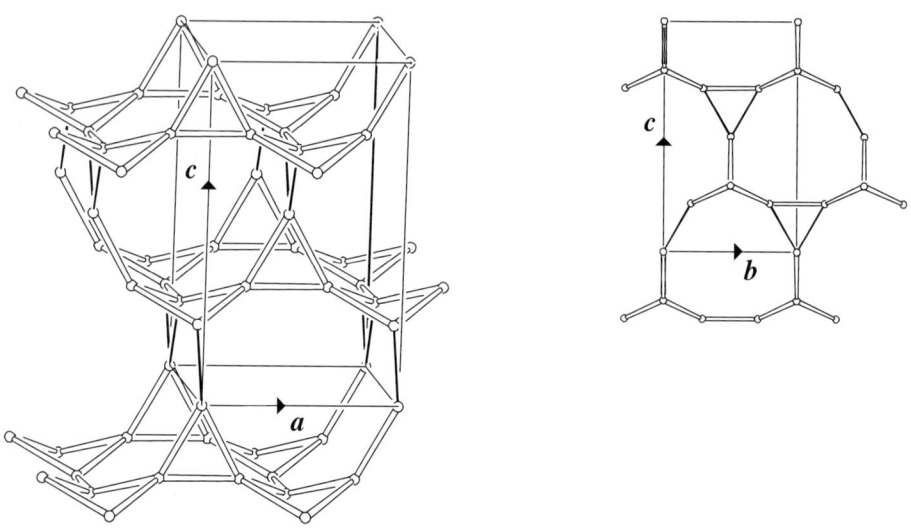

Fig. 2.   Connection mode (and unit cell content) viewed along **b** (left) and unit cell content projected along **a** (right).

## 3. Channels and/or cages

Intersecting 9-ring channels are parallel to <100>. The 9-ring channels (Figure 3) are topologically equivalent to those in **LOV**, **RSN** and **VSV**. The channels are interconnected along $c$ through 8-rings perpendicular to $c$ as depicted in Figure 3.

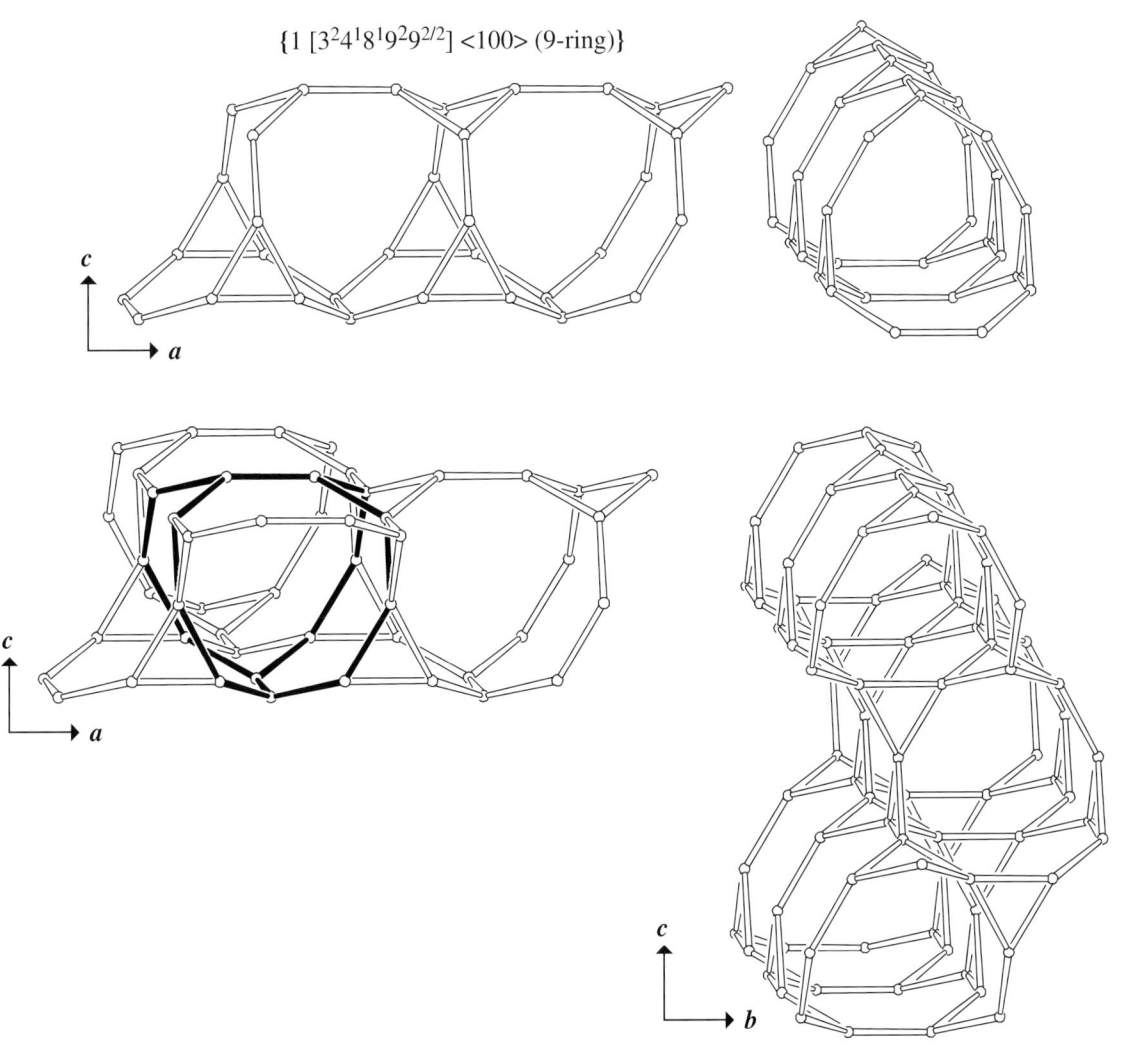

$\{1\ [3^2 4^1 8^1 9^2 9^{2/2}] <100>\ (9\text{-ring})\}$

Fig. 3.   Top: 9-ring channel viewed perpendicular to the channel axis (left) and along the channel axis (right). Bottom left: intersecting 9-ring channels have $[9^4]$-cavities (in bold) in common. Bottom right: channels are interconnecting through 8-rings perpendicular to $c$.

# NAT  Building Scheme

## 1. Periodic Building Unit

Tetragonal **NAT** can be built using the fibrous chain (or natrolite-chain) as one-dimensional PerBU. The chain is composed of units of 5 T atoms (bold in Figure 1). These T5-units ($[4^3]$-cages, or $4=1$ units) are related along *c* by pure translations. (See also **EDI** and **THO**.)

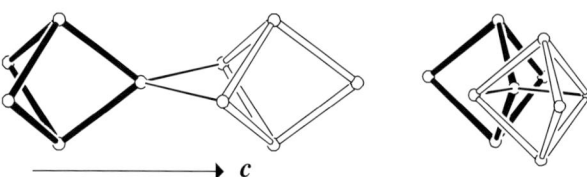

Fig. 1.   PerBU viewed perpendicular to the fibrous chain axis (left), and down the chain axis (right).

## 2. Connection mode

Neighboring PerBUs, related along $a(b)$ by a rotation of 90° about the chain axis followed by a shift of $\frac{1}{2}a$ ($\frac{1}{2}b$), are connected along $a(b)$ as shown in Figure 2.

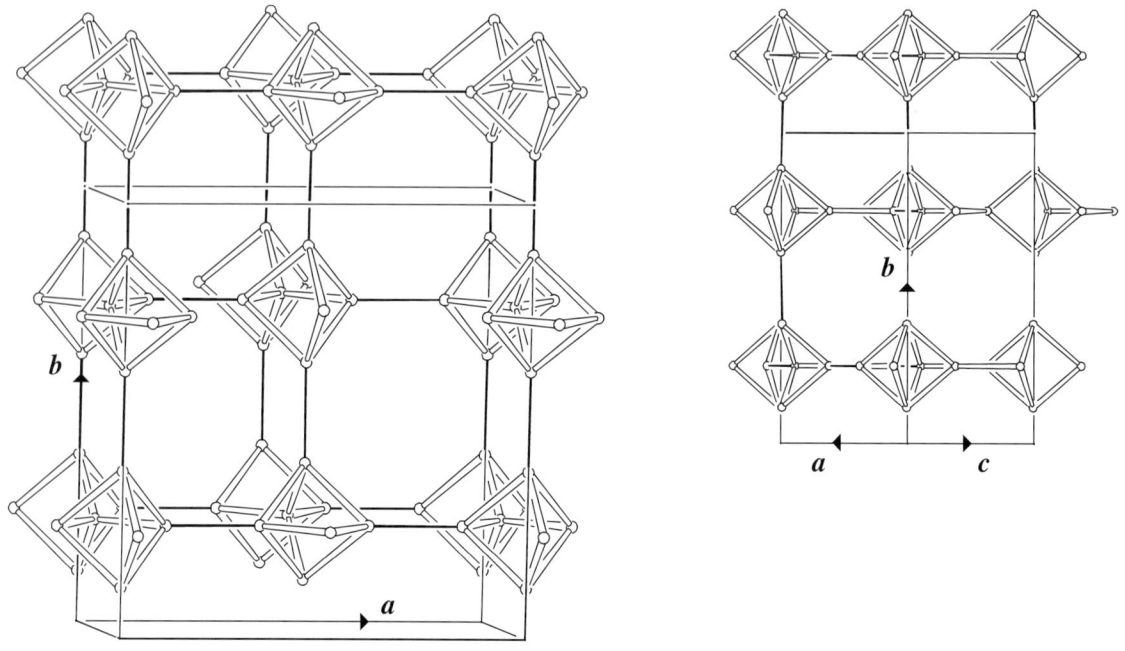

Fig. 2.   Connection mode (and unit cell content) viewed along the fibrous chain axis *c* (left) and unit cell content projected along <101> (right).

## 3. Channels and/or cages

Intersecting 8- and 9-ring channels (Figure 3) are parallel to <101> and $c$, respectively.

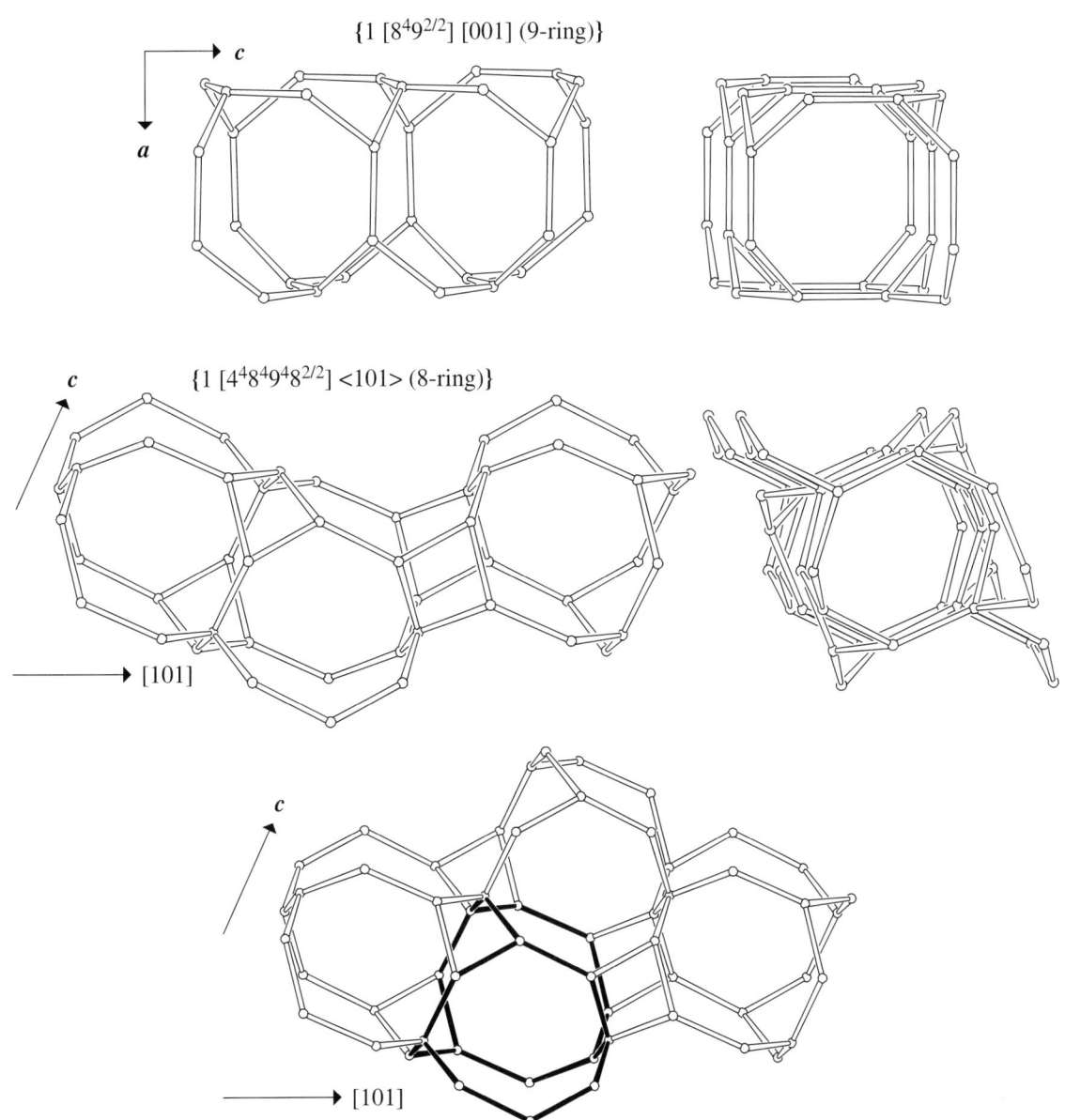

{1 [$8^49^{2/2}$] [001] (9-ring)}

{1 [$4^48^49^48^{2/2}$] <101> (8-ring)}

Fig. 3.   9-Ring channel parallel to $c$ (top) and 8-ring channel parallel to <101> (middle) viewed perpendicular to the channel axis (left) and along the channel axis (right). Bottom: intersecting 9-ring channels have [$8^49^2$]-cavities in common. One intersection in bold.

# NES

## Building Scheme

### 1. Periodic Building Unit

**NES** can be built using building units composed of 17 T atoms: two 5-1 units and a 5-ring (Figure 1). The two-dimensional PerBU is obtained when T17-units, related along *b* by a screw rotation of 180° about *b* and related along *c* by a screw rotation of 180° about *c*, are connected into the *bc* layer shown in Figure 1. (Compare with PerBUs in **EUO**, **IWV** and **NON**.)

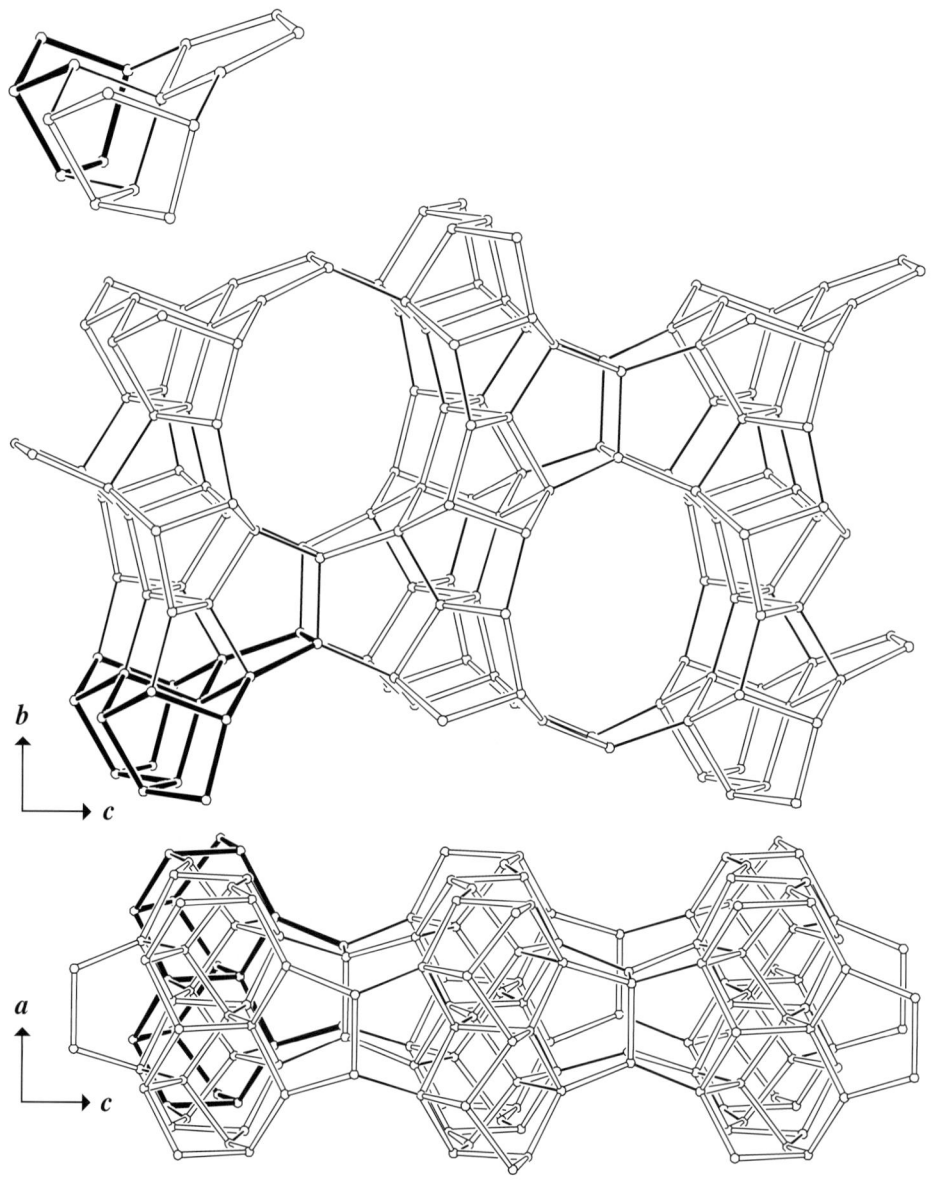

Fig. 1. Finite building unit (top) viewed along *a* and PerBU viewed along *a* (middle), and along *b* (bottom; one T17-unit in bold).

# Building Scheme

## 2. Connection mode

Neighboring PerBUs, related along $a$ by a shift of $\frac{1}{2}c$ (or $\frac{1}{2}b$), are connected along $a$ as shown in Figure 2.

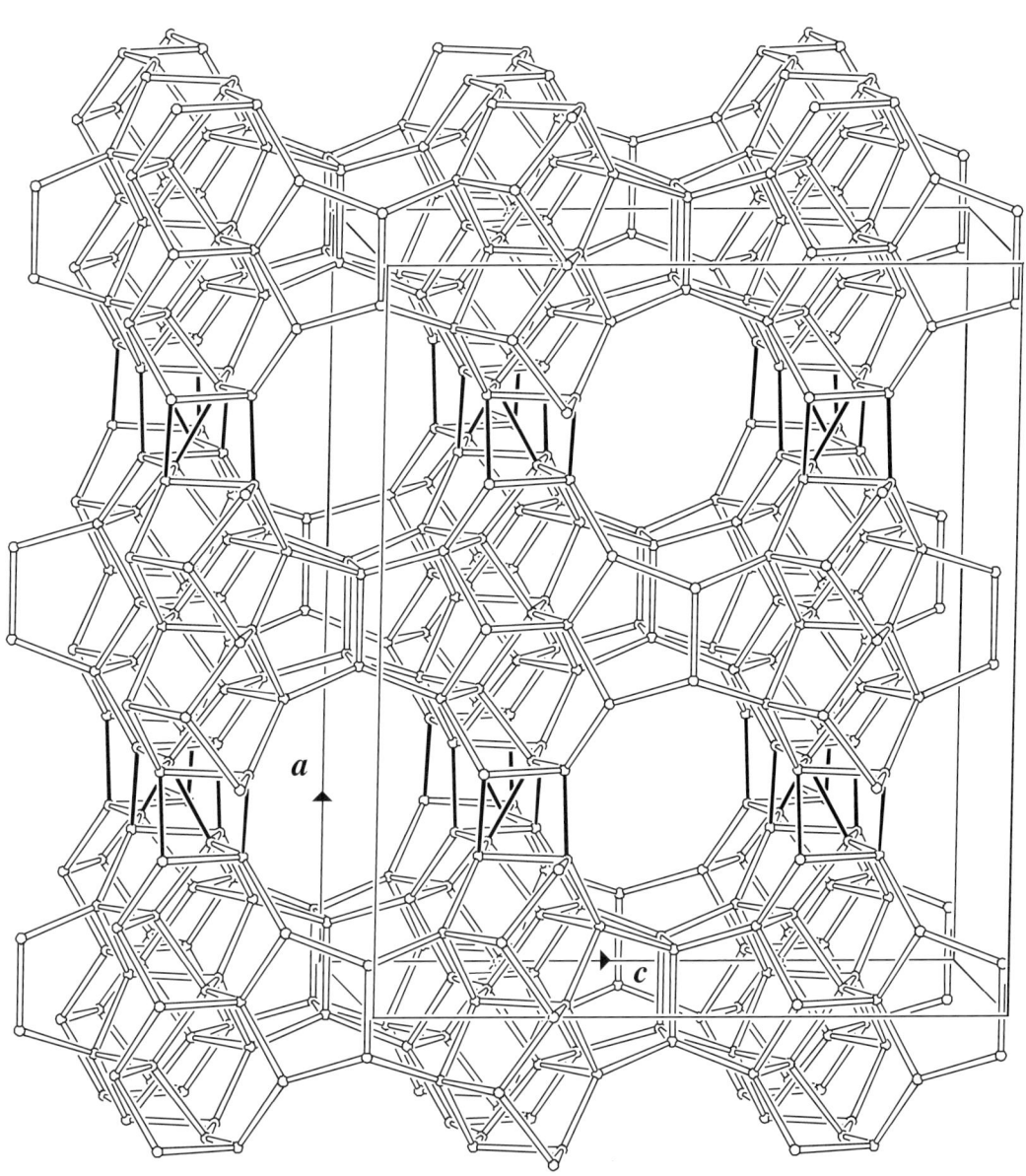

Fig. 2a. Connection mode (and unit cell content) viewed along $b$.

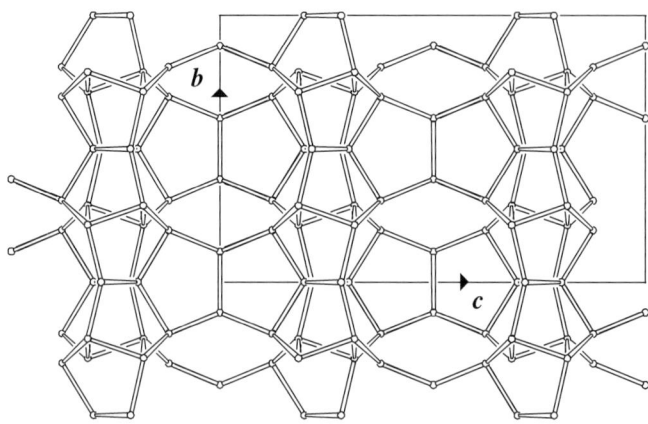

Fig. 2b.　Unit cell content projected along *a*.

## 3. Channels and/or cages

Double-cavities are connected through common 10-rings into 10-ring channels parallel to *b* as illustrated in Figure 3. The 10-ring channels are interconnecting through 12-rings in the double-cavity.

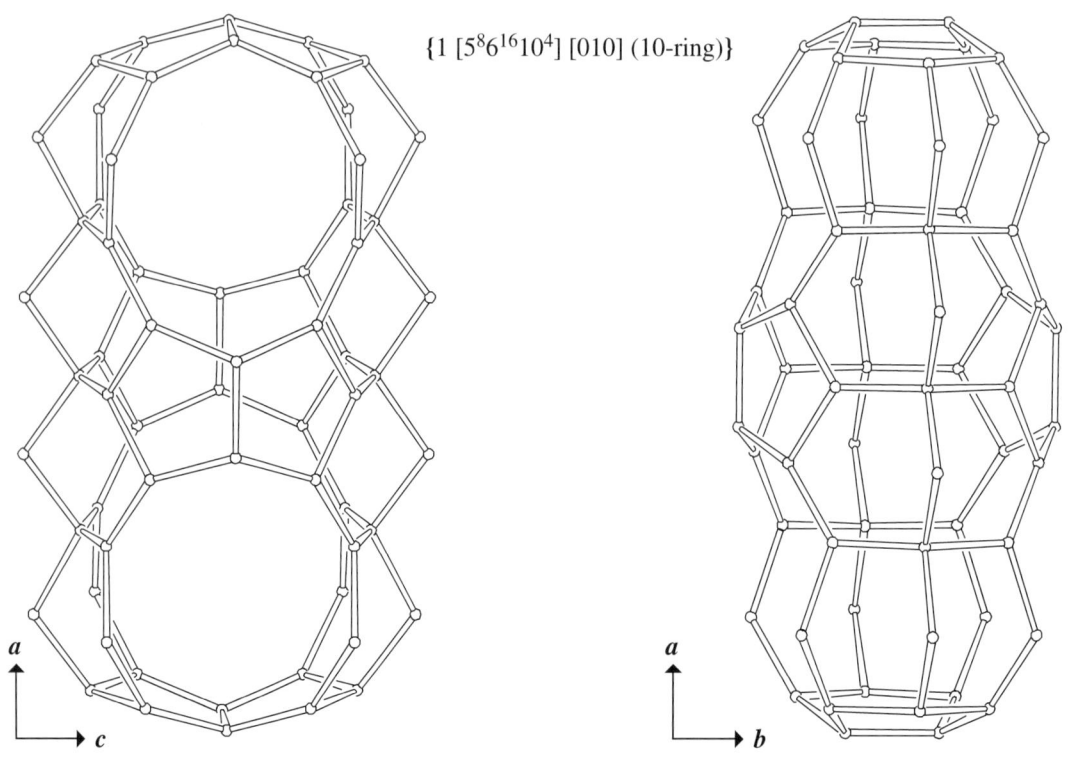

{1 [$5^8 6^{16} 10^4$] [010] (10-ring)}

Fig. 3a.　Double-cavity viewed along *b* (left) and along *c* (right).

# Cage/Channel

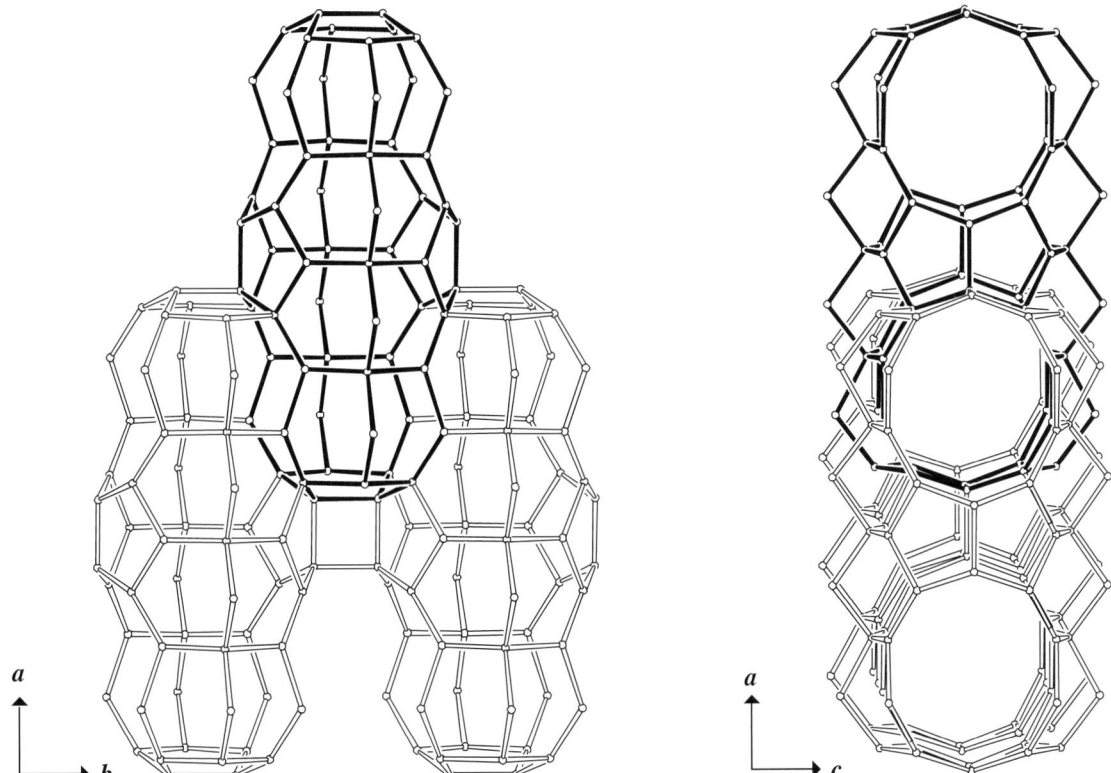

Fig. 3b. Fused double-cavities, viewed along *c* (left) and along *b* (right), form 10-ring channels along *b*. Pairs of 10-ring channels are interconnecting along *a* through 12-rings in the cavity.

# NON

## Building Scheme

### 1. Periodic Building Unit

**NON** can be built using building units composed of 11 T atoms: a 5-ring and a 6-ring (Figure 1). The two-dimensional PerBU is obtained when T11-units, related along **c** by a screw rotation of 180° about **c** and related along **a** by a screw rotation of 180° about **a**, are connected into the **ac** layer shown in Figure 1. (Compare with the PerBUs in **EUO**, **IWV** and **NES**.)

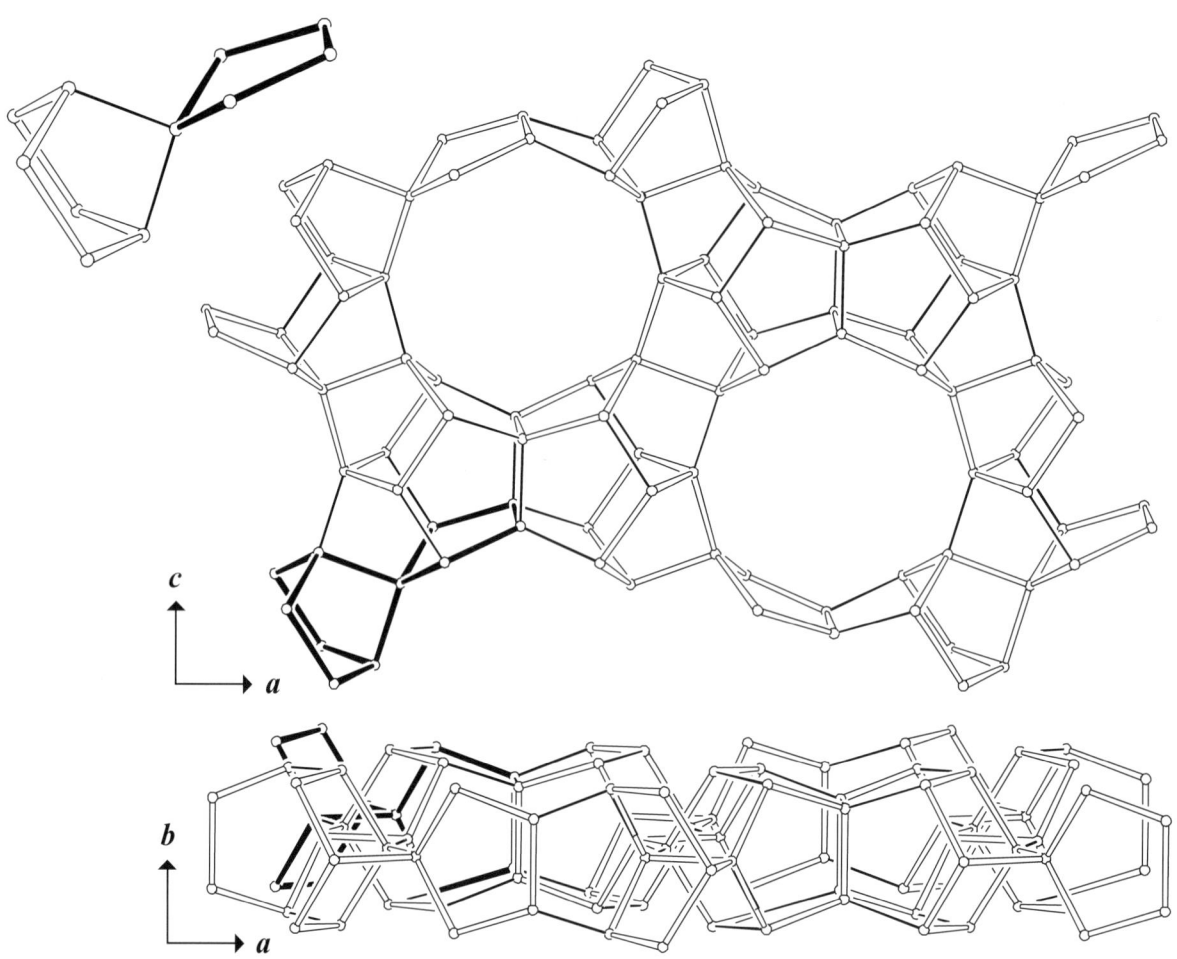

Fig. 1.    Finite building unit viewed along **b** (top left) and PerBU viewed along **b** (middle) and along **c** (bottom). One T11-unit in bold.

### 2. Connection mode

Neighboring PerBUs, related along **b** by a shift of $\frac{1}{2}$**a** (or $\frac{1}{2}$**c**), are connected along **b** as shown in Figure 2.

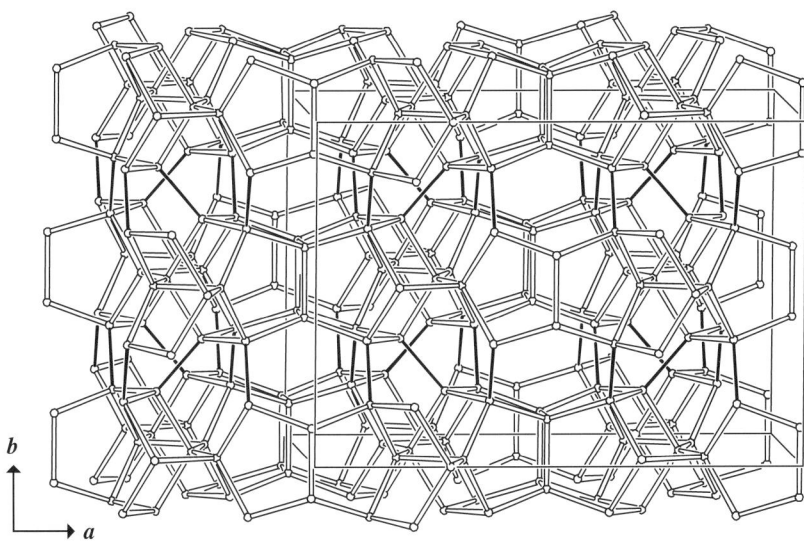

Fig. 2.  Connection mode (and unit cell content) viewed along **c**.

## 3.  Channels and/or cages

The cage (Figure 3) has apertures formed by 5- and 6-rings only.

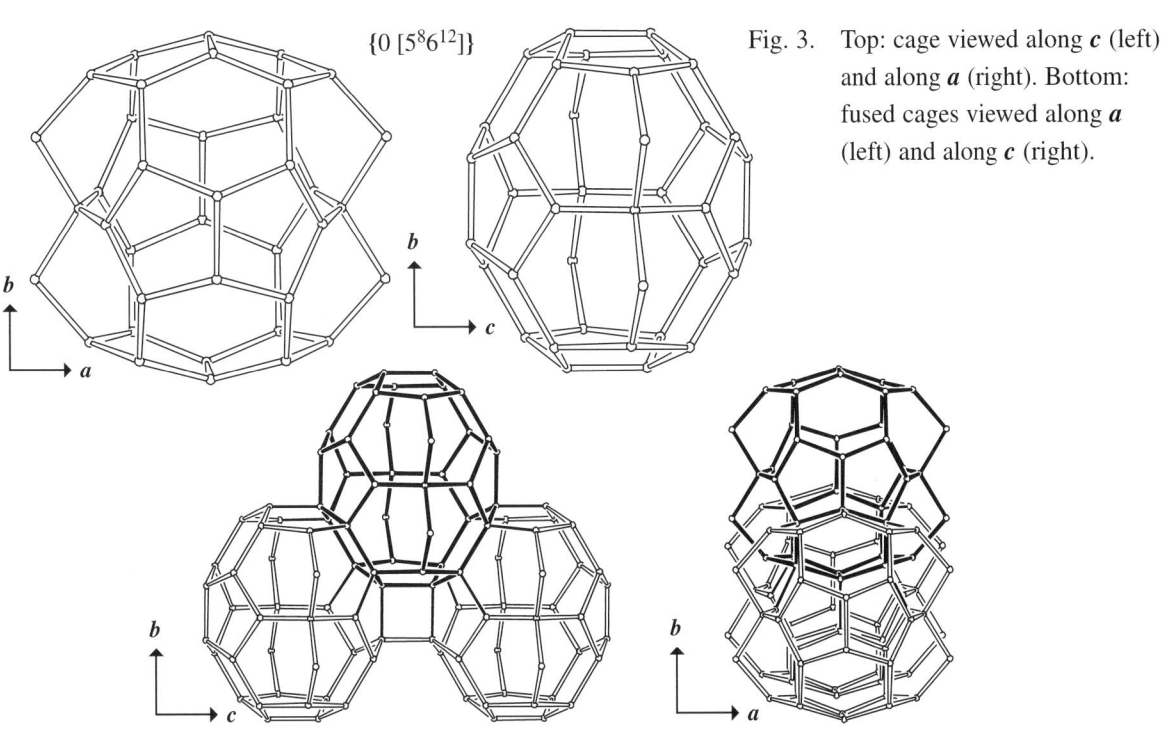

{0 [$5^8 6^{12}$]}

Fig. 3.  Top: cage viewed along **c** (left) and along **a** (right). Bottom: fused cages viewed along **a** (left) and along **c** (right).

# NPO — Building Scheme

## 1. Periodic Building Unit

Hexagonal **NPO** can be built using the zigzag chain (bold in Figure 1, left) parallel to $c$. The one-dimensional PerBU is obtained when three zigzag chains are connected into a column of $[3^26^3]$-cages sharing 3-rings. The repeat unit of the PerBU is composed of 6 T atoms (bold in Figure 1, right).

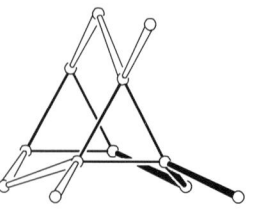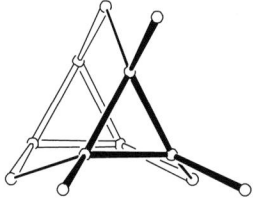

Fig. 1. PerBU constructed from three zigzag chains (left) and PerBU constructed from T6-units consisting of a 3-ring with three additional single T atoms (right) viewed along the chain axis $c$.

## 2. Connection mode

Neighboring PerBUs, related along $a$ and $b$ by pure translations, are connected into the three-dimensional framework type as illustrated in Figure 2.

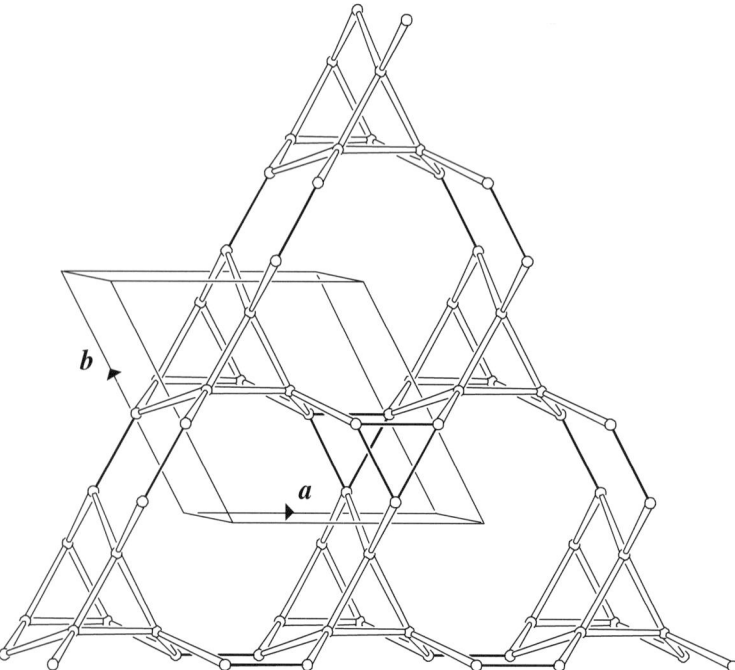

Fig. 2. Connection mode (and unit cell content) viewed along $c$.

## 3. Channels and/or cages

Non-interconnecting 12-ring channels are parallel to *c* (Figure 3). The channel is topologically equivalent to the 12-ring channels in **ATO** and **CAN**. The pore wall consists of fused 6-rings. The framework consists of fused 12-ring channels.

$$\{1\ [6^6 12^{2/2}]\ [001]\ (12\text{-ring})\}$$

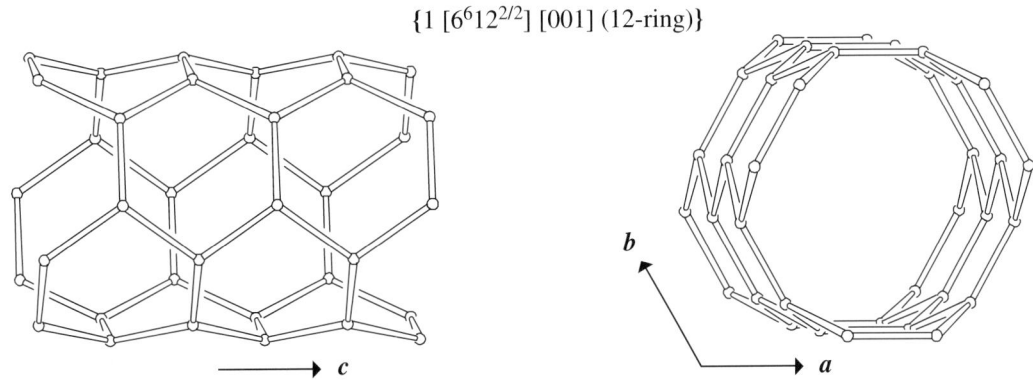

Fig. 3.  Channel viewed perpendicular to the channel axis (left) and along the channel axis (right).

# NSI         Building Scheme

## 1. Periodic Building Unit

**NSI** can be built using the zigzag chain (bold in Figure1) running parallel to **b**. Three zigzag chains are connected into an infinite building unit. The repeat unit of the infinite building unit consists of a 5-1 unit (bold in Figure 1). A two-dimensional PerBU is obtained when infinite building units, related along **a** by a shift of $\frac{1}{2}(a+b)$, are connected along **a** into the **ab** layer (Figure 1). (Compare with **BIK** and **CAS**.)

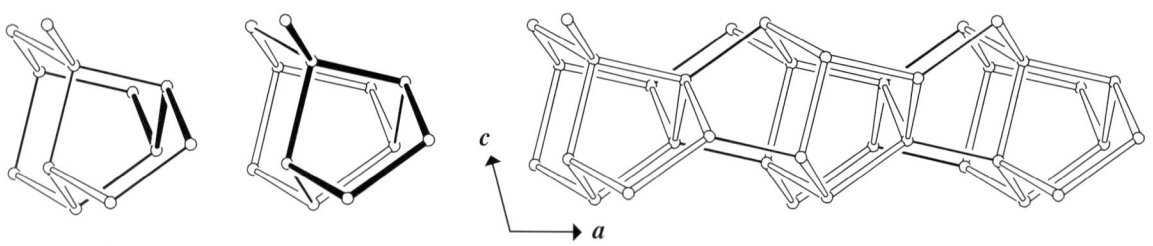

Fig. 1.    Infinite building unit, constructed from three zigzag chains (left) and from 5-1 units (middle) and PerBU (right) viewed along the chain axis **b**.

## 2. Connection mode

Neighboring PerBUs, related along **c** by a pure translation, are connected along **c** through 6-rings and 8-rings as illustrated in Figure 2.

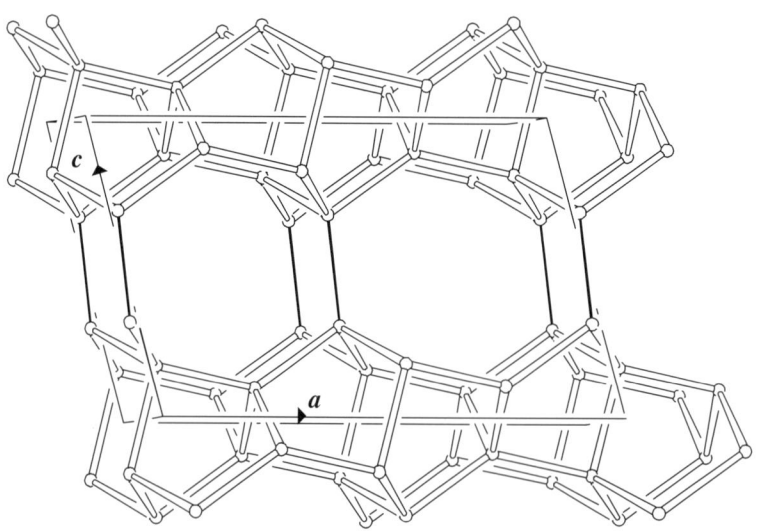

Fig. 2.    Connection mode (and unit cell content) viewed along **b**.

### 3. Channels and/or cages

Non-interconnecting 8-ring channels are parallel to $b$ (Figure 3). The channel is topologically equivalent to the channel in **BIK**.

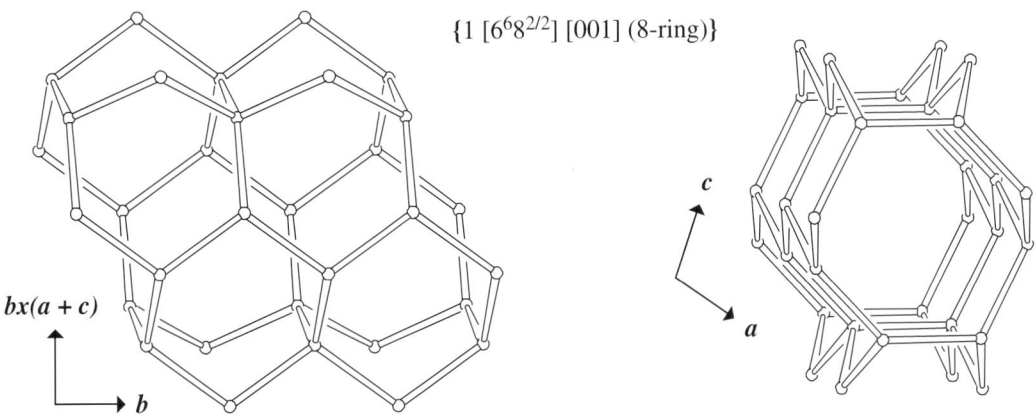

$\{1 \ [6^6 8^{2/2}] \ [001] \ (8\text{-ring})\}$

Fig. 3.   Channel viewed perpendicular to the channel axis (left) and along the channel axis (right).

# OBW  Building Scheme

## 1. Periodic Building Unit

Tetragonal **OBW** can be built using units of 38 T atoms. The T38-unit consists of twelve 3-rings that are connected around a 4-fold axis parallel to $c$ and two additional T atoms (Be). Six (fused) 8-rings are formed. A two-dimensional PerBU is obtained when T38-units, related along $a$ and $b$ by pure translations, are connected into the $ab$ layer as shown in Figure 1. The two additional T atoms connect the PerBUs in three dimensions.

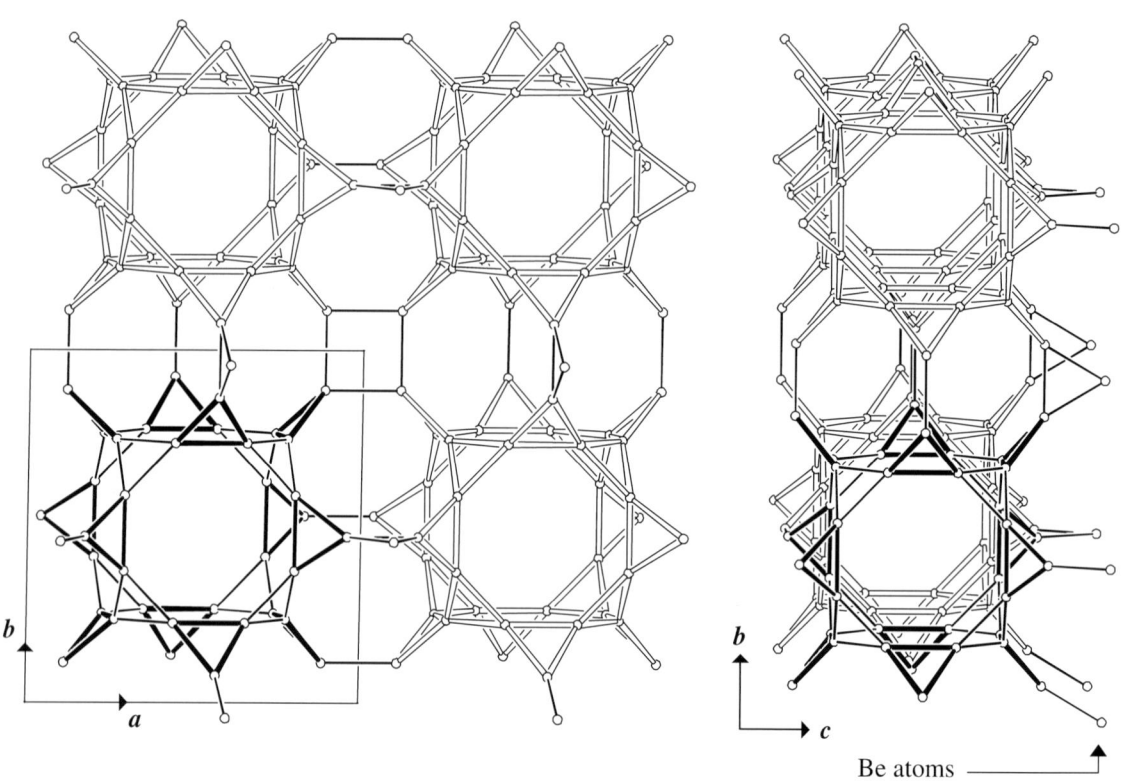

Be atoms

Fig. 1.  PerBU viewed along $c$ (left) and along $a$ (or $b$) (right). The Be atoms (4 per unit cell), connecting the PerBU to its neighboring PerBUs (see Figure 2, Connection mode), are also shown.

## 2. Connection mode

Neighboring PerBUs, related along $c$ by a shift of $\frac{1}{2}(a+b+c)$, are connected along $c$ through spiro-5 rings (with the Be atom as the central atom) as illustrated in Figure 2.

# Building Scheme

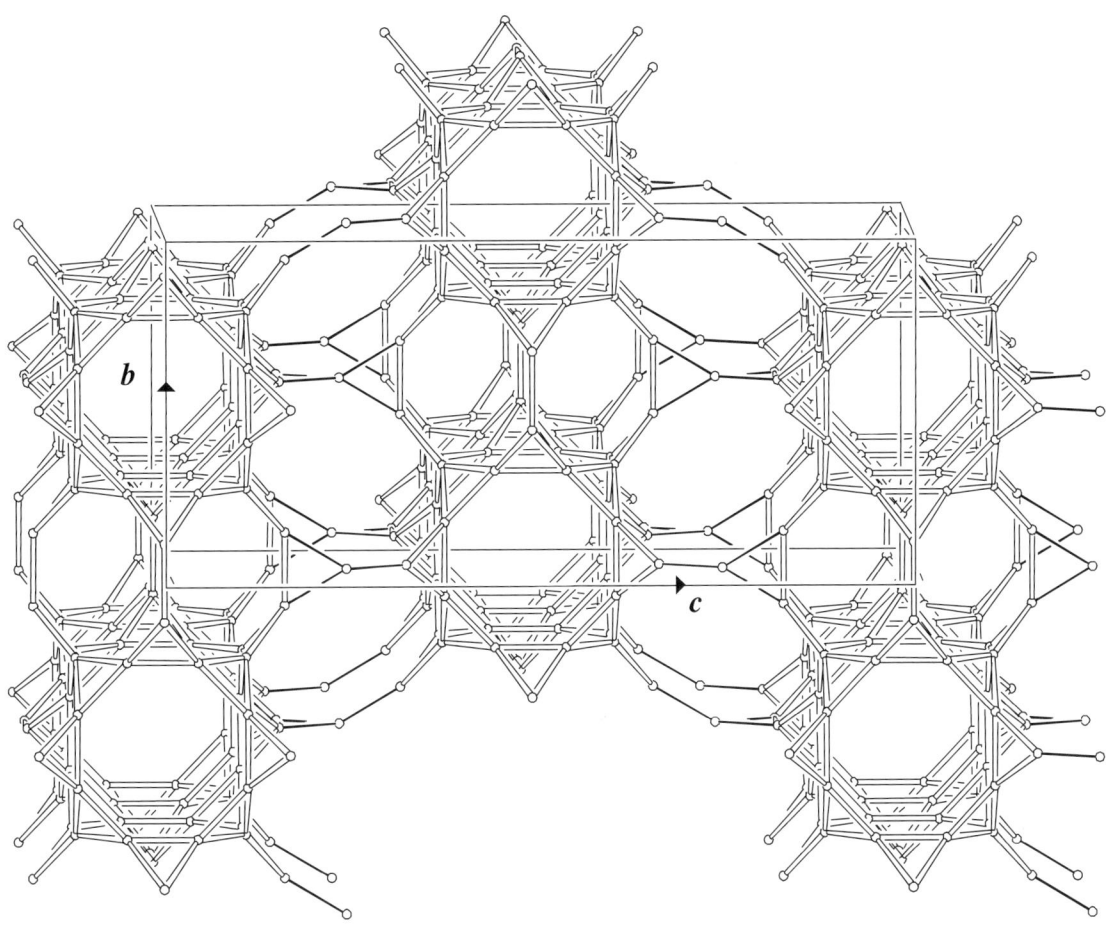

Fig. 2a. Connection mode (and unit cell content) viewed along **a**. The PerBUs are connected along **c** through spiro-5-rings with Be atoms as central atoms.

# OBW

## Cage/Channel

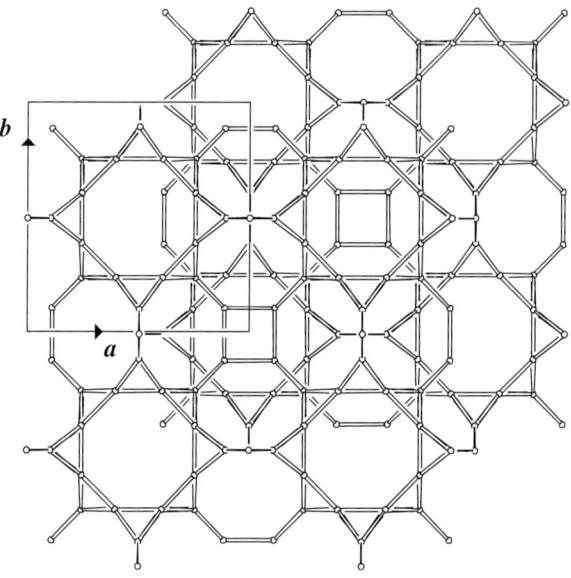

Fig. 2b.   Unit cell content projected along **c**.

## 3. Channels and/or cages

8-Ring channels are parallel to <100> and 10-ring channels are parallel to <110>. The channel systems are interconnecting through common 8-rings. The two types of channel intersections (cavities) are shown in Figure 3 together with the connection of the cavities.

Cavity 1:
$\{3\ [3^{12}4^18^310^4]\ <110>\ (10\text{-ring}),\ [001]\ (8\text{-ring})\}$

Cavity 2:
$\{3\ [3^88^6]\ <100>\ (8\text{-ring}),[001]\ (8\text{-ring}\}$

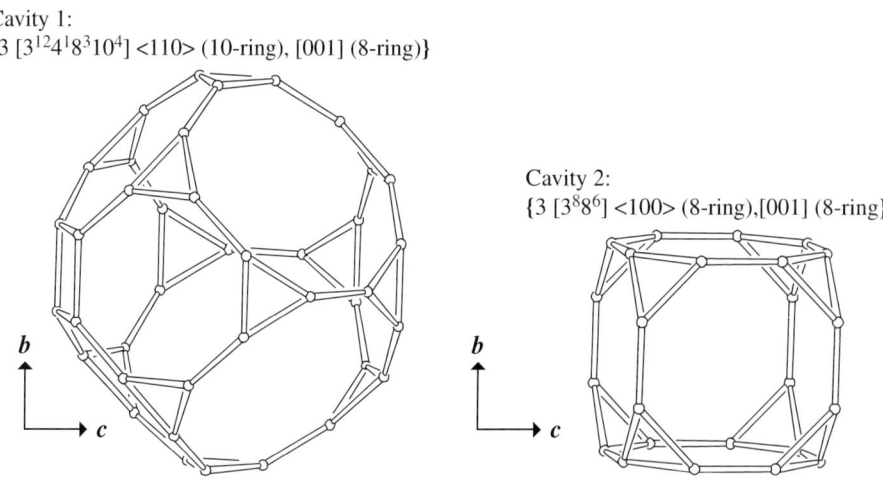

Fig. 3a.   Cavity 1 (left) and cavity 2 (right) viewed along **a** (or **b**).

# Cage/Channel

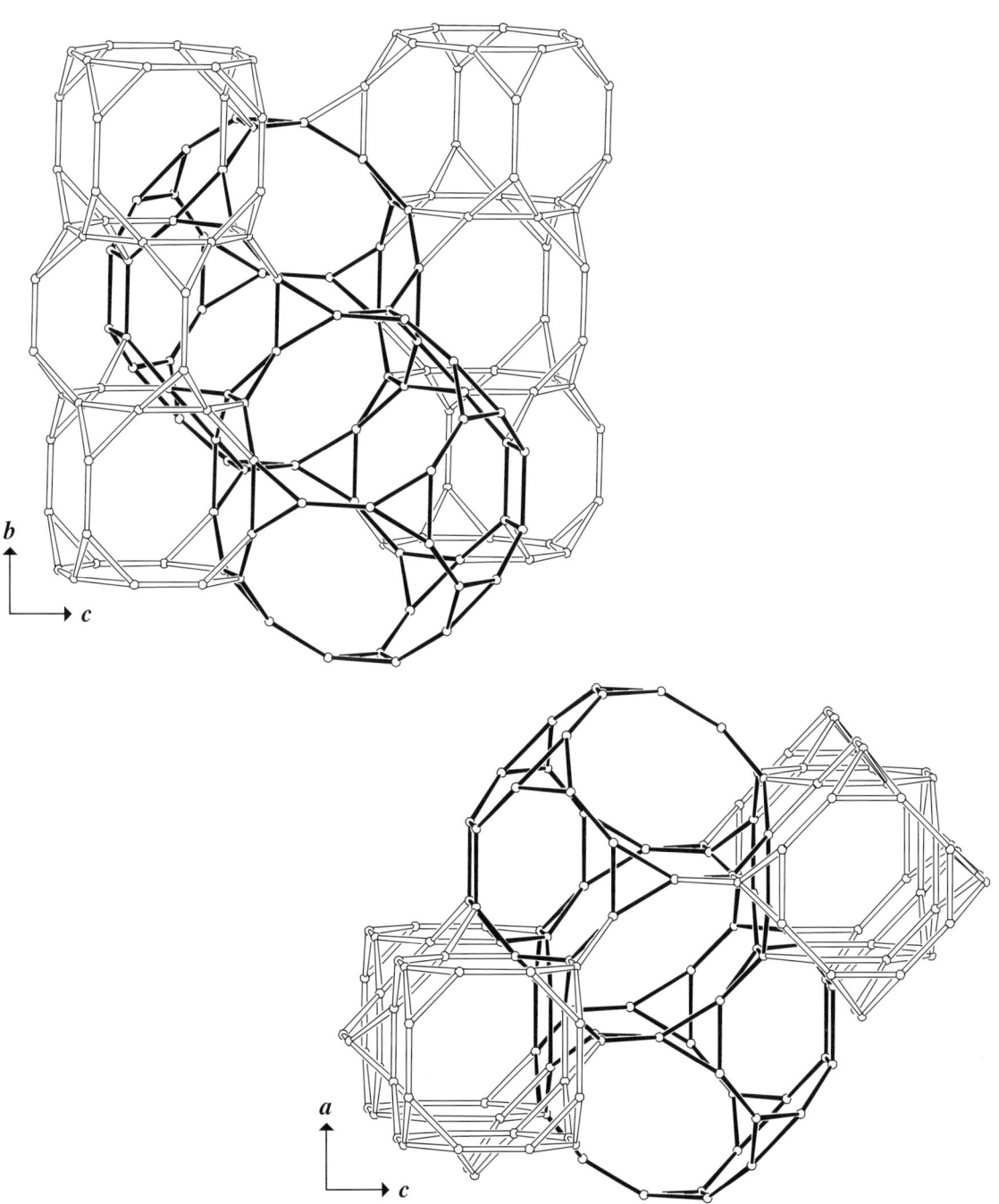

Fig. 3b.  Connection of the cavities viewed along *a* (or *b*, left) and along the 8-ring channel axis parallel to *b* (or *a*, right). Cavity 1 in bold.

# OFF                                  Building Scheme

## 1. Periodic Building Unit

The two-dimensional PerBU of **OFF** consists of a hexagonal array of non-connected planar 6-rings (bold in Figure 1), which are related along *a* and *b* by pure translations. The 6-rings are centered at (0,0) in the *ab* layer. This position is usually called the **A** position. **OFF** belongs to the ABC-6 family (see Introduction).

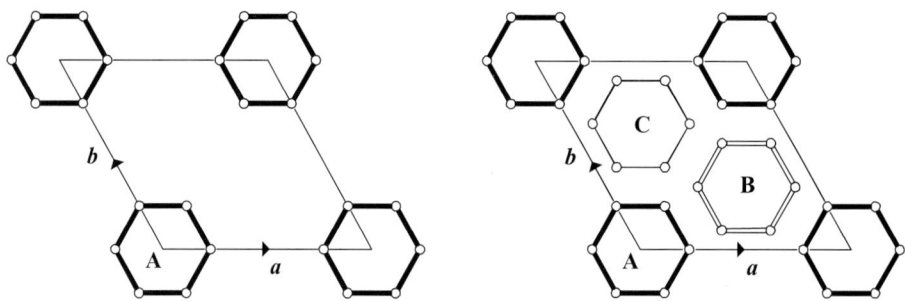

Fig. 1.   PerBU (left) and definition of 6-ring positions used in the stacking modes (right).

## 2. Connection mode

Neighboring PerBUs can be connected along *c* through tilted 4-rings in three different ways (see Introduction). In **OFF** all three connection modes between the PerBUs are observed. **can** Cages are formed. (See Figure 2; see also **ERI**.)

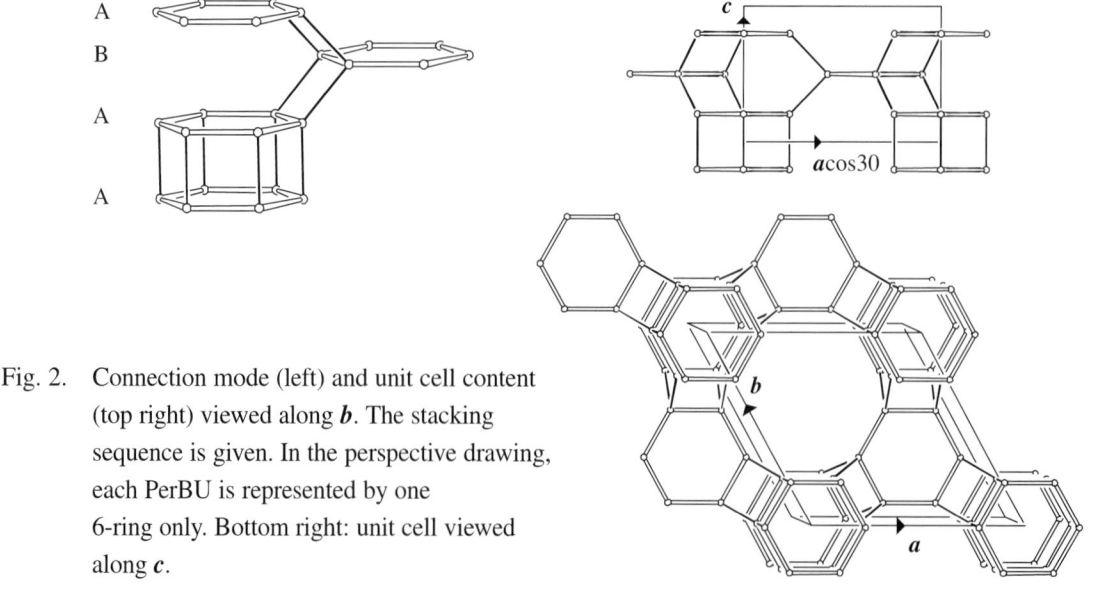

Fig. 2.   Connection mode (left) and unit cell content (top right) viewed along *b*. The stacking sequence is given. In the perspective drawing, each PerBU is represented by one 6-ring only. Bottom right: unit cell viewed along *c*.

## 3. Channels and/or cages

Interconnecting 12-ring channels, equivalent to the 12-ring channel in **MOZ** (Figure 3), are parallel to *c*. Two-dimensional 8-ring channels are parallel to <110>. **OFF** can also be built from *can* cages (see also **ERI**, **LTL** and **MOZ**). A column of (fused) *gmel* cavities (bold in Figure 3) is formed when *can* cages are connected around a 3-fold axis at $(\frac{2}{3}, \frac{1}{3})$. The *gmel* cavity is also present in **AFT**, **AFX**, **EAB**, **EON**, **GME** and **MAZ**. A column of *can* cages is formed when the 12-ring channels are connected around the 3-fold axis at (0,0). The *can* cage is also present in **AFG**, **CAN**, **ERI**, **FAR**, **FRA**, **GIU**, **LIO**, **LOS**, **LTL**, **LTN**, **MAR**, **MOZ**, **SAT**, **SBS**, **SBT**, **TOL** and **-WEN**.

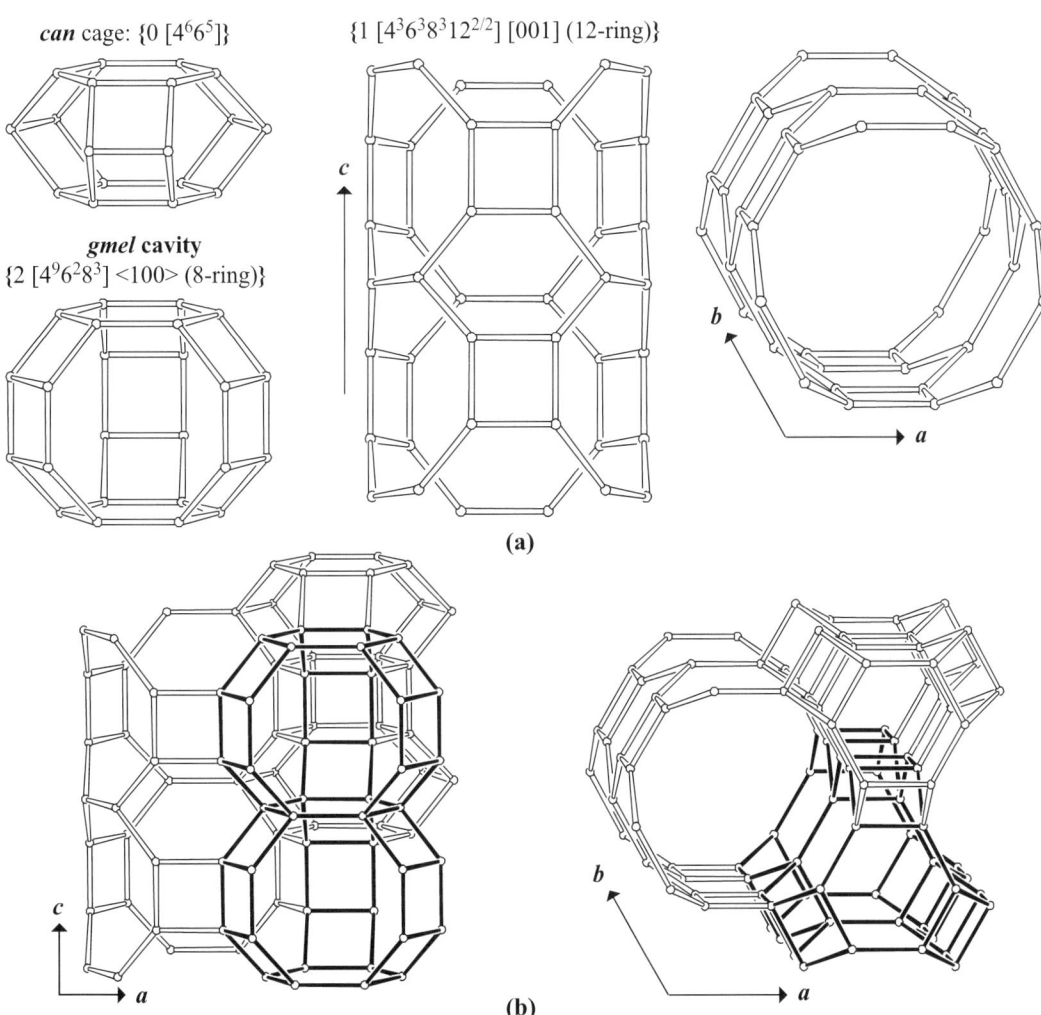

Fig. 3.    (a) *can* Cage (top left), *gmel* cavity (bottom left) viewed along <120> and 12-ring channel viewed perpendicular to the channel axis (middle) and along the channel axis (right); (b) Connection of channel, *can* cages and *gmel* cavities (in bold) viewed along <120> (left) and along *c* (right).

# OSI                     Building Scheme

## 1. Periodic Building Unit

Tetragonal **OSI** can be built using the zigzag chain (one in bold in Figure 1) parallel to *c*. The one-dimensional PerBU is obtained when eight zigzag chains are connected into a channel with a 12-ring aperture. The channel wall consists of fused 6-rings. The repeat unit consists of 16 T atoms: a 12-ring and 4 T atoms (bold in Figure 1). (See also **VET**.)

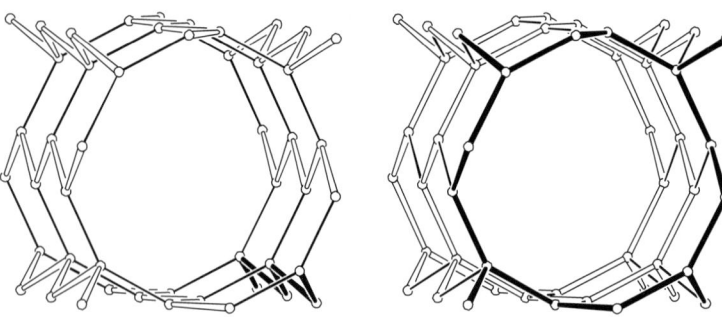

Fig. 1.   PerBU, constructed from eight zigzag chains (left) and from T16-units (right), viewed along *c*.

## 2. Connection mode

Neighboring PerBUs, related along *a* (and *b*) by a shift of $\frac{1}{2}(a+b+c)$, are connected along *a* (and *b*) through 4- and 6-rings as illustrated in Figure 2. The connection mode between four direct neighboring PerBUs shows a 4-fold rotation axis through the central 4-ring.

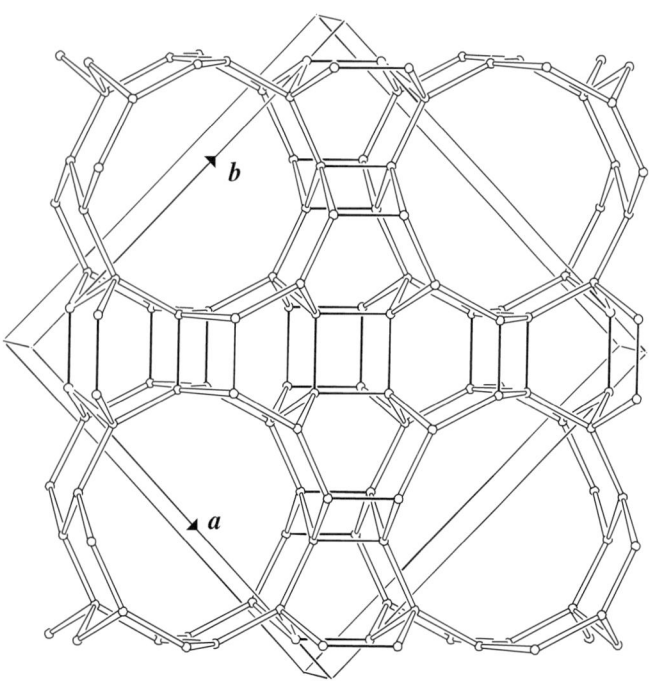

Fig. 2.   Connection mode (and unit cell content) viewed along *c*. Only two repeat units of the PerBUs are drawn for clarity.

### 3. Channels and/or cages

Non-interconnecting 12-ring channels are parallel to *c*. The channel wall consist of fused 6-rings as shown in Figure 3. The channel is topologically equivalent to the 12-ring channel in **VET**.

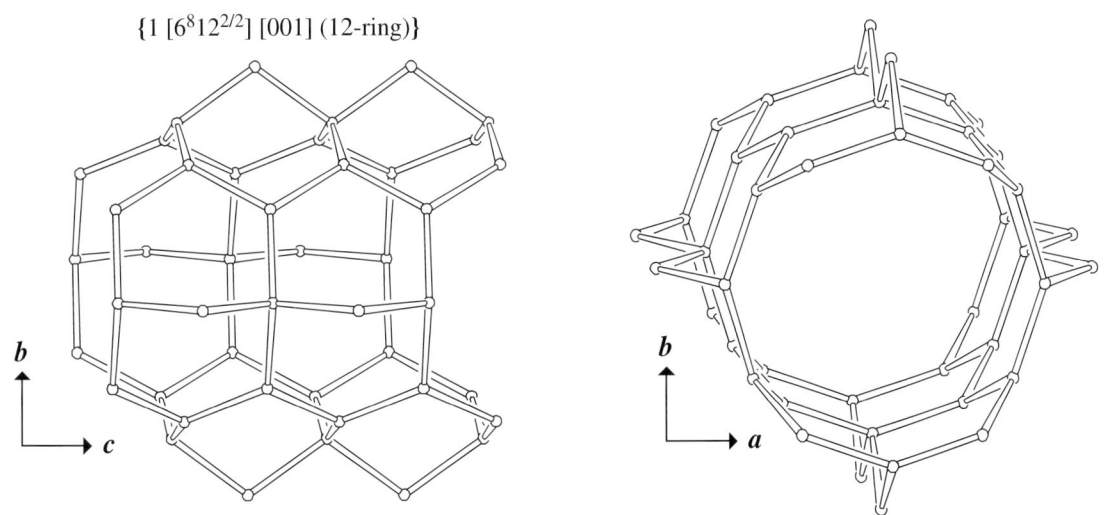

$\{1 \, [6^8 12^{2/2}] \, [001] \, (12\text{-ring})\}$

Fig. 3.    Channel viewed perpendicular to the channel axis (left) and along the channel axis (right).

# OSO

## Building Scheme

### 1. Periodic Building Unit

Hexagonal **OSO** can be built using units of 9 T atoms. The T9-unit consists of four fused 3-rings (or a 3-1 unit and a spiro-5 unit; bold in Figure 1). The two-dimensional PerBU is obtained when T9-units, related along **a** and **b** by pure translations, are connected into the **ab** layer depicted in Figure 1.

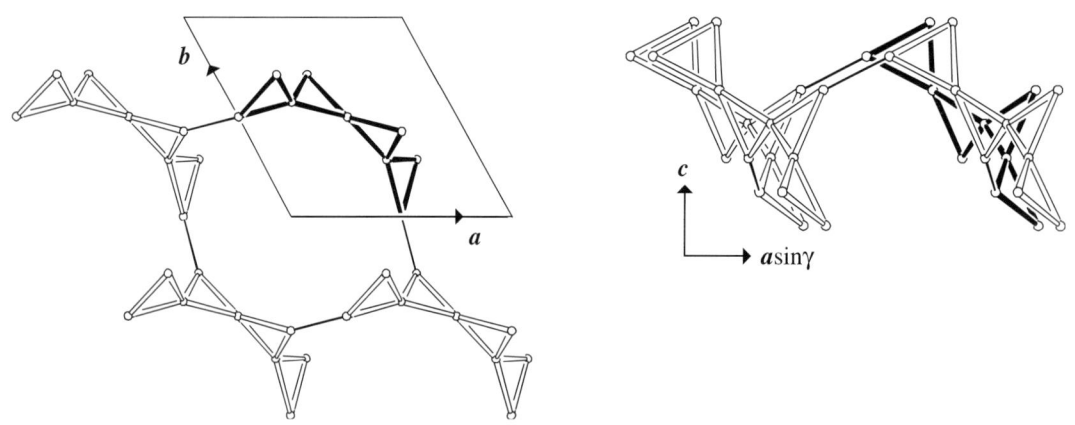

Fig. 1.   PerBU viewed along **c** (left) and along **b** (right).

### 2. Connection mode

Neighboring PerBUs, related along **c** by a pure translation, are connected along **c** through 3-rings as shown in Figure 2.

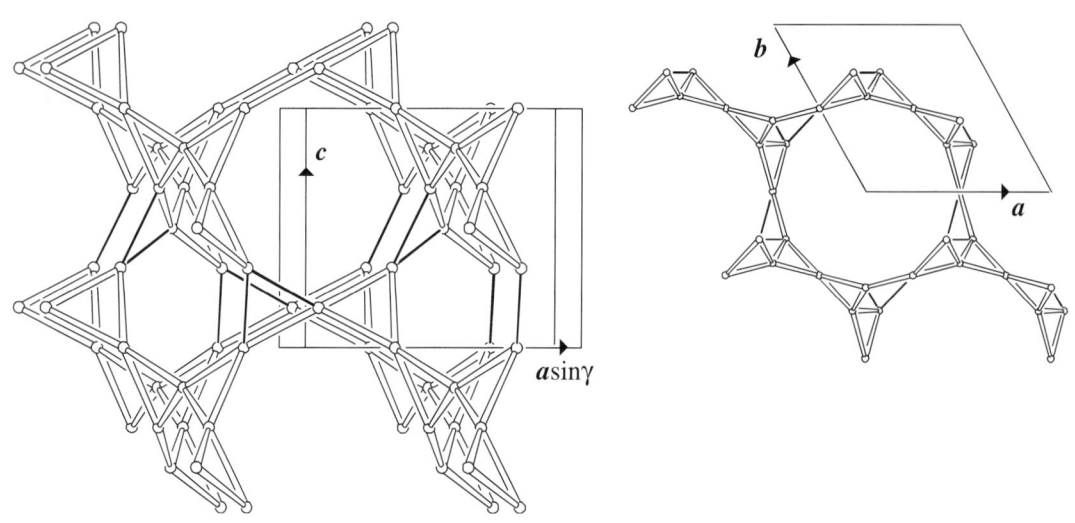

Fig. 2.   Connection mode (and unit cell content) viewed along **b** (left) and unit cell content projected along **c** (right).

## 3. Channels and/or cages

14-Ring channels, parallel to *c*, intersect with 8-ring channels parallel to <100>. The channel intersection is shown in Figure 3 together with the connection of intersections.

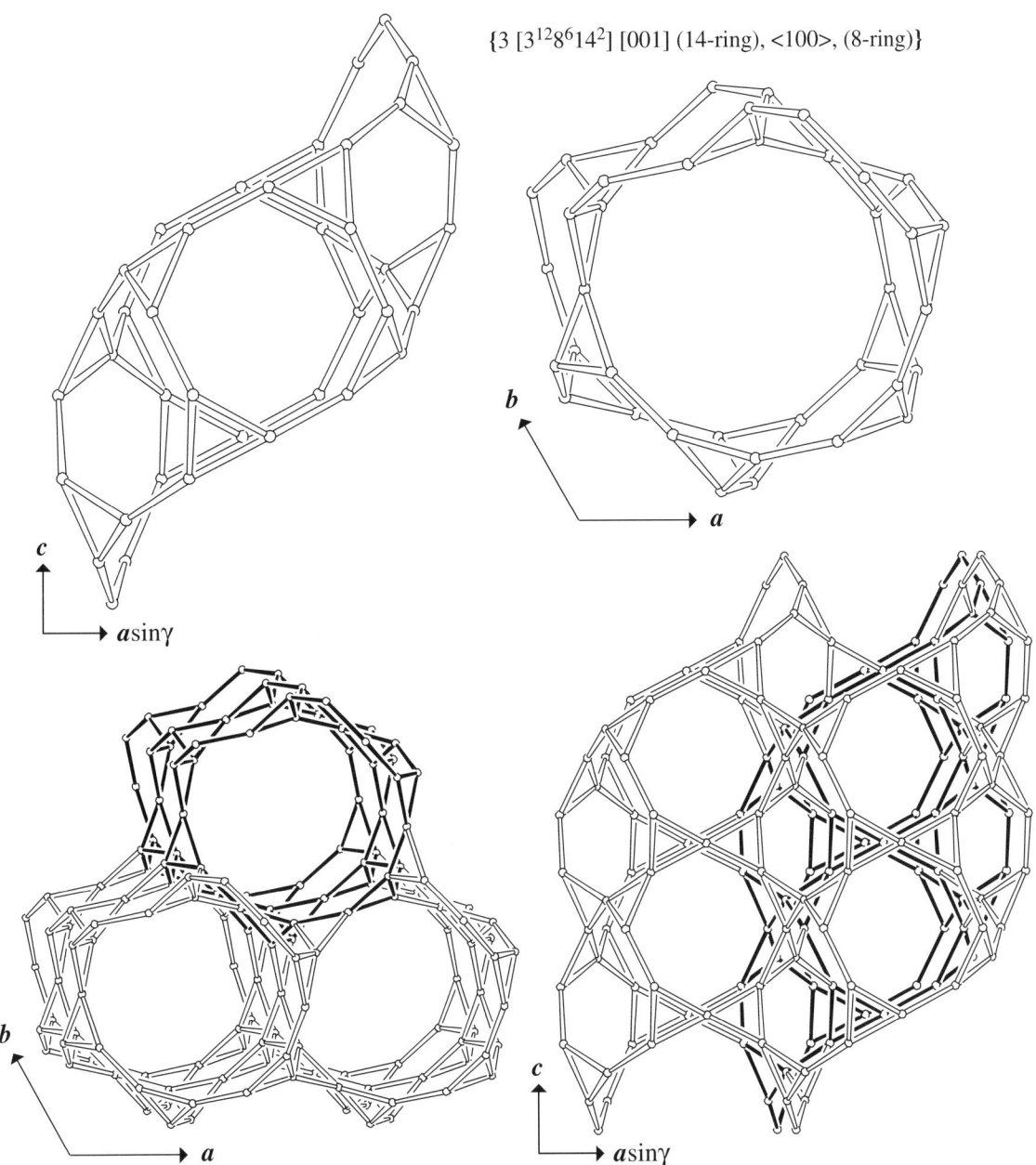

{3 [3$^{12}$8$^6$14$^2$] [001] (14-ring), <100>, (8-ring)}

Fig. 3. Top: channel intersection viewed along <010> (left) and along the 14-ring channel axis parallel to *c* (right). Bottom: connection of intersections viewed along *c* (left) and along the 8-ring channel axis parallel to <010> (right). One 14-ring channel in bold.

# OWE

## Building Scheme

### 1. Periodic Building Unit

**OWE** can be built using double 4-rings with one disconnected edge (a $[4^4 6^1]$-cage; bold in Figure 1). T8-units, related along **a** by a screw rotation of 180° about **a**, are connected into chains parallel to **a**. Neighboring chains, related along **c** by a pure translation, are linked into the **ac** layer. The two-dimensional PerBU is depicted in Figure 1.

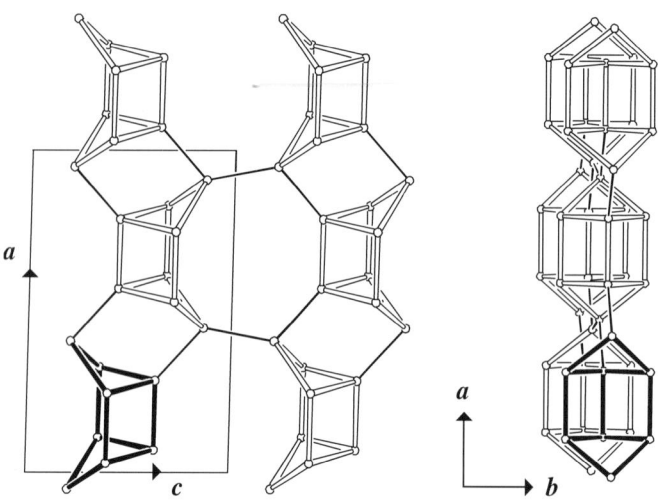

Fig. 1. PerBU viewed along **b** (left) and along **c** (right).

### 2. Connection mode

Neighboring PerBUs, related along **b** by a pure translation, are connected along **b** through 4-rings. Double saw chains and intersecting 8-ring channels parallel to **a** and **b** are formed (Figure 2).

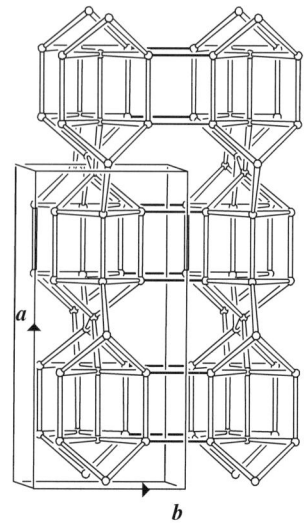

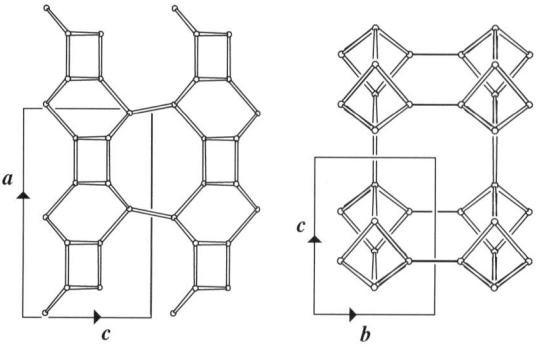

Fig. 2. Connection mode (and unit cell content) viewed along **c** (left) and unit cell content projected along **b** (middle) and along **a** (right).

## 3. Channels and/or cages

Intersecting 8-ring channels are parallel to *a* and *b*. The two different channels and their interconnection are depicted in Figure 3.

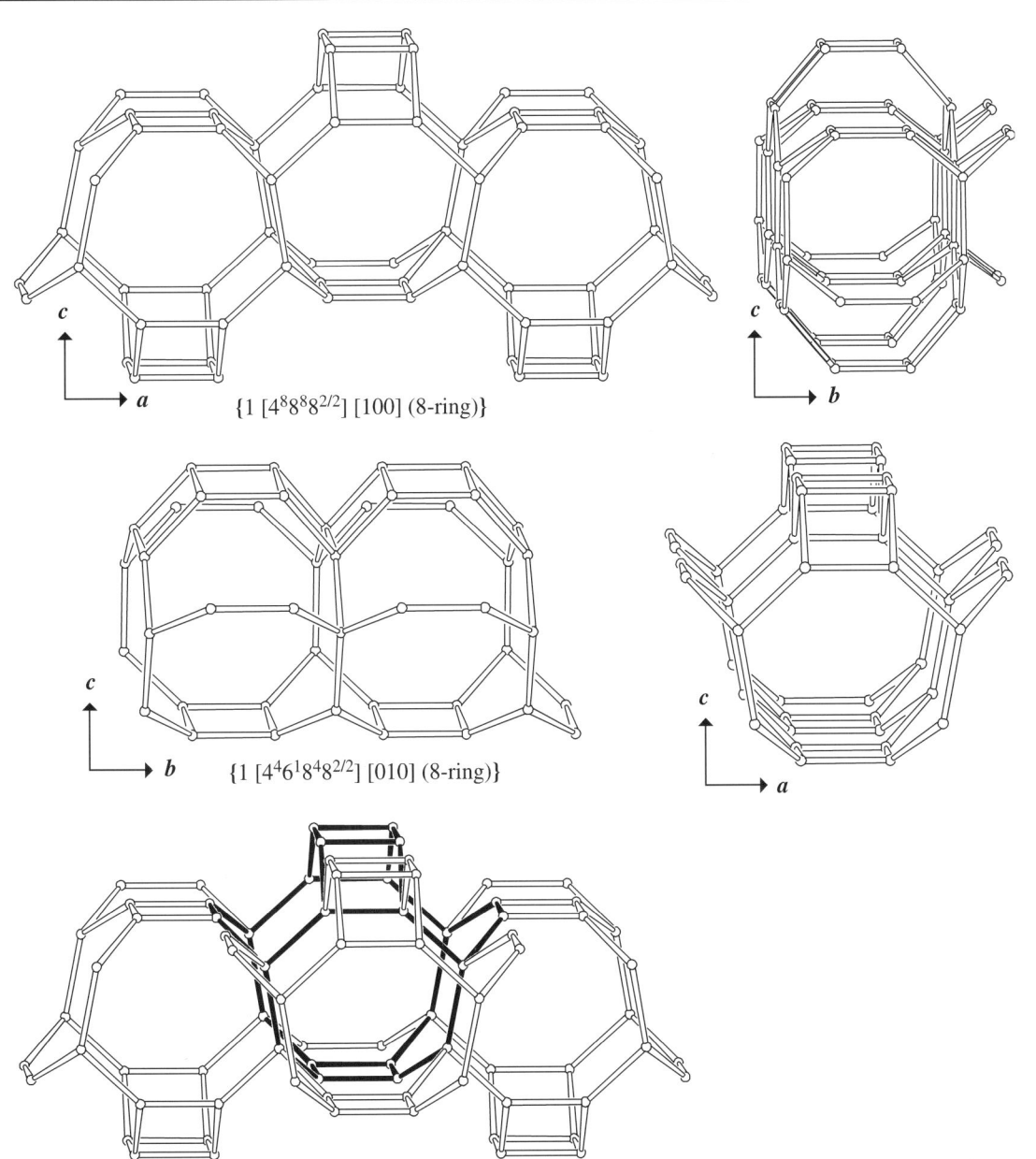

**Fig. 3.** 8-Ring channel parallel to *a* (top) and 8-ring channel parallel to *b* (middle) viewed perpendicular to the channel axis (left) and along the channel axis (right). Bottom: intersecting 8-ring channels share [4⁴8⁶]-cavities. One intersection in bold.

# -PAR                   Building Scheme

## 1. Periodic Building Unit

The interrupted framework of **-PAR** can be built using units of 16 T atoms. This T16-unit (bold in Figure 1) consists of four 4-rings that are connected in such a way that a 6-ring chair is formed. The two-dimensional PerBU is obtained when neighboring T16-units, related along $a$ (and $b$) by a shift of $\frac{1}{2}(a+b)$, are connected into the $ab$ layer as shown in Figure 1.

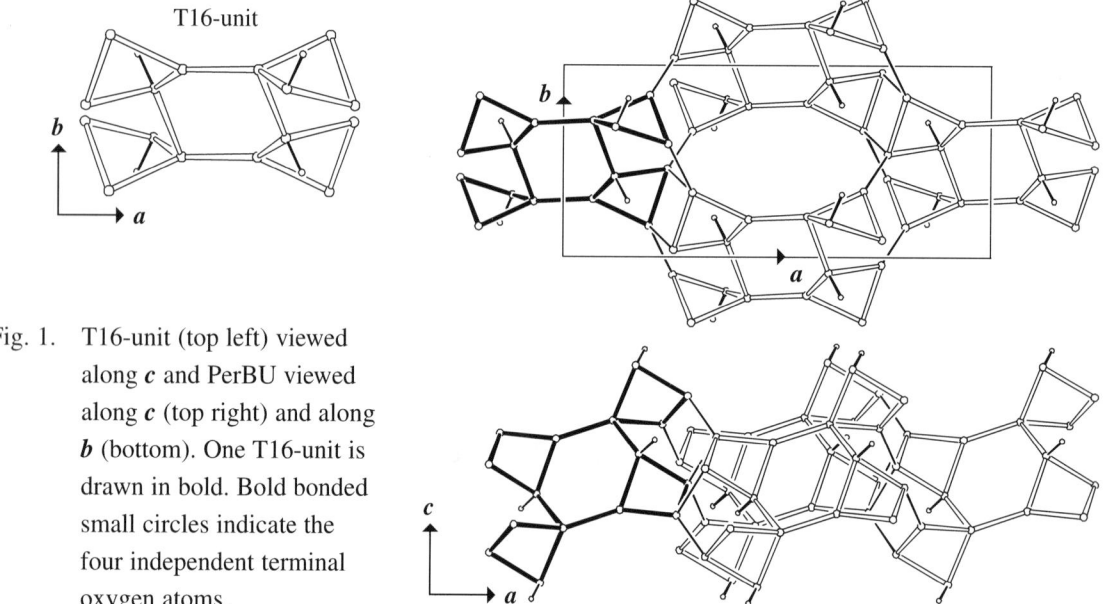

Fig. 1.  T16-unit (top left) viewed along $c$ and PerBU viewed along $c$ (top right) and along $b$ (bottom). One T16-unit is drawn in bold. Bold bonded small circles indicate the four independent terminal oxygen atoms.

## 2. Connection mode

Neighboring PerBUs, related by pure translations along $c$, are connected along $c$ as shown in Figure 2.

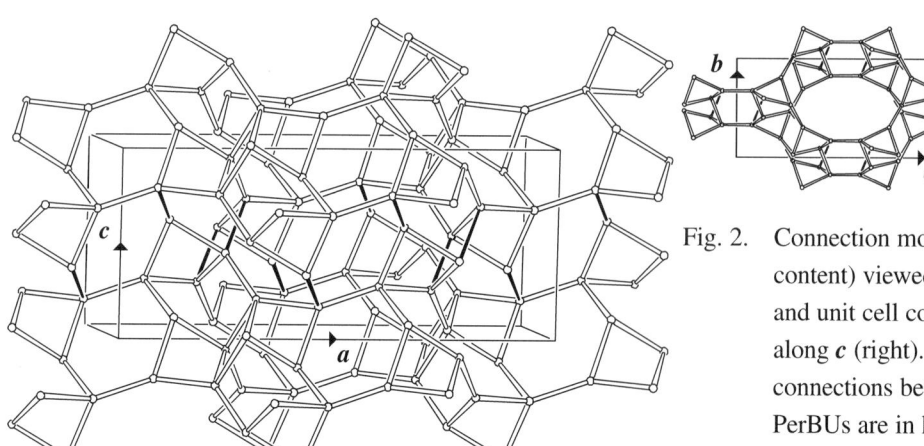

Fig. 2.  Connection mode (and unit cell content) viewed along $b$ (left) and unit cell content projected along $c$ (right). For clarity, T–T connections between the PerBUs are in heavy bold and oxygen atoms are left out.

## 3. Channels and/or cages

10-Ring channels are parallel to *c*. The 8-rings in the channel wall are strongly corrugated as is illustrated in Figure 3.

{1 [$4^8 6^8 8^4 10^{2/2}$] [001] (10-ring)}

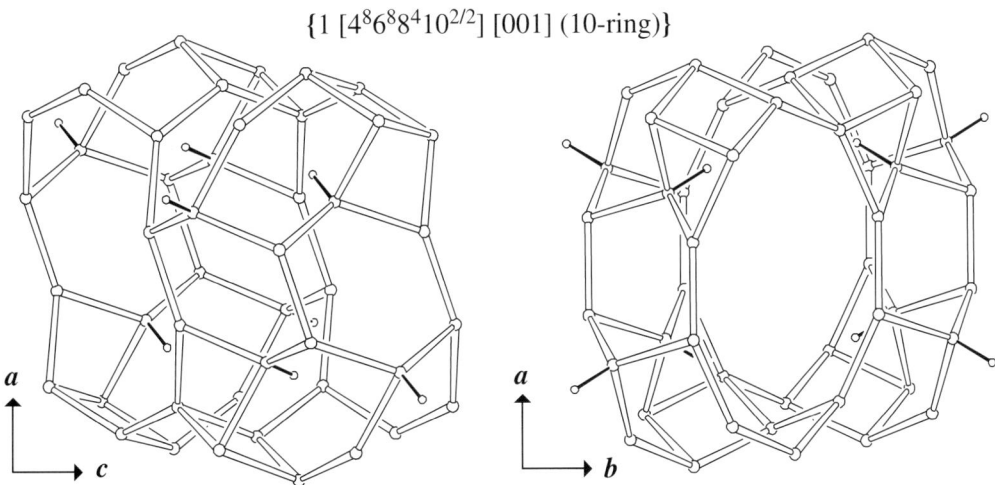

Fig. 3.  Channel viewed perpendicular to the channel axis (left) and along the channel axis (right).

# PAU                    Building Scheme

## 1. Periodic Building Unit

In cubic **PAU** three types of cavities can be distinguished. The ***rho*** cavity (Figure 1a) consists of 48 T atoms: six 8-rings, eight 6-rings or twelve 4-rings. The second cavity (cavity2) contains 64 T atoms: eight 8-rings or sixteen 4-rings connected into a double 24-ring "capped" on both sides with an 8-ring (Figure 1a; 8-rings in bold). The ***mer*** cavity is equal to cavity2 without the "side-pockets" and contains 32 T atoms: four 8-rings or eight 4-rings as shown in Figure 1a. The cavities are connected into two different building units. The first building unit (BU1; Figure 1b) consists of six cavities2, which are connected to a ***rho*** cavity through double 8-rings (and contains $48 + 6 \times 64 = 432$ T atoms). The second building unit (BU2; Figure 1b) consists of six ***mer*** cavities, which are connected to a ***rho*** cavity through double 8-rings (and contains $48 + 6 \times 32 = 240$ T atoms). Neighboring BU1s, related by pure translations along the cube axes, are connected into cubic faces through double 8-rings as shown in Figure 1c. The PerBU is obtained when the "empty spaces" in the BU1-cubic faces are filled with BU2 units. The BU1 and BU2 units, shifted with respect to each other over $\frac{1}{2}(a + b + c)$, are connected through 4-rings as shown in Figure 1c and d.

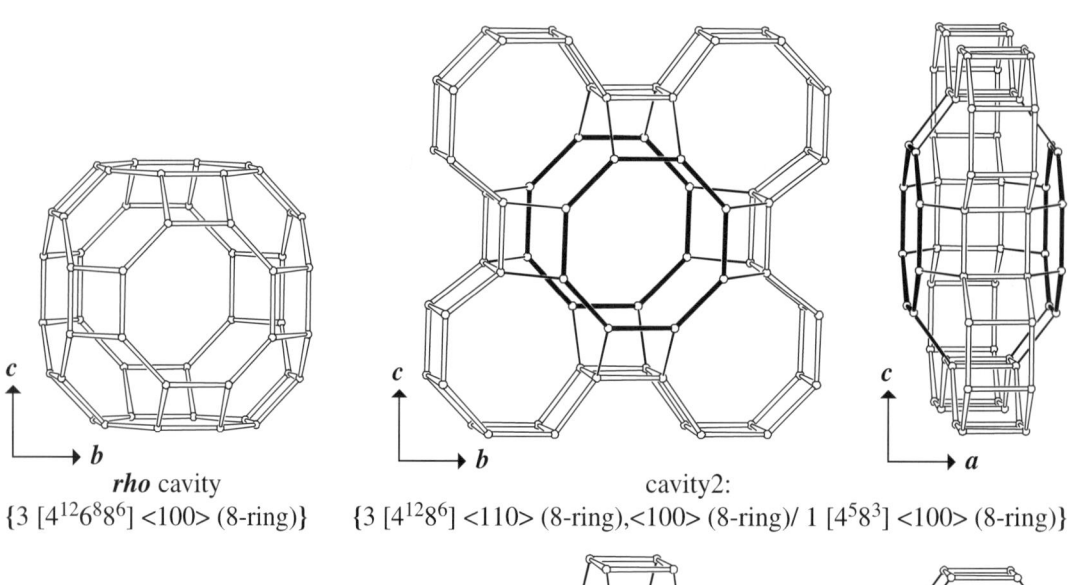

***rho*** cavity
{3 [$4^{12}6^86^6$] <100> (8-ring)}

cavity2:
{3 [$4^{12}8^6$] <110> (8-ring),<100> (8-ring)/ 1 [$4^58^3$] <100> (8-ring)}

Fig. 1a. ***rho*** Cavity viewed along ***a*** (top left), cavity2 viewed along ***a*** and along ***b*** (top middle and top right) and ***mer*** cavity viewed along ***a*** and along ***b*** (bottom left and bottom right).

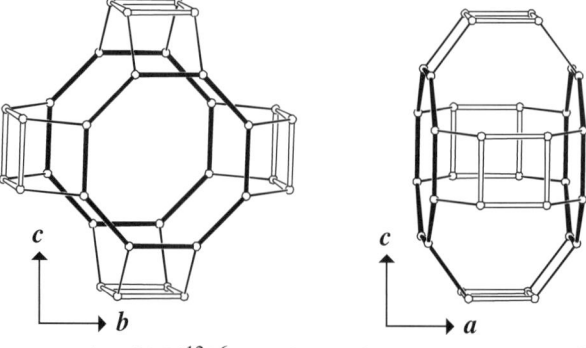

***mer*** cavity: {3 [$4^{12}8^6$] <110> (8-ring), <100> (8-ring)}

# Building Scheme

# PAU

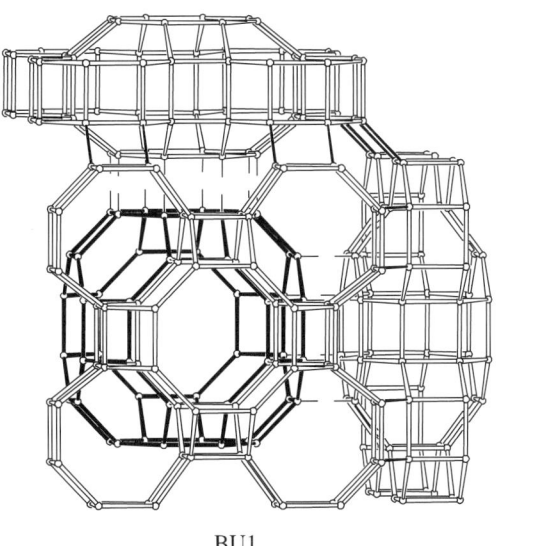

BU1

BU2

Fig. 1b.   BU1 (left) and BU2 (right) viewed along a cubic axis. For clarity, only three of the six cavities2 (left) and three of the six *mer* cavities (right) are attached to a (bold) *rho* cavity; double 8-ring connections to the *rho* cavity are dashed and the additional connections (4-rings and a 6-ring) between cavities2 are in light bold.

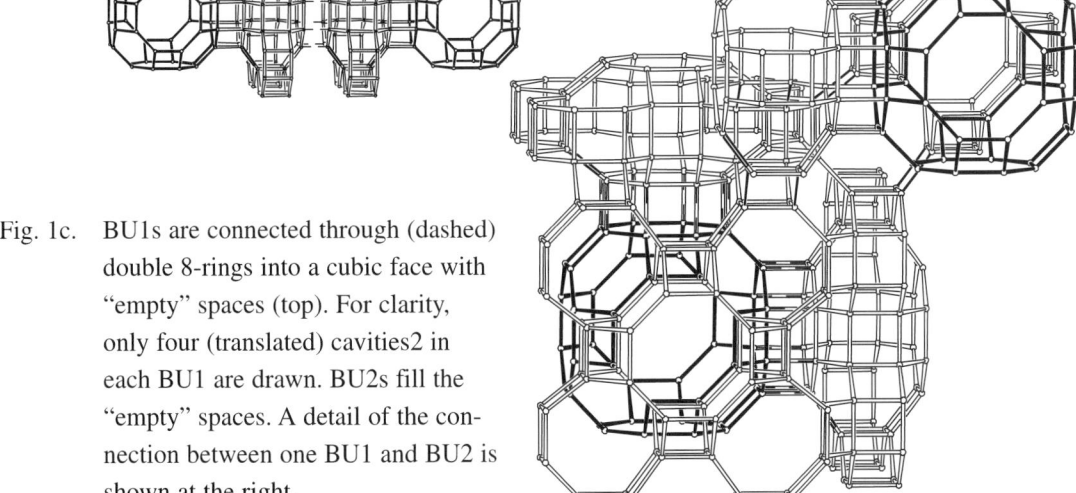

Fig. 1c.   BU1s are connected through (dashed) double 8-rings into a cubic face with "empty" spaces (top). For clarity, only four (translated) cavities2 in each BU1 are drawn. BU2s fill the "empty" spaces. A detail of the connection between one BU1 and BU2 is shown at the right.

# PAU

## Building Scheme

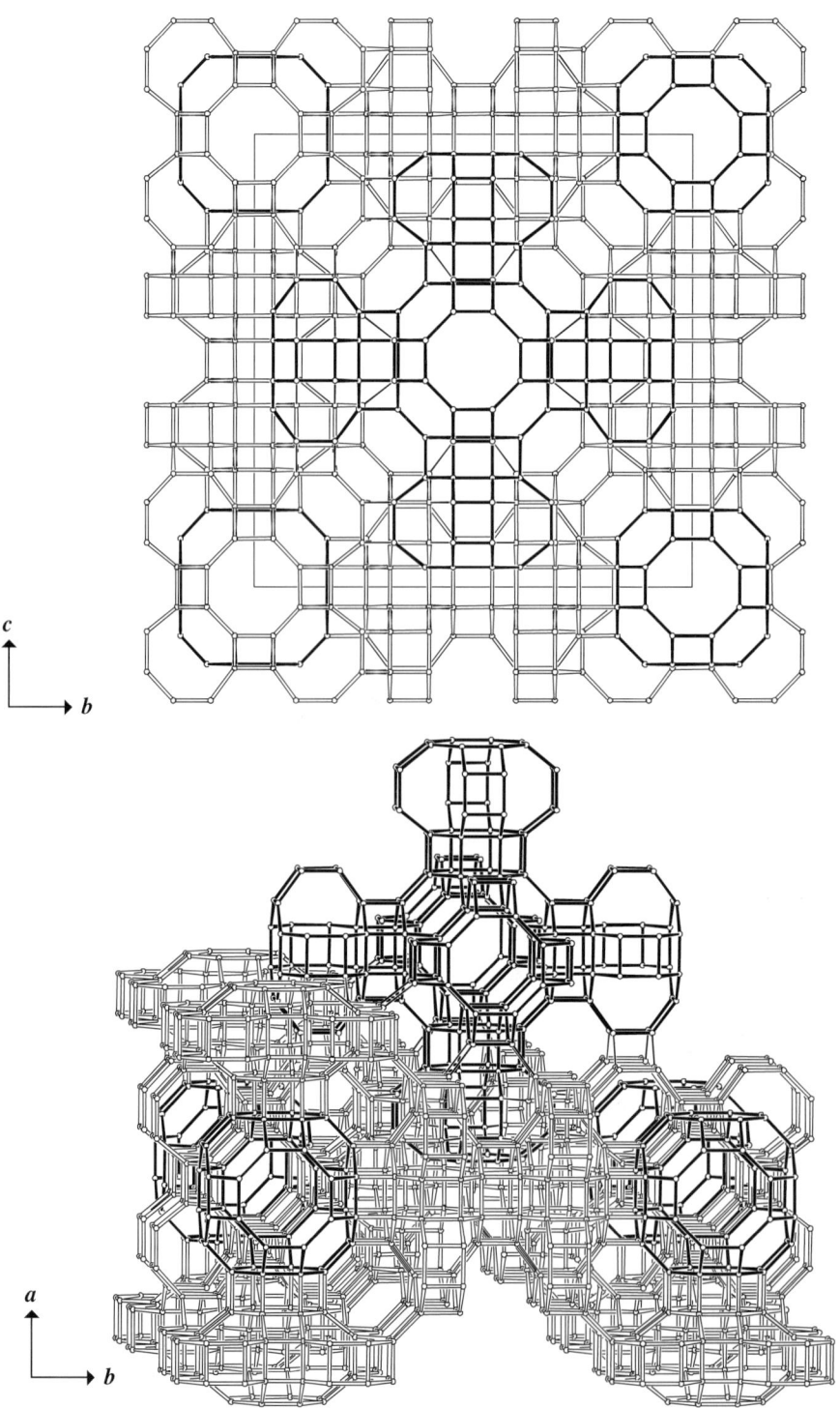

Fig. 1d.   PerBU in **PAU** viewed along *a* (top) and along *c* (bottom). For clarity, not all cavities2 are drawn. Bu2 and *rho* cavities are in bold.

## 2. Connection mode

Neighboring PerBUs are related by pure translations along **a**. The connection modes in the **ab**- and **ac**-planes are equivalent to those shown in Figure 1 for the **bc** plane.

## 3. Channels and/or cages

Two systems of equal, non-interconnecting 8-ring channels are parallel to <100> (Figure 2). In both channel systems, intersecting 8-ring channels are parallel to the cell axes. The channel intersection is equal to the **rho** cavity. (Extended) **mer** cavities connect the **rho** cavities into 8-ring channels. The **mer** cavity is also present in **KFI**, **MER** and **MOZ**. The **rho** cavity is also observed in **-CLO**, **KFI**, **LTA**, **LTN**, **RHO**, **TSC** and **UFI**. Cavities are shown in Figure 1a.

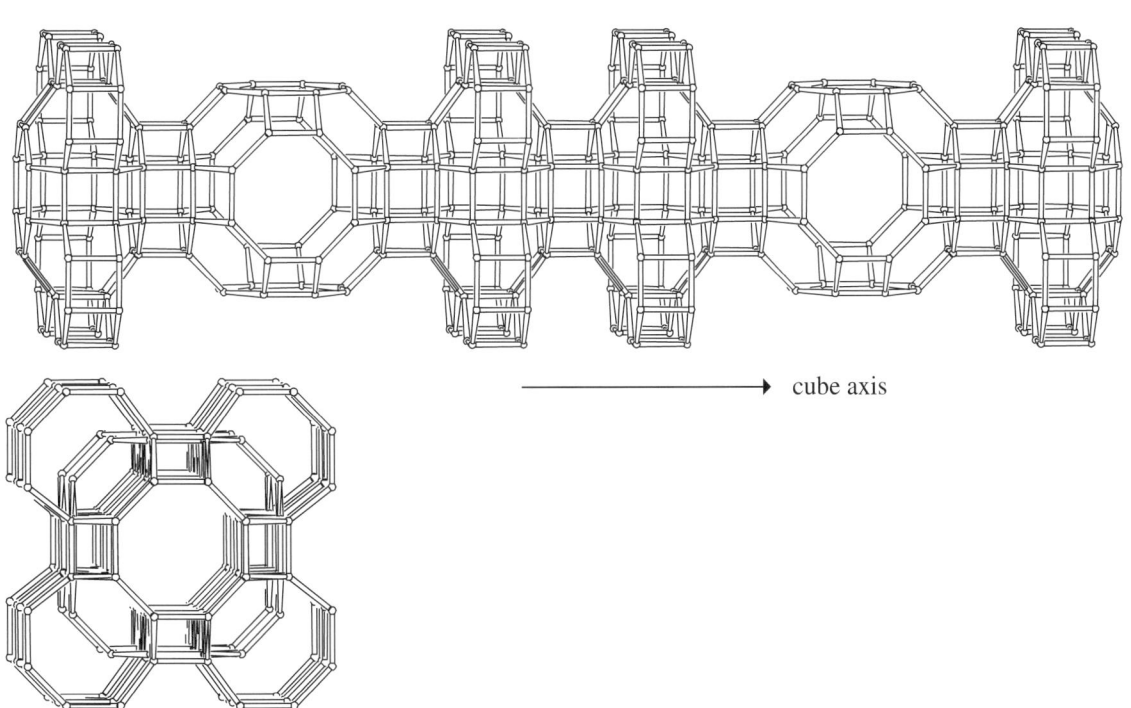

cube axis

Fig. 2.    8-Ring channel parallel to <100> viewed perpendicular to the channel axis (top) and along the channel axis (bottom).

# PHI                    Building Scheme

## 1. Periodic Building Unit

**PHI** can be built using the 4-ring and the crankshaft chain (bold in Figure1) running parallel to *a*. A one-dimensional PerBU is obtained when two crankshaft chains and 4-rings are connected into a channel with an 8-ring aperture. The repeat unit of the PerBU consists of a 6-fold (1,2,3,4,5,6)-connected double 8-ring (bold in Figure 1). (Compare with **SIV**.)

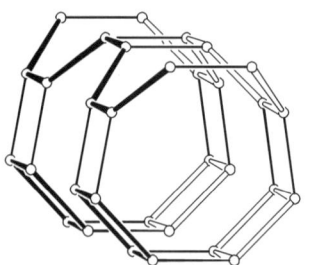 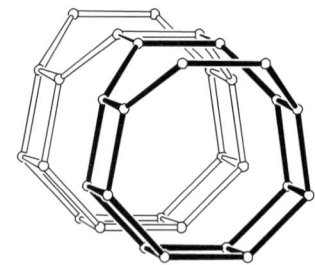

Fig. 1.   PerBU constructed from crankshaft chains and 4-rings (or solely from 4-rings; left) and from 6-fold connected double 8-rings (right) viewed along *a*.

## 2. Connection mode

Neighboring PerBUs, related along *b* (and *c*) by a rotation of 180° about *a* followed by a shift of $\frac{1}{2}(a + b + c)$, are connected along *b* (and *c*) through double-crankshaft chains as shown in Figure 2.

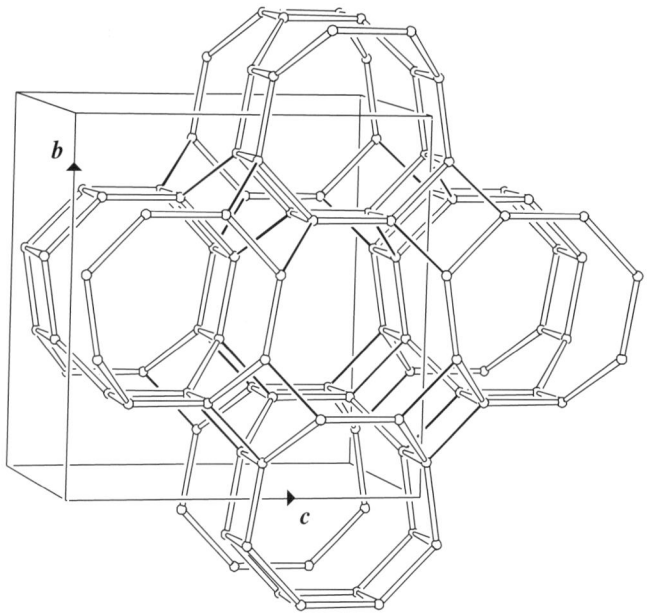 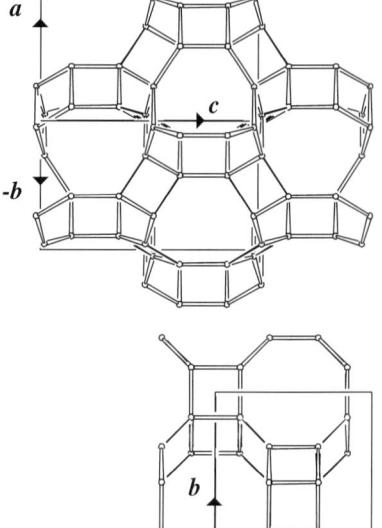

Fig. 2.   Connection mode (and unit cell content) viewed along *a* (left) and unit cell content projected along [1–10] (top right) and along *c* (bottom right). For clarity, only $1\frac{1}{2}$ repeat unit along *a* of each PerBU is drawn.

## Cage/Channel

### 3. Channels and/or cages

Intersecting 8-ring channels are parallel to $c$, $a$ and [1–10] (and [110]), respectively. The three different types of channels and their interconnection are shown in Figure 3. The 8-ring channel parallel to $a$ is (topologically) equivalent to (one of) the channels in **ATT**, **GIS** and **SIV**.

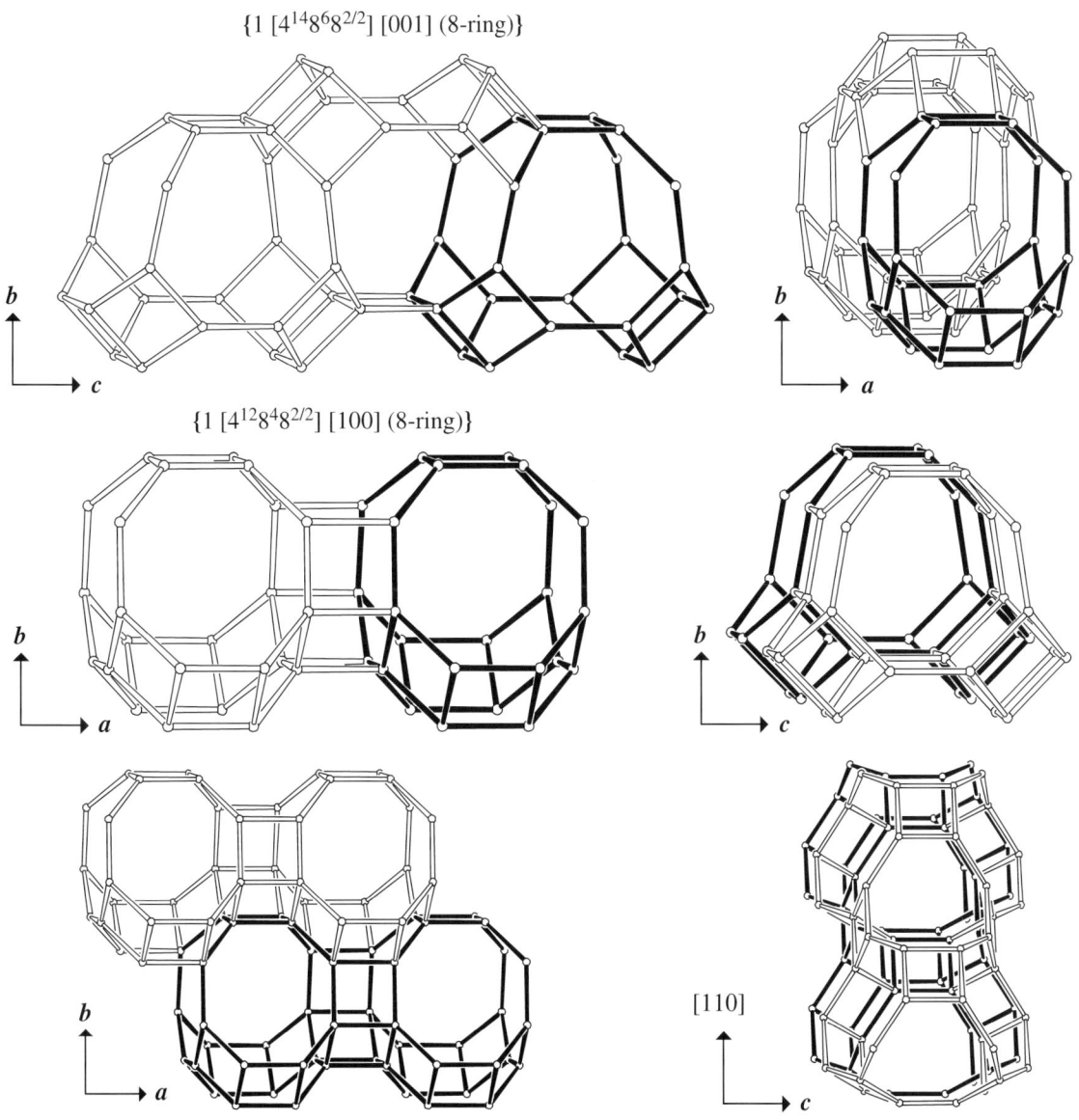

Fig. 3.  8-Ring channel parallel to $c$ (top) and 8-ring channel parallel to $a$ (middle) viewed perpendicular to the channel axis (left) and along the channel axis (right). One channel intersection, a $[4^78^5]$-cavity, in bold. Bottom: channels parallel to $a$ (one channel in bold) are connected into channels parallel to [1–10] and [110]. View along $c$ (left) and along [−110] (right).

# PON **Building Scheme**

## 1. Periodic Building Unit

**PON** can be built using units of 12 T atoms. The T12-unit (bold in Figure 1) consists of five fused 4-rings (or two 4-2 units). A two-dimensional PerBU is obtained when T12- units, related along *a* and *b* by pure translations, are connected into the *ab* layer shown in Figure 1.

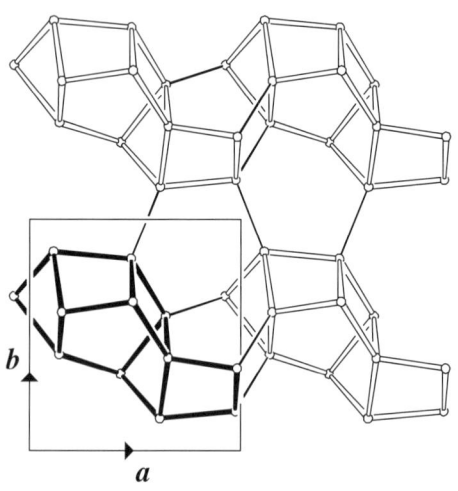
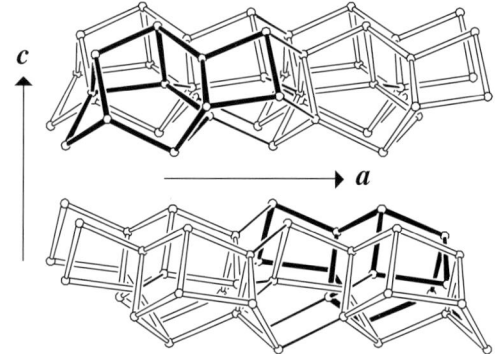

Fig. 1.   PerBU viewed along *c* (left) and along *b* (right). The PerBUs, depicted at the right are related by a rotation of 180° about the plane normal parallel to *c*.

## 2. Connection mode

Neighboring PerBUs, related along *c* by a screw rotation of 180° about *c*, are connected along *c* through 6- and 10-rings as depicted in Figure 2.

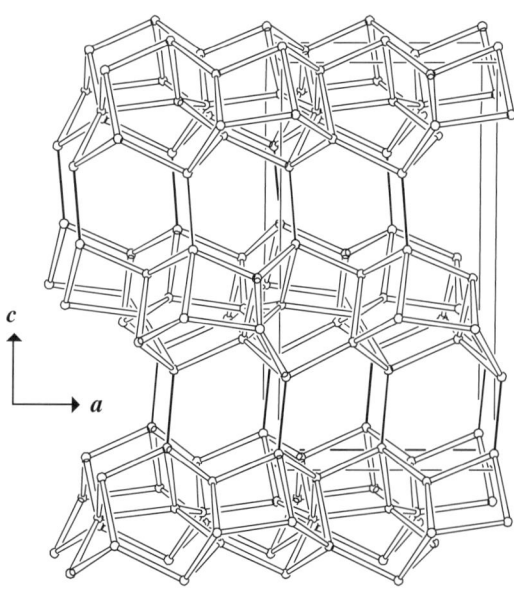
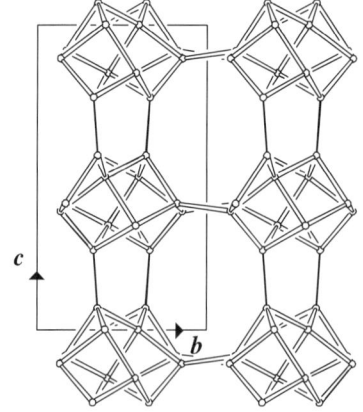

Fig. 2.   Connection mode (and unit cell content) viewed along *b* (left) and projection of the unit cell content along *a* (right).

## Cage/Channel

### 3. Channels and/or cages

Non-interconnecting 10-ring channels are parallel to *a* as depicted in Figure 3.

$\{1 \ [4^4 6^{12} 10^{2/2}] \ [100] \ (10\text{-ring})\}$

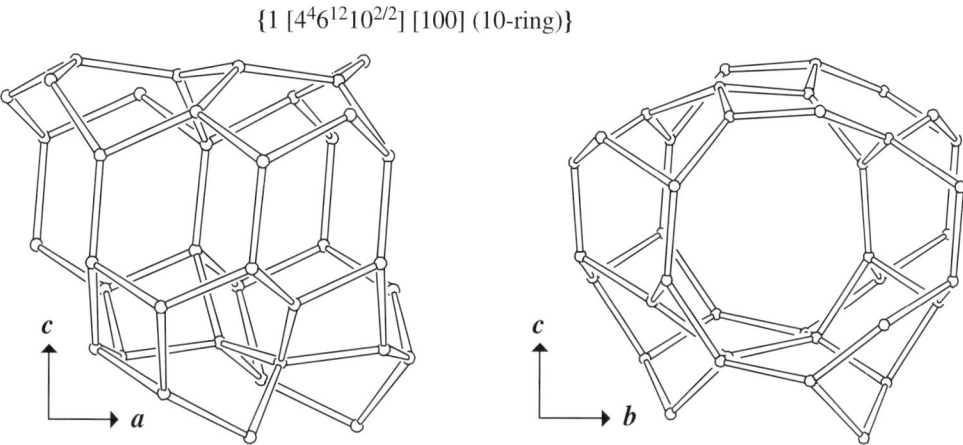

Fig. 3.　Channel viewed perpendicular to the channel axis (left) and along the channel axis (right).

# RHO                    Building Scheme

## 1. Periodic Building Unit

Cubic **RHO** can be built using a zero-dimensional PerBU, the *rho* cavity, composed of 48 T atoms (twelve 4-rings, eight 6-rings or six 8-rings; Figure 1). (See also **KFI**.)

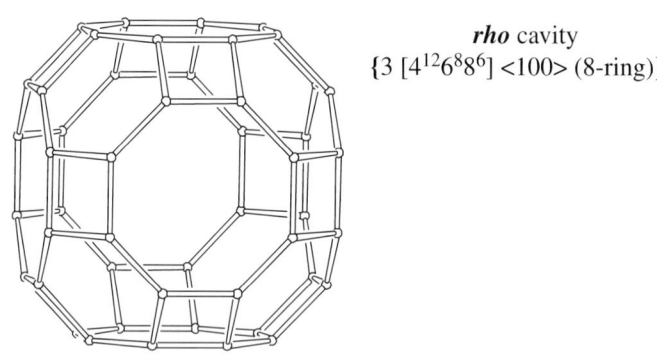

*rho* cavity
$\{3\ [4^{12}6^88^6]\ <100>\ (8\text{-ring})\}$

Fig. 1.   PerBU (*rho* cavity) viewed along a cube axis.

## 2. Connection mode

The three-dimensional framework is obtained when PerBUs, related along the cube axes by pure translations, are connected through double 8-rings. Another *rho* cavity is formed at the center of the cube. The connection mode in a cubic face is illustrated in Figure 2. An alternative PerBU of **RHO** is the double 8-ring.

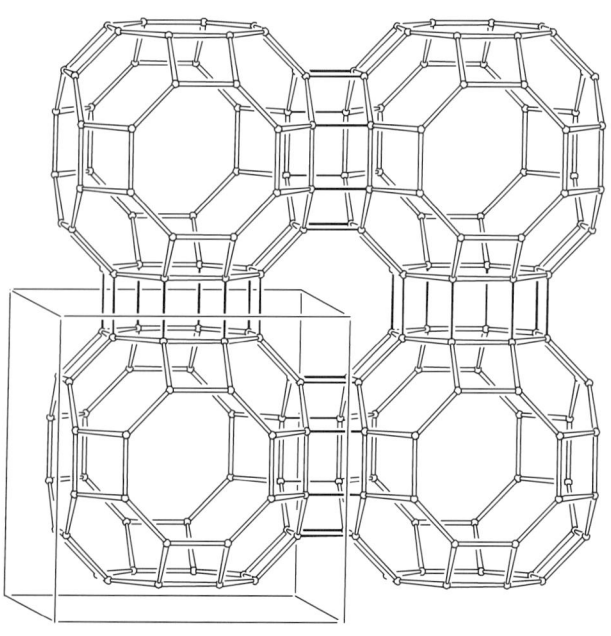

Fig. 2.   Connection mode (and unit cell content) viewed along a cube axis.

## 3. Channels and/or cages

Two equal, non-interconnecting three-dimensional systems of 8-ring channels are parallel to <100>. In both channel systems, intersecting 8-ring channels are parallel to the cell axes (Figure 3). The channel intersection is equal to the *rho* cavity. The *rho* cavity is also observed in **-CLO**, **KFI**, **LTA**, **LTN**, **PAU**, **TSC** and **UFI**. The *rho* cavity is shown in Figure 1.

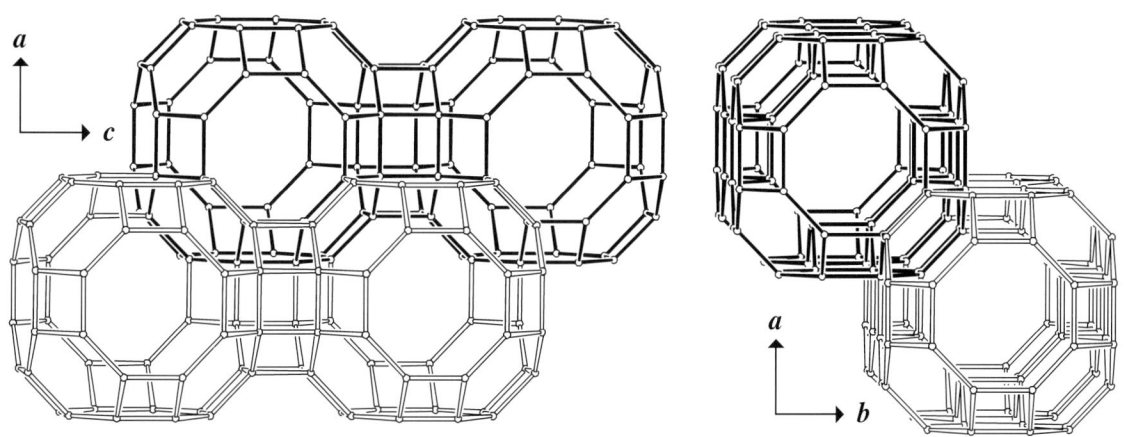

Fig. 3.    Two non-interconnected (equal) systems of 8-ring channels parallel to <100> viewed perpendicular to the channel axis (left) and along the channel axis (right).

# -RON

# Building Scheme

## 1. Periodic Building Unit

The interrupted framework of tetragonal **-RON** can be built using units of 28 T atoms. The T28-unit consists of four 3-rings (bold in Figure 1, left) and two 6-[1,1] units (or a $[4^2 6^4]$-cage with four additional T atoms). The one-dimensional PerBU is obtained when T28-units, related along $c$ by pure translations, are connected along $c$ into a column of $[4^2 6^4]$-cages sharing a 4-ring with (substituted) 3-rings attached to the column as illustrated in Figure 1 (right).

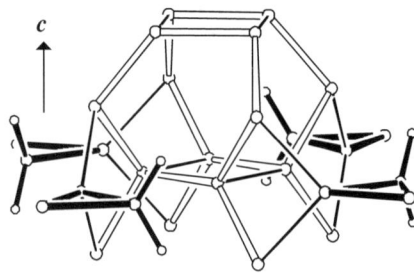
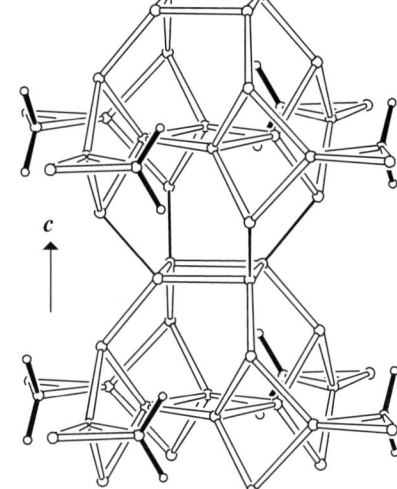

Fig. 1.   Left: T28-unit viewed perpendicular to $c$. In each 3-ring, one T atom is connected to two terminal oxygen atoms (indicated by bold bonded small circles). Right: PerBU viewed perpendicular to $c$.

## 2. Connection mode

Neighboring PerBUs, related by a shift of $\frac{1}{2}(a + b + c)$, are connected through 4-rings (Figure 2).

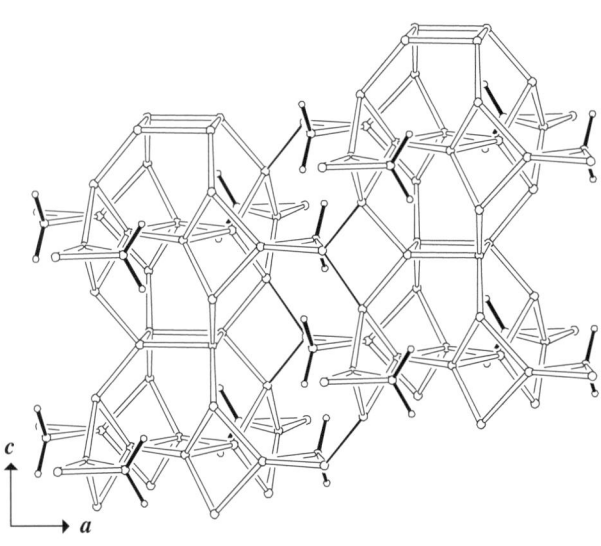

Fig. 2a.   Connection mode viewed along $b$.

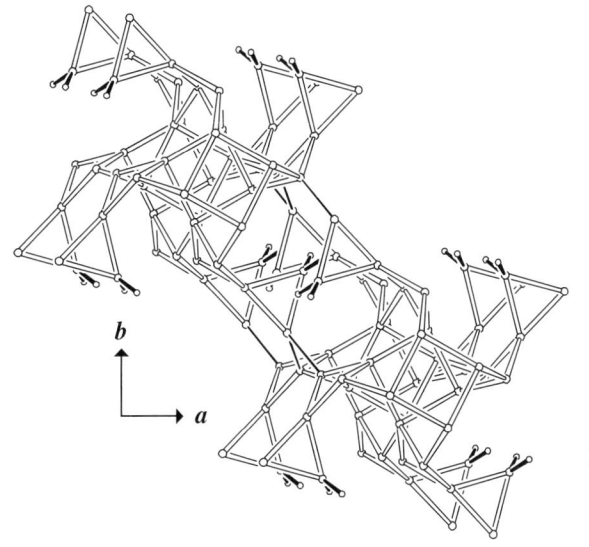

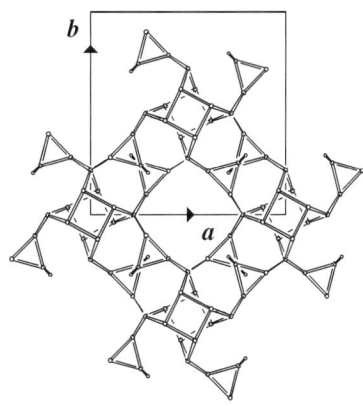

Fig. 2b.    Connection mode viewed along $c$ (left) and unit cell content projected along $c$ (right).

## 3. Channels and/or cages

     12-Ring channels are parallel to $c$ as shown in Figure 3. Terminal oxygen atoms and neighboring 12-ring channels block the entrance to the 10-ring windows perpendicular to <110> (see Figure 3).

$\{1\ [3^8 4^{12} 10^4 12^{2/2}]\ [001]\ (12\text{-ring})\}$

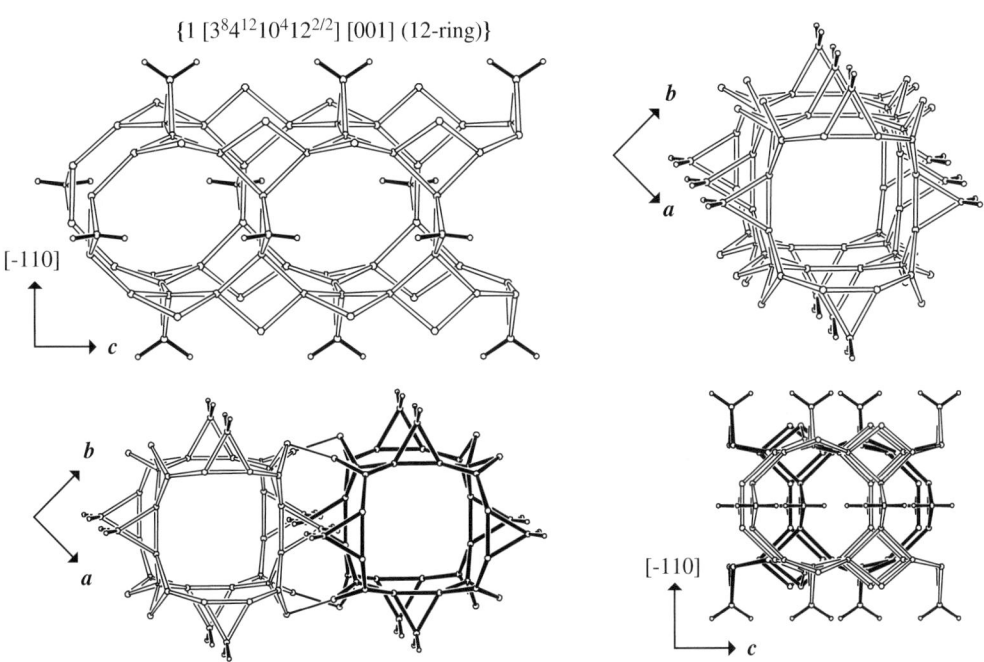

Fig. 3.    Top: 12-ring channel viewed along [110] (left) and along $c$ (right). Bottom: linkage of the channels along [110] viewed along $c$ (left) and along [110] (right).

# RRO                     Building Scheme

## 1. Periodic Building Unit

**RRO** can be built using T9-units: a double 4-ring with an additional bridging T atom (a $[4^2 5^2 6^1]$-cage, or 4-4 = 1 unit; bold in Figure 1). T9-units, related along $c$ by a rotation of 180° about an axis parallel to $b$ and passing through the bridging T atom, are connected into chains parallel to $c$. The two-dimensional PerBU is obtained when neighboring chains, related along $a$ by a pure translation, are connected through (fused) 4- and 5-rings into the $ac$ layer depicted in Figure 1. (See also **HEU** and **STI**.)

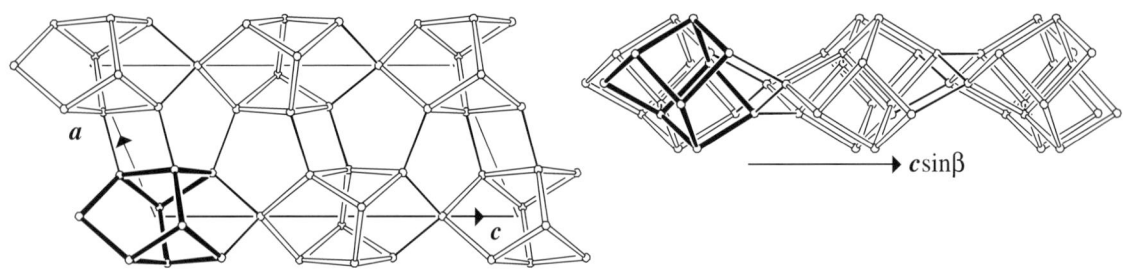

Fig. 1.    PerBU viewed along $b$ (left) and along $a$ (right).

## 2. Connection mode

Neighboring PerBUs, related along $b$ by a pure translation, are connected along $b$ through 8- and 10-rings as depicted in Figure 2.

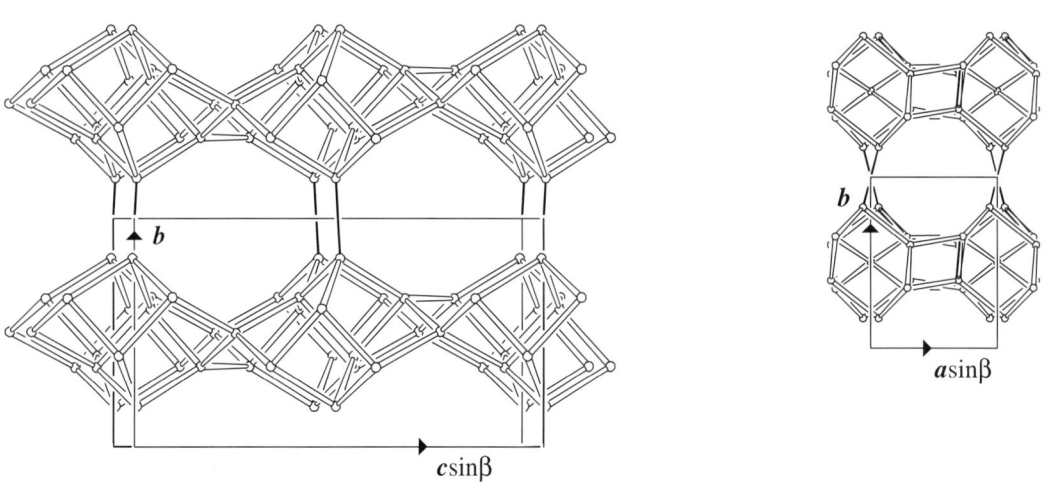

Fig. 2.    Connection mode (and unit cell content) viewed along $a$ (left) and unit cell content projected along $c$ (right).

## 3. Channels and/or cages

Intersecting 8-ring and 10-ring channels are parallel to $c$ and $a$, respectively. One channel intersection, equal to the repeat unit in the channel parallel to $a$, is drawn in bold (Figure 3).

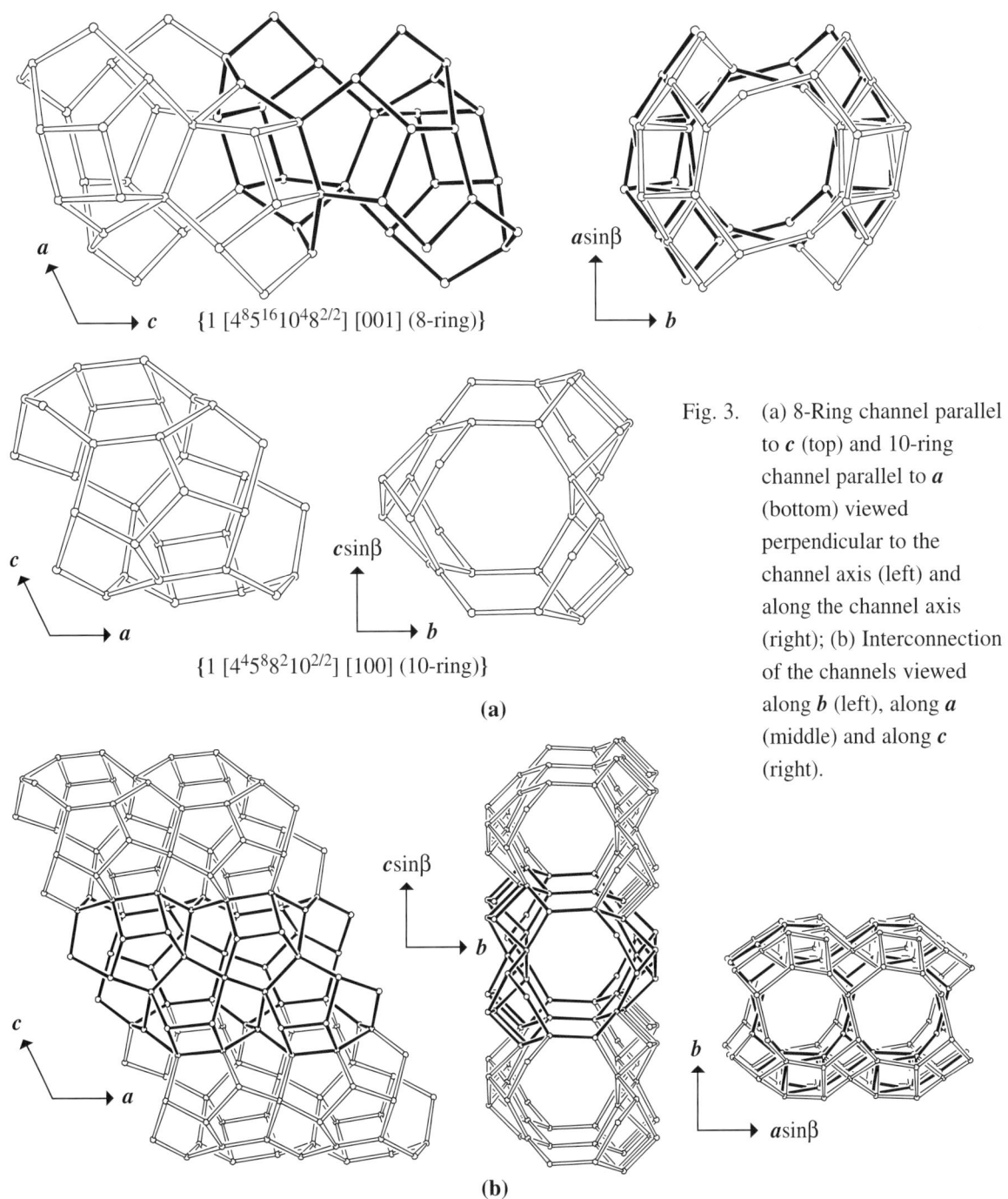

$\{1\ [4^85^{16}10^48^{2/2}]\ [001]\ (8\text{-ring})\}$

$\{1\ [4^45^88^210^{2/2}]\ [100]\ (10\text{-ring})\}$

**(a)**

Fig. 3. (a) 8-Ring channel parallel to $c$ (top) and 10-ring channel parallel to $a$ (bottom) viewed perpendicular to the channel axis (left) and along the channel axis (right); (b) Interconnection of the channels viewed along $b$ (left), along $a$ (middle) and along $c$ (right).

**(b)**

# RSN        Building Scheme

## 1. Periodic Building Unit

**RSN** can be built using units of 9 T atoms: two 4-rings connected through a single T atom (bold in Figure 1). The two-dimensional PerBU, composed of T9-units related along $a$ and $b$ by pure translations, is equal to the layer depicted in Figure 1. (See also **LOV** and **VSV**.)

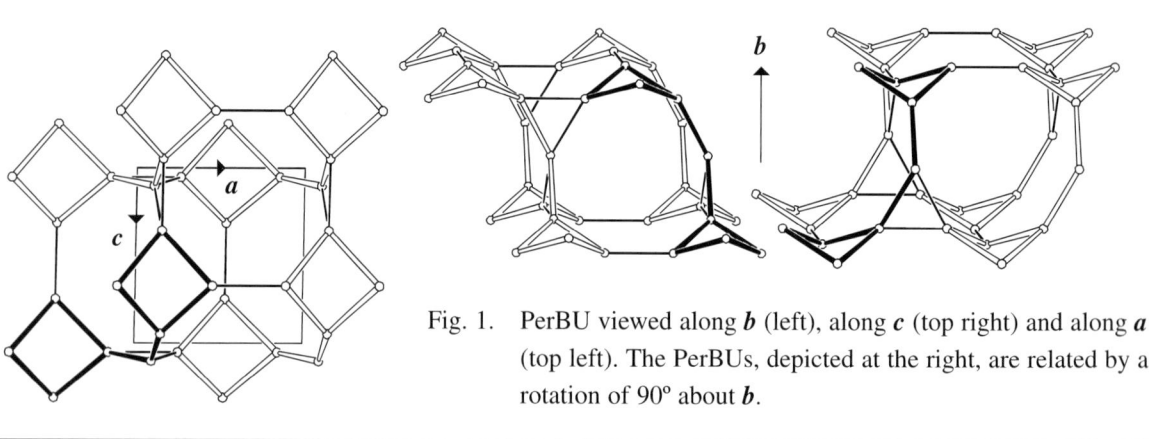

Fig. 1.    PerBU viewed along $b$ (left), along $c$ (top right) and along $a$ (top left). The PerBUs, depicted at the right, are related by a rotation of 90° about $b$.

## 2. Connection mode

Neighboring PerBUs are related along $b$ by (approximately) a rotation of 90° about $b$ and a shift of $\frac{1}{4}b$ followed by a lateral shift along $a$ and $c$ of (simultaneously) zero and zero (connection mode (**1**)), and zero and $\frac{1}{2}a$ (connection mode (**2**)) as illustrated in Figure 2.

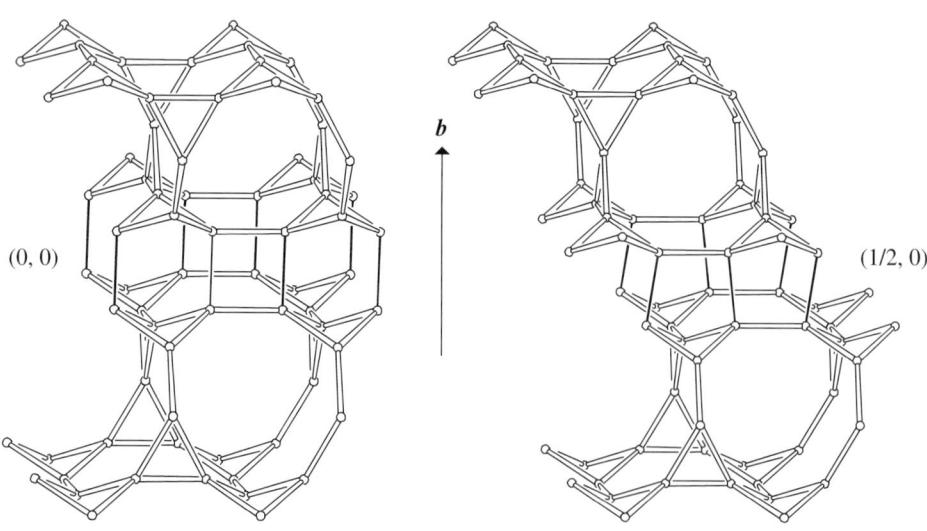

Fig. 2a.    Left: connection mode (**1**). The PerBUs are connected through 4- and 6-rings. Right: connection mode (**2**). The PerBUs are connected through 5-rings. The lateral shifts between neighboring PerBUs are given in fractions of $a$ and $c$.

# Building Scheme

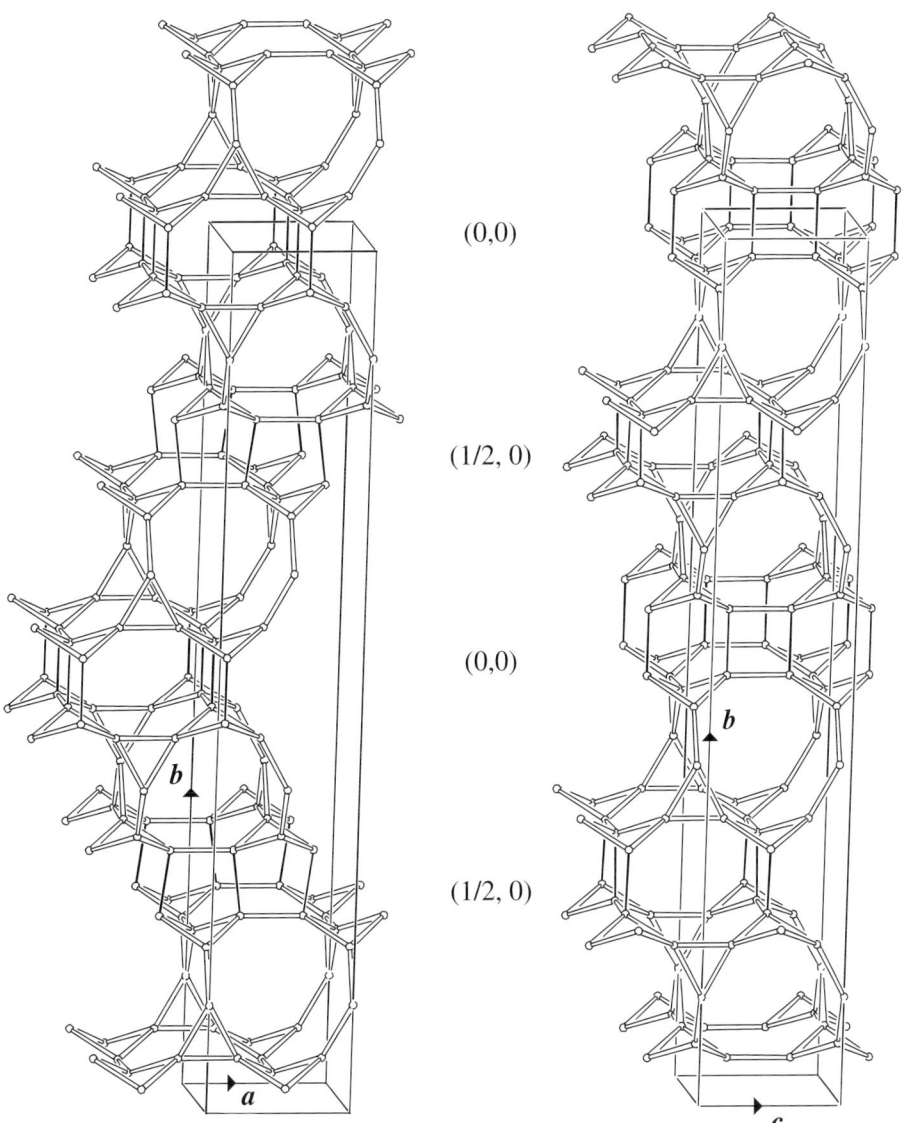

(0,0)

(1/2, 0)

(0,0)

(1/2, 0)

Fig. 2b. Unit cell content viewed along **c** (left) and along **a** (right). The lateral shifts between neighboring PerBUs, related by a rotation of 90° about **c**, are given in the drawings in fractions of **a** and **c**.

### 3. Channels and/or cages

Two (different) types of interconnecting 8-ring channels are parallel to $a$ and $c$, respectively (Figure 3). The 8-ring channels parallel to $c$ are topologically equivalent to (one of) those in **DFT**, **LOV** and **LTL**, and the 8-ring channels parallel to $a$ to those in **EON**, **MAZ**, **MON**, **MOR** and **VSV**. Intersecting 9-ring channels are parallel to $a$ and $c$. The 9-ring channels (Figure 3) are topologically equivalent to those in **NAB**, **LOV** and **VSV**. The channels are interconnected along $b$ through 8-rings perpendicular to $b$ as can be seen in Figure 3b on next page.

$\{1\ [4^2 6^2 8^2 8^{2/2}]\ [001]\ (8\text{-ring})\}$

$\{1\ [5^4 8^2 8^{2/2}]\ [100]\ (8\text{-ring})\}$

$\{1\ [3^2 4^1 8^1 9^2 9^{2/2}]\ [100]\ (9\text{-ring}),\ [001]\ (9\text{-ring})\}$

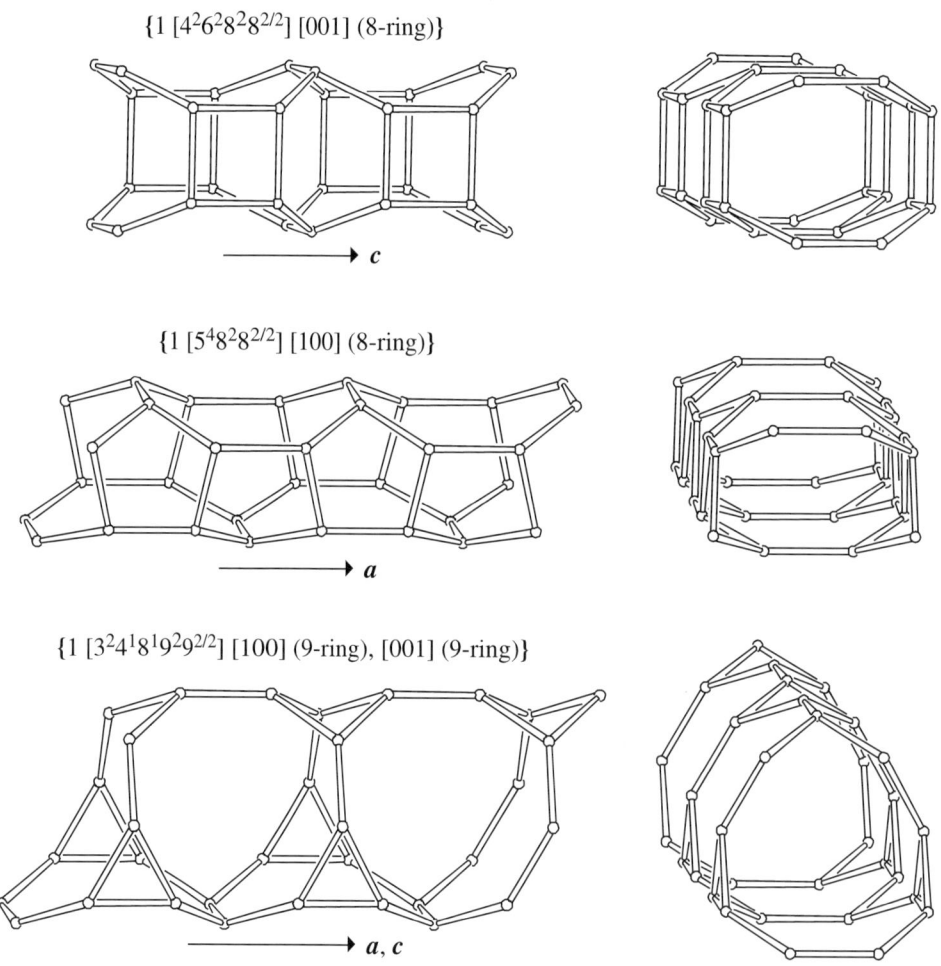

Fig. 3a. 8-Ring channels (top and middle) and 9-ring channel (bottom) viewed perpendicular to the channel axis (left) and along the channel axis (right).

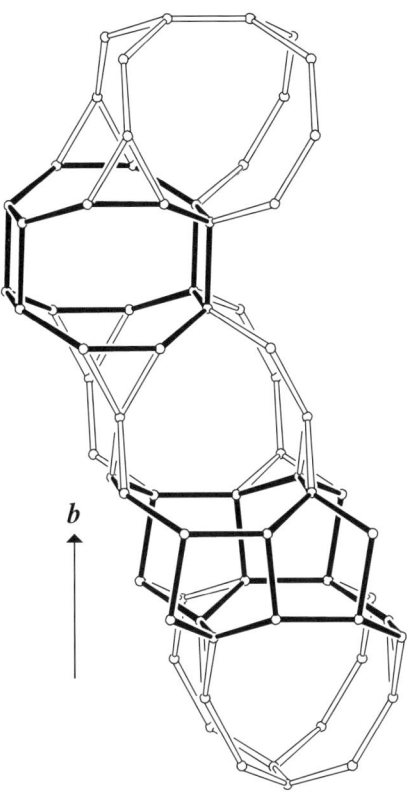

Fig. 3b. The 8-ring and 9-ring channels are interconnected along **b** through common 8-rings (the two types of 8-ring channels are in bold). Only one repeat unit in the channels is drawn for clarity. Intersecting 9-ring channels parallel to **a** and **c** have [9⁴]-cavities in common. (See **NAB**.)

# RTE                     Building Scheme

## 1. Periodic Building Unit

**RTE** can be built using T12-units consisting of 2-fold (1,4)-connected double 6-rings. The one-dimensional PerBU is the chain obtained when T12-units (one bold in Figure 1), related along *c* by pure translations, are connected along *c* through 4-rings. Alternatively, the chain can be built using 5-1 units as is also illustrated in Figure 1. In this building scheme T12-units will be used.

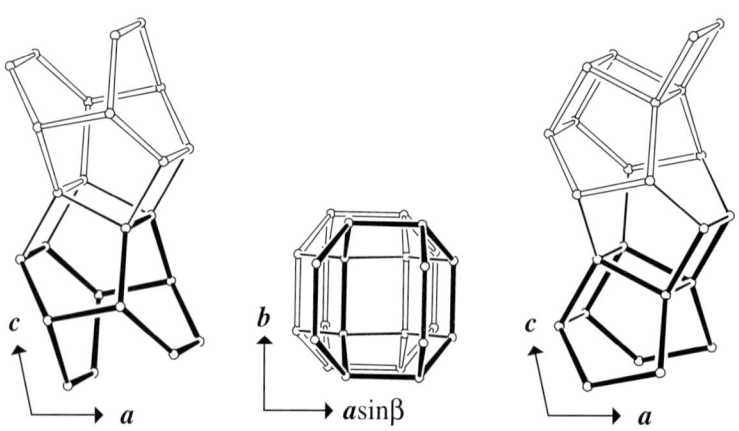

Fig. 1.  PerBU built from T12-units viewed along *b* (left) and down *c* (middle) and PerBU built from 5-1 units viewed along *b* (right).

## 2. Connection mode

Neighboring PerBUs, related along *a* (and *b*) by a shift of $\frac{1}{2}(a+b)$, are connected along *a* (and *b*) through 4-rings into the three-dimensional framework of **RTE** depicted in Figure 2.

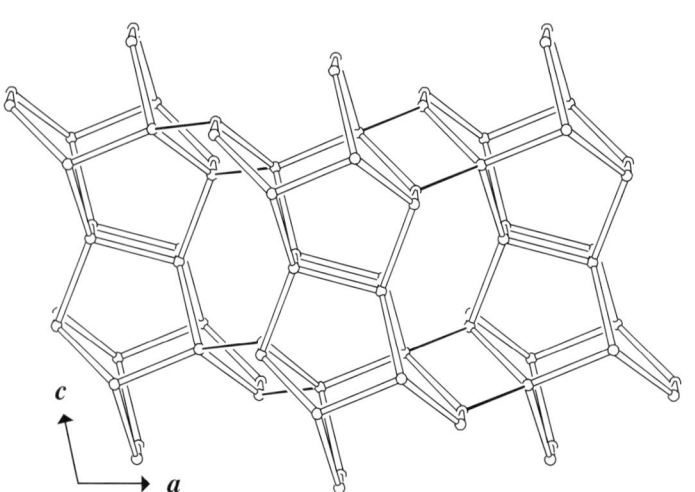

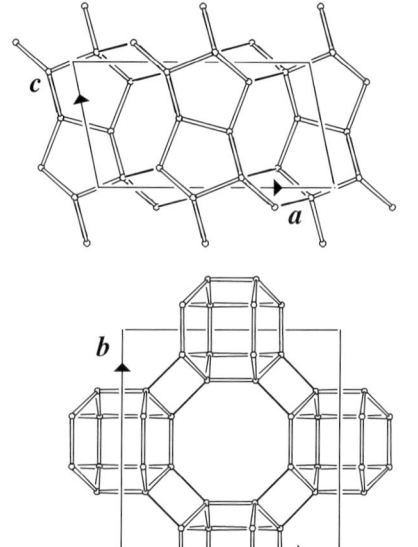

Fig. 2.  Connection mode viewed along *b* (left) and unit cell content projected along *b* (top right) and along *c* (bottom right). In the perspective drawing only three PerBUs are drawn for clarity.

## 3. Channels and/or cages

Non-interconnecting 8-ring channels, parallel to $c$, consist of cavities that are connected along $c$ through common 8-rings (Figure 3).

$\{1\ [4^{6}5^{4}6^{6}8^{2/2}]\ [001]\ (8\text{-ring})\}$

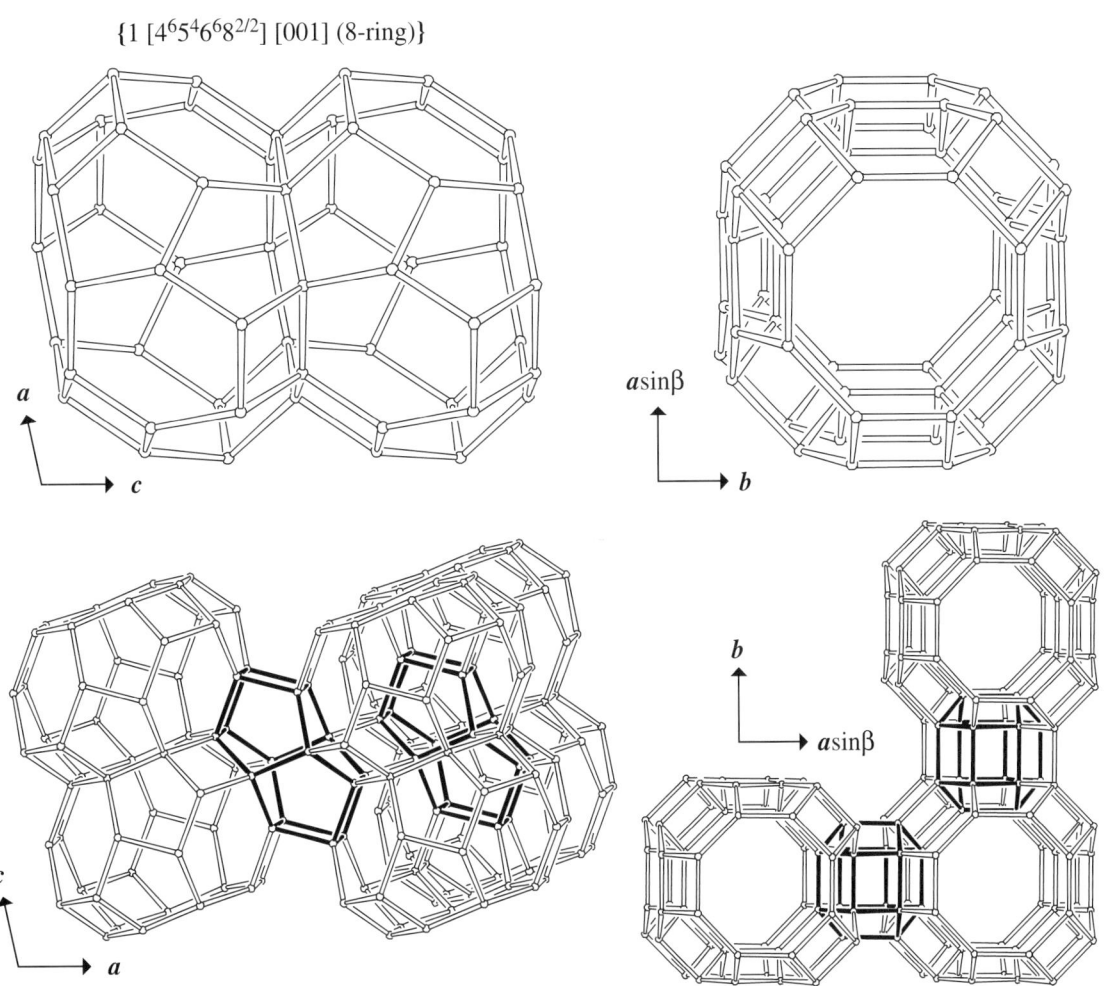

Fig. 3. Top: channel viewed perpendicular to the channel axis (left) and along the channel axis (right). Bottom: connection of channels viewed along $b$ (left) and along $c$ (right). The inter-connecting $[4^{4}5^{4}6^{2}]$-cages (also present in **RUT**) are in bold.

# RTH    Building Scheme

## 1. Periodic Building Unit

**RTH** can be built using units of 16 T atoms consisting of two sets of three (fused) 4-rings that are related by a rotation of 90° about **b**. T16-units (one in bold in Figure 1), related along **a** by a pure translation, are connected along **a** into a chain of [4⁴5⁴]-cages sharing a 4-ring. The two-dimensional PerBU is obtained when chains, related along **c** by a pure translation, are linked along **c** through 6-rings into the *ac* layer shown in Figure 1. (See also **ITE**.)

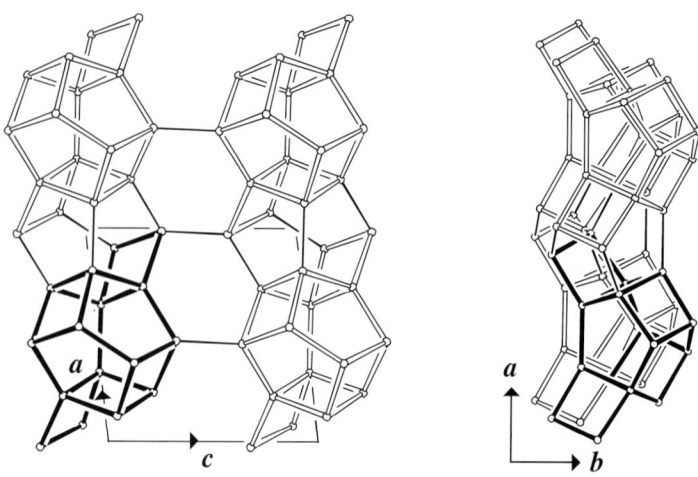

Fig. 1.   PerBU viewed along **b** (left) and along **c** (right).

## 2. Connection mode

Neighboring PerBUs, related by a shift of $\frac{1}{2}(\boldsymbol{a}+\boldsymbol{b})$, are connected along **b** through 4-rings (Figure 2).

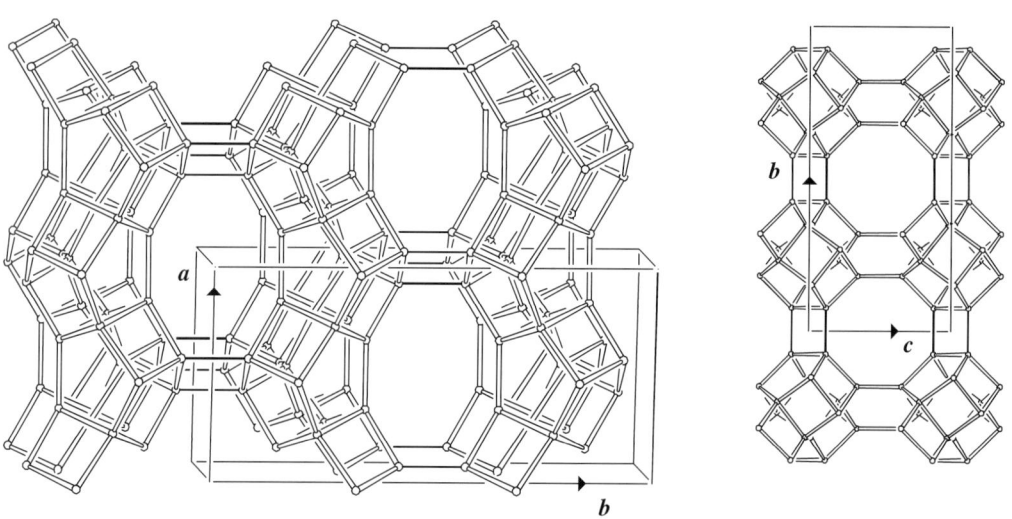

Fig. 2.   Connection mode (and unit cell content) viewed along **c** (left) and unit cell content projected along **a** (right).

### 3. Channels and/or cages

Intersecting (different) 8-ring channels are parallel to **a** and **c**. The intersection, topologically equivalent to the intersection in **ITE**, is depicted in Figure 3 together with the linkage of intersections.

$\{2\ [4^65^86^48^4]\ [100]\ (8\text{-ring}),\ [001]\ (8\text{-ring})\}$

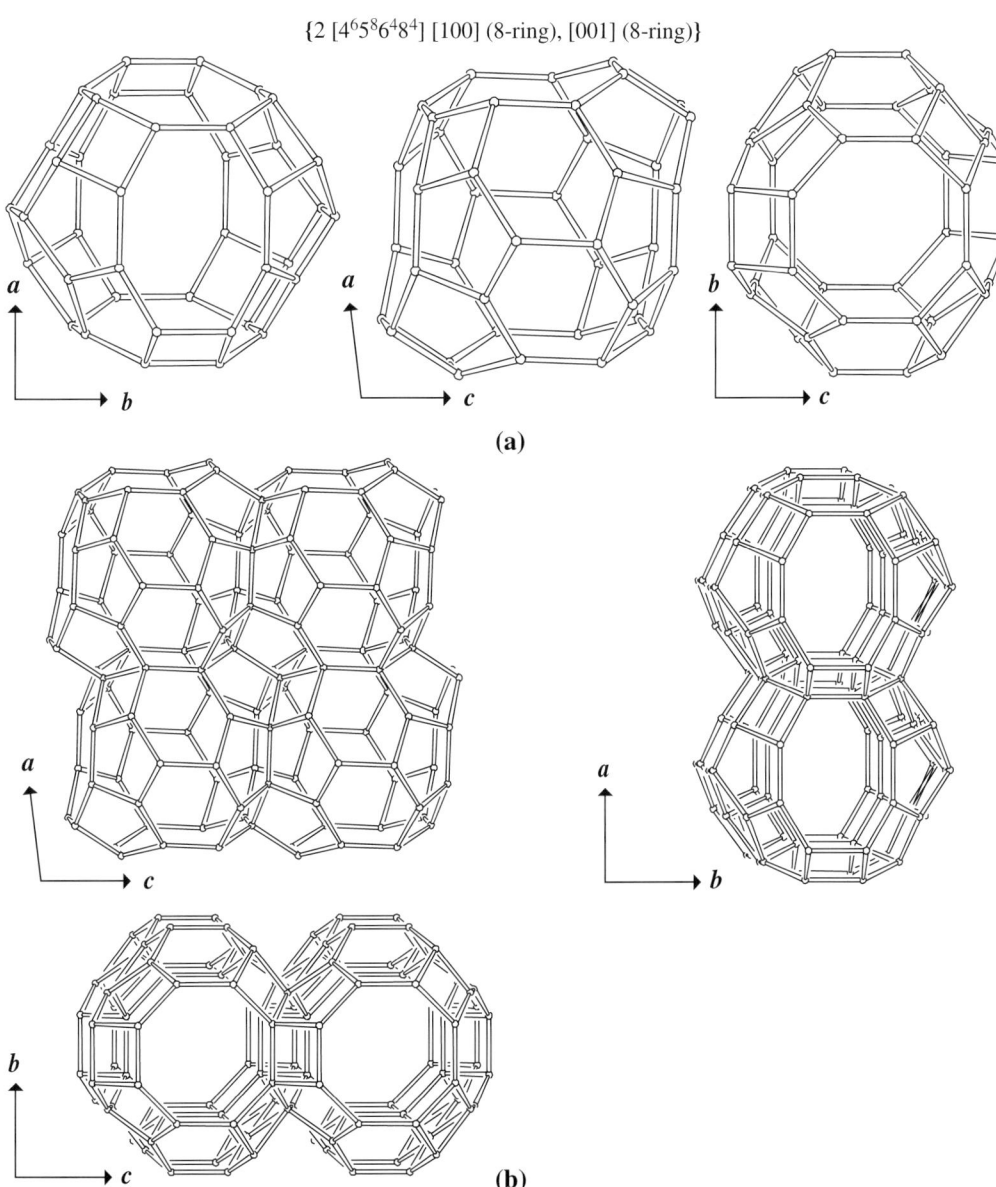

Fig. 3. (a) Channel intersection viewed along (from left to right) **c**, **b** and **a**. (b) Linkage of intersections into 8-ring channels parallel to **a** and **c** viewed along **b** (top left), along the 8-ring channel axis parallel to **c** (top right) and along the 8-ring channel axis parallel to **a** (bottom).

# RUT                     Building Scheme

## 1. Periodic Building Unit

**RUT** can be built using units of 18 T atoms. The T18-units consist of 2-fold (1,4)-connected double 6-rings linked to a single 6-ring in such a way that 4- and 5-rings are formed. The one-dimensional PerBU is the chain obtained when T18-units (one bold in Figure 1), related by a pure translation along [101], are connected along [101] through 4-rings. Alternatively, the PerBU can be built from $[4^4 5^4 6^2]$-cages connected through two additional T atoms (see Figure 1).

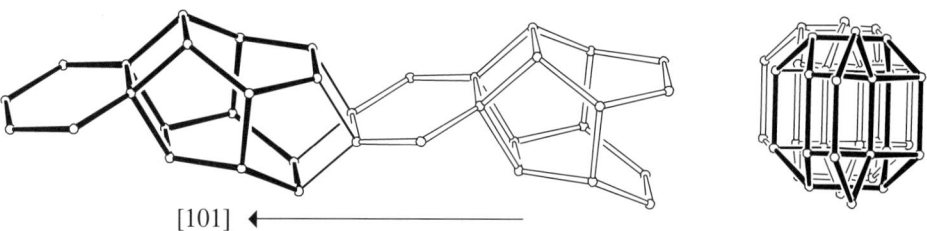

[101]

Fig. 1.    PerBU built from T18-units viewed along **b** (left) and down [101] (right).

## 2. Connection mode

Neighboring PerBUs, related along **a** (and **b**) by a shift of $\frac{1}{2}(a+b)$, are connected along **a** (and **b**) through 4-, 5- and 6-rings into the three-dimensional framework as depicted in Figure 2.

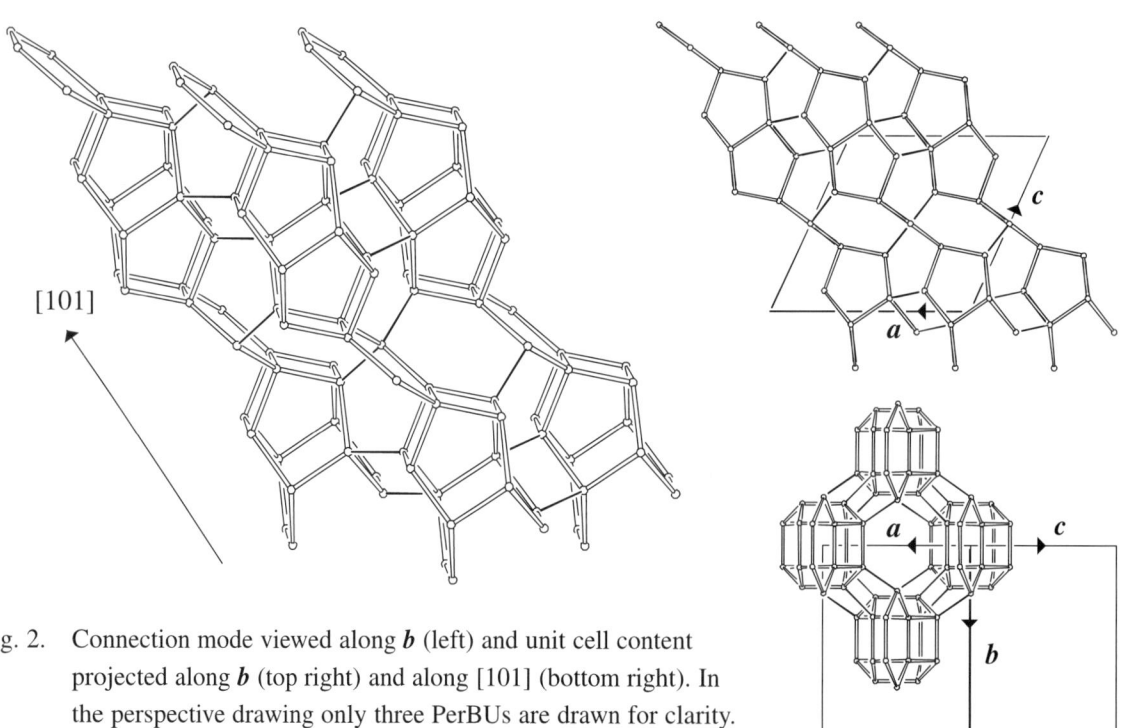

[101]

Fig. 2.    Connection mode viewed along **b** (left) and unit cell content
projected along **b** (top right) and along [101] (bottom right). In
the perspective drawing only three PerBUs are drawn for clarity.

### 3. Channels and/or cages

The cage in **RUT** is shown in Figure 3. Apertures of "channels" are formed by 6-rings only.

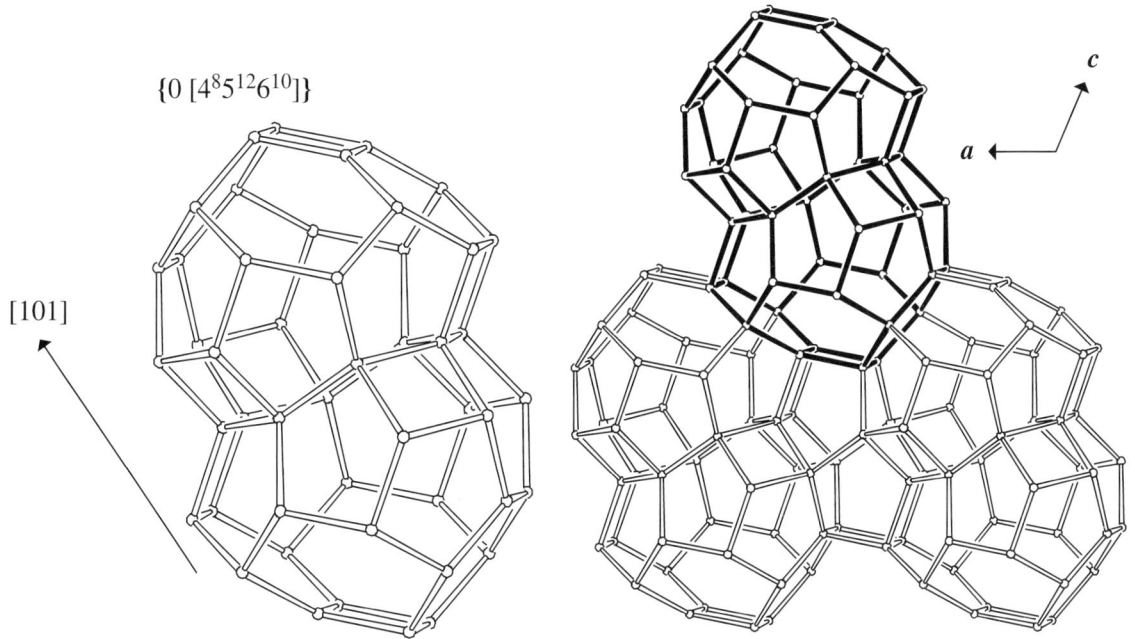

$\{0 \, [4^8 5^{12} 6^{10}]\}$

[101]

Fig. 3.  Cage viewed along **b** (left) and linkage of cages along **a** and **c** viewed along **b** (right). The interconnecting $[4^4 5^4 6^2]$-cage along **a** is also present in **RTE**.

# RWR       Building Scheme

## 1. Periodic Building Unit

The building units of 8 T atoms in tetragonal **RWR** are composed of four fused 5-rings (a $[5^4]$-cage; bold in Figure 1). The two-dimensional PerBU is obtained when these T8-units, related along *a* and *b* by pure translations, are connected into a layer with a tetragonal repeat unit (Figure 1). Infinite saw chains along *a* and *b* (bold in Figure 1) are formed.

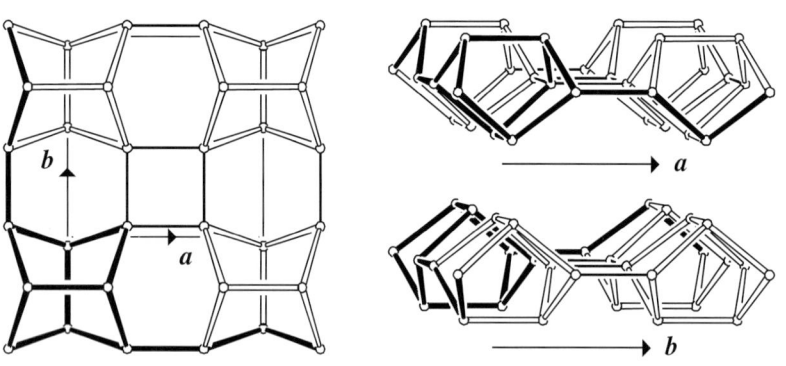

Fig. 1.    Parallel projection of the PerBU along *c* (left; one T8-unit and two saw chains in bold) and PerBU viewed along *b* (top right) and along *a* (bottom right). The PerBUs shown at the right are related by a rotation of 90° about the plane normal parallel to *c*.

## 2. Connection mode

Neighboring PerBUs, related along *c* by a rotation of 90° about *c* and a shift of $\frac{1}{2}\boldsymbol{a}$ or $\frac{1}{2}\boldsymbol{b}$ (denoted as (1/2, 0) or (0, 1/2)), are connected along *c* through 6- and 8-rings as shown in Figure 2.

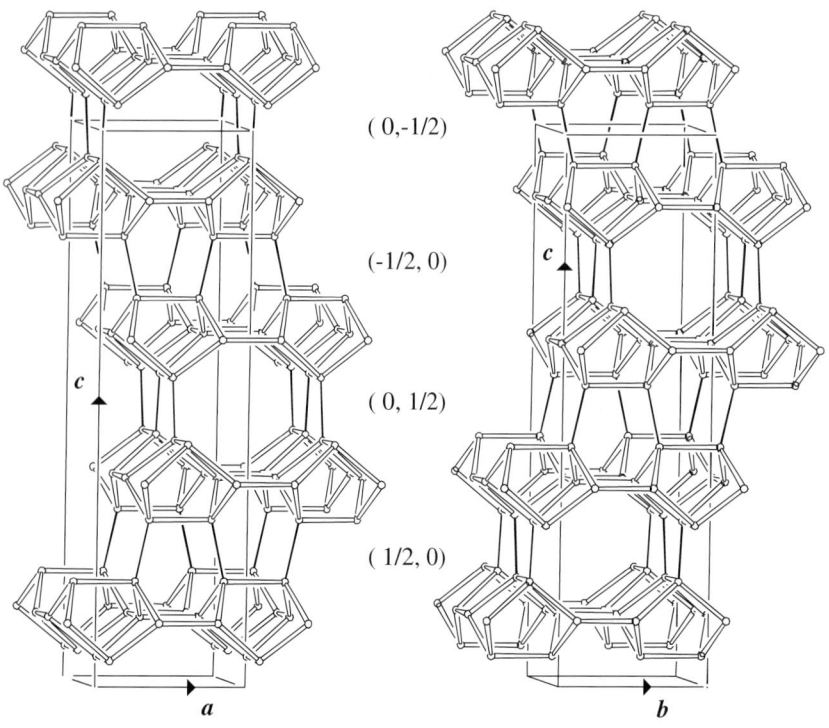

( 0,-1/2)

(-1/2, 0)

( 0, 1/2)

( 1/2, 0)

Fig. 2.    Connection mode (and unit cell content) viewed along *b* (left) and along *a* (right). The connection codes are given in fractions of *a* and *b*.

## 3. Channels and/or cages

Non-interconnecting (equal) 8-ring channels are parallel to <100>. Channel and fusion of channels is illustrated in Figure 3.

$\{1 \ [4^2 5^4 6^6 8^{2/2}] \ <100> \ (8\text{-ring})\}$

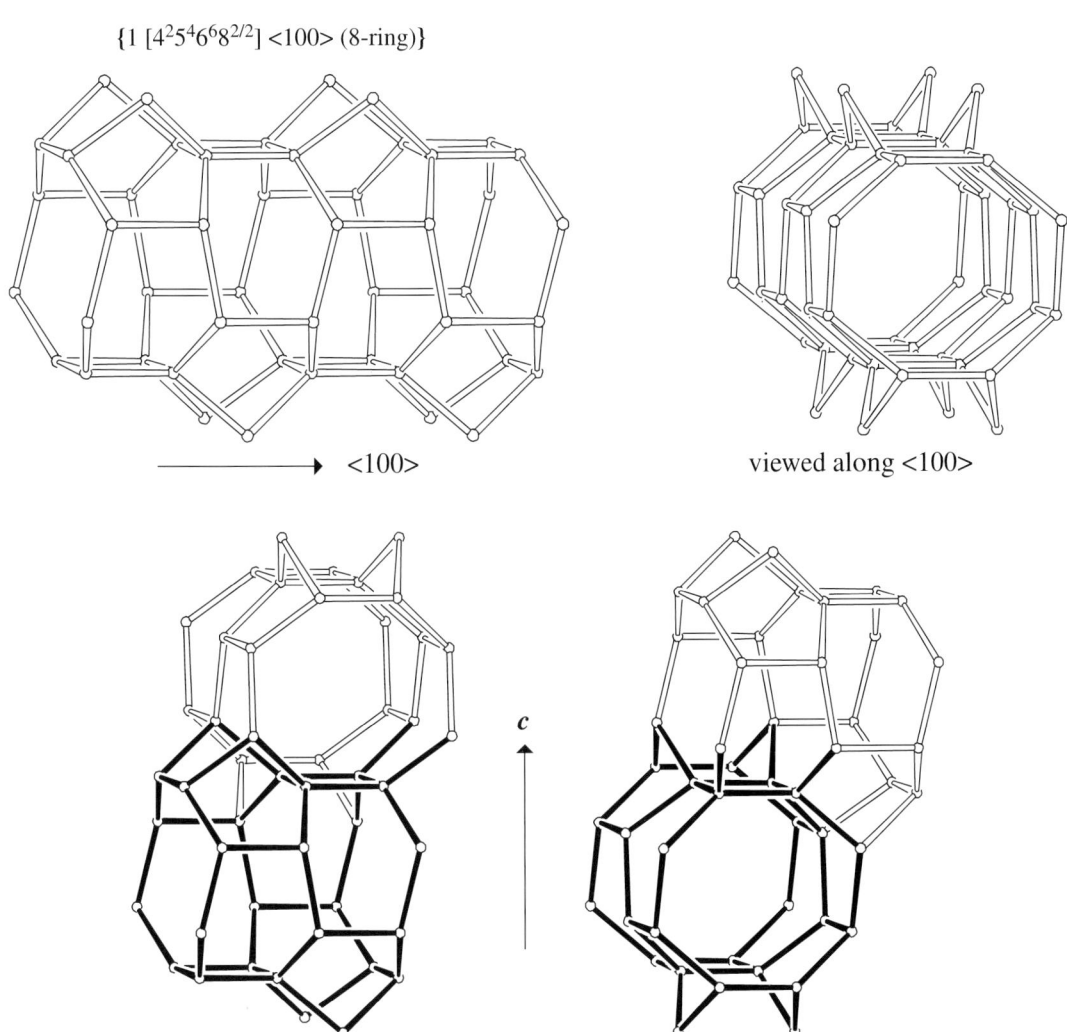

Fig. 3.    Top: 8-ring channel parallel to <100> viewed perpendicular to <100> (left) and along <100> (right). Bottom: connected channels viewed perpendicular to <100> and along <100> (right).

# RWY                    Building Scheme

## 1. Periodic Building Unit

Cubic **RWY** can be built using units of 4 T atoms: four fused 3-rings connected into a tetrahedron of T atoms (a 3*1 unit or [3⁴]-cage). The PerBU, periodic in zero dimensions, is obtained when T4-units (one in bold in Figure 1) are linked about a 4-fold axis as shown in Figure 1.

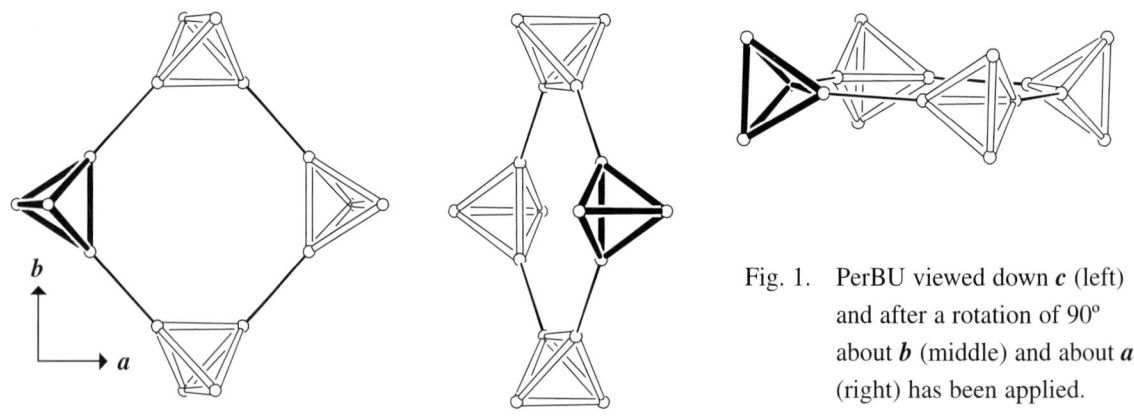

Fig. 1.  PerBU viewed down **c** (left) and after a rotation of 90° about **b** (middle) and about **a** (right) has been applied.

## 2. Connection mode

Three neighboring PerBUs, related by a rotation of 90° about **a** or **b** (or by the cubic 3-fold axis) are connected through single T–T connections as illustrated in Figure 2.

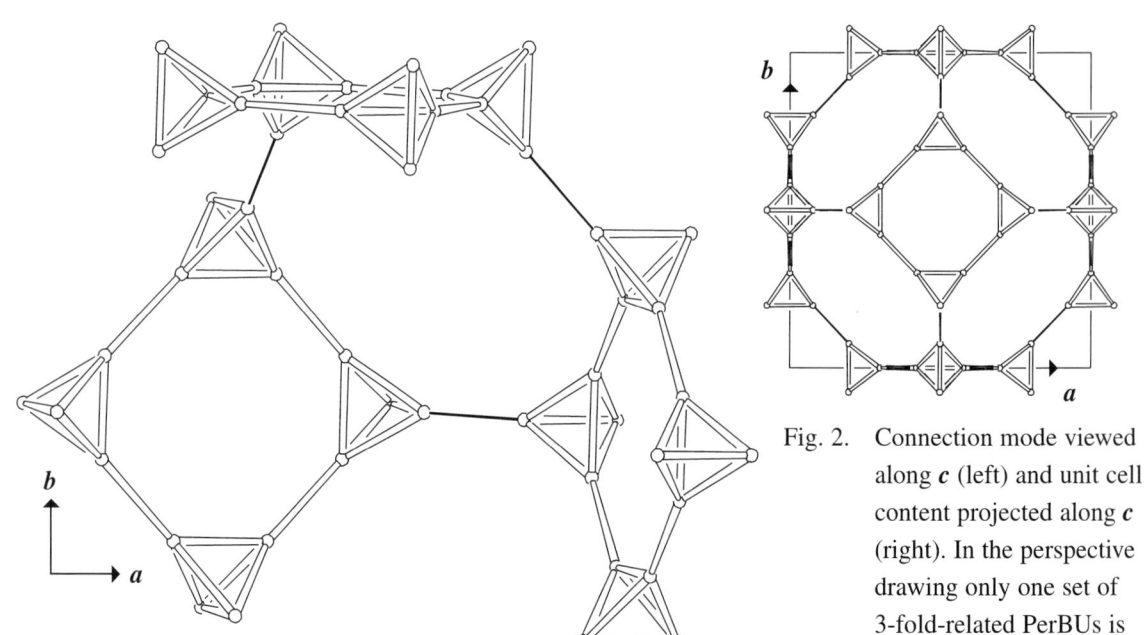

Fig. 2.  Connection mode viewed along **c** (left) and unit cell content projected along **c** (right). In the perspective drawing only one set of 3-fold-related PerBUs is shown for clarity.

## 3. Channels and/or cages

Intersecting 8-ring and 12-ring channels are parallel to <100> and <111>, respectively. The channel intersection is shown in Figure 3 together with the linkage of intersections along <100> (through common 8-rings) and along <111> (through common 12-rings) into 8-ring and 12-ring channels.

$\{3\ [3^{24}8^612^8]\ <100>\ (8\text{-ring}),\ <111>\ (12\text{-ring})\}$

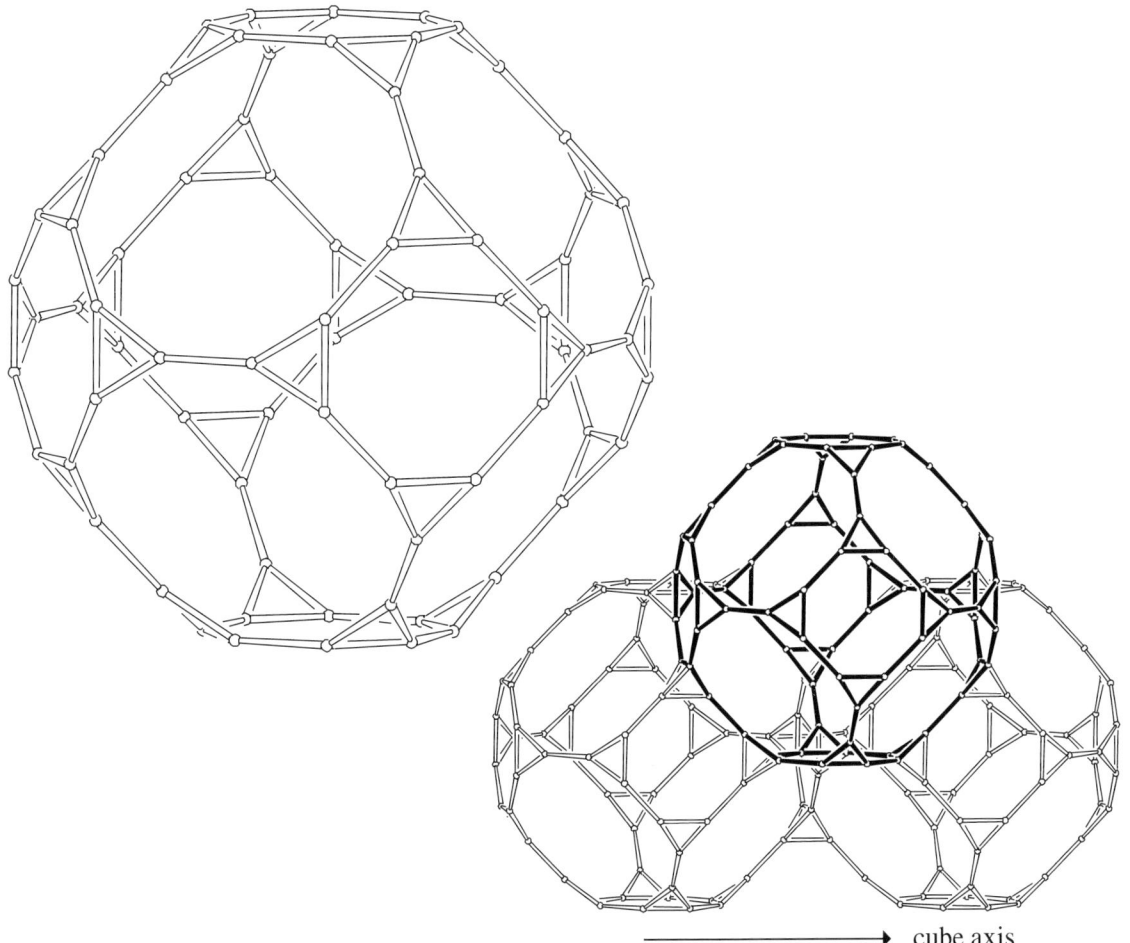

cube axis

Fig. 3.   Channel intersection (left) and linked intersections (right) viewed approximately along <001>.

# SAO                    **Building Scheme**

## 1. Periodic Building Unit

Tetragonal **SAO** can be built using units of 28 T atoms consisting of four 4-rings which are connected tetrahedral to two 4-fold (1,2,4,5)-connected 6-ring "boats" (Figure 1; in bold). The two-dimensional PerBU is obtained when T28-units, related along **a** and **b** by pure translations, are connected into the **ab** layer shown in Figure 1.

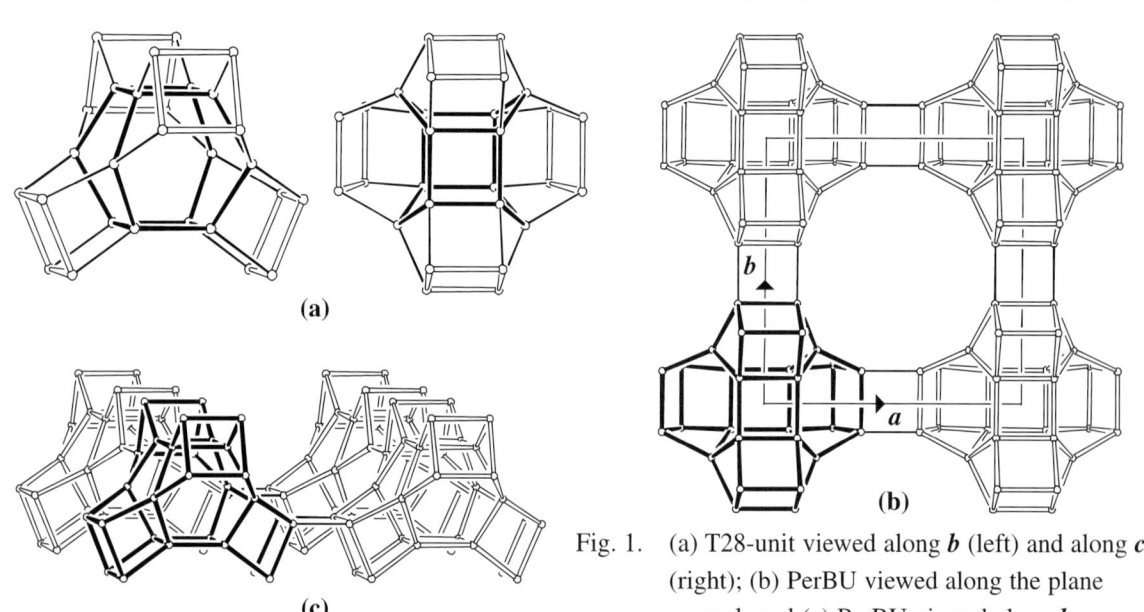

Fig. 1.   (a) T28-unit viewed along **b** (left) and along **c** (right); (b) PerBU viewed along the plane normal; and (c) PerBU viewed along **b**.

## 2. Connection mode

Neighboring PerBUs, related by a shift of $\frac{1}{2}(a+b+c)$, are connected along **c** as shown in Figure 2.

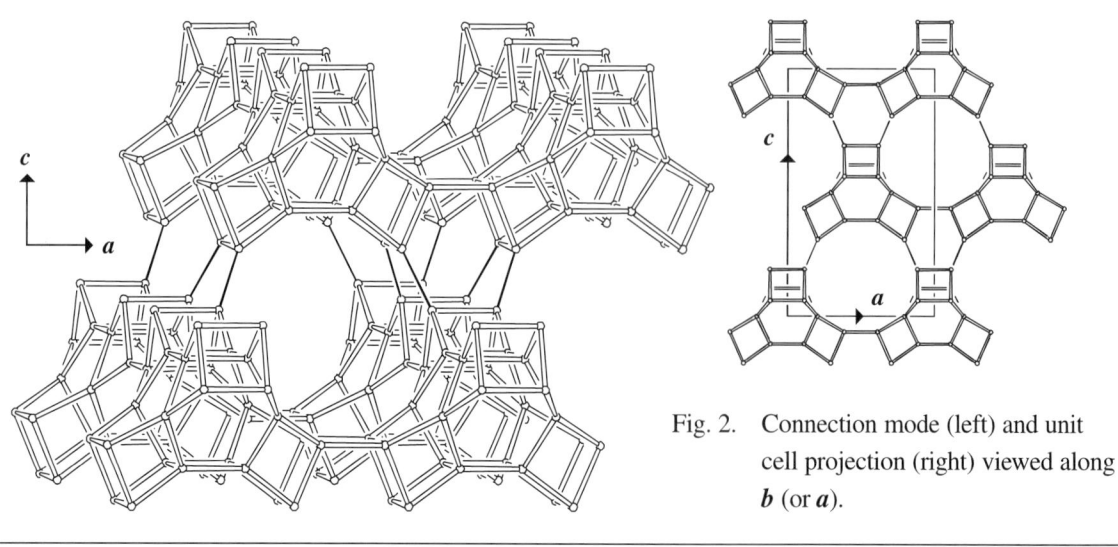

Fig. 2.   Connection mode (left) and unit cell projection (right) viewed along **b** (or **a**).

# Cage/Channel                                    **SAO**

## 3. Channels and/or cages

Interconnecting (equal) 12-ring channels are parallel to <100>. The channel and the interconnection of channels are depicted in Figure 3. Diffusion through the 16-ring windows perpendicular to *c* is obstructed as can be seen in Figure 3.

$\{1\ [4^{7}6^{4}12^{1}16^{1}12^{2/2}]\ <010>\ (12\text{-ring})\}$

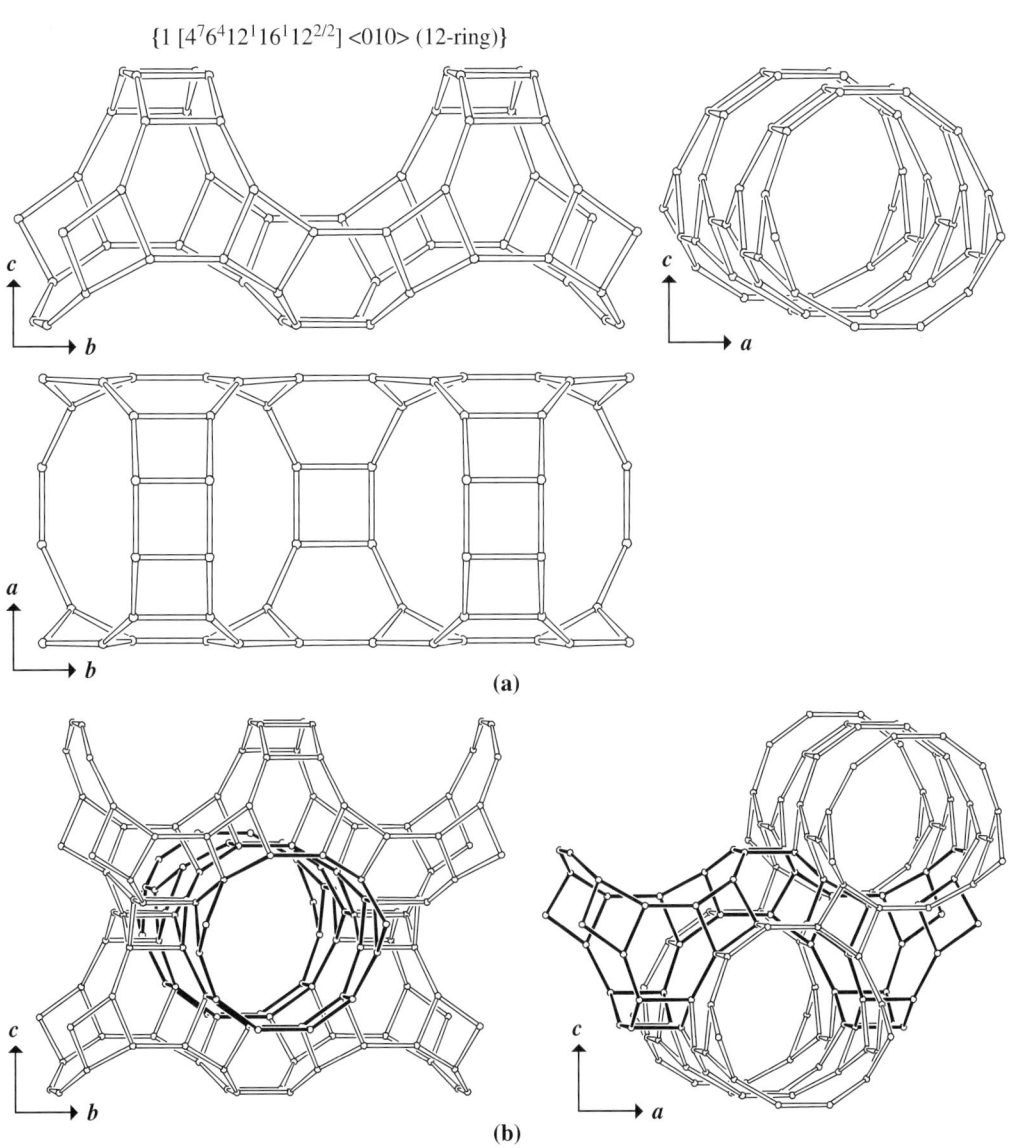

Fig. 3.   (a) 12-Ring channel viewed perpendicular to the tetragonal *c* axis (top left and top right) and along the tetragonal axis (bottom); (b) interconnection of channels viewed along *a* (left) and along *b* (right).

# SAS  Building Scheme

## 1. Periodic Building Unit

Tetragonal **SAS** can be built using as one-dimensional PerBU the chain depicted in Figure 1. The chain is composed of units of 16 T atoms (in bold) related along $c$ by a pure translation. The T16-unit consists of two 2-fold (1,2)-connected double 4-rings or, alternatively, of a double 6-ring with two "handles" (or two 6-2 units).

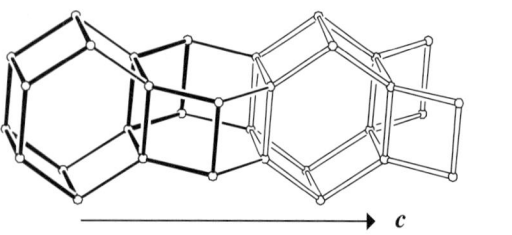 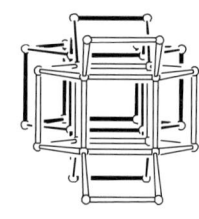

Fig. 1.  PerBU viewed perpendicular to the chain axis $c$ (left) and down $c$ (right).

## 2. Connection mode

Neighboring PerBUs are connected along a 4-fold axis parallel to $c$ through 6-rings (Figure 2).

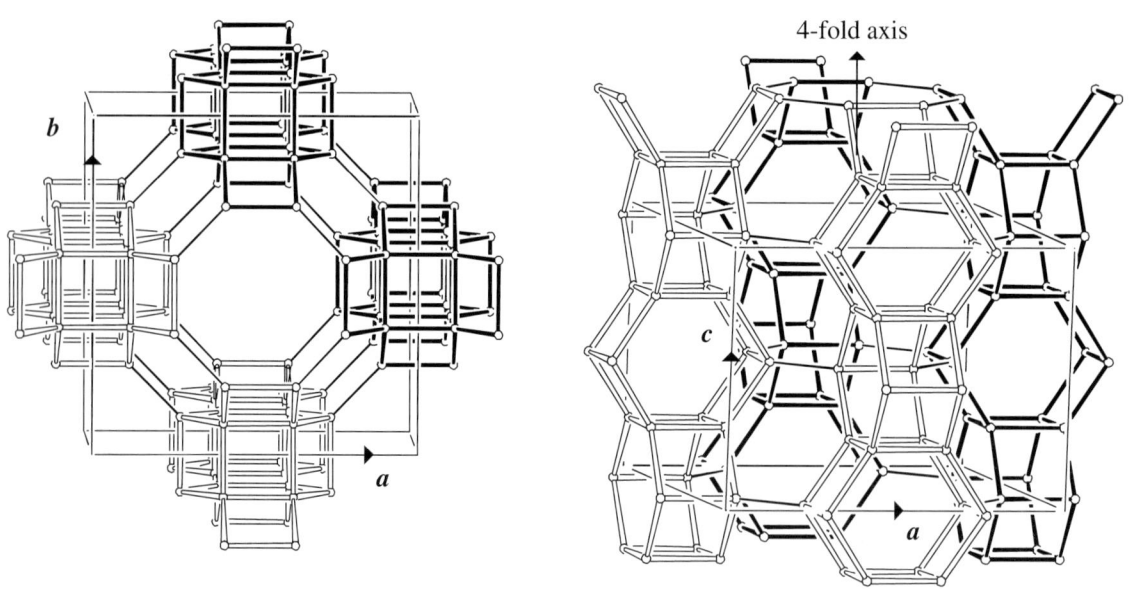

Fig. 2.  Connection mode (and unit cell content) viewed along the 4-fold axis parallel to $c$ (left) and perpendicular to $c$ (right). In the drawings two PerBUs are in bold for clarity. Contracted **rho** cavities with 8-ring windows are formed.

## 3. Channels and/or cages

Non-interconnecting 8-ring channels are parallel to *c*. The channel is composed of contracted *rho* cavities that are connected through common 8-rings along *c*. The cavities are connected along <100> through double 6-rings. Cavity and interconnection of cavities is shown in Figure 3.

Contracted *rho* cavity: $\{1\ [4^8 6^{12} 8^2]\ [001]\ (8\text{-ring})\}$

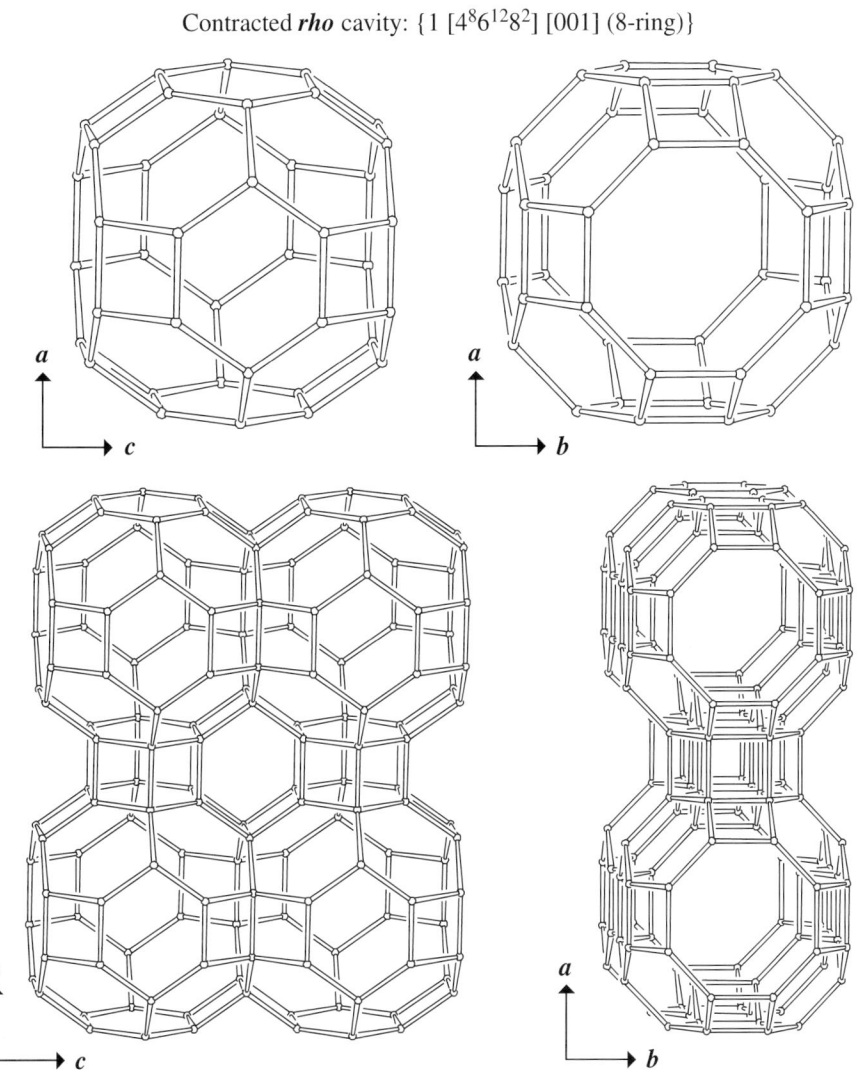

Fig. 3.    Top: contracted *rho* cavity viewed along *b* (or *a*; left) and along *c* (right). Bottom: interconnection of cavities viewed along *b* (or *a*; left) and along *c* (right).

# SAT　　　　　　　Building Scheme

## 1. Periodic Building Unit

The two-dimensional PerBU of **SAT** consists of a hexagonal array of non-connected planar 6-rings (bold in Figure 1), which are related along *a* and *b* by pure translations. The 6-rings are centered at (0,0) in the *ab* layer. This position is usually called the **A** position. **SAT** belongs to the ABC-6 family (see Introduction).

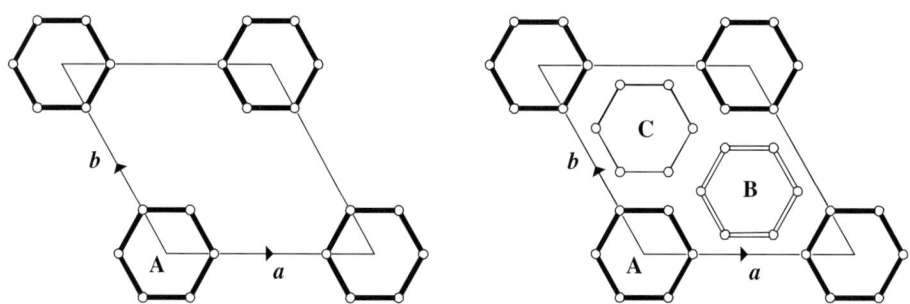

Fig. 1.　PerBU (left) and definition of 6-ring positions used in the stacking modes (right).

## 2. Connection mode

Neighboring PerBUs can be connected along *c* through tilted 4-rings in three different ways (see Introduction). In **SAT** all three connection modes between the PerBUs are observed (Figure 2).

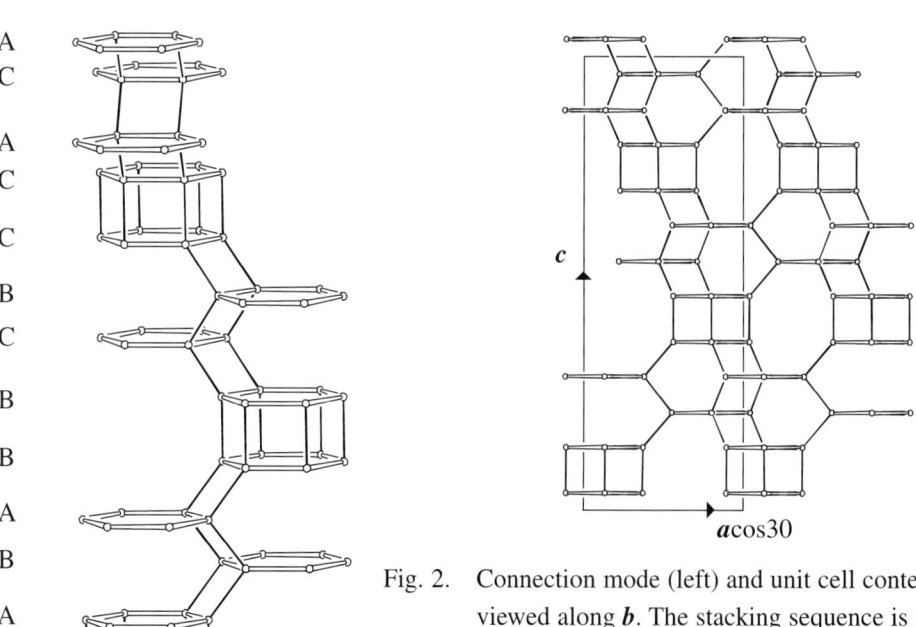

Fig. 2.　Connection mode (left) and unit cell content (right) viewed along *b*. The stacking sequence is given. In the perspective drawing, each PerBU is represented by one 6-ring only.

### 3. Channels and/or cages

The two types of cavities are depicted in Figure 3. The ***can*** cage is also present in **AFG**, **CAN**, **ERI**, **FAR**, **FRA**, **GIU**, **LIO**, **LOS**, **LTN**, **LTL**, **MAR**, **MOZ**, **OFF**, **SBS**, **SBT**, **TOL** and **-WEN**. Interconnecting two-dimensional 8-ring channel systems are perpendicular to *c*.

***can* cage:**
$\{0\,[4^66^5]\}$

***sat* cavity:**
$\{2\,[4^{12}6^86^6]\,<100>\,(8\text{-ring})\}$

Fused cages

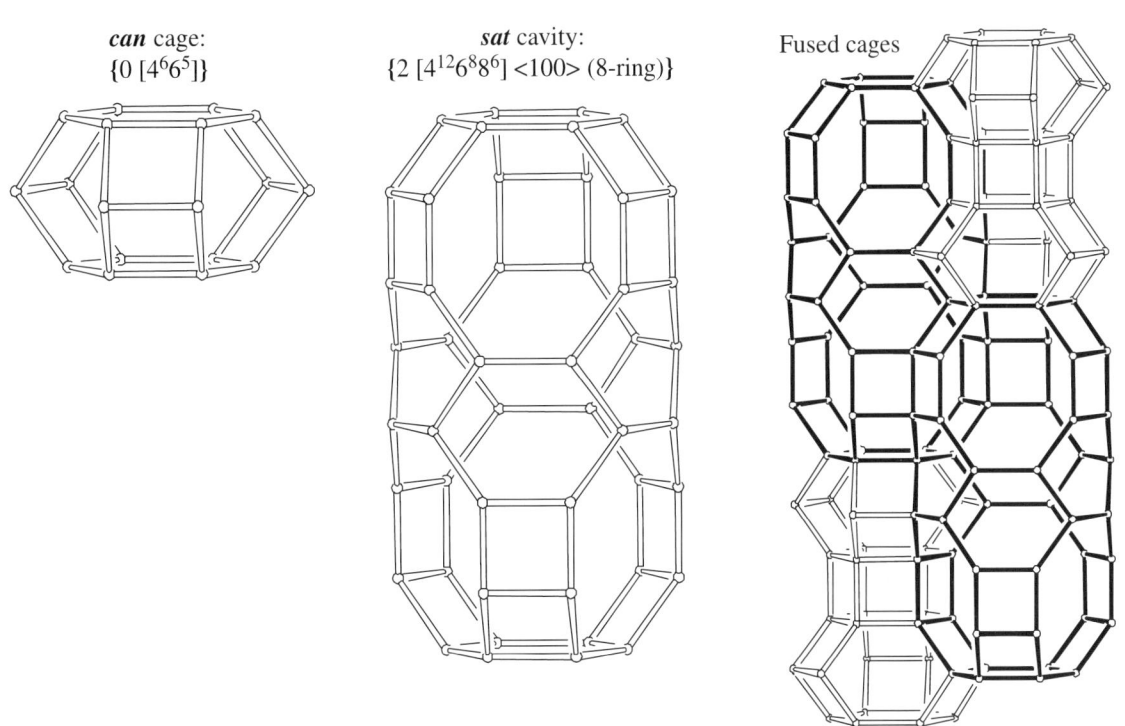

Fig. 3.  Cavities viewed along <120> (left and middle). Two-dimensional channels with 8-ring apertures are interconnected along *c* through 12-rings in the ***sat*** cavities leading to a three-dimensional channel system (***sat*** cavities in bold; right).

# SAV                    Building Scheme

## 1. Periodic Building Unit

The two-dimensional PerBU in tetragonal **SAV** is the double 6-ring layer depicted in Figure 1. Double 6-rings (one in bold), related along **a** and **b** by screw rotations of 180° about **a** and **b**, are connected into the **ab** layer through 4-rings. (See also **KFI**; compare with **AEI**.)

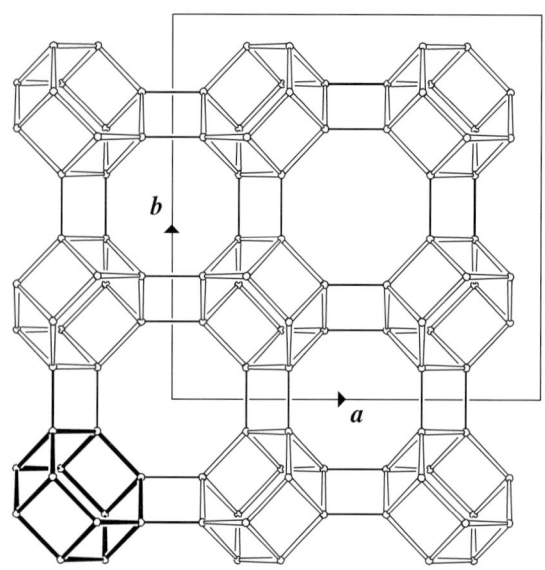

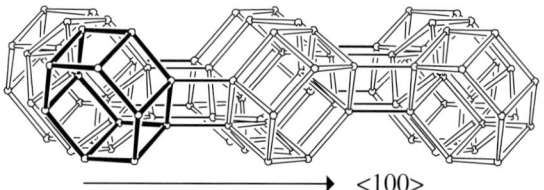

Fig. 1.    PerBU viewed along **c** (left) and along <010> (right).

## 2. Connection mode

Neighboring PerBUs, related along **c** by a pure translation, are connected along **c** through 4-rings as depicted in Figure 2.

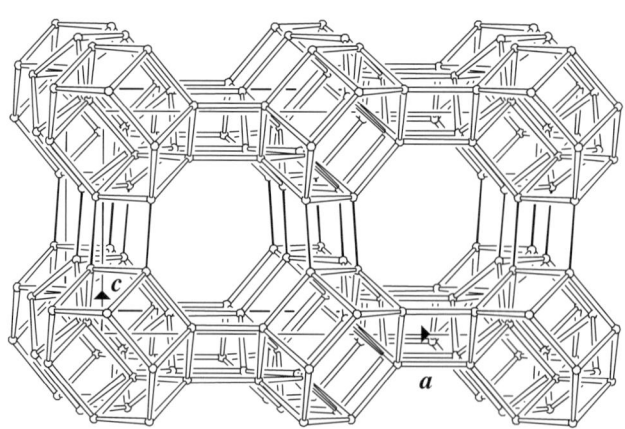

Fig. 2.    Connection mode (and unit cell content) viewed along <010>. The projection of the cell content along **c** is equal to Figure 1.

## 3. Channels and/or cages

Intersecting 8-ring channels are parallel to <100> and *c*. The channel intersections (cavity 1 and the "truncated" *rho* cavity) are depicted in Figure 3. Two types of interconnecting 8-ring channels parallel to *c*, and a third type of interconnecting 8-ring channels parallel to <100> are formed when the cavities are connected through common 8-rings.

Cavity1:
{3 [$4^{12}8^6$] <100> (8-ring),
[001] (8-ring)}

"Truncated" *rho* cavity:
{3 [$4^{12}6^48^6$] <100> (8-ring),
[001] (8-ring)}

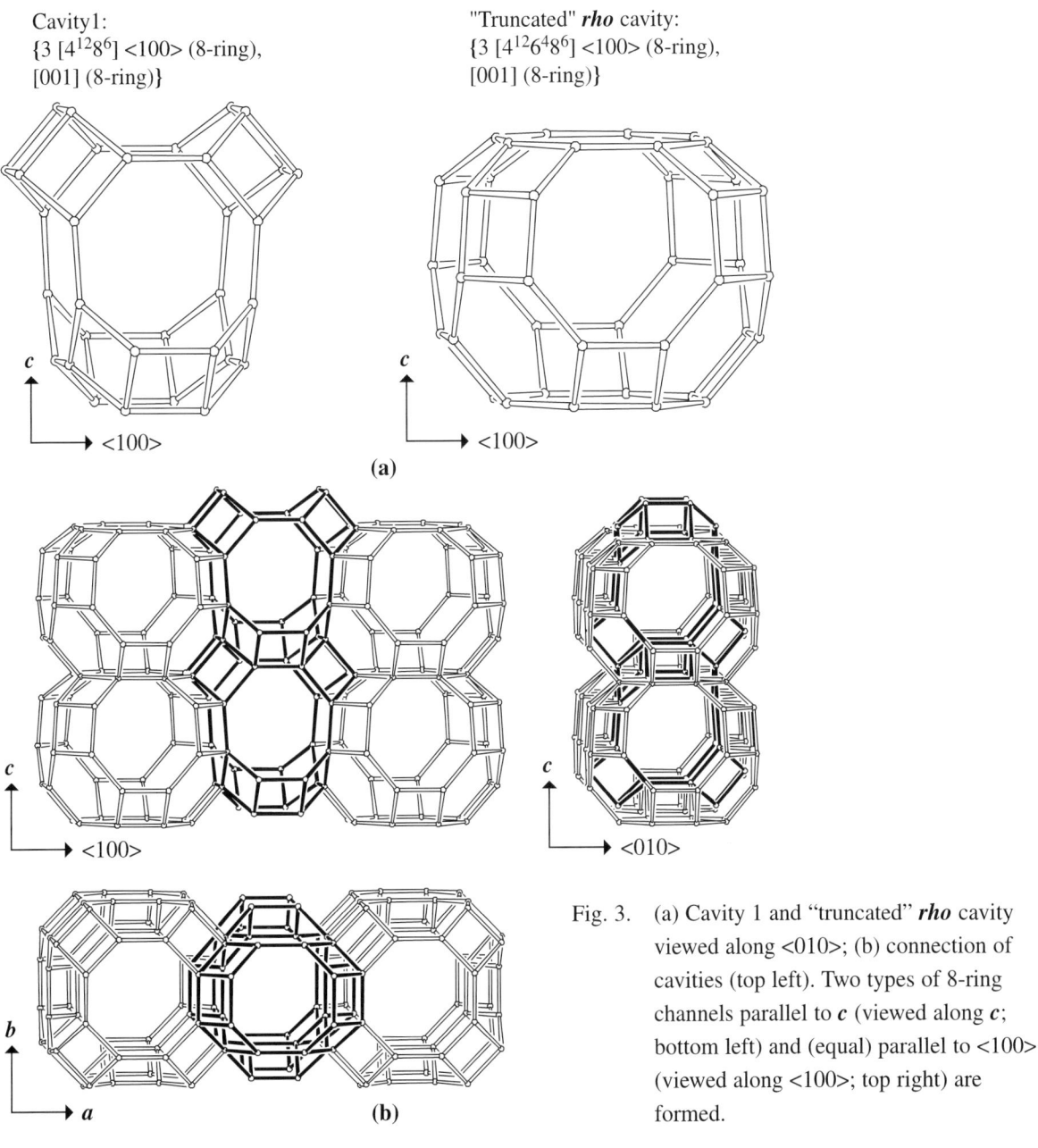

Fig. 3.  (a) Cavity 1 and "truncated" *rho* cavity viewed along <010>; (b) connection of cavities (top left). Two types of 8-ring channels parallel to *c* (viewed along *c*; bottom left) and (equal) parallel to <100> (viewed along <100>; top right) are formed.

# SBE        Building Scheme

## 1. Periodic Building Unit

Tetragonal **SBE** can be built using the *sbe* cavity, composed of 64 T atoms, as zero-dimensional PerBU. The cavity consists of eight 8-rings (one in bold) connected around the 4-fold axis parallel to *c* through 4-rings. 12-Rings are formed as shown in Figure 1.

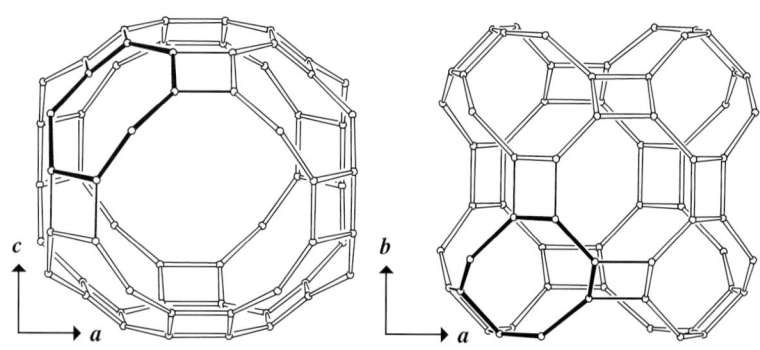

$\{3 [4^{12}8^{10}12^4]$
$<100> (12\text{-ring}),$
$[001] (8\text{-ring})\}$

Fig. 1.  PerBU viewed perpendicular to the 4-fold axis (left) and along the 4-fold axis (right).

## 2. Connection mode

PerBUs, related along *a* and *b* by pure translations and along *c* by a shift of $\frac{1}{2}(a+b+c)$, are connected through 4-rings into the three-dimensional framework as shown in Figure 2.

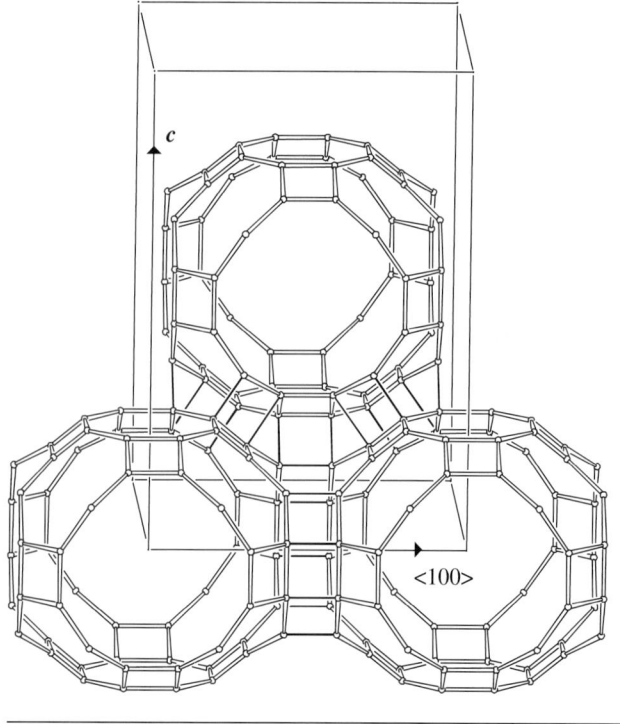

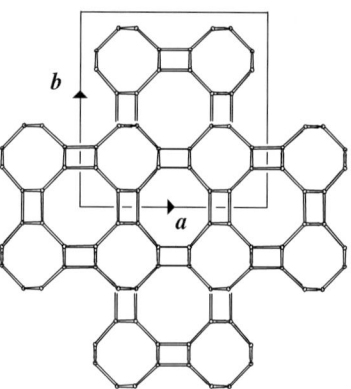

<100>

Fig. 2.  Connection mode (and unit cell content) viewed along <010> (left; only three PerBUs are drawn for clarity) and unit cell content projected along *c* (right).

### 3. Channels and/or cages

12-Ring channels are parallel to <100> and 8-ring channels are parallel to *c*. The channel intersection is equal to the *sbe* cavity depicted in Figure 3. The linkage of complete *sbe* cavities generate *atn* cavities connected through double 8-rings. The location of the "double" *atn* cavities within the framework is illustrated in Figure 3. The *atn* cavity is also present in **ATN**.

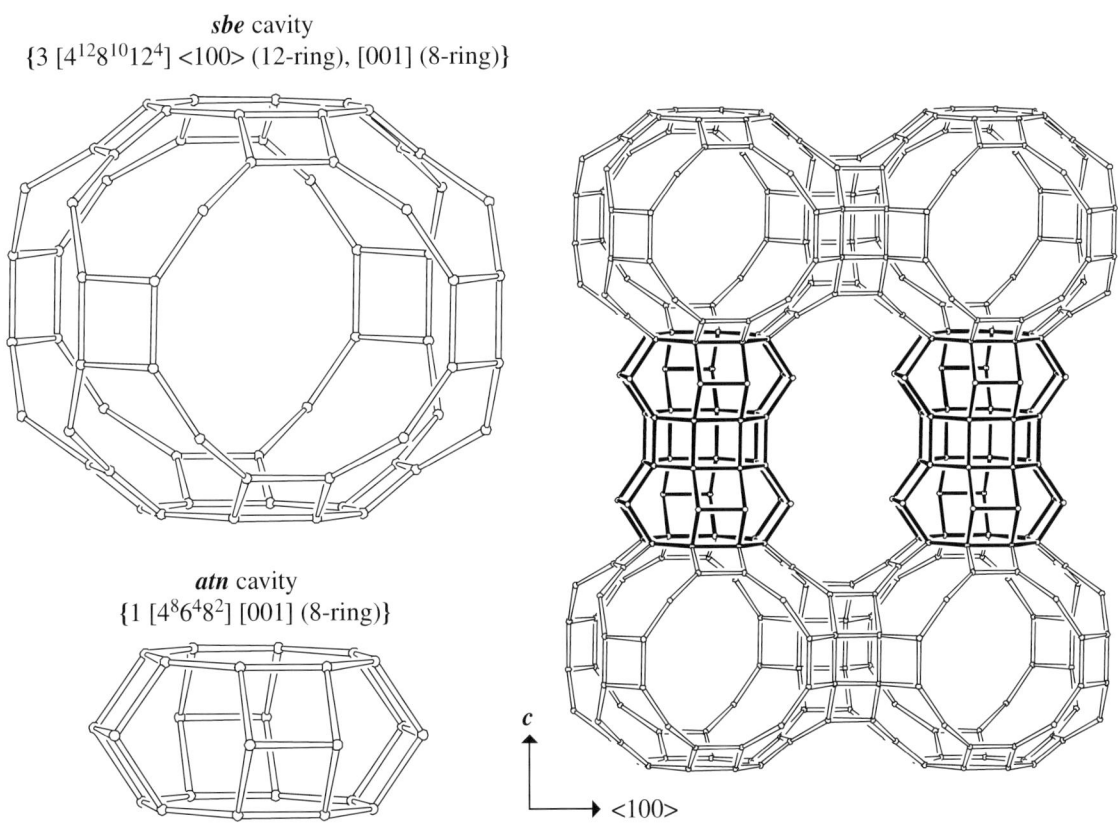

*sbe* cavity
$\{3 \, [4^{12}8^{10}12^{4}] \, <100> \, (12\text{-ring}), \, [001] \, (8\text{-ring})\}$

*atn* cavity
$\{1 \, [4^{8}6^{4}8^{2}] \, [001] \, (8\text{-ring})\}$

Fig. 3.   *sbe* Cavity (top left) and *atn* cavity (bottom left). Right: *sbe* cavities are connected into 12-ring channels parallel to <100> and (through "double" *atn* cavities) into 8-ring channels parallel to *c*.

# SBS

## Building Scheme

### 1. Periodic Building Unit

Hexagonal **SBS** can be built using the *sbs* cavity, composed of 48 T atoms, as zero-dimensional PerBU. The cavity consists of six 8-rings (one in bold) connected through 4-rings around the 3-fold axis parallel to *c*. 12-Rings are formed as shown in Figure 1.

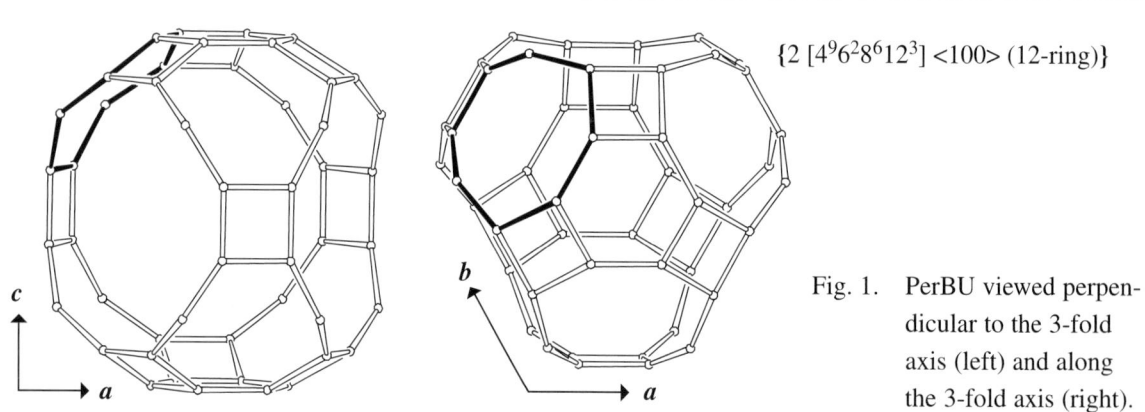

$\{2\ [4^9 6^2 8^6 12^3]\ <100>\ (12\text{-ring})\}$

Fig. 1. PerBU viewed perpendicular to the 3-fold axis (left) and along the 3-fold axis (right).

### 2. Connection mode

Neighboring PerBUs, related along *a* and *b* by pure translations and along *c* by a rotation of 60° about *c* followed by a shift of $\frac{1}{3}(-a+b)$, are connected through (deformed) 4-rings between 12- and 8-rings, respectively, as shown in Figure 2.

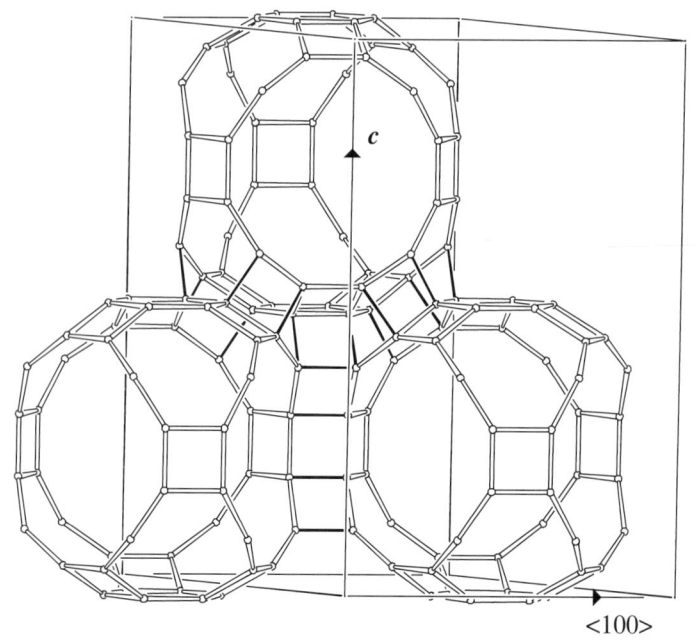

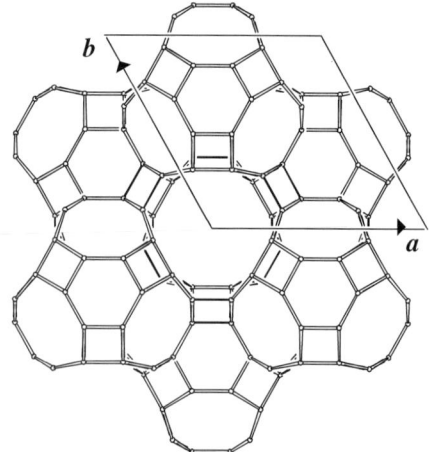

Fig. 2. Connection mode (and unit cell content) viewed along <120> (left; only three PerBUs are drawn for clarity) and unit cell content projected along *c* (right).

### 3. Channels and/or cages

12-Ring channels parallel to *c* are interconnected through the intersection (the *sbs* cavity) of 12-ring channels parallel to <100>. The *sbs* cavities are interconnected through (deformed) 8-ring windows (see also Figure 2). Channel, *sbs* cavity and their linkage are depicted in Figure 3. *can* Cages are formed when complete *sbs* cavities are connected. The *can* cage is also present in **AFG**, **CAN**, **ERI**, **FAR**, **FRA**, **GIU**, **LIO**, **LOS**, **LTL**, **LTN**, **MAR**, **MOZ**, **OFF**, **SAT**, **SBT**, **TOL** and **-WEN**.

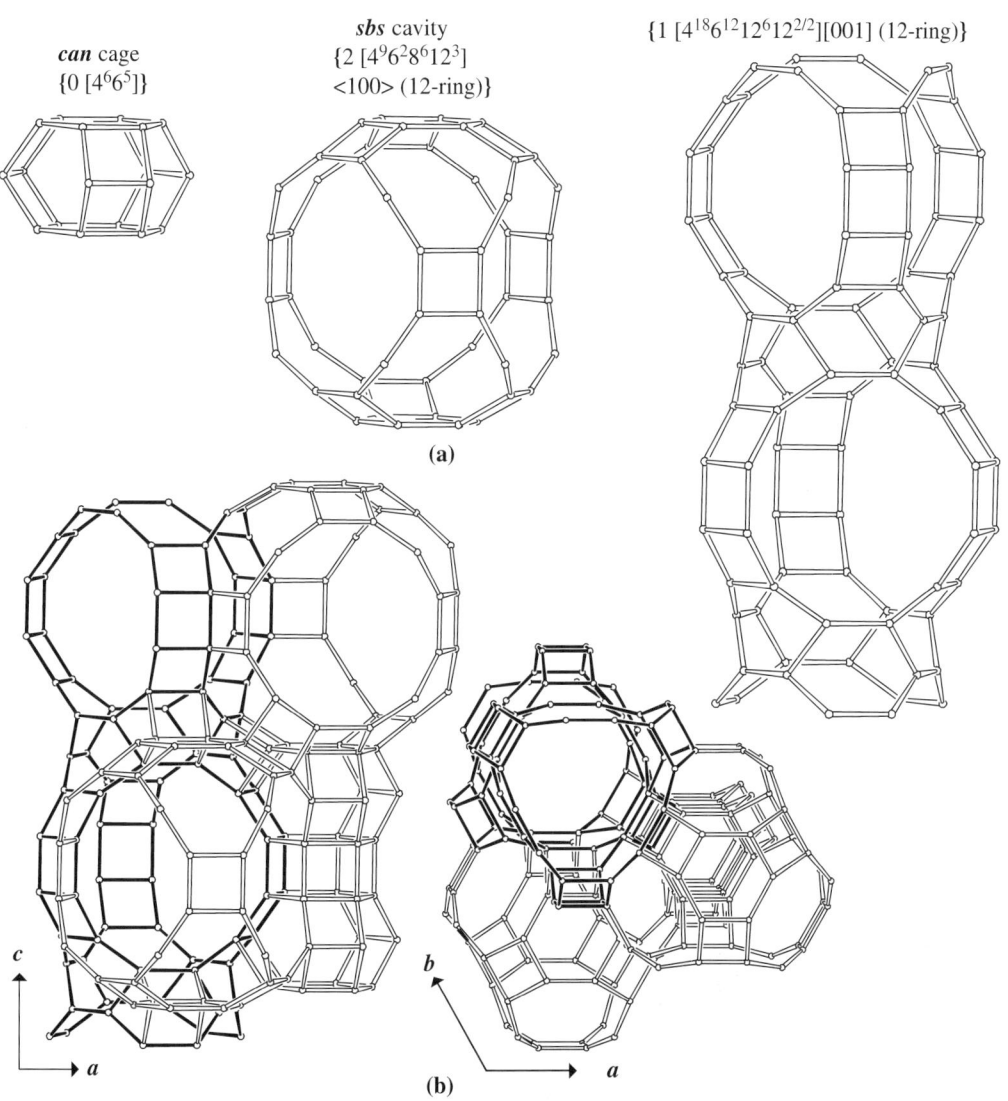

Fig. 3.   (a) From left to right: *can* cage, *sbs* cavity and 12-ring channel parallel to *c* viewed along <120>;
(b) linkage of channel (bold), *sbs* cavities and *can* cages viewed along <120> (left) and along the 12-ring channel axis parallel to *c* (right).

# SBT                      Building Scheme

## 1. Periodic Building Unit

Trigonal **SBT** can be built using the ***can*** cage with "handles" consisting of 24 T atoms (or six 4-rings; Figure 1). The two-dimensional PerBU is obtained when these cages are linked through 4-rings into the hexagonal layer depicted in Figure 1. 8-Rings are formed.

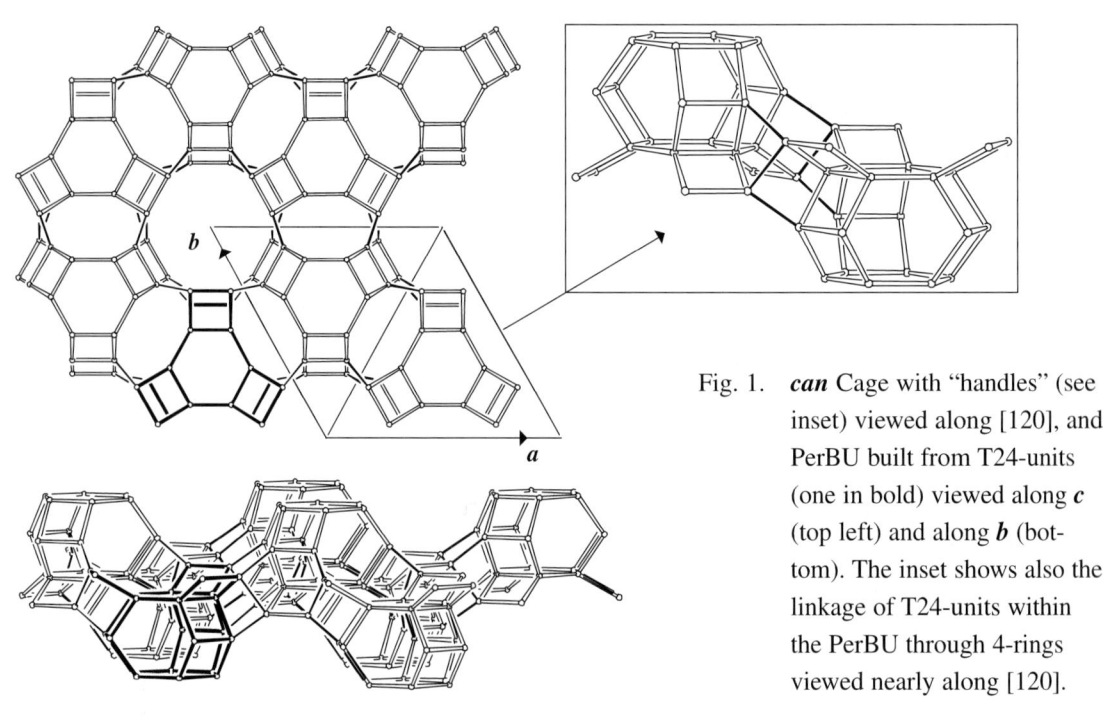

Fig. 1.   ***can*** Cage with "handles" (see inset) viewed along [120], and PerBU built from T24-units (one in bold) viewed along ***c*** (top left) and along ***b*** (bottom). The inset shows also the linkage of T24-units within the PerBU through 4-rings viewed nearly along [120].

## 2. Connection mode

Neighboring PerBUs, related along ***c*** by a shift of $\frac{1}{3}(-\boldsymbol{a}+\boldsymbol{b}+\boldsymbol{c})$, are connected along ***c*** through double 6-rings as shown in Figure 2.

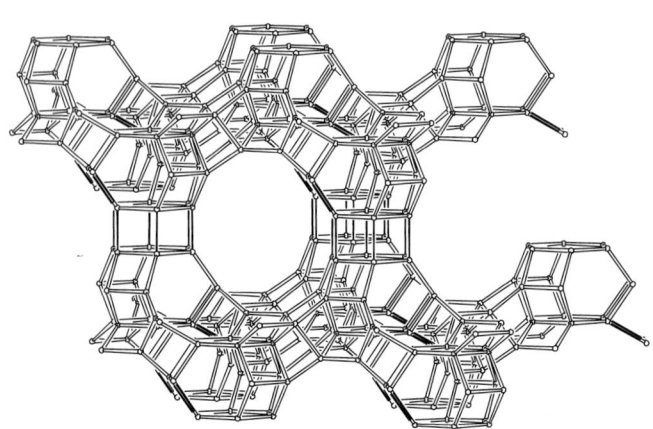

Fig. 2a.   Connection mode viewed along ***b***.

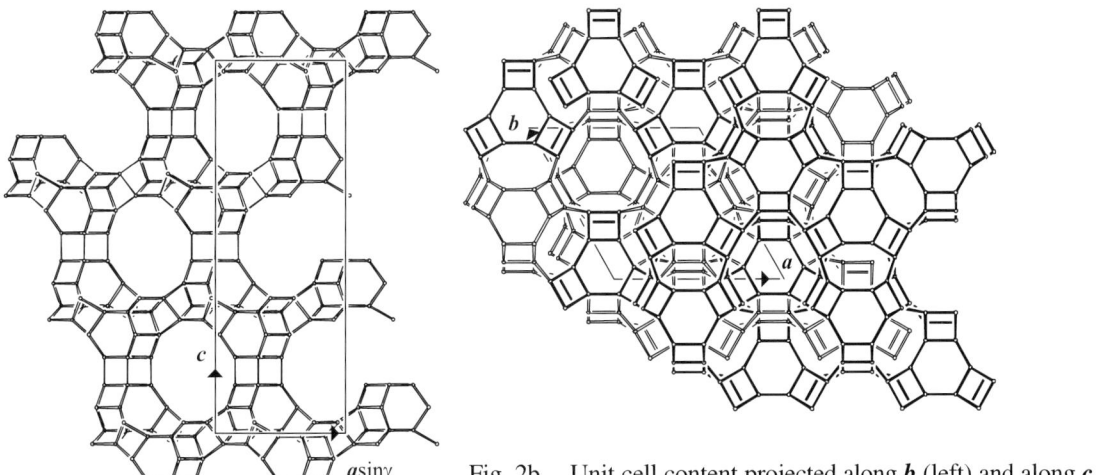

Fig. 2b.    Unit cell content projected along **b** (left) and along **c** (right).

### 3. Channels and/or cages

Two-dimensional intersecting 12-ring channel systems are perpendicular to **c**. The channel intersection is equal to a "double" cavity (the **sbt** cavity). The channel systems are interconnecting along **c** through 12-rings in the **sbt** cavities as shown in Figure 3. **can** Cages, formed when **sbt** cavities are connected around a 3-fold axis parallel to **c**, are also present in **AFG**, **CAN**, **ERI**, **FAR**, **FRA**, **GIU**, **LIO**, **LOS**, **LTL**, **LTN**, **MAR**, **MOZ**, **OFF**, **SAT**, **SBS**, **TOL** and **-WEN**.

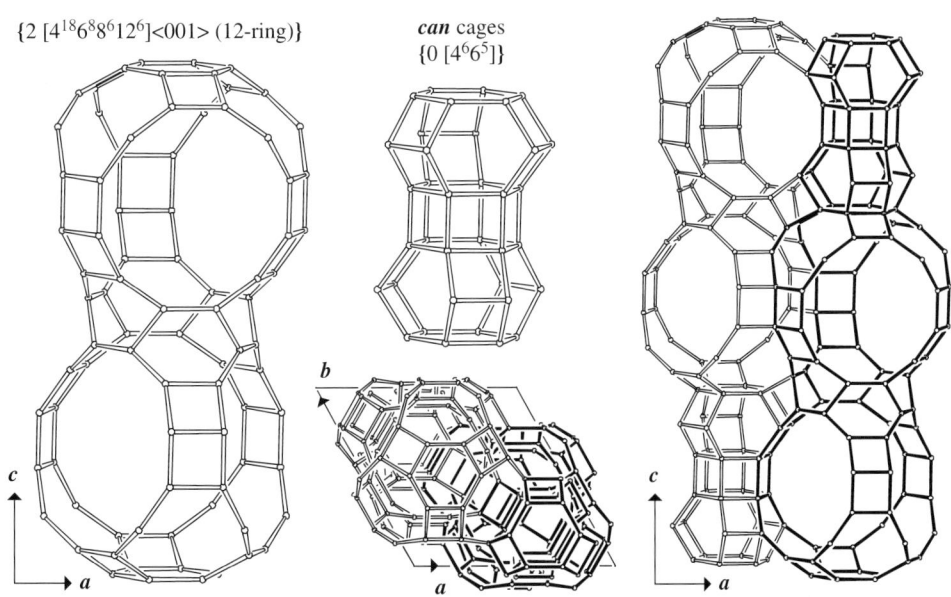

$\{2\ [4^{18}6^{8}8^{6}12^{6}]{<}001{>}\ (12\text{-ring})\}$     **can** cages $\{0\ [4^{6}6^{5}]\}$

Fig. 3.   **sbt** Cavity (left), "double" **can** cage (top middle) viewed along <120> and connection of cages and cavities viewed along <120> (right) and along **c** (bottom middle).

# SFE                    Building Scheme

## 1. Periodic Building Unit

**SFE** can be built using the zigzag chain parallel to **b** (bold in Figure 1). Seven zigzag chains are connected into an infinite building unit. The repeat unit consists of 5-1 and 5-3 units (bold in Figure 1). A two-dimensional PerBU is obtained when infinite building units, related along **c** by a pure translation, are connected through double zigzag chains into the **bc** layer (Figure 1). (See also **SSY**.)

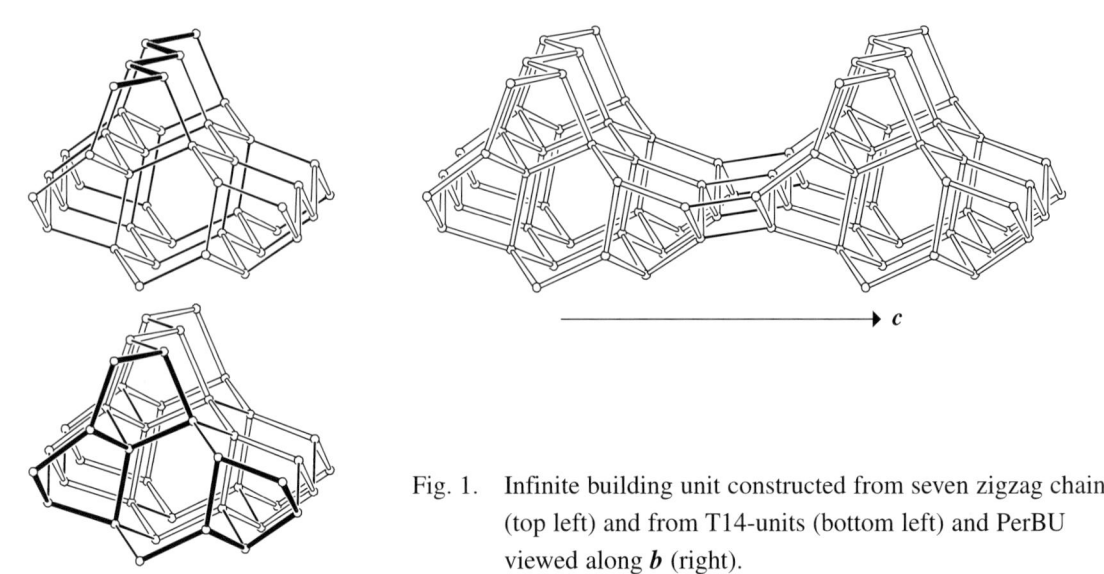

Fig. 1.   Infinite building unit constructed from seven zigzag chains (top left) and from T14-units (bottom left) and PerBU viewed along **b** (right).

## 2. Connection mode

Neighboring PerBUs, related along **a** by a pure translation, are connected along **a** through (fused) 5- and 6-rings as shown in Figure 2.

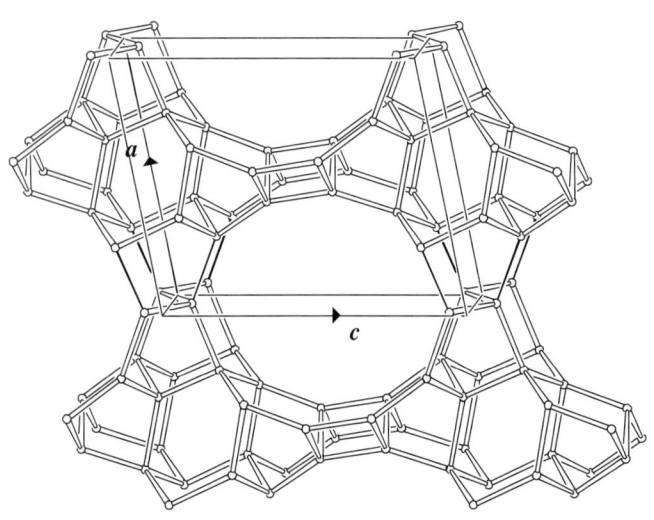

Fig. 2.   Connection mode (and unit cell content) viewed along **b**. Only two repeat units along **b** of each PerBU are drawn for clarity.

## 3. Channels and/or cages

Non-interconnecting 12-ring channels (Figure 3) are parallel to *b*. The channel is topologically equivalent to the 12-ring channel in **SSY**.

{1 [$4^4 6^8 12^{2/2}$] [010] (12-ring)}

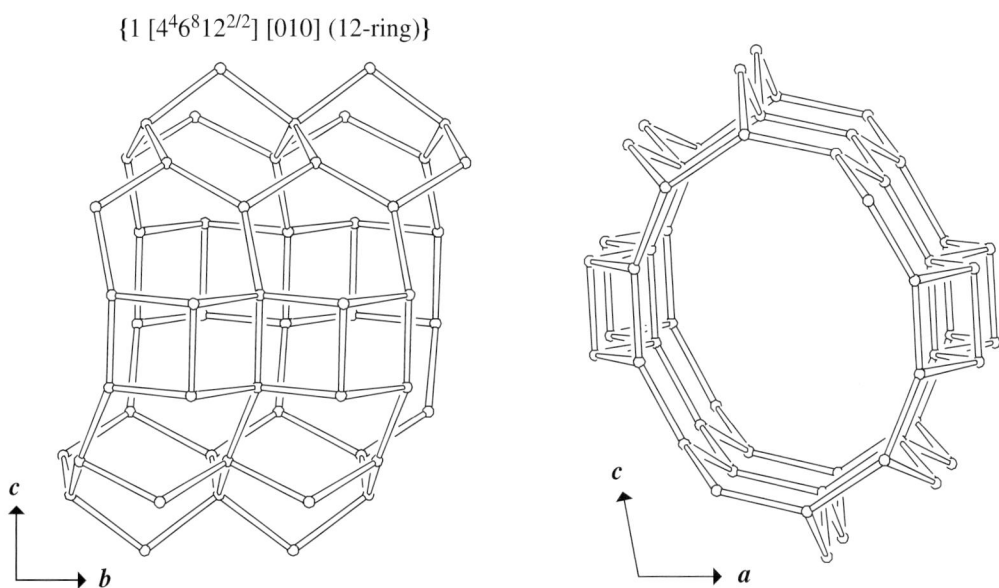

Fig. 3.    Channel viewed perpendicular to the channel axis (left) and along the channel axis (right).

# SFF

## Building Scheme

### 1. Periodic Building Unit

The two-dimensional PerBU in **SFF**, composed of units of 16 T atoms, is equal to the *ac* layer shown in Figure 1. The T16-unit consists of two 5-3 units (bold in Figure 1). T16-units, related along *a* and *c* by pure translations, are linked through 4-rings and finite zigzag chains, respectively. (See also **STF**.)

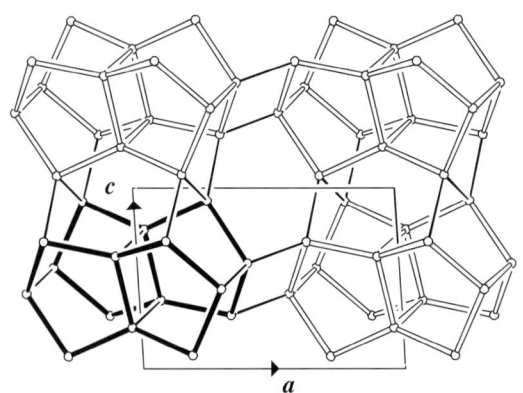
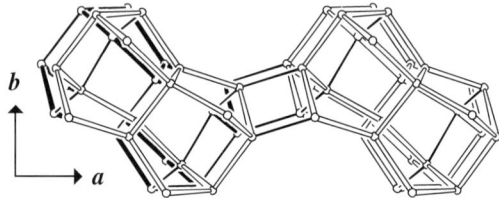

Fig. 1.   PerBU viewed along *b* (left) and along *c* (right).

### 2. Connection mode

Neighboring PerBUs, related along *b* by a screw rotation of 180° about *b*, are connected along *b* through 4-rings as shown in Figure 2.

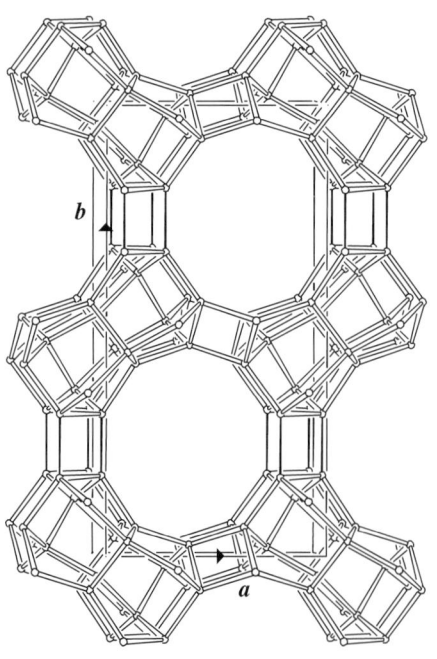

Fig. 2.   Connection mode (and unit cell content) viewed along *c*.

### 3. Channels and/or cages

Non-interconnecting 10-ring channels are parallel to *c*. The cavity, depicted in Figure 3, shows the repeat unit in the channel. The channel is topologically equivalent to the 10-ring channel in **STF**.

{1 [$4^4 5^8 6^6 10^2$][001] (10-ring)}

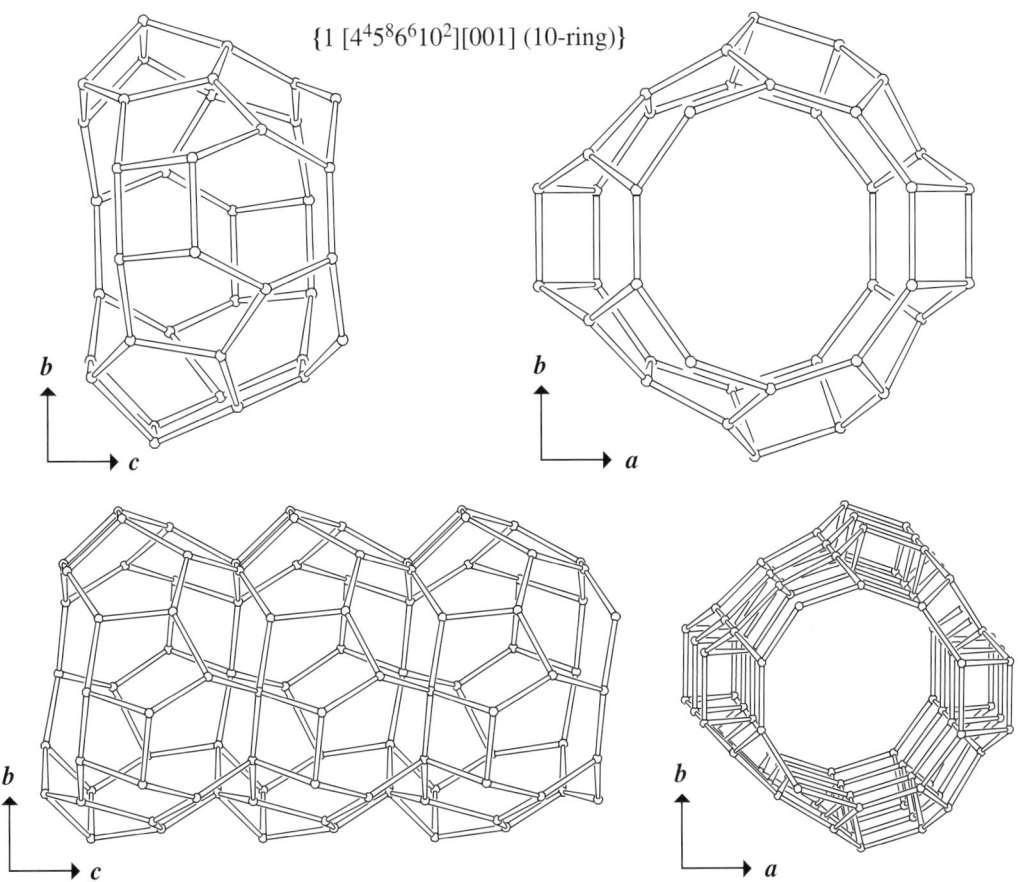

Fig. 3.   Cavity (top) and channel (bottom) viewed perpendicular to the channel axis (left) and along the channel axis (right).

# SFG

## Building Scheme

### 1. Periodic Building Unit

**SFG** can be built using units of 37 T atoms (bold in Figure 1): five finite "zigzag" chains (each containing 5 T atoms) are connected around a 5-fold axis into a $[5^2 6^5]$-"double cage" (see inset Figure 1). Three additional 4-rings are linked to the "double cage". T37-units, related along **a** by a screw rotation of 180° about **a** and along **c** by a pure translation, are connected into the **ac** layer by a system of (fused) 6- and 7-rings. The two-dimensional PerBU equals this **ac** layer.

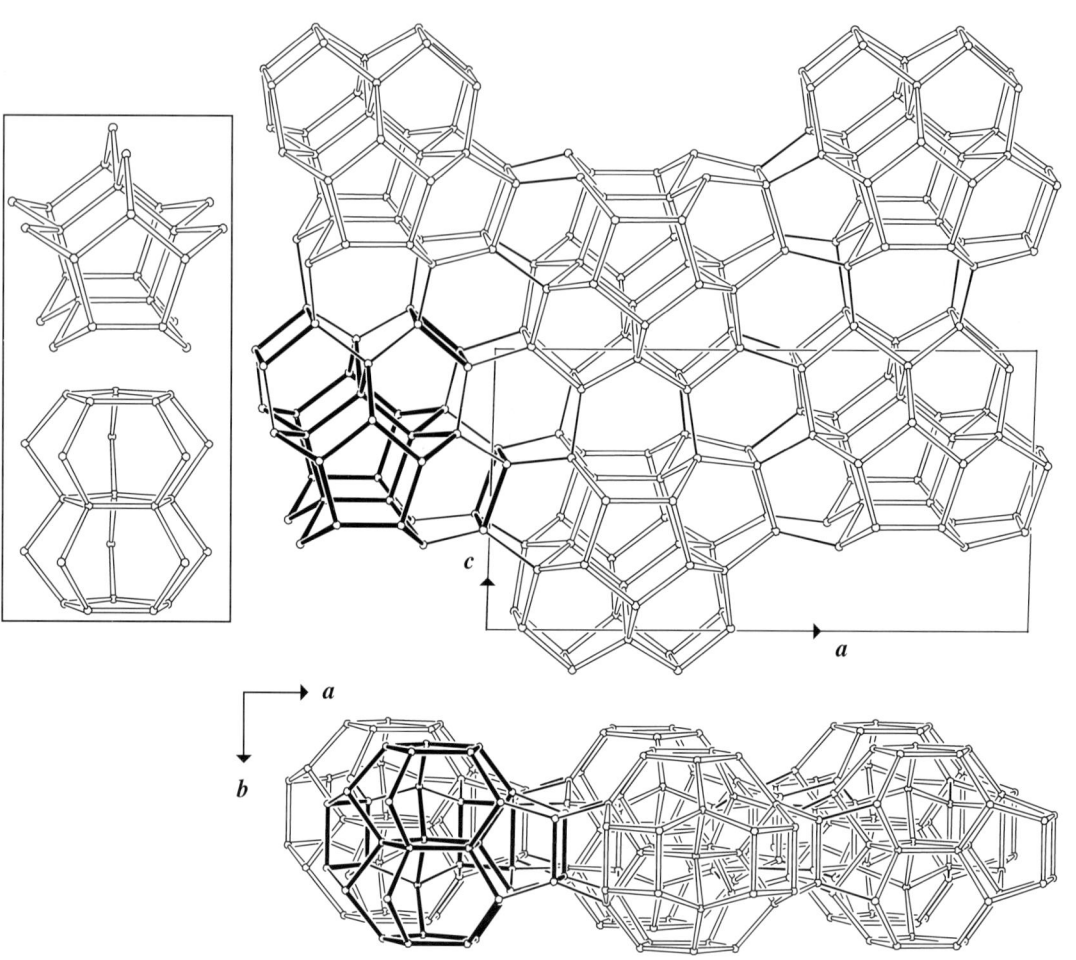

Fig. 1. PerBU viewed along **b** (top) and along **c** (bottom). The inset shows the $[5^2 6^5]$-"double cage" viewed along **b** (top) and along **c** (bottom).

### 2. Connection mode

Neighboring PerBUs, related along **b** by a pure translation, are connected along **b** through double 5-rings as shown in Figure 2 on next page.

# Cage/Channel **SFG**

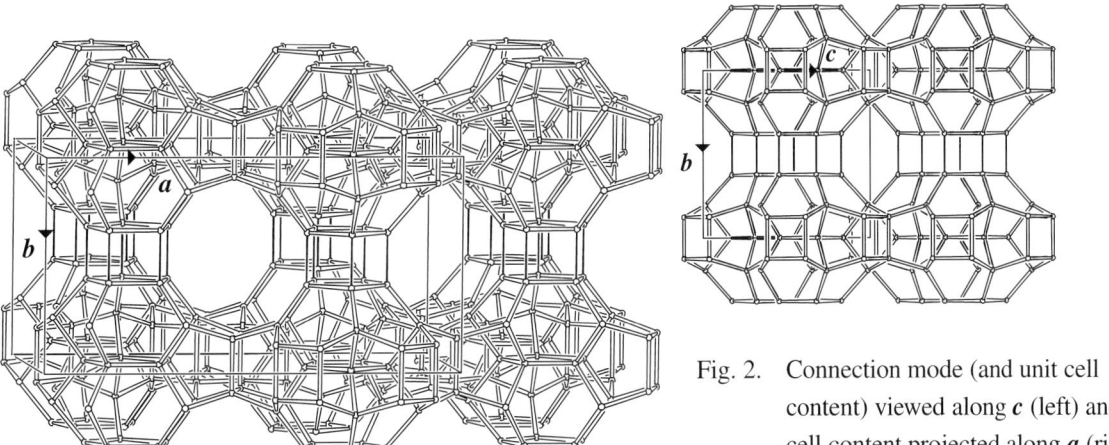

Fig. 2. Connection mode (and unit cell content) viewed along *c* (left) and unit cell content projected along *a* (right).

## 3. Channels and/or cages

Intersecting straight and sinusoidal 10-ring channels are parallel to *c* and *a*, respectively. The channel intersection, equal to the repeat unit in the straight channel, is shown in Figure 3 (top). Straight channels are connected (through double 5- and 7-rings) into sinusoidal channels parallel to *a* as is also illustrated in Figure 3. The two-dimensional channel systems are interconnecting through 7-rings perpendicular to *b*.

{1 [$4^4 6^{12} 10^4$] [001] (10-ring)}

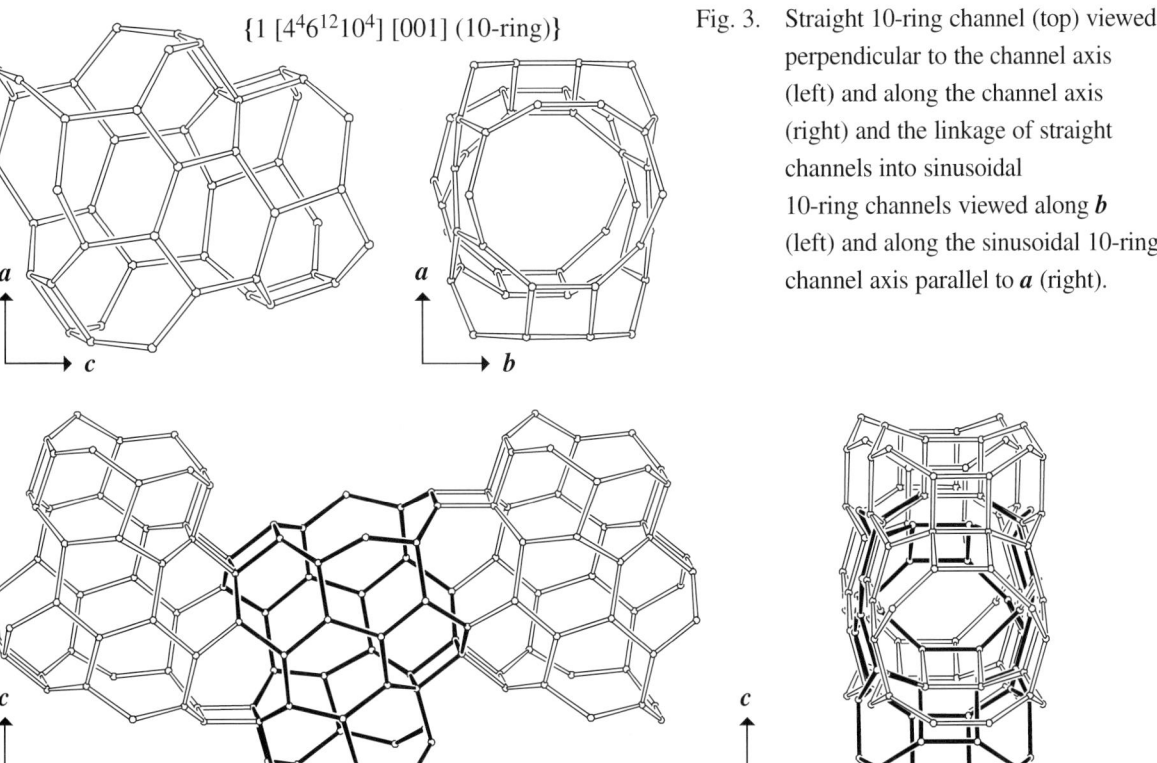

Fig. 3. Straight 10-ring channel (top) viewed perpendicular to the channel axis (left) and along the channel axis (right) and the linkage of straight channels into sinusoidal 10-ring channels viewed along *b* (left) and along the sinusoidal 10-ring channel axis parallel to *a* (right).

# SFH                              **Building Scheme**

## 1. Periodic Building Unit

**SFH** can be built using the zigzag chain (bold in Figure 1) running parallel to *a*. Eight zigzag chains are connected into an infinite building unit. The repeat unit of this building unit is composed of two 5-3 units (bold in Figure 1). A two-dimensional PerBU is obtained when infinite building units, related along *b* by a translation of $\frac{1}{2}(a+b)$, are connected along *b* through 4-rings.

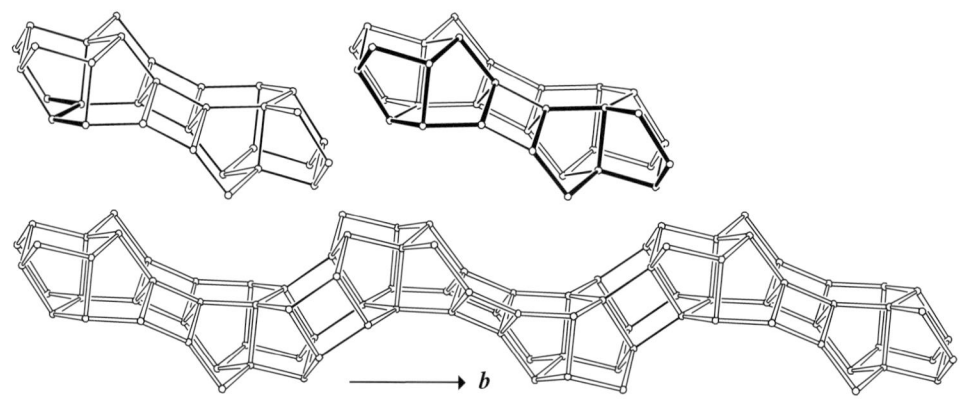

Fig. 1.   Infinite building unit (top), constructed from zigzag chains (left) and from T16-units (right), and PerBU (bottom) viewed along *a*.

## 2. Connection mode

Neighboring PerBUs, related along *c* by a screw rotation of 180° about *c*, are connected along *c* through 6-rings as depicted in Figure 2.

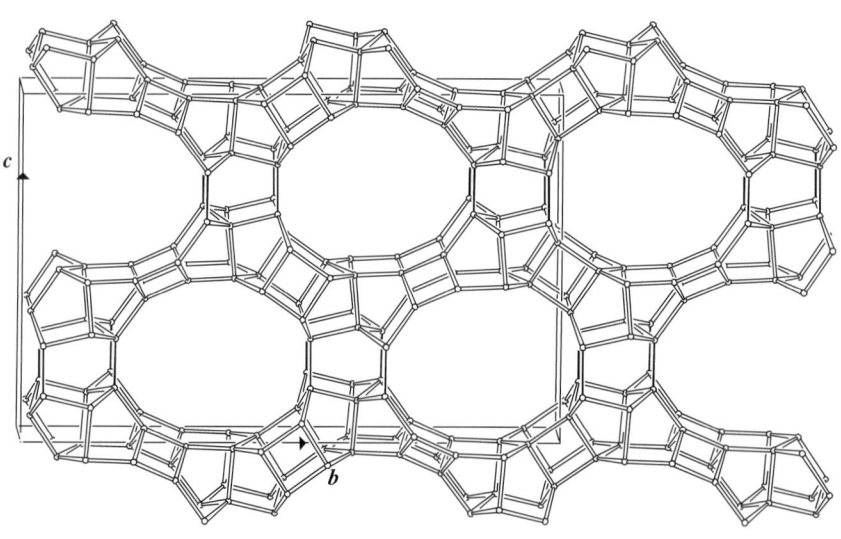

Fig. 2.   Connection mode (and unit cell content) viewed along *a*.

## 3. Channels and/or cages

Non-interconnecting 14-ring channels (Figure 3) are parallel to $a$. The channel is topologically equivalent to the channel in **SFN**.

{1 [$4^4 6^8 14^{2/2}$] [100] (14-ring)}

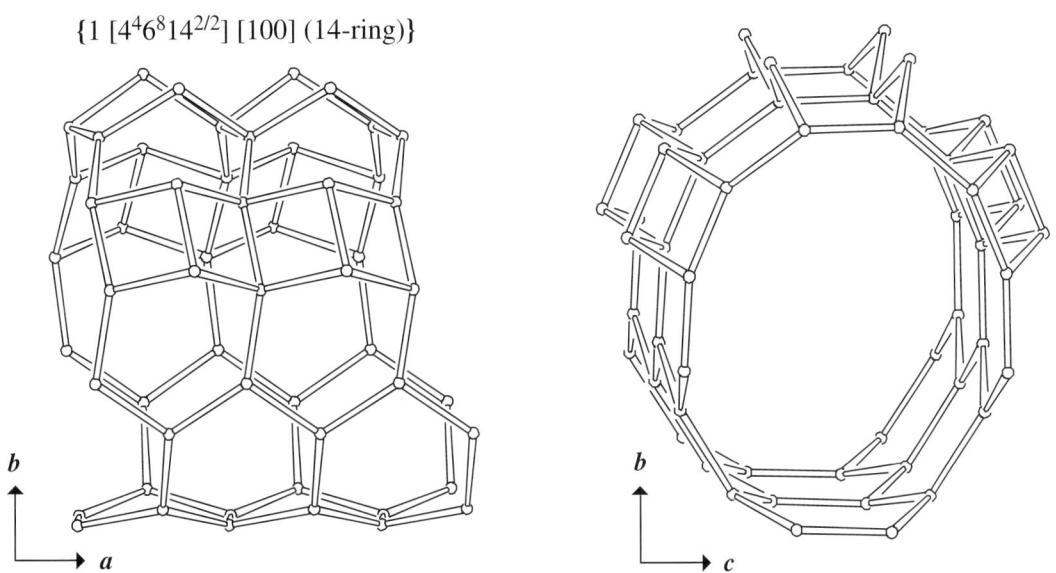

Fig. 3.    Channel viewed perpendicular to the channel axis (left) and along the channel axis (right).

# SFN                                   Building Scheme

## 1. Periodic Building Unit

**SFN** can be built using the zigzag chain (bold in Figure 1) running parallel to **b**. Eight zigzag chains are connected into an infinite building unit. The repeat unit of this building unit is composed of two 5-3 units (bold in Figure 1). A two-dimensional PerBU is obtained when infinite building units, related along **a** by a translation of $\frac{1}{2}(a + b)$, are connected along **a** through 4-rings.

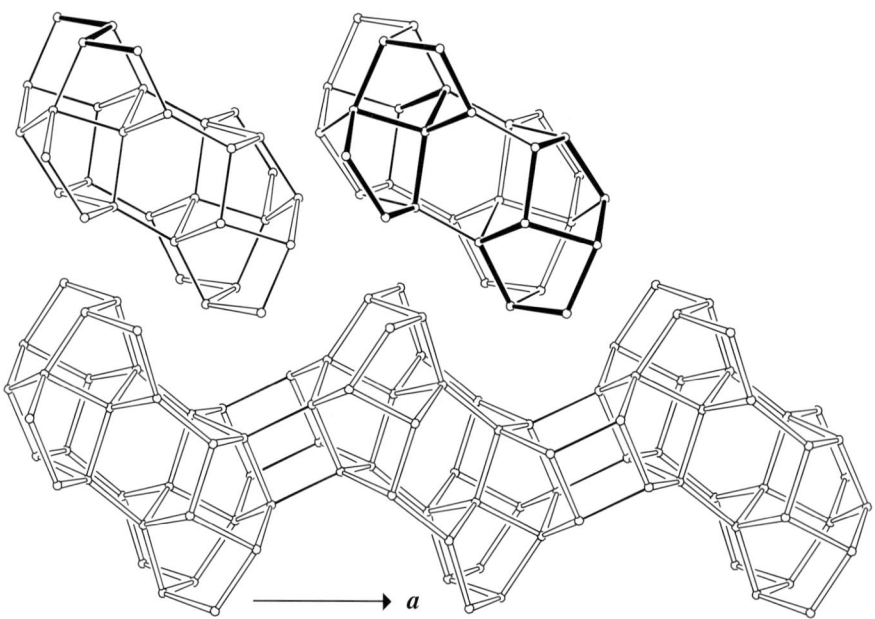

Fig. 1.  Infinite building unit (top), constructed from eight zigzag chains (left) and from T16-units (right), and PerBU (bottom) viewed along the zigzag chain axis parallel to **b**.

## 2. Connection mode

Neighboring PerBUs, related along **c** by a pure translation, are connected along **c** through 4-rings as depicted in Figure 2 on next page.

# SFN

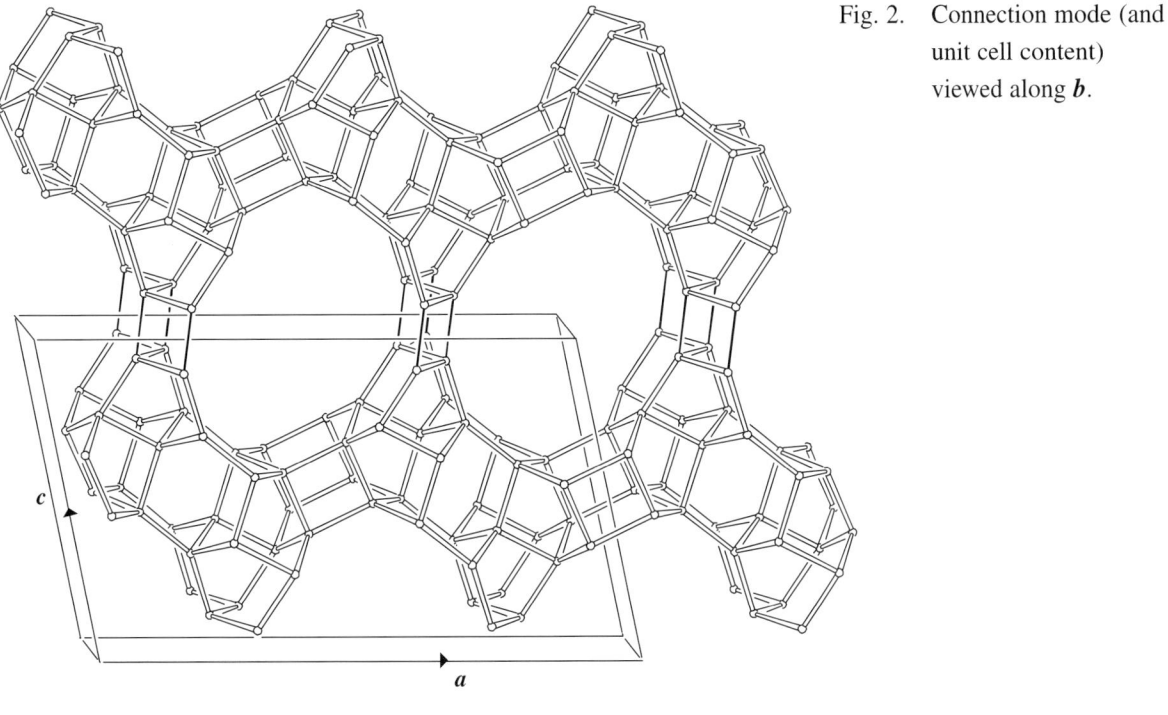

Fig. 2. Connection mode (and unit cell content) viewed along **b**.

## 3. Channels and/or cages

Non-interconnecting 14-ring channels (Figure 3) are parallel to **b**. The channel is topologically equivalent to the channel in **SFH**.

{1 [$4^4 6^8 14^{2/2}$] [010] (14-ring)}

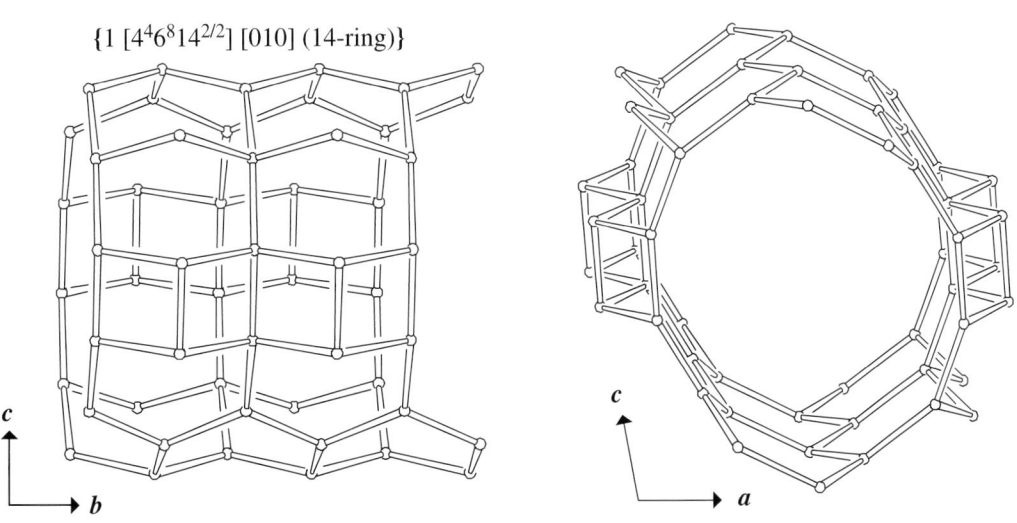

Fig. 3. Channel viewed perpendicular to the channel axis (left) and along the channel axis (right).

# SFO

## Building Scheme

### 1. Periodic Building Unit

**SFO** can be built using double 4-rings with one disconnected edge (a [$4^46^1$]-cage; bold in Figure 1). These T8-units, related along *c* by a pure translation, are connected into chains parallel to *c*. The two-dimensional PerBU is obtained when neighboring chains, related by a screw rotation of 180° about *b*, are linked into the undulating *bc* layer (Figure 1). (See also **AFR**; compare with **ZON**.)

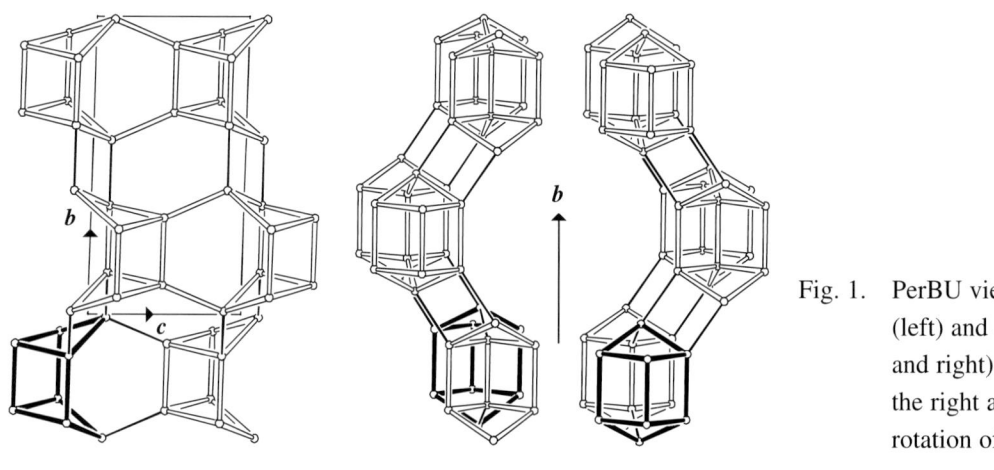

Fig. 1. PerBU viewed along *a* (left) and along *c* (middle and right). The PerBUs at the right are related by a rotation of 180° about *b*.

### 2. Connection mode

Neighboring PerBUs, related along *a* by a rotation of 180° about *b*, are connected along *a* through 4-rings as illustrated in Figure 2.

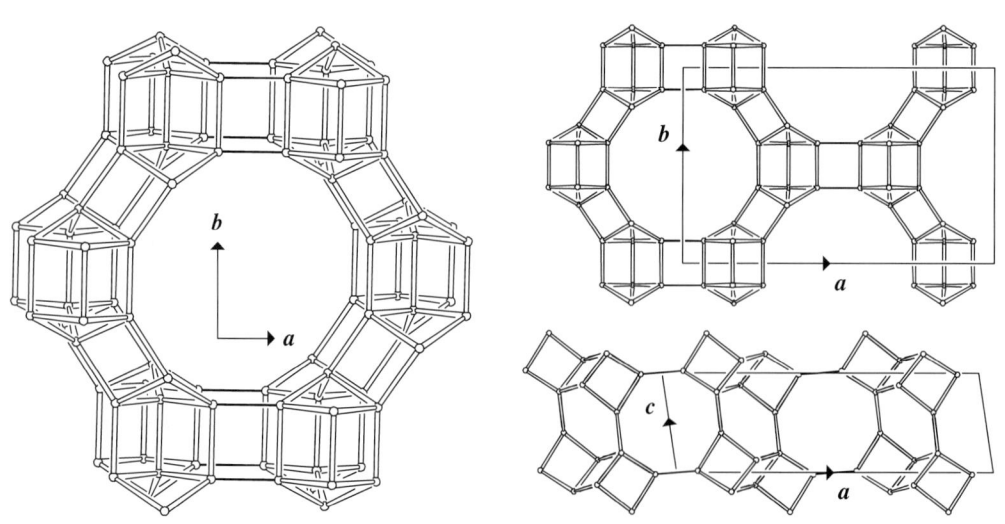

Fig. 2. Connection mode viewed along *c* (left) and unit cell content projected along *c* (top) and along *b* (bottom).

## 3. Channels and/or cages

Intersecting 8-ring and 12-ring channels are parallel to *b* and *c*, respectively. The channel intersection, equal to the repeat unit of the channel along *c*, and linked intersections are depicted in Figure 3. The channel intersection is topologically equivalent to the channel intersection in **AFR**.

$\{2 \ [4^{10}6^48^212^2] \ [010] \ (8\text{-ring}), \ [001] \ (12\text{-ring})\}$

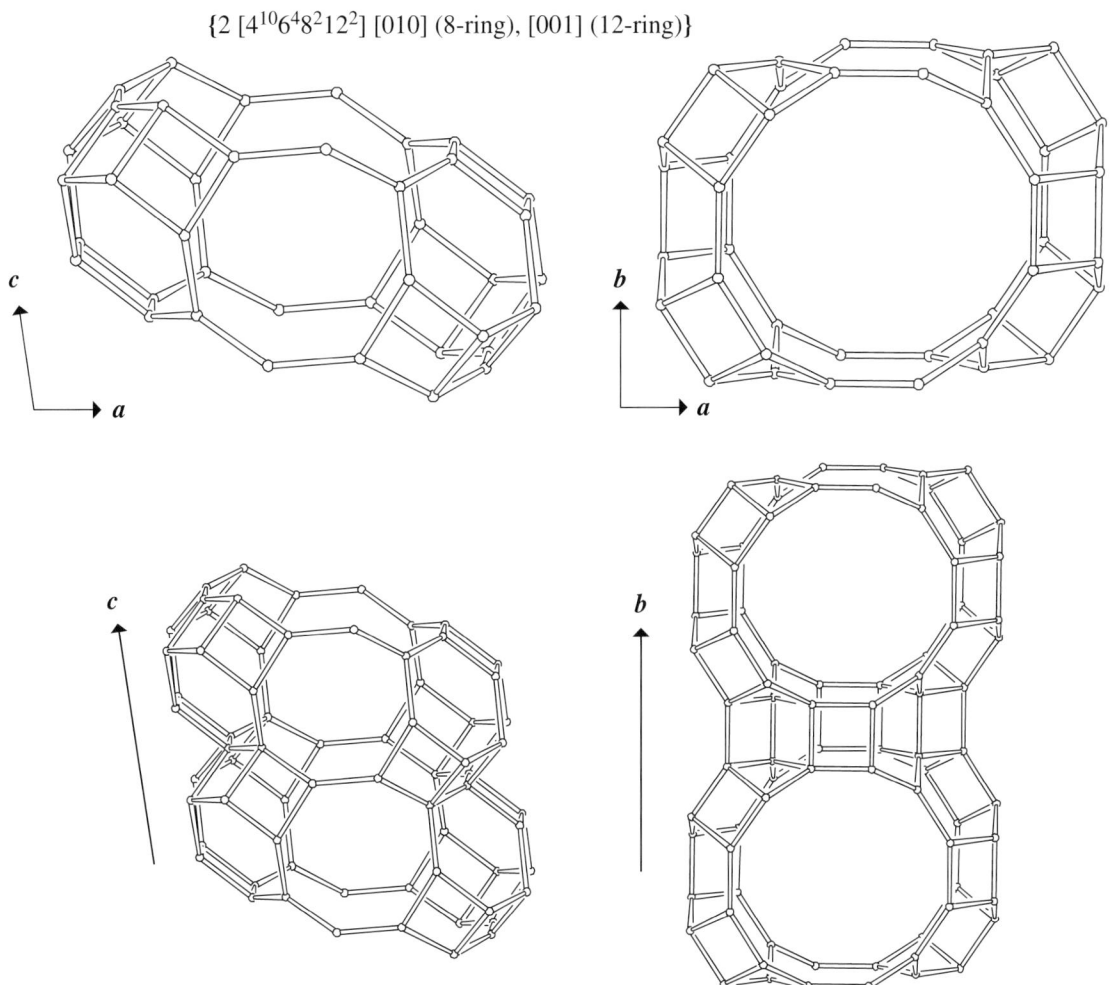

Fig. 3.    Top: channel intersection viewed along *b* (left) and along *c* (right). Bottom: 12-ring channel parallel to *c* viewed along *b* (left) and 8-ring channel parallel to *b* viewed along *c* (right).

# SGT                    Building Scheme

## 1. Periodic Building Unit

Tetragonal **SGT** can be built using units of 32 T atoms. Four 5-3 units (one unit in bold in Figure 1) are connected through (fused) 4- and 5-rings into T32-units. A two-dimensional PerBU is obtained when T32-units, related along $b$ by a pure translation and along [111] by a translation of $\frac{1}{2}(a+b+c)$, are connected into the undulating $bc$ layer through 4-, 5- and 6-rings as shown in Figure 1. [$4^35^6$]-Cages are formed.

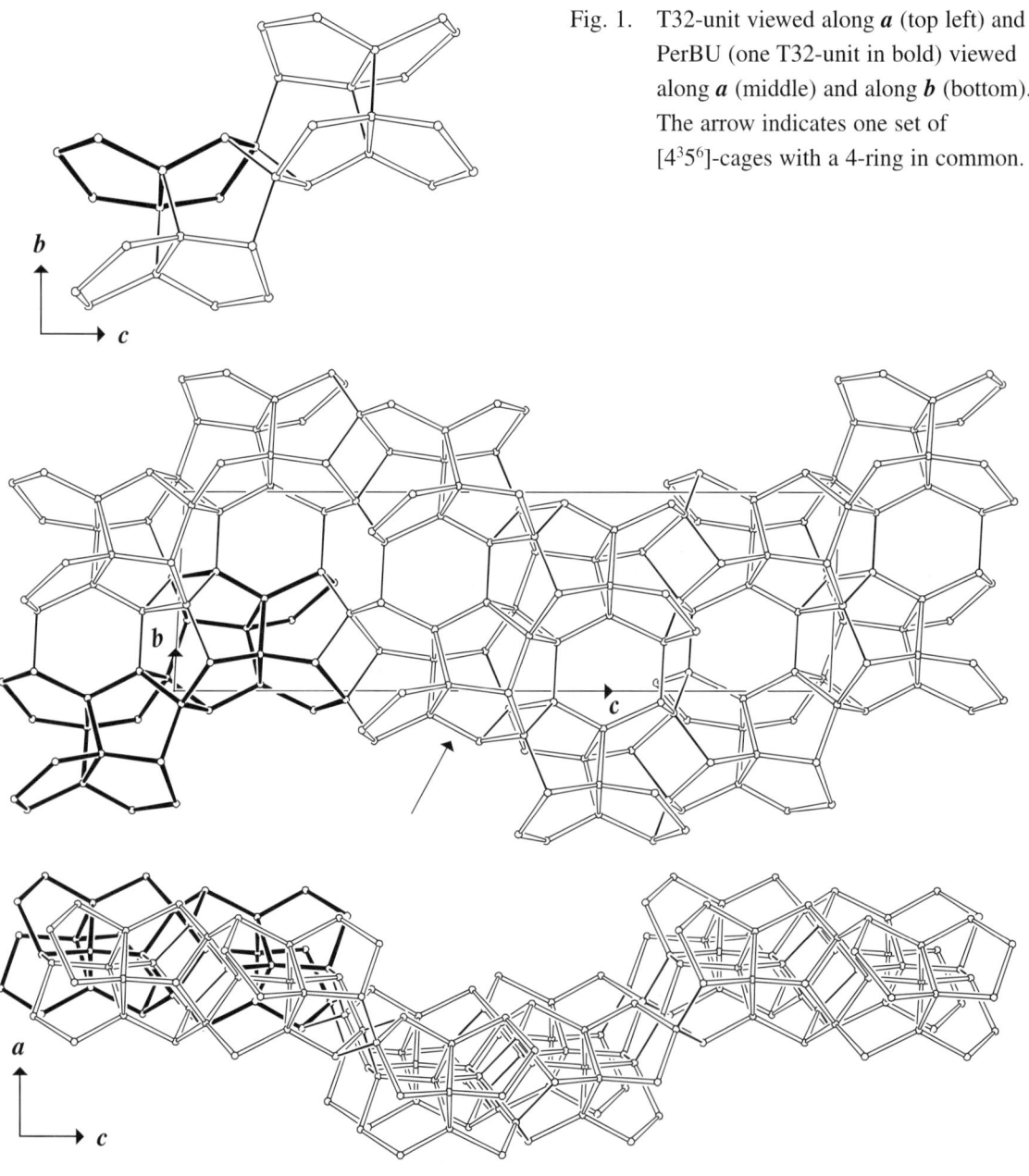

Fig. 1. T32-unit viewed along $a$ (top left) and PerBU (one T32-unit in bold) viewed along $a$ (middle) and along $b$ (bottom). The arrow indicates one set of [$4^35^6$]-cages with a 4-ring in common.

## 2. Connection mode

Neighboring PerBUs, related along *a* by a pure translation, are connected through 4-, 5- and 6-rings as shown in Figure 2. Small $[4^35^6]$-cages and large $[5^{12}6^8]$-cages are formed. The connection modes along *a* and *b* are equal as can be seen from Figures 1 and 2.

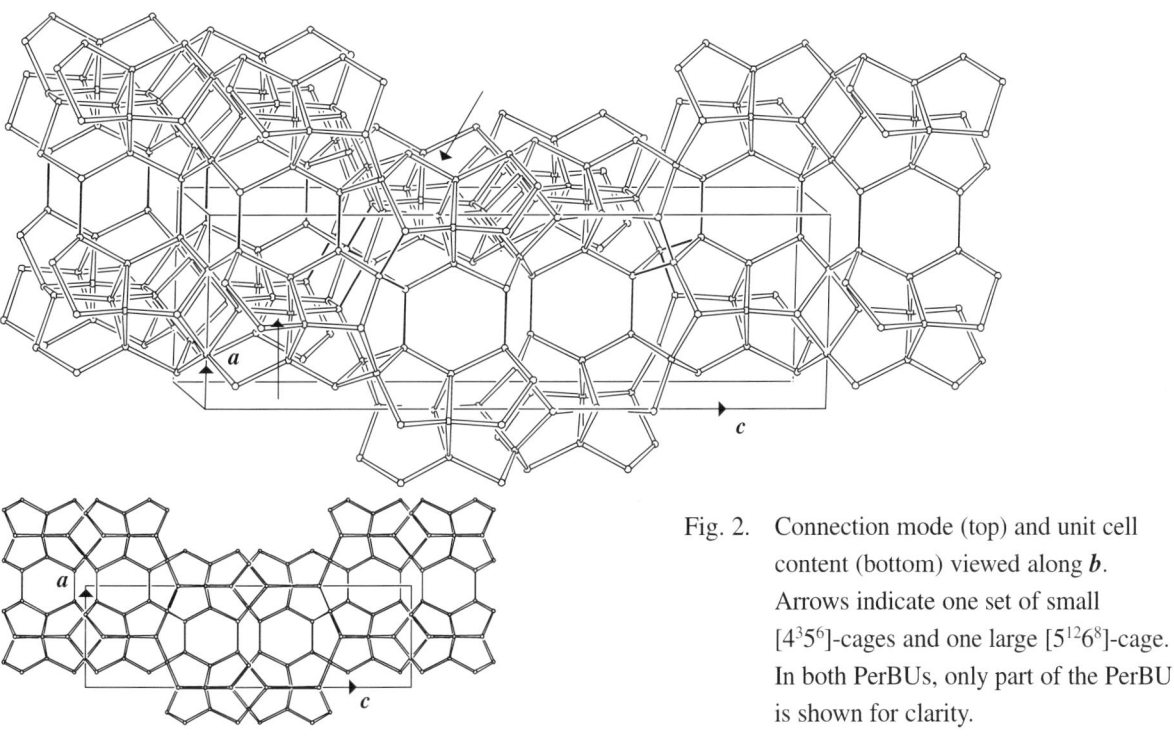

Fig. 2.  Connection mode (top) and unit cell content (bottom) viewed along *b*. Arrows indicate one set of small $[4^35^6]$-cages and one large $[5^{12}6^8]$-cage. In both PerBUs, only part of the PerBU is shown for clarity.

## 3. Channels and/or cages

The two types of cages are shown in Figure 3. Apertures are formed by 6-rings only.

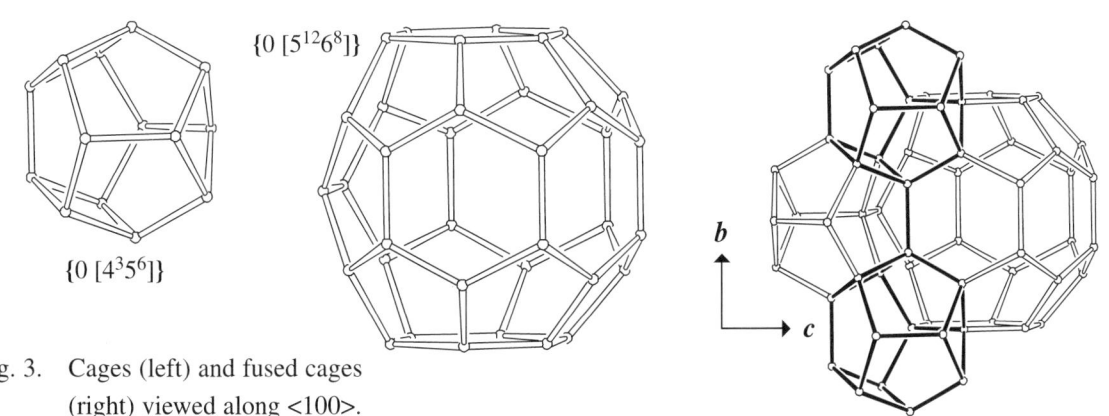

$\{0 \ [5^{12}6^8]\}$

$\{0 \ [4^35^6]\}$

Fig. 3.  Cages (left) and fused cages (right) viewed along <100>.

# SIV                          Building Scheme

## 1. Periodic Building Unit

**SIV** can be built using units of 32 T atoms: a 6-fold (1,2,3,4,5,6)-connected double 8-ring (see **PHI**) and a 4-fold (1,2,3,4)-connected double 8-ring (see **GIS**) linked along *c* through 4-rings. T32-units are connected along *a* through 4-rings into a one-dimensional PerBU depicted in Figure 1 (one T32-unit in bold). Double crankshaft chains parallel to *a* are formed.

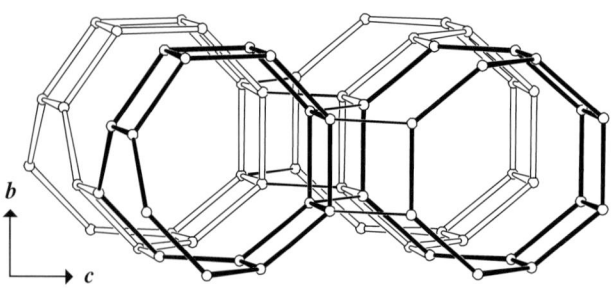

Fig. 1.   PerBU, constructed from 6- and 4-fold connected double 8-rings, viewed along *a*.

## 2. Connection mode

Neighboring PerBUs, related along *b* by a pure translation and along *c* by a screw rotation of 180° about *c*, are connected along *b* and *c* through double crankshaft chains as shown in Figure 2.

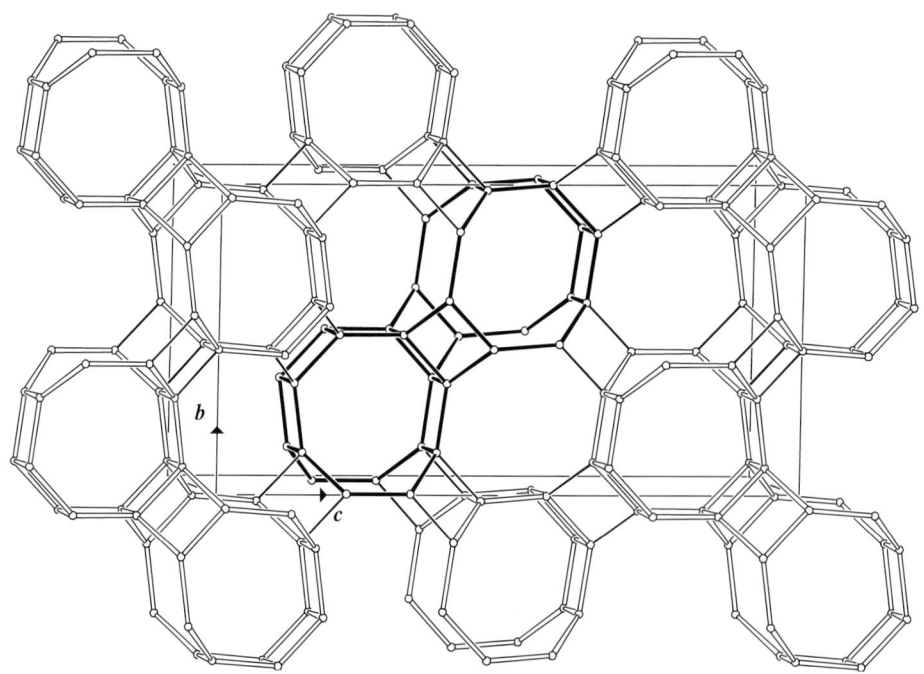

Fig. 2.   Connection mode (and unit cell content) viewed along *a*. For clarity, only one repeat unit along *a* of each PerBU is shown.

### 3. Channels and/or cages

Two types of interconnecting 8-ring channels (with the same topology) are parallel to $a$. The first type is equivalent to the 8-ring channel in **PHI** and the other type to the 8-ring channel in **ATT** and **GIS**. The channels and their interconnection are shown in Figure 3.

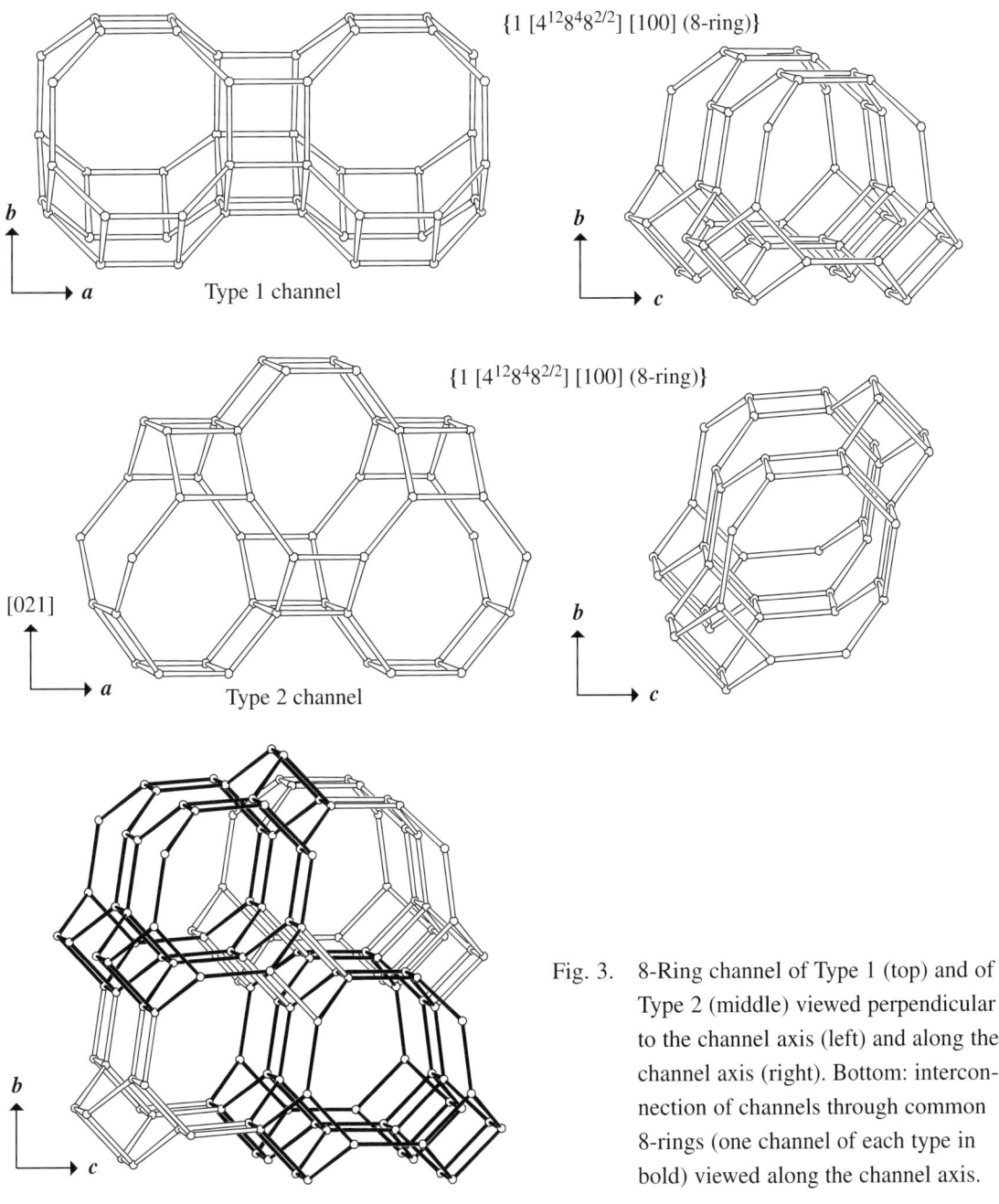

{1 [$4^{12}8^48^{2/2}$] [100] (8-ring)}

Type 1 channel

{1 [$4^{12}8^48^{2/2}$] [100] (8-ring)}

Type 2 channel

Fig. 3.    8-Ring channel of Type 1 (top) and of Type 2 (middle) viewed perpendicular to the channel axis (left) and along the channel axis (right). Bottom: interconnection of channels through common 8-rings (one channel of each type in bold) viewed along the channel axis.

# SOD                                     Building Scheme

## 1. Periodic Building Unit

The two-dimensional PerBU of **SOD** consists of a hexagonal array of non-connected planar 6-rings (bold in Figure 1), which are related along **a** and **b** by pure translations. The 6-rings are centered at (0,0) in the **ab** layer. This position is usually called the **A** position. **SOD** belongs to the ABC-6 family (see Introduction).

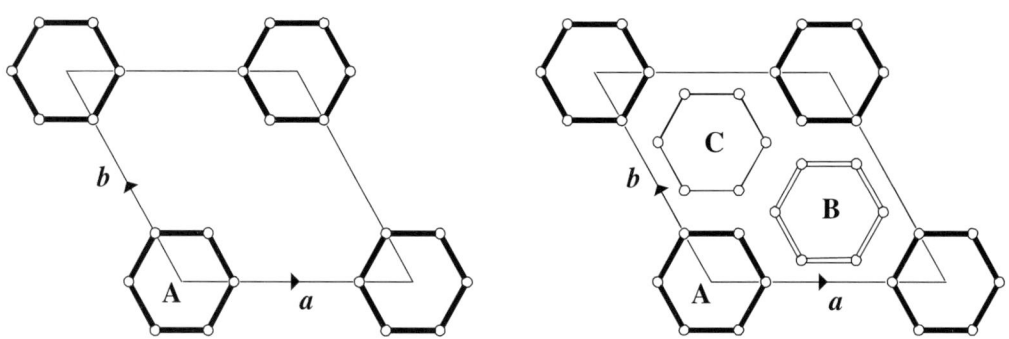

Fig. 1.   PerBU (left) and definition of 6-ring positions used in the stacking modes (right).

## 2. Connection mode

Neighboring PerBUs are connected along **c** through tilted 4-rings by connection modes (**1**) and (**2**) (see Introduction) as illustrated in Figure 2.

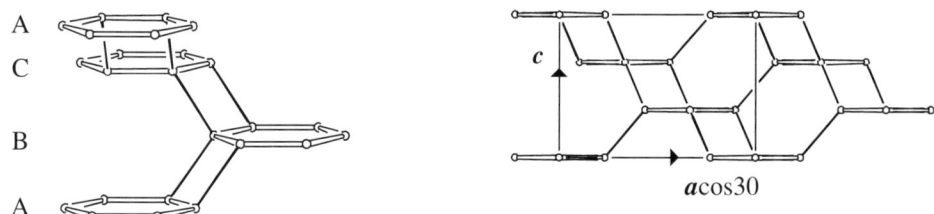

Fig. 2.   Connection mode (left) and unit cell content (right) viewed along **b**. The stacking sequence is given. In the perspective drawing, each PerBU is represented by one 6-ring only.

### 3. Channels and/or cages

The sodalite cage (or *sod* cage) is depicted in Figure 3. The *sod* cage is also found in **EMT**, **FAU**, **FRA**, **GIU**, **LTA**, **LTN**, **MAR** and **TSC**. Apertures of "channels" are formed by 6-rings only.

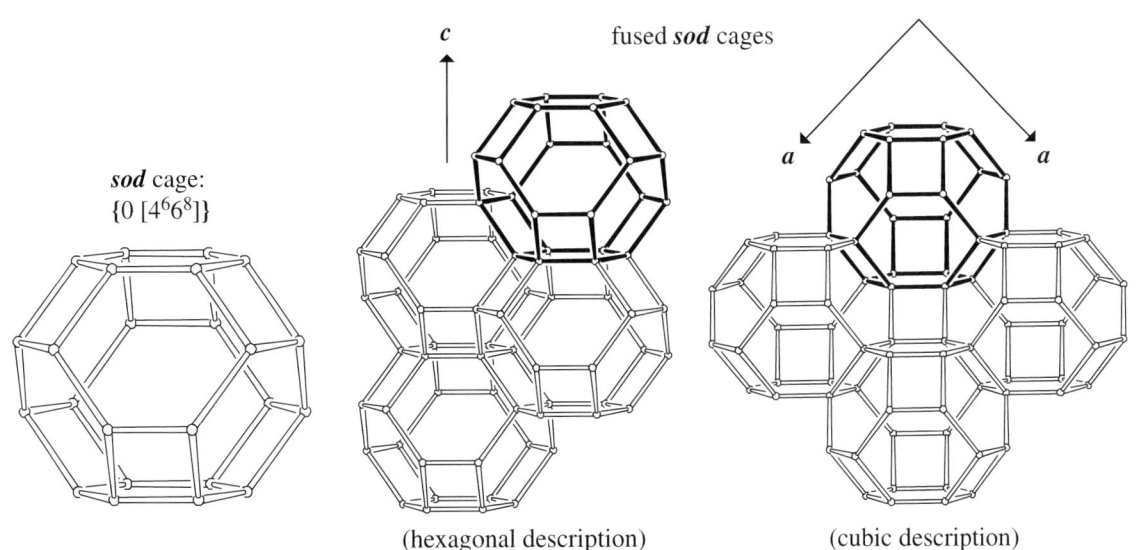

*sod* cage:
$\{0\ [4^6 6^8]\}$

fused *sod* cages

(hexagonal description)　　　　(cubic description)

Fig. 3.　*sod* Cage (left) and fused cages in hexagonal description (middle) and in cubic description (right).

# SOS Building Scheme

## 1. Periodic Building Unit

**SOS** can be built using double 6-rings with two disconnected edges (or two 4-2 units; bold in Figure 1). The one-dimensional PerBU is obtained when T12-units, related along $b$ by a pure translation, are linked into chains along $b$ through 3-rings sharing an edge.

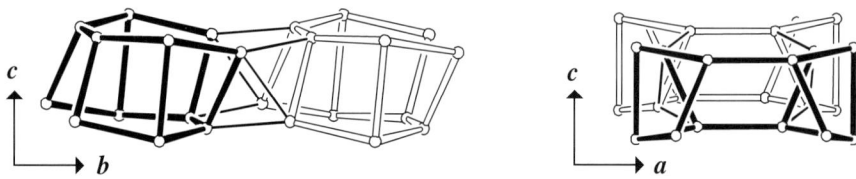

Fig. 1.   PerBU viewed along $a$ (left) and along $b$ (right).

## 2. Connection mode

Neighboring PerBUs, related along $a$ (and $c$) by a rotation of 180° about $b$ followed by a shift of $\frac{1}{2}(a+c)$, are connected as shown in Figure 2. 12-Ring channels parallel to $b$ are formed.

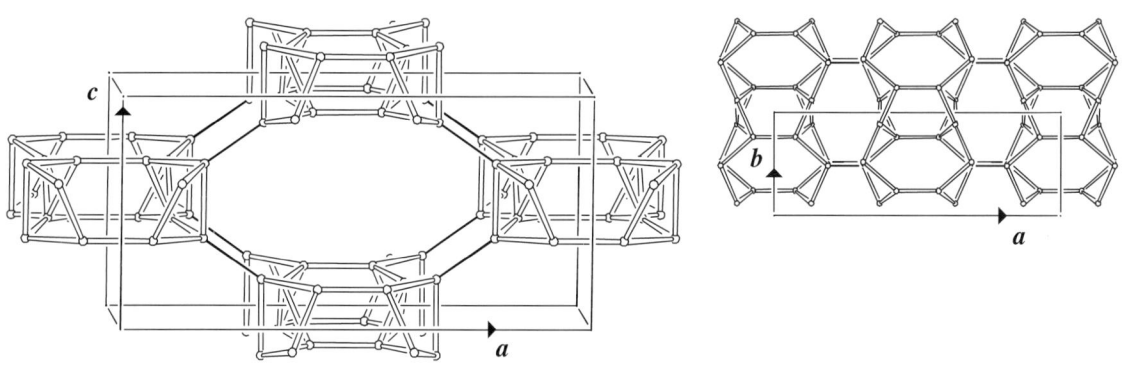

Fig. 2.   Connection mode viewed along $b$ and unit cell content viewed along $b$ (top right) and along $c$ (bottom right).

### 3. Channels and/or cages

Interconnecting 12-ring channels are parallel to **b**. The channels are interconnected by sharing 8-rings (Figure 3).

$\{1\ [3^4 4^4 8^6 12^{2/2}]\ [010]\ (12\text{-ring})\}$

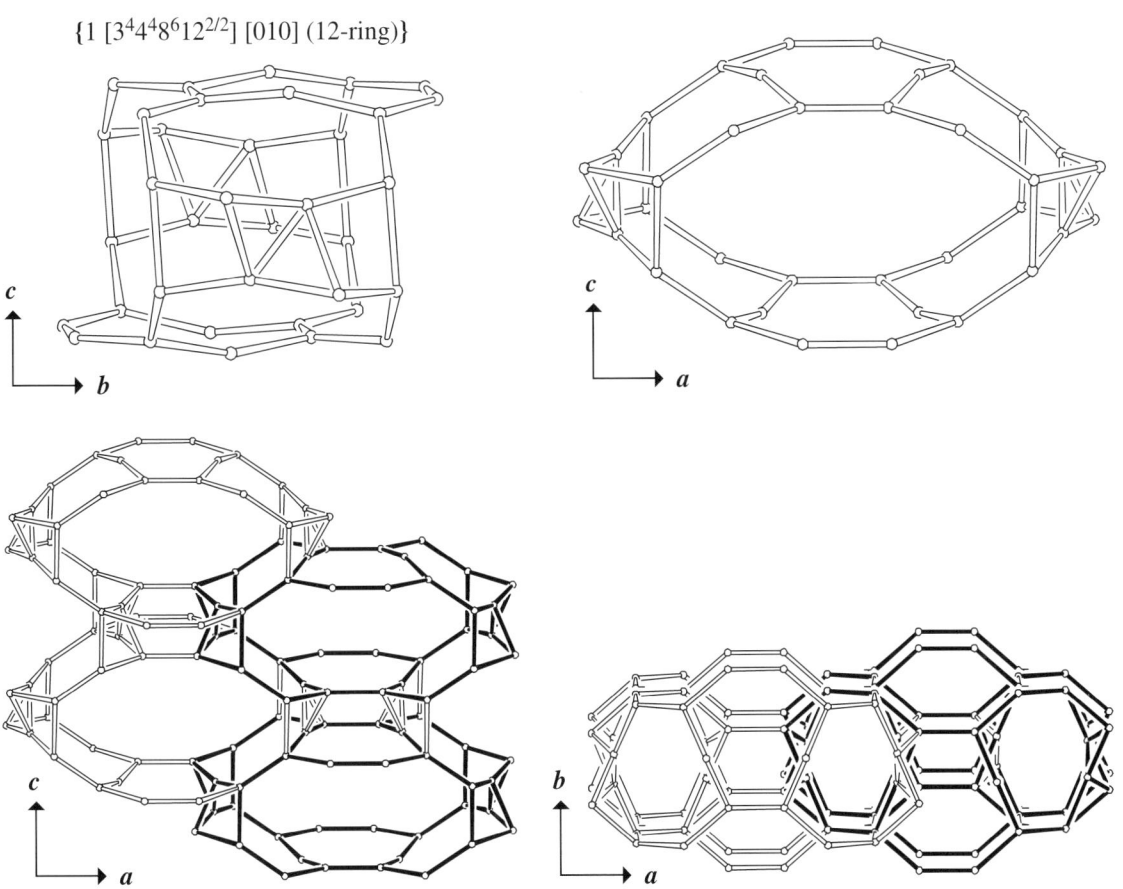

Fig. 3. Top: repeat unit in the 12-ring channel viewed perpendicular to the channel axis (left) and along the channel axis (right). Bottom: interconnecting 12-ring channels (bottom) viewed along **b** (left) and along **c** (right).

# SSY        Building Scheme

## 1. Periodic Building Unit

**SSY** can be built using the zigzag chain running parallel to **c** (bold in Figure 1). Seven zigzag chains are connected into an infinite building unit. A two-dimensional PerBU is obtained when infinite building units, related along **b** by a pure translation, are connected through double zigzag chains into the **bc** layer shown in Figure 1 (see also **SFE**).

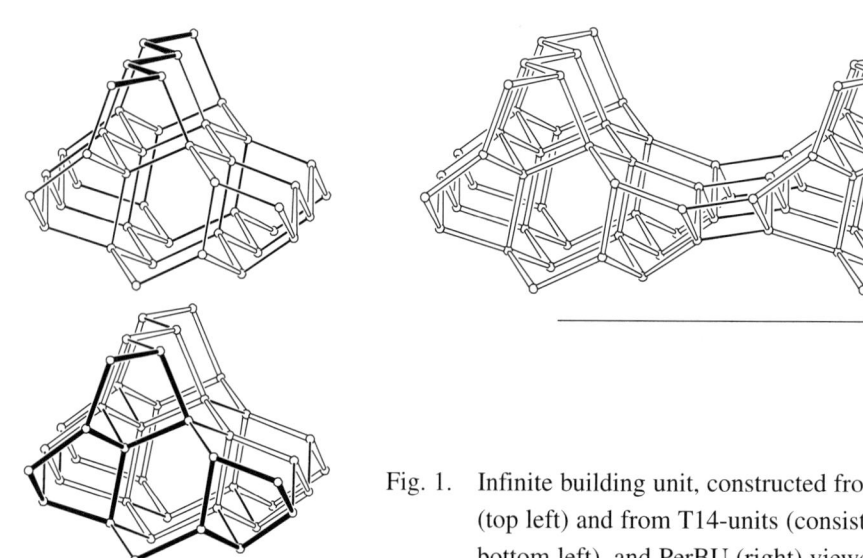

Fig. 1.  Infinite building unit, constructed from seven zigzag chains (top left) and from T14-units (consisting of 5-1 and 5-3 units; bottom left), and PerBU (right) viewed along **c**.

## 2. Connection mode

Neighboring PerBUs, related along **a** by a screw rotation of 180° about **a**, are connected along **a** through (fused) 5- and 6-rings as shown in Figure 2.

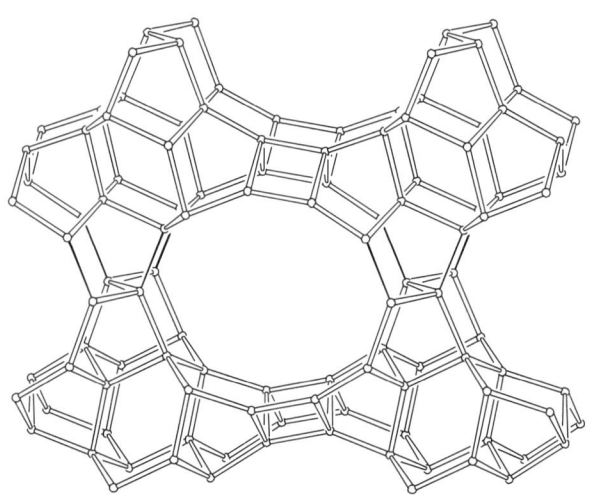

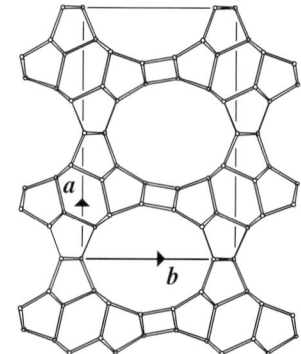

Fig. 2.  Connection mode (left) and unit cell content (right) viewed along **c**.

## 3. Channels and/or cages

Non-interconnecting 12-ring channels (Figure 3) are parallel to $c$. The channel is topologically equivalent to the 12-ring channel in **SFE**.

$\{1\,[4^4 6^8 12^{2/2}][001]\,(12\text{-ring})\}$

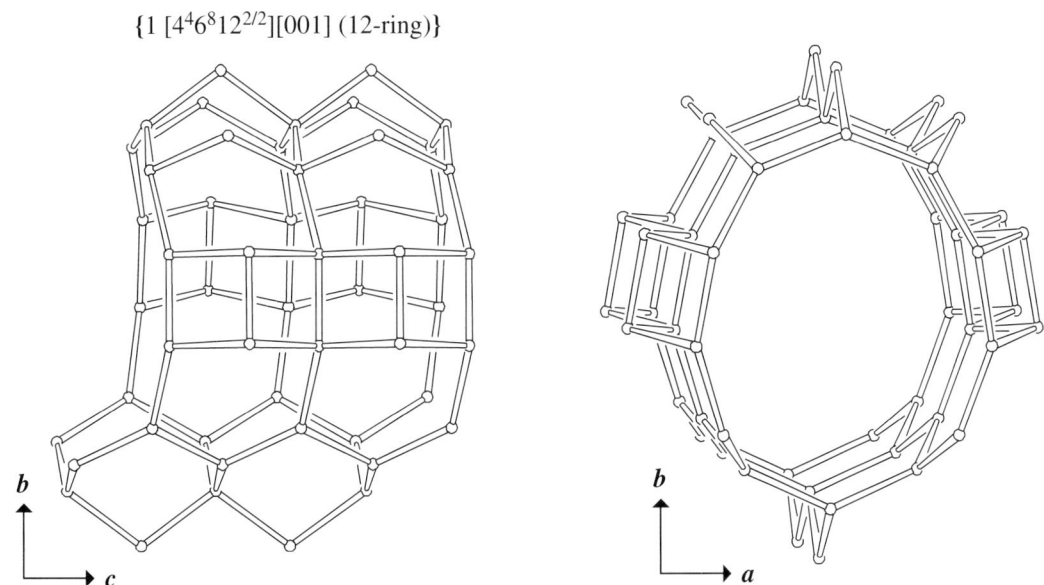

Fig. 3.    Channel viewed perpendicular to the channel axis (left) and along the channel axis (right).

# STF                    **Building Scheme**

## 1. Periodic Building Unit

The two-dimensional PerBU in **STF** is composed of units of 16 T atoms and equals the *ac* layer shown in Figure 1. The T16-unit consists of two 5-3 units (bold in Figure 1). T16-units, related along *a* and *c* by pure translations, are linked through 4-rings and finite zigzag chains, respectively (see also **SFF**).

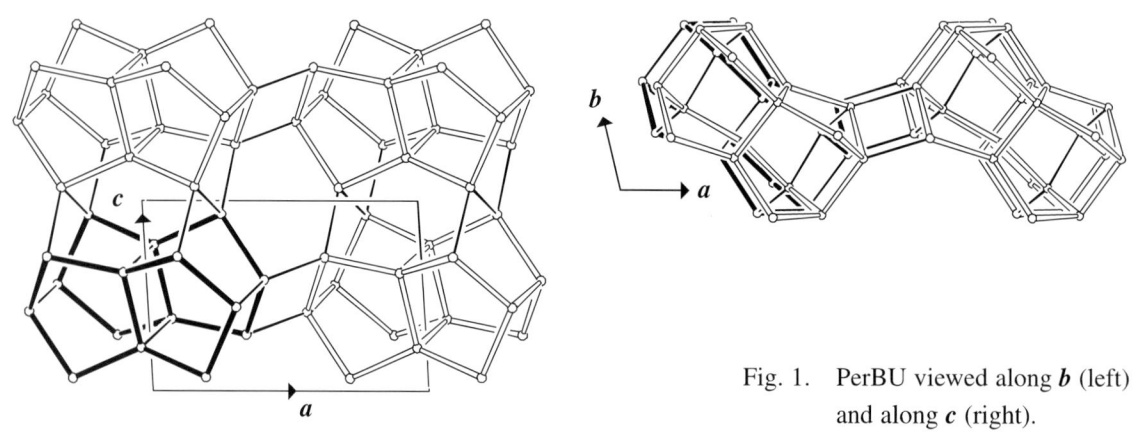

Fig. 1.    PerBU viewed along *b* (left) and along *c* (right).

## 2. Connection mode

Neighboring PerBUs, related along *b* by a pure translation, are connected along *b* through 4-rings as illustrated in Figure 2.

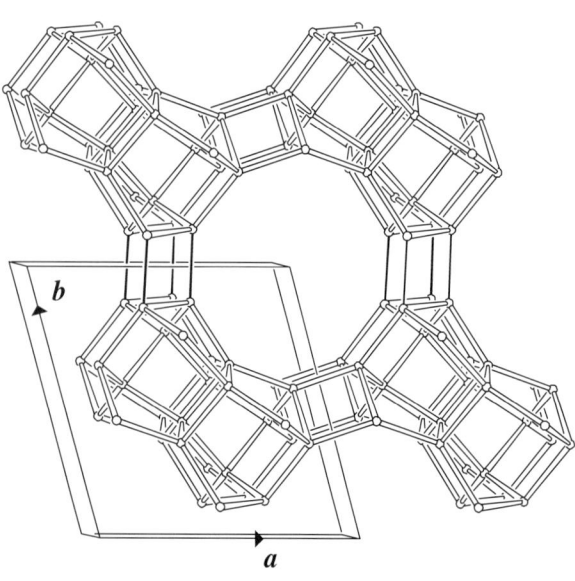

Fig. 2.    Connection mode (and unit cell content) viewed along *c*.

## 3. Channels and/or cages

Non-interconnecting 10-ring channels are parallel to *c*. The cavity, depicted in Figure 3, shows the repeat unit in the channel. The channel is topologically equivalent to the 10-ring channel in **SFF**.

$\{1 \ [4^4 5^8 6^6 10^2][001] \ (10\text{-ring})\}$

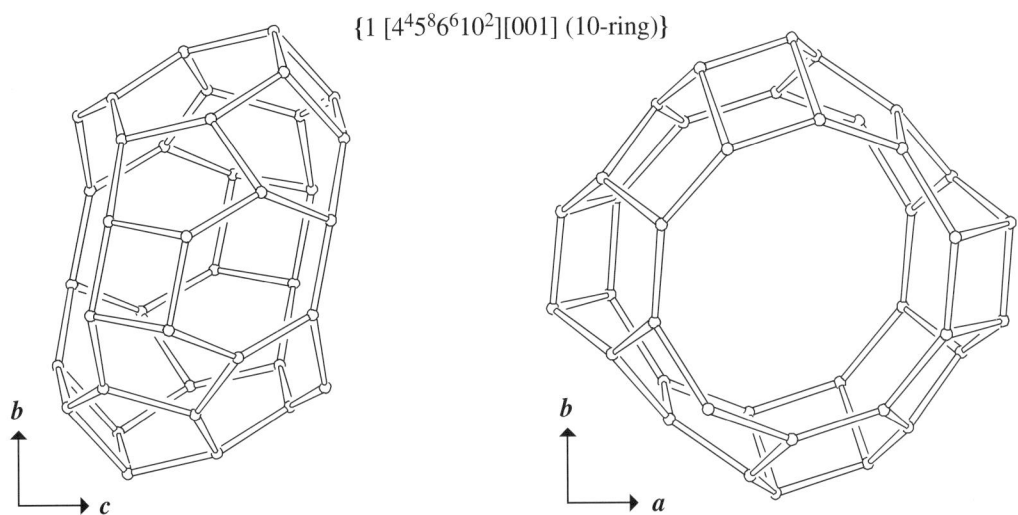

Fig. 3.    Channel viewed perpendicular to the channel axis (left) and along the channel axis (right).

# STI                    Building Scheme

## 1. Periodic Building Unit

The two-dimensional PerBU in **STI** is composed of units of 9 T atoms (one in bold in Figure 1). The T9-unit consists of a 2-fold (1,3)-connected double 4-ring and an additional bridging T atom (a $[4^25^26^1]$-cage, or 4–4 = 1 unit). T 9-units, related along **b** by a rotation of 180° about an axis parallel to **b** and passing through the bridging T atom, are connected into a chain along **b** through the bridging T atoms. Neighboring chains, related along **a** by a shift of $\frac{1}{2}(\boldsymbol{a}+\boldsymbol{b})$, are connected along **a** through 4-rings as shown in Figure 1. (See also **HEU** and **RRO**.)

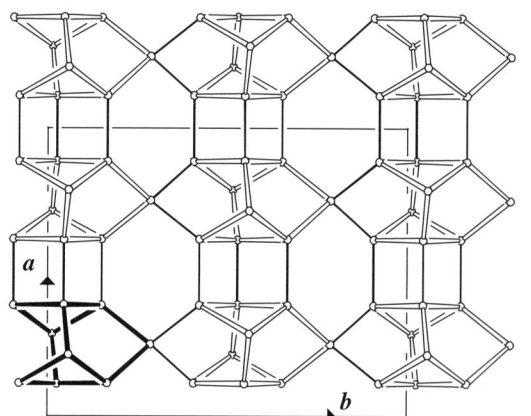
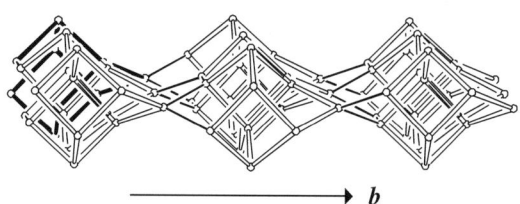

Fig. 1.   PerBU composed of T9-units viewed along **c** (left) and along **a** (right).

## 2. Connection mode

Neighboring PerBUs, related along **c** by a shift of $\frac{1}{2}(\boldsymbol{b}+\boldsymbol{c})$, are connected as shown in Figure 2.

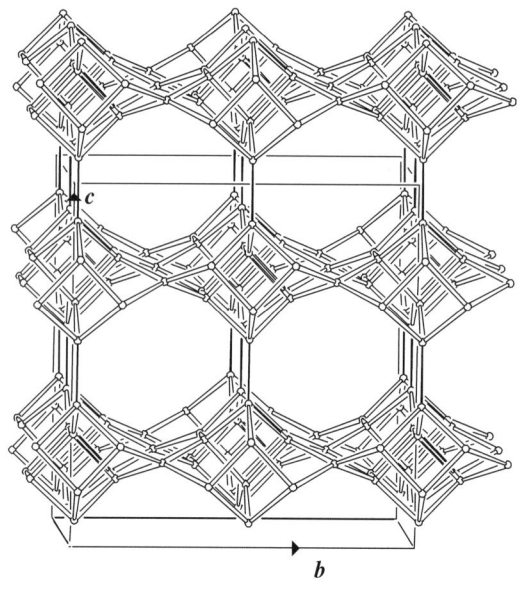
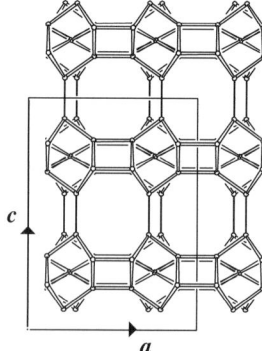

Fig. 2.   Connection mode (and unit cell content) viewed along **a** (left) and unit cell content projected along **b** (right).

## 3. Channels and/or cages

Intersecting 10-ring and 8-ring channels are parallel to *a* and *b*, respectively. The channel intersection is shown in Figure 3 together with the connection of the intersections into channels.

{2 [$4^2 5^4 6^2 8^2 10^2$] [100] (10-ring), [010] (8-ring)}

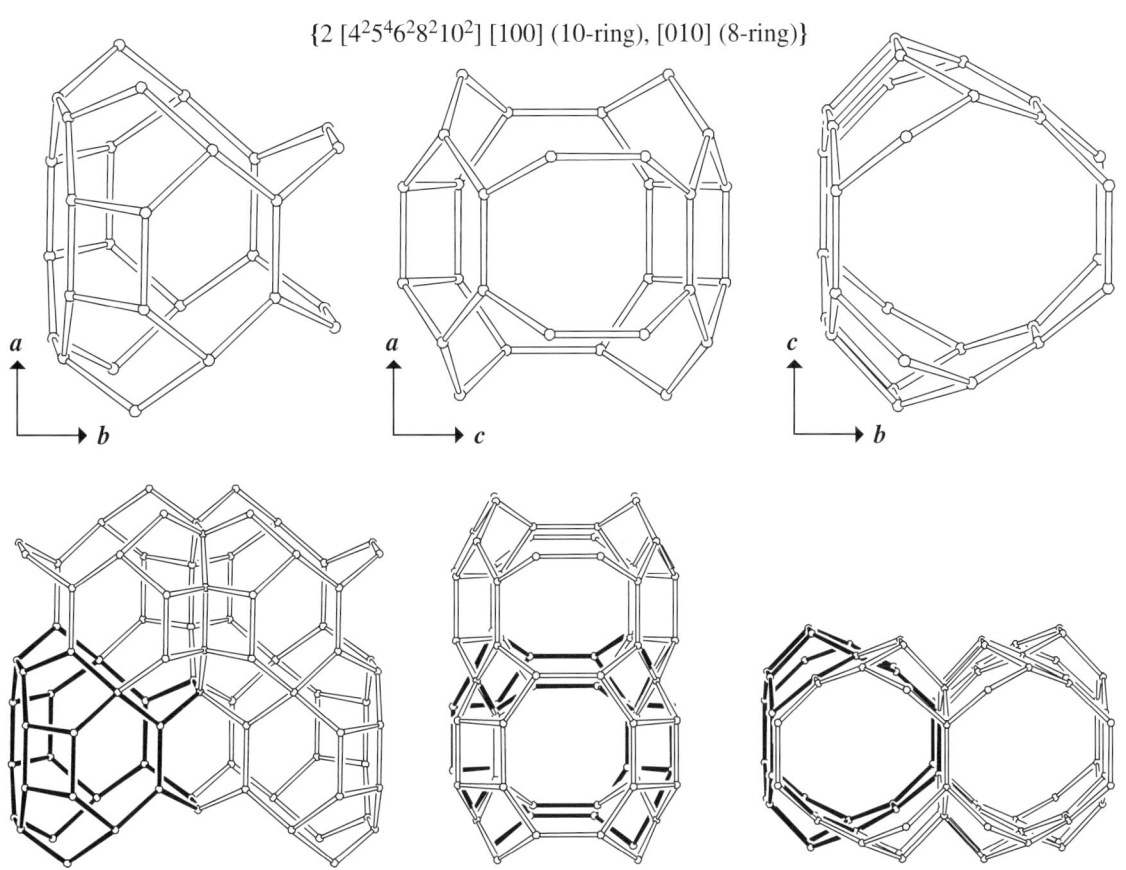

Fig. 3.    Channel intersection (top) and connection of intersections into channels (bottom) viewed (from left to right) along *c*, along the 8-ring channel axis parallel to *b* and along the 10-ring channel axis parallel to *a*.

# STT                    Building Scheme

## 1. Periodic Building Unit

The two-dimensional PerBU in **STT**, composed of units of 16 T atoms, is equal to the ***ac*** layer shown in Figure 1. The T16-unit consists of two 5-3 units (bold in Figure 1). Two "nearest neighbor" T16-units are related by inversion centers; the two "next-nearest" T16-units are related along ***a*** and ***c*** by pure translations. The T16-units are linked in the ***ac*** plane through (finite) single and double zigzag chains, 4- and 6-rings (Figure 1).

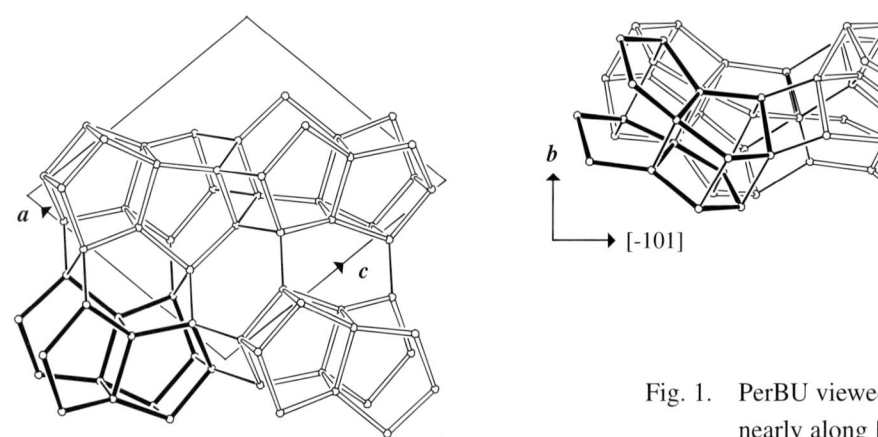

Fig. 1.   PerBU viewed along ***b*** (left) and nearly along [101] (right).

## 2. Connection mode

Neighboring PerBUs, related along ***b*** by a screw rotation of 180° about ***b***, are connected along ***b*** through 4-rings as shown in Figure 2.

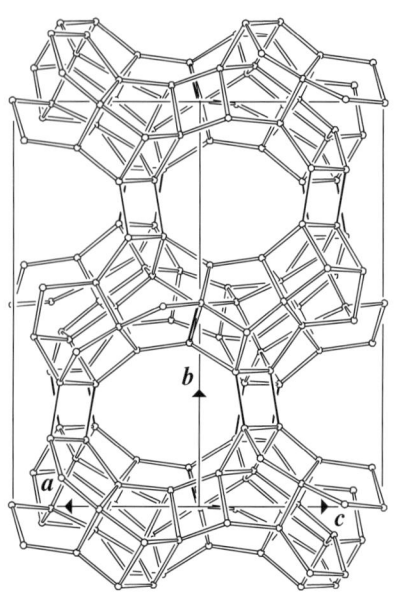
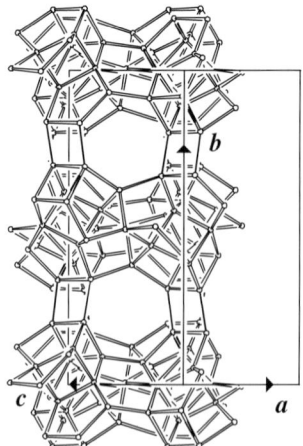

Fig. 2.   Connection mode (and cell content) viewed along [101] (left) and along [–101] (right).

## 3. Channels and/or cages

Intersecting 7-ring and 9-ring channels are parallel to [–101] and [101], respectively. The channel intersection is depicted in Figure 3. Intersections are linked through common 7- and 9-rings.

{2 [$4^6 5^6 6^5 7^2 9^2$][-101] (7-ring), [101] (9-ring)}

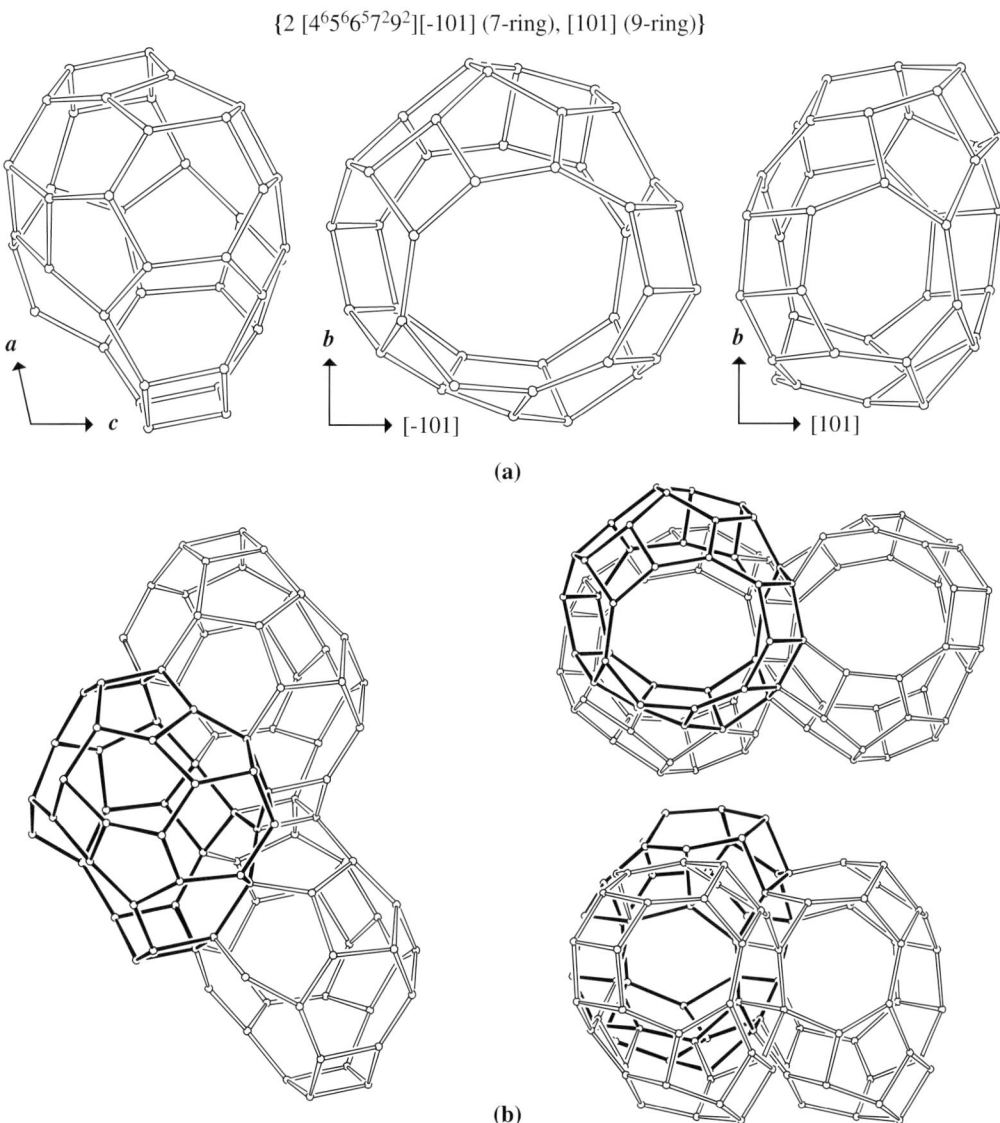

**(a)**

**(b)**

Fig. 3.  (a) Channel intersection viewed along **b** (left), [101] (middle) and along [–101] (right).
(b) Connection of intersections into channels parallel to [101] and [–101] viewed along **b** (left), along the 9-ring channel axis parallel to [101] (top right) and along the 7-ring channel axis parallel to [–101] (bottom right).

# SZR

# Building Scheme

## 1. Periodic Building Unit

**SZR** can be built using the $[6^8]$-cage (Figure 1) built from three 6-ring boats (one in bold). The $[6^8]$-cage is also observed in **MSO**. The one-dimensional PerBU is obtained when $[6^8]$-cages, related along $c$ by a pure translation, are connected along $c$ through double 6-rings as shown in Figure 1. Saw chains (one in bold) are formed. Large spheres indicate K atoms.

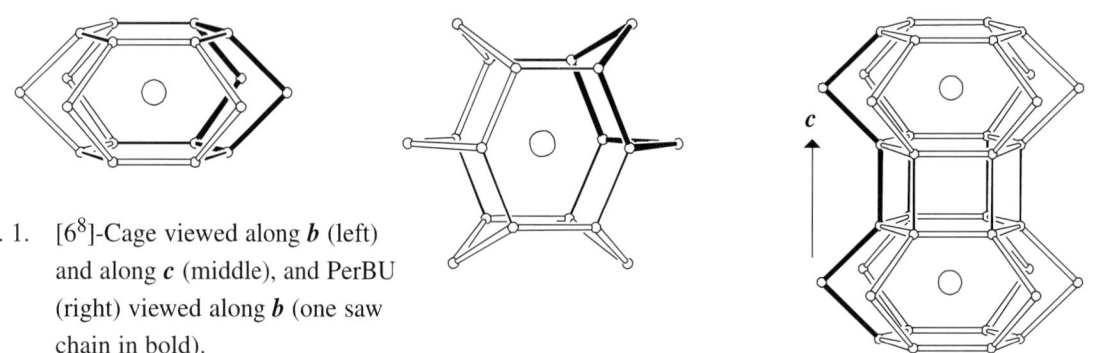

Fig. 1.   $[6^8]$-Cage viewed along $b$ (left) and along $c$ (middle), and PerBU (right) viewed along $b$ (one saw chain in bold).

## 2. Connection mode

Neighboring PerBUs, related along $a$ (and $b$) by a shift of $\frac{1}{2}(a+b)$, are connected along $a$ (and $b$) through 5-rings (Figure 2).

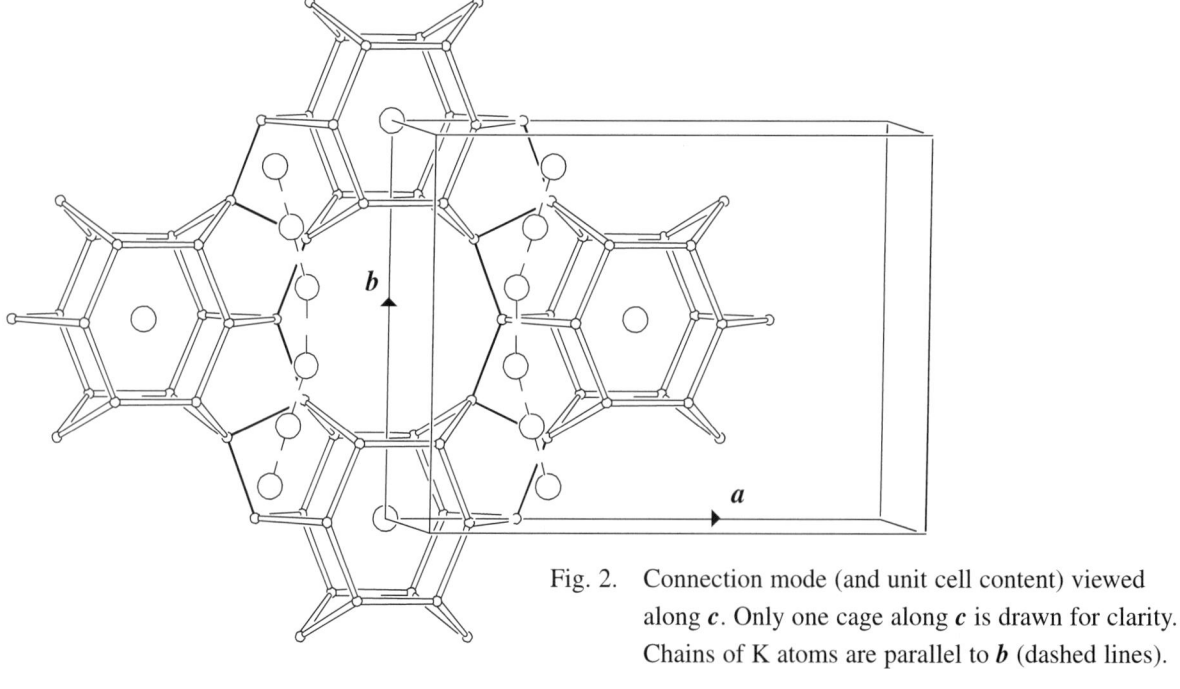

Fig. 2.   Connection mode (and unit cell content) viewed along $c$. Only one cage along $c$ is drawn for clarity. Chains of K atoms are parallel to $b$ (dashed lines).

## 3. Channels and/or cages

One-dimensional 10-ring channels are parallel to $c$. Transport of matter between the 10-ring channels is impossible because the 8-ring windows are blocked by potassium chains (which cannot be removed) in 8-ring channels parallel to $b$. One repeat unit of each channel is shown in Figure 3.

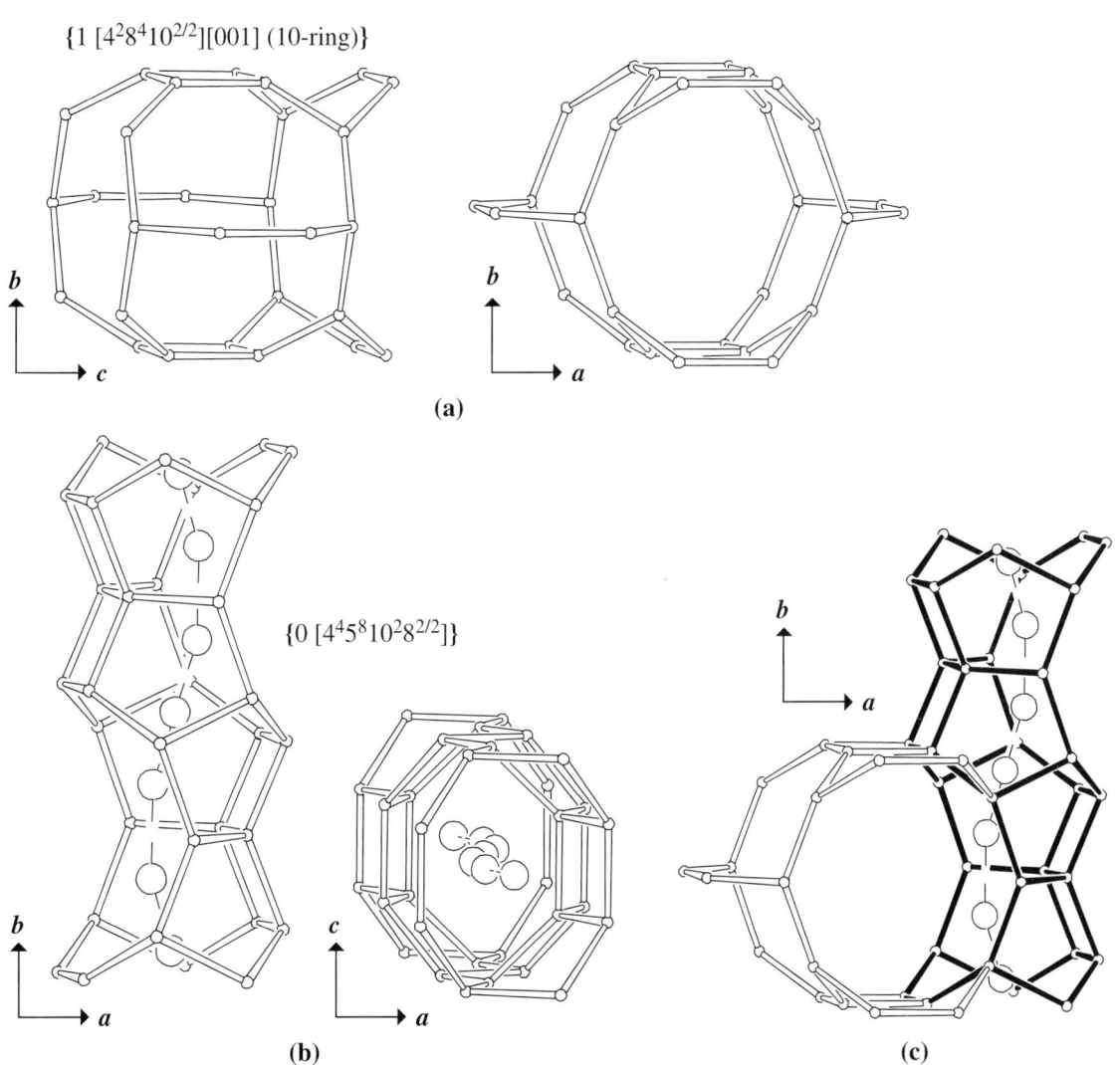

Fig. 3.　(a) 10-Ring channel viewed along $a$ (left) and along $c$ (right); (b) 8-ring channel (K chain dashed) viewed along $c$ (left) and along b (right); and (c) linked channels viewed along $c$ (8-ring channel in bold).

# TER **Building Scheme**

## 1. Periodic Building Unit

The one-dimensional PerBU is composed of units of 20 T atoms: two [$4^2 5^4$]-cages sharing an edge and two additional T atoms (or four 4-1 units; bold in Figure 1). The PerBU is equal to the strip, extending infinitely along $c$, shown in Figure 1. Neighboring T20-units in the strip, related along $c$ by a screw rotation of 180° about $c$, are connected along $c$ by (fused) 6-rings and 10-rings.

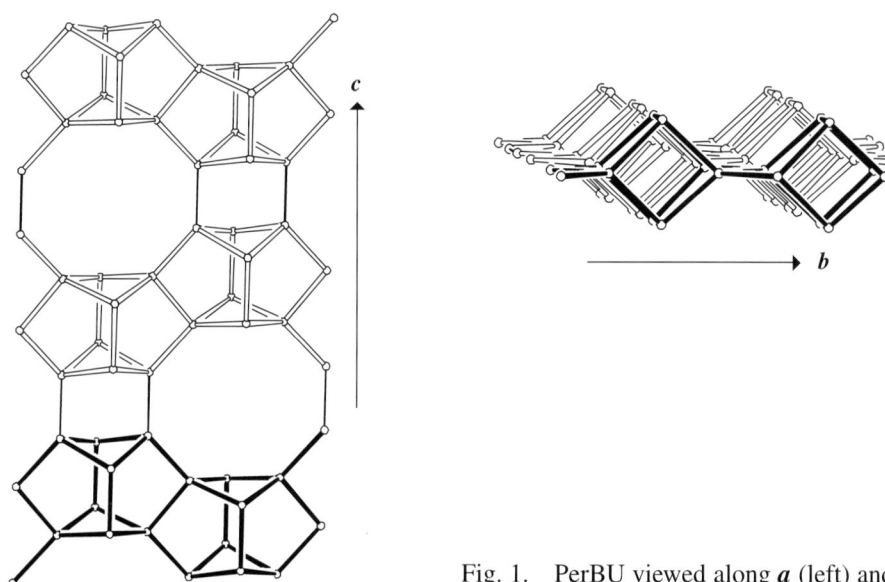

Fig. 1.   PerBU viewed along $a$ (left) and along $c$ (right).

## 2. Connection mode

Neighboring PerBUs, related along $a$ (and $b$) by a shift of $\frac{1}{2}(a+b)$, are connected along $a$ (and $b$) through 6-rings as shown in Figure 2. 10-Ring channels parallel to $a$ and $c$ are formed.

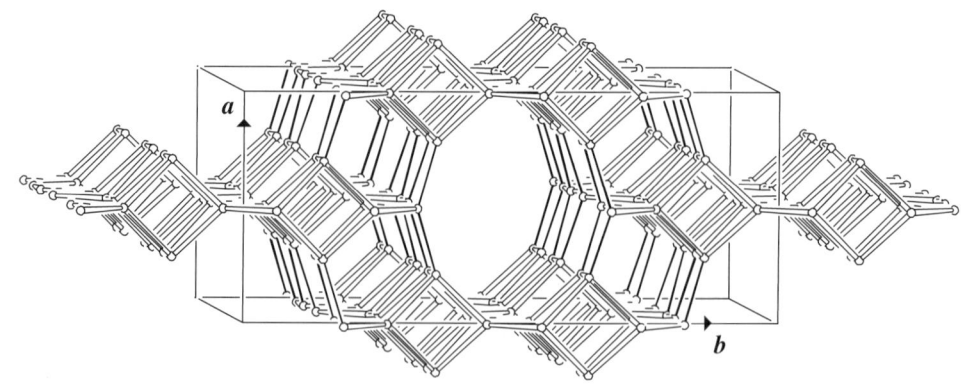

Fig. 2a.   Connection mode (and cell content) viewed along $c$.

## Cage/Channel

# TER

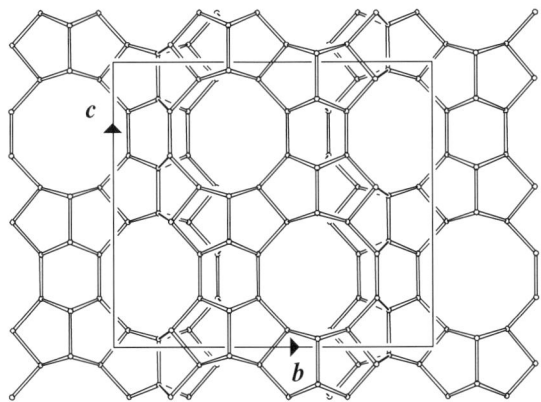

Fig. 2b.   Unit cell content projected along **a**.

## 3. Channels and/or cages

Intersecting 10-ring channels are parallel to **a** and **c**. The repeat unit of the channel parallel to **a** is shown in Figure 3 together with the linkage of these channels into 10-ring channels parallel to **c**.

$\{1\ [4^2 6^8 10^2 10^{2/2}]\ [100]\ (10\text{-ring})\}$

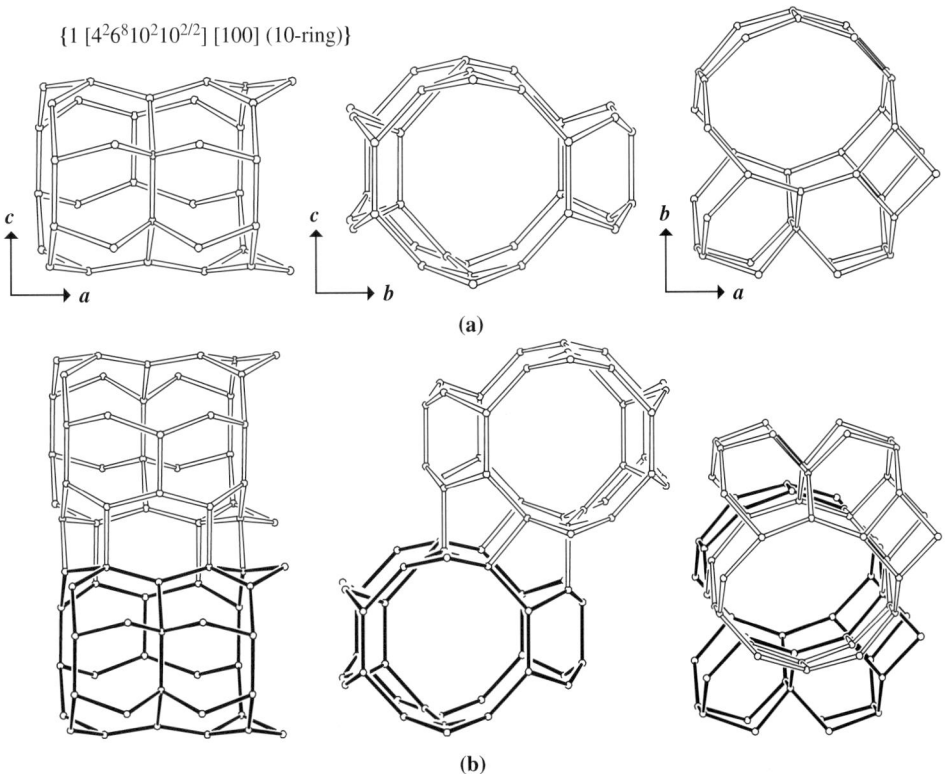

Fig. 3.   (a) 10-Ring channel parallel to **a** viewed (from left to right) along **b**, **a** and **c**; (b) Channels parallel to **a** are connected into 10-ring channels parallel to **c**. View along **b** (left), along the 10-ring channel axis parallel to **a** (middle) and parallel to **c** (right).

# THO

## Building Scheme

### 1. Periodic Building Unit

**THO** can be built using the fibrous chain (or natrolite-chain) as one-dimensional PerBU. The chain is composed of units of 5T atoms (bold in Figure 1). These T5-units ($[4^3]$-cages or $4=1$ units) are related along $c$ by pure translations. (See also **EDI** and **NAT**.)

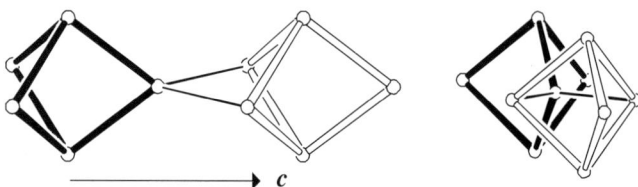

Fig. 1.    PerBU viewed perpendicular to the chain axis (left) and down the chain axis (right).

### 2. Connection mode

Neighboring PerBUs, related along $a$ by a rotation of 90° about the chain axis and a shift of $\frac{1}{2}$ $a$ and along $b$ by a pure translation, are connected along $a$ and $b$ as depicted in Figure 2.

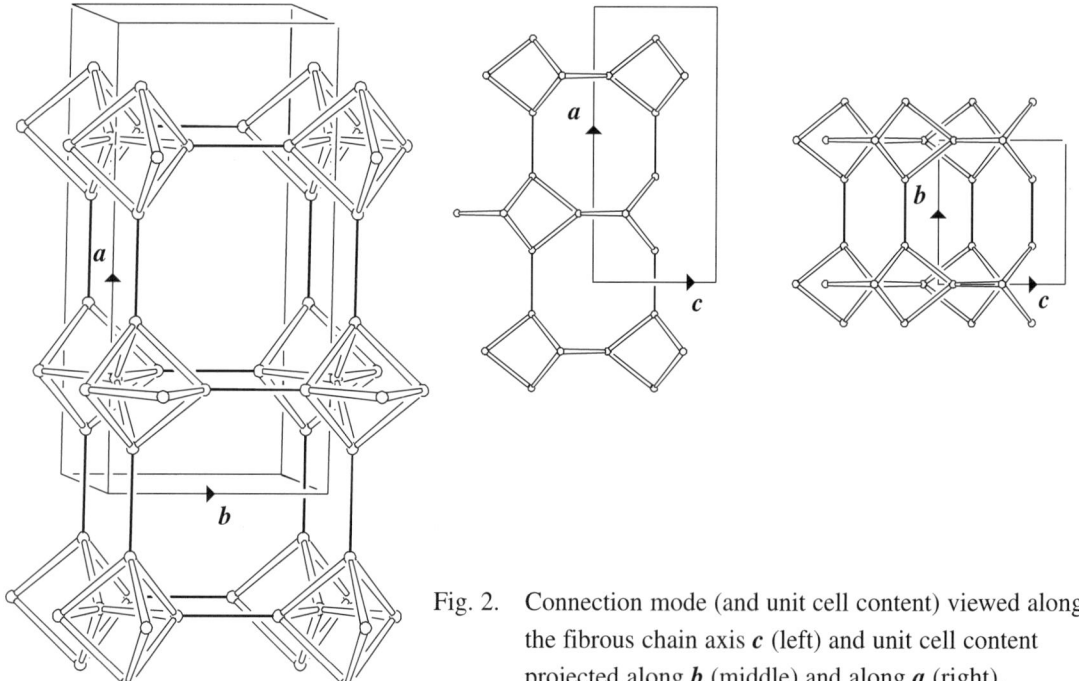

Fig. 2.    Connection mode (and unit cell content) viewed along the fibrous chain axis $c$ (left) and unit cell content projected along $b$ (middle) and along $a$ (right).

### 3. Channels and/or cages

Intersecting 8-ring channels are parallel to **b** and **c**. The channels are topologically equivalent to the channels in **EDI**. The repeat unit in both channels is depicted in Figure 3. The channels are interconnecting through common 8-rings. Sinusoidal 8-ring channels parallel to **a** are formed as illustrated in Figure 3.

$\{1\ [8^4 8^{2/2}]\ [001]\ (8\text{-ring})\}$

$\{1\ [4^2 8^4 8^{2/2}]\ [010]\ (8\text{-ring})\}$

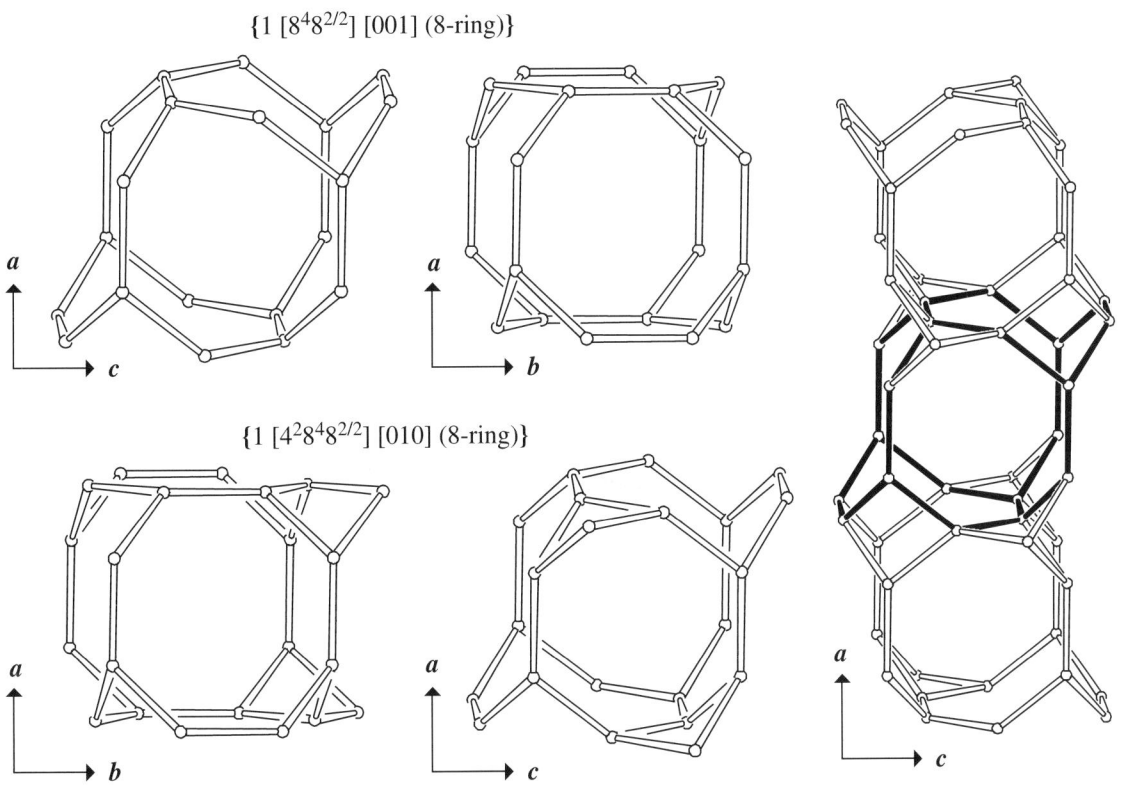

Fig. 3.    8-Ring channels, parallel to **c** (top) and to **b** (bottom), viewed perpendicular to the channel axis (left) and along the channel axis (right). At the far right: channels are connected through common 8-rings into sinusoidal channels parallel to **a**. The intersecting channels have [8⁶] cavities in common. One intersection in bold.

# TOL Building Scheme

## 1. Periodic Building Unit

The two-dimensional Periodic Building Unit (PerBU) of **TOL** consists of a hexagonal array of non-connected planar 6-rings (bold in Figure 1), which are related by pure translations along *a* and *b*. The 6-rings are centered at (0,0) in the *ab* layer. This position is usually called the **A** position. **TOL** belongs to the ABC-6 family (see Introduction).

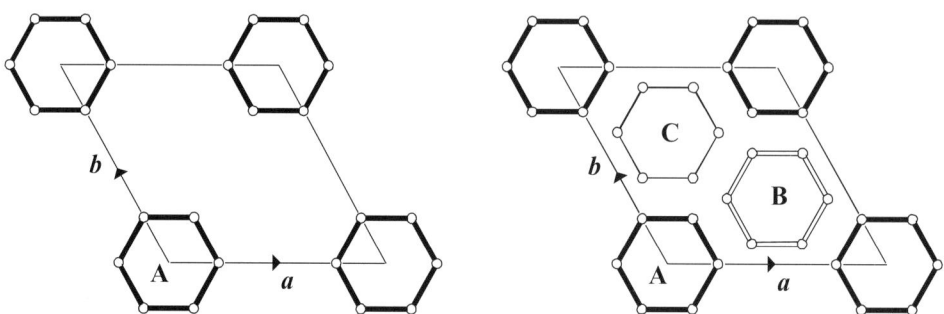

Fig. 1. PerBU (left) and definition of 6-ring positions used in the stacking modes (right).

## 2. Connection mode

Neighboring PerBUs can be connected through tilted 4-rings along *c* in three different ways (see Introduction). In **TOL**, only connection modes **(1)** and **(2)** between the PerBUs are observed.

A
C

A
C

A

B

A

C

A

B

A

B

A

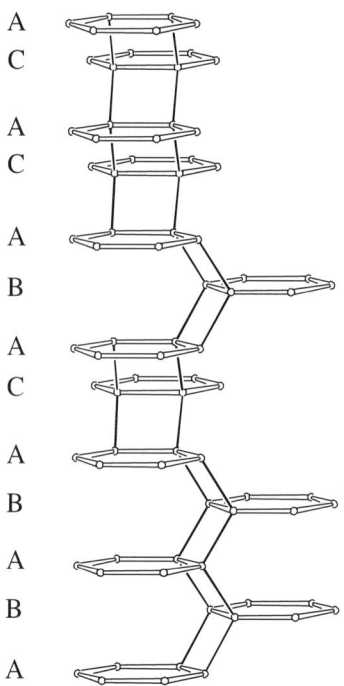

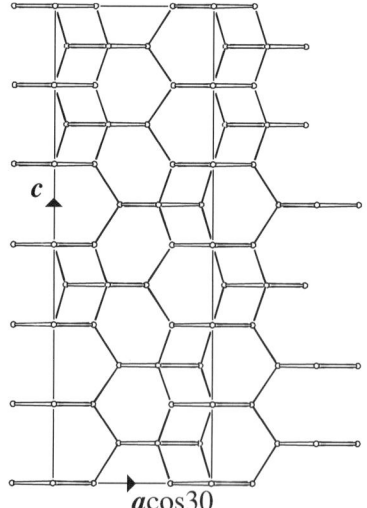

Fig. 2. Perspective drawing (left) and projection of the unit cell content (right) along *b*. The stacking sequence is given. In the perspective drawing each PerBU is represented by one 6-ring only.

## 3. Channels and/or cages

The three types of cages in **TOL** are depicted in Figure 3. The *can* cage is also present in **AFG**, **CAN**, **ERI**, **FAR**, **FRA**, **GIU**, **LIO**, **LOS**, **LTL**, **LTN**, **MAR**, **MOZ**, **OFF**, **SAT**, **SBS**, **SBT** and **-WEN**. The *los* cage is also found in **FRA**, **LIO** and **LOS**. Finally, the *lio* cage is also found in **AFG**, **FAR**, **LIO** and **MAR**. Apertures of "channels" are formed by 6-rings only.

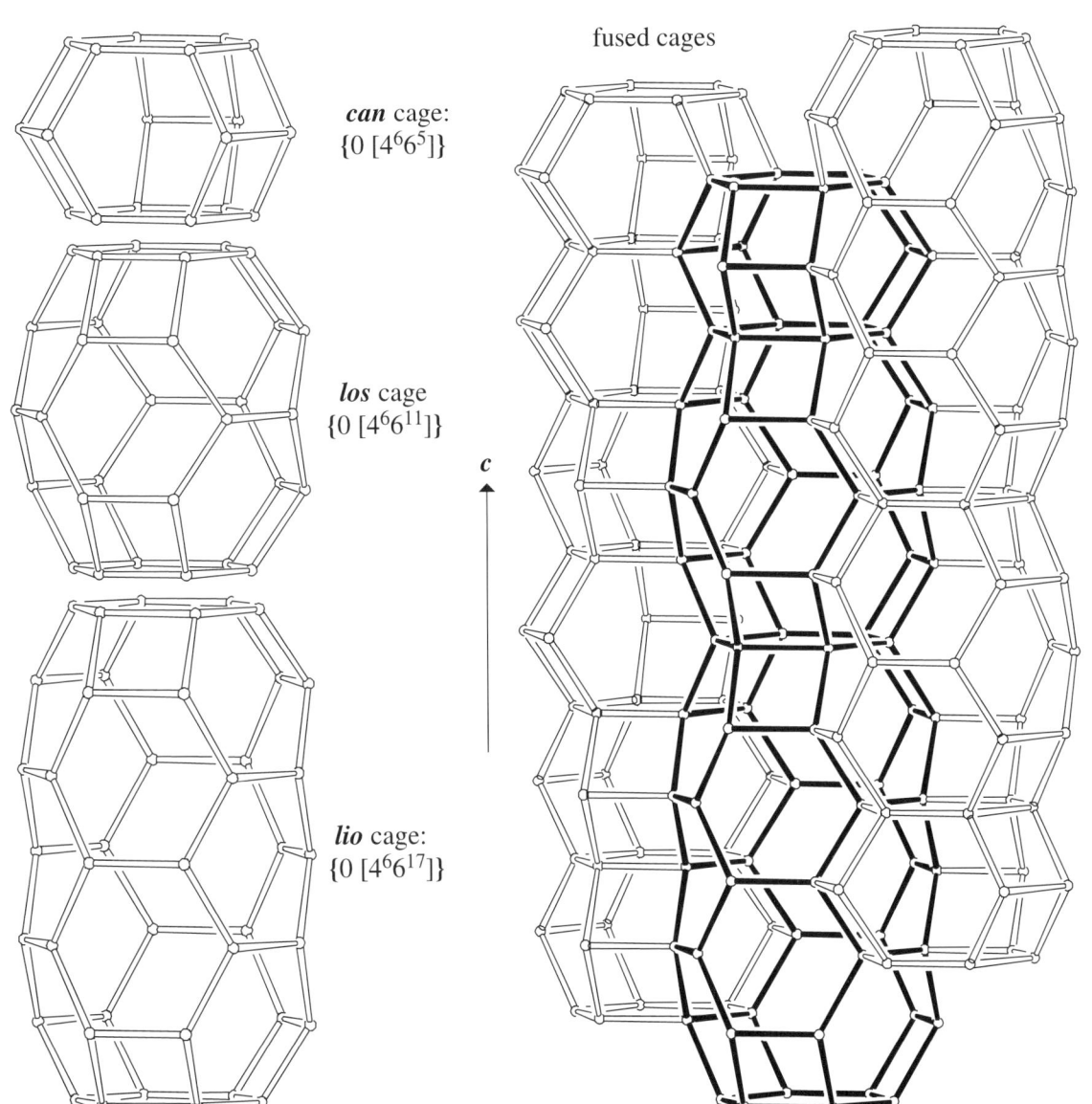

fused cages

*can* cage:
$\{0\ [4^6 6^5]\}$

*los* cage
$\{0\ [4^6 6^{11}]\}$

*lio* cage:
$\{0\ [4^6 6^{17}]\}$

$c$

Fig. 3.   *can* Cage (top left), *los* cage (middle left), *lio* cage (bottom left) and connection of cages (right) as viewed perpendicular to $c$.

# TON — Building Scheme

## 1. Periodic Building Unit

**TON** can be built using the zigzag chain (one bold in Figure 1, left). Six zigzag chains form an infinite building unit (Figure 1, left). The repeat unit of this infinite building unit consists of two 5–1 units (bold in Figure 1, middle). The two-dimensional PerBU is obtained when infinite building units, related along [110] by a shift of $\frac{1}{2}(\boldsymbol{a}+\boldsymbol{b})$, are connected into the layer shown in Figure 1 (right). (See also **MTT**.)

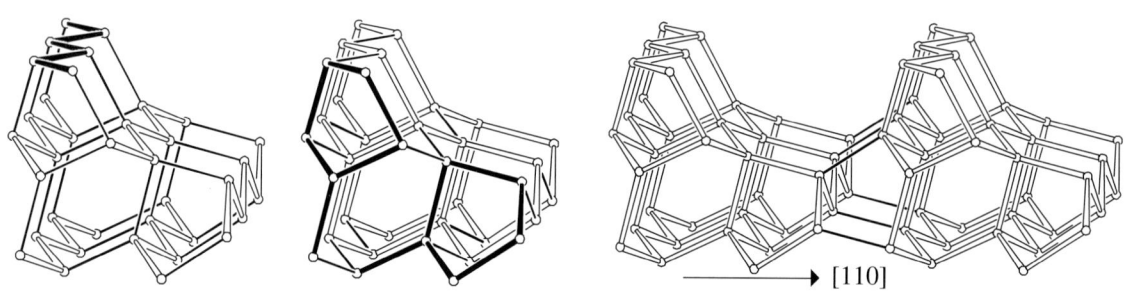

Fig. 1.  Infinite building unit, constructed from six zigzag chains (left) and from 5-1 units (middle), and PerBU (right) viewed along the zigzag chain axis.

## 2. Connection mode

Neighboring PerBUs, related along [1-10] by a shift of $\frac{1}{2}(\boldsymbol{a}-\boldsymbol{b})$, are connected along [1-10] through (fused) 5- and 6-rings as shown in Figure 2.

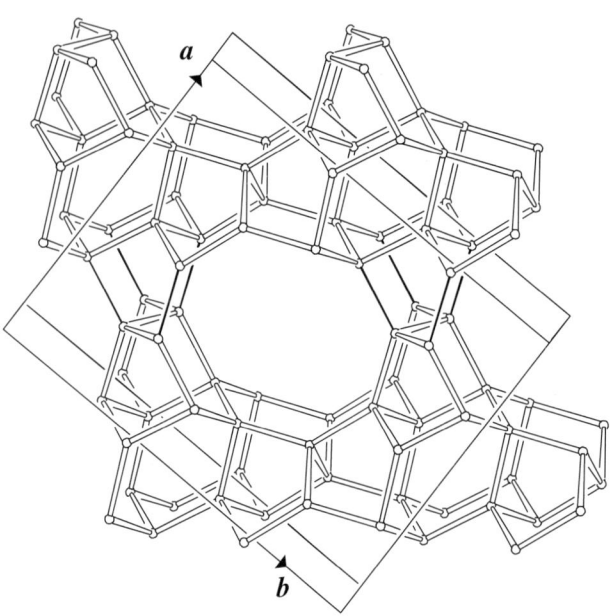

Fig. 2.  Connection mode (and unit cell content) viewed along $\boldsymbol{c}$. Only two repeat units along $\boldsymbol{c}$ of each PerBU are drawn for clarity.

## 3. Channels and/or cages

One-dimensional non-interconnecting 10-ring channels are parallel to *a*. The channel, depicted in Figure 3, is topologically equivalent to the channel in **MTT**.

$$\{1\ [6^8 10^{2/2}]\ [001]\ (10\text{-ring})\}$$

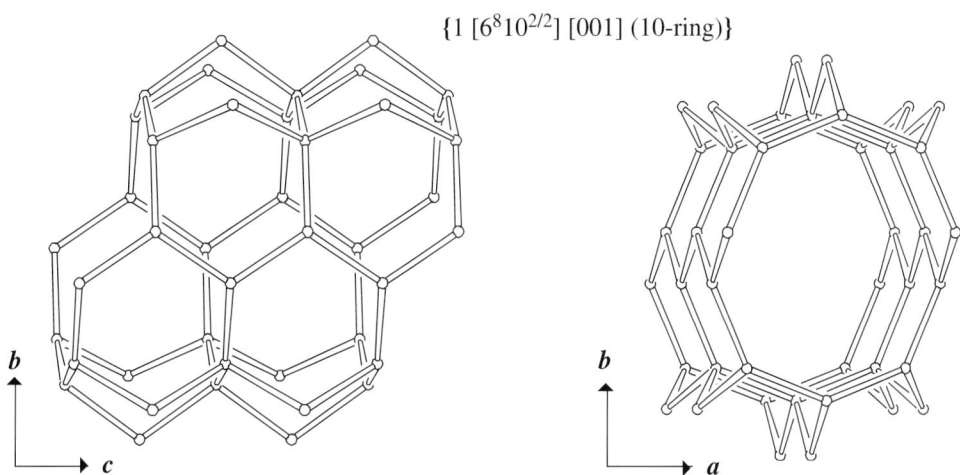

Fig. 3.　Channel viewed perpendicular to the channel axis (left) and along the channel axis (right).

# TSC

# Building Scheme

## 1. Periodic Building Unit

Cubic **TSC** can be built using the tschoertnerite (***tsch***) cavity consisting of twenty-four 4-rings (or twelve 8-rings) that are connected as shown in Figure 1 (left). A two-dimensional PerBU is obtained when these cavities, related along ***a*** (and ***b***) by a shift of $\frac{1}{2}(\boldsymbol{a}+\boldsymbol{b})$, are connected along ***a*** (and ***b***) through double 8-rings (D8Rs) as illustrated in Figure 1 (right).

***tsch*** cavity
{3 [$4^{24}6^88^{18}$] <100> (8-ring), <110> (8-ring)}

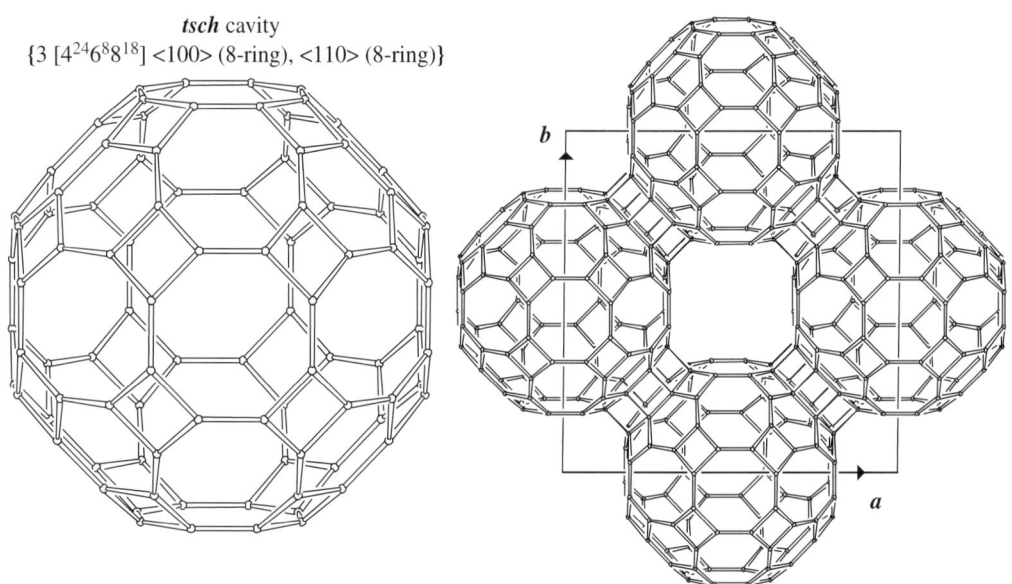

Fig. 1.   ***tsch*** Cavity (left) and PerBU (right).

## 2. Connection mode

Neighboring PerBUs, related along ***c*** by a shift of $\frac{1}{2}\boldsymbol{a}$ (or $\frac{1}{2}\boldsymbol{b}$), are connected through D8Rs (Figure 2).

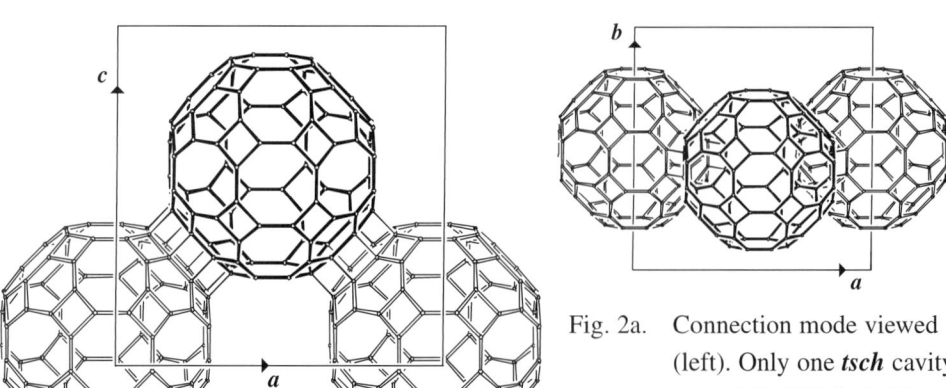

Fig. 2a.   Connection mode viewed along ***b*** (left). Only one ***tsch*** cavity of the second PerBU (in bold) and two of the first PerBU are drawn for clarity.

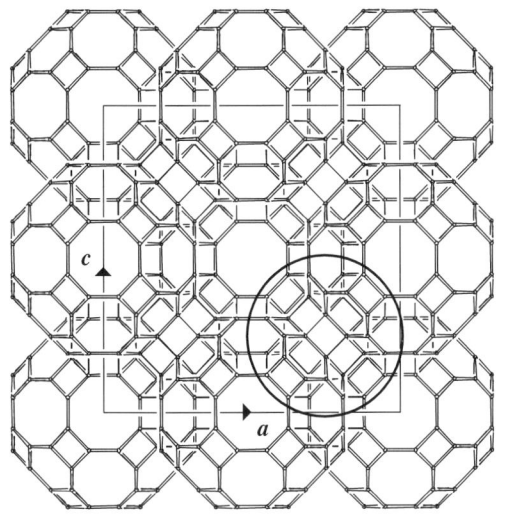

Fig. 2b.   Unit cell content projected along a cubic axis. **TSC** can also be built using units consisting of four double 6-rings (48 T atoms) that are tetrahedral coordinated around the center of the *sod* cage (circled in the Figure).

## 3. Channels and/or cages

Intersecting 8-ring channels are parallel to <100> and to <011>. The channel intersections are shown in Figures 1 and 3. The linkage of *rho*- and *tsch*-cavities into 8-ring channels parallel to <100> is also illustrated in Figure 3. The connection of *tsch* cavities into 8-ring channels parallel to <110> is shown in Figure 1. The *rho* cavity is also observed in **-CLO**, **KFI**, **LTA**, **LTN**, **PAU**, **RHO** and **UFI**. The *sod* cage is also found in **EMT**, **FAU**, **FRA**, **GIU**, **LTA**, **LTN**, **MAR** and **SOD**.

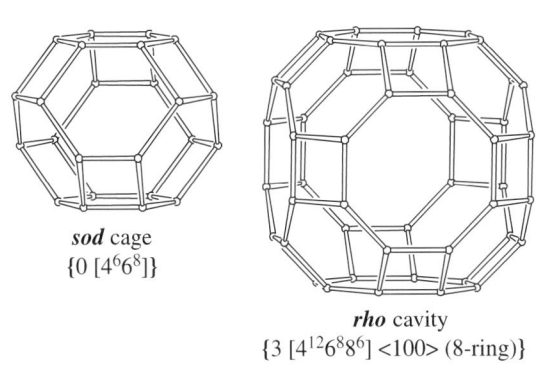

*sod* cage
{0 [4⁶6⁸]}

*rho* cavity
{3 [4¹²6⁸8⁶] <100> (8-ring)}

Fig. 3.   *sod* Cage and *rho* cavity (the *tsch* cavity is shown in Figure 1) viewed along <100> (top), and linkage of cavities into 8-ring channel parallel to <100> (bottom) viewed perpendicular to the channel axis (left) and along the channel axis (right). The linkage with *sod* cages (in bold) is added.

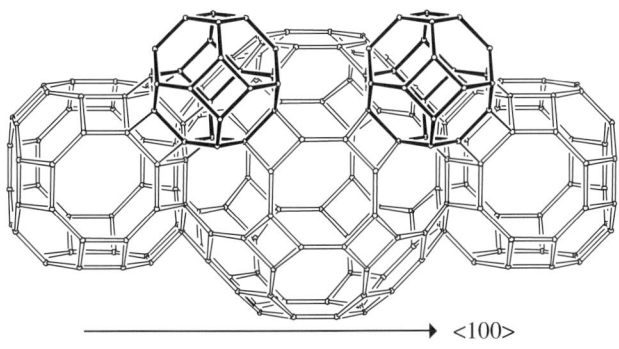

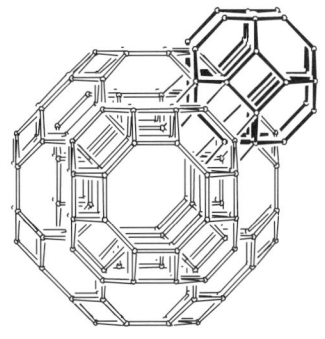

<100>

# TUN     Building Scheme

## 1. Periodic Building Unit

**TUN** can be built using left- and right-handed units of 24 T atoms (one bold in Figure 1). The T24-unit consists of four 5-1 units, or four "finite" zigzag chains (each containing 4 T atoms) and four dimers. Left- and right-handed T24-units form left- and right-handed columns parallel to $c$, respectively. The T24-units in the column are related by a rotation of 180° about $b$. Left- and right-handed columns are connected into the two dimensional PerBU. The PerBU equals the $bc$ layer shown in Figure 1b.

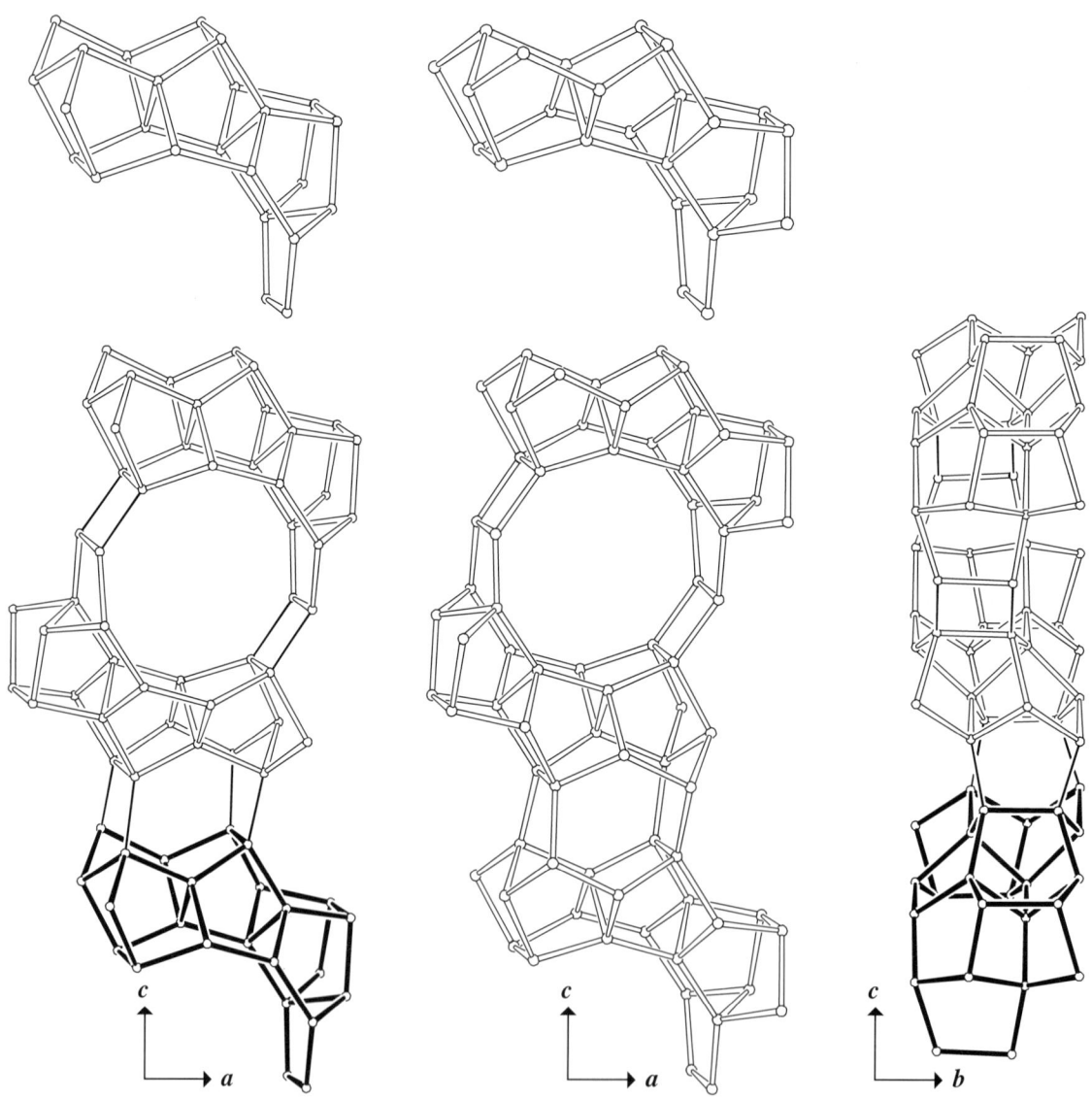

Fig. 1a.    Left- and right-handed T24-units (top) and the polar columns obtained (bottom). Columns viewed along $b$ (left and middle), and one column viewed along $a$ (right). One T24-unit in bold.

# Building Scheme

# TUN

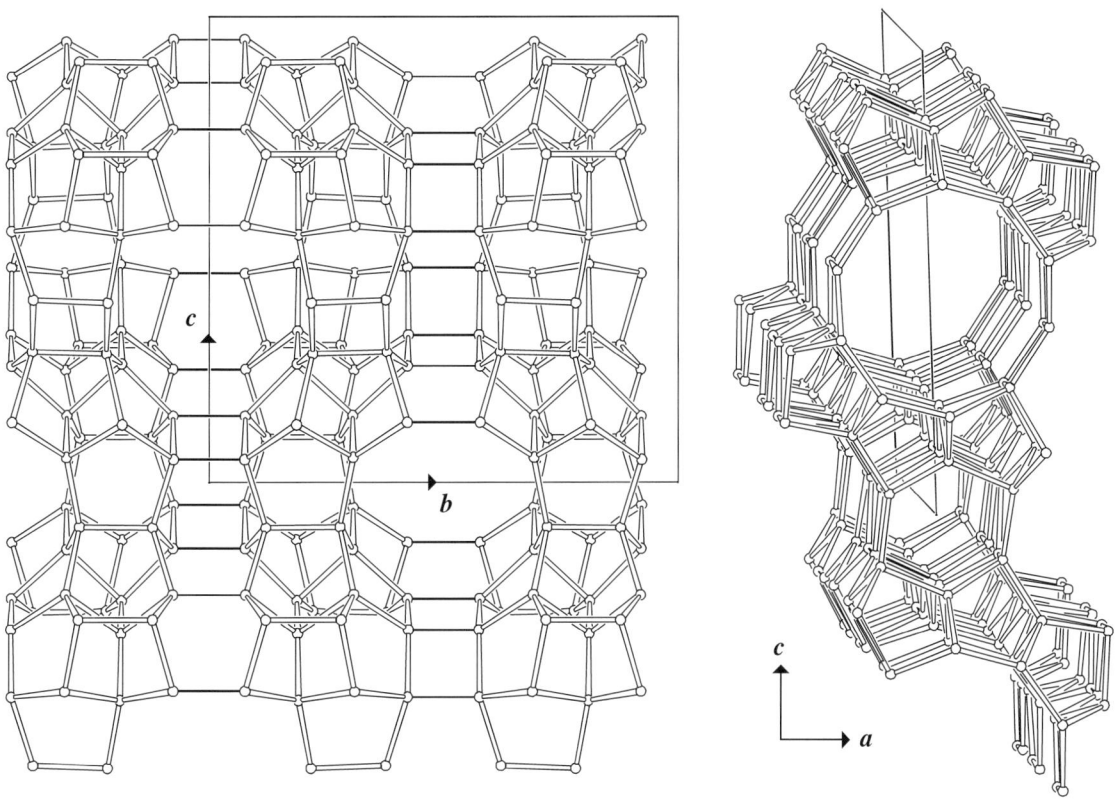

Fig. 1b.   Left- and right-handed columns are connected along **b** through 4- and 6-rings into the PerBU viewed along **a** (left) and along **b** (right).

## 2. Connection mode

Neighboring PerBUs, related along $a$ by a shift of $\frac{1}{2}(a+b)$, are connected along $a$ through 5-rings as depicted in Figure 2.

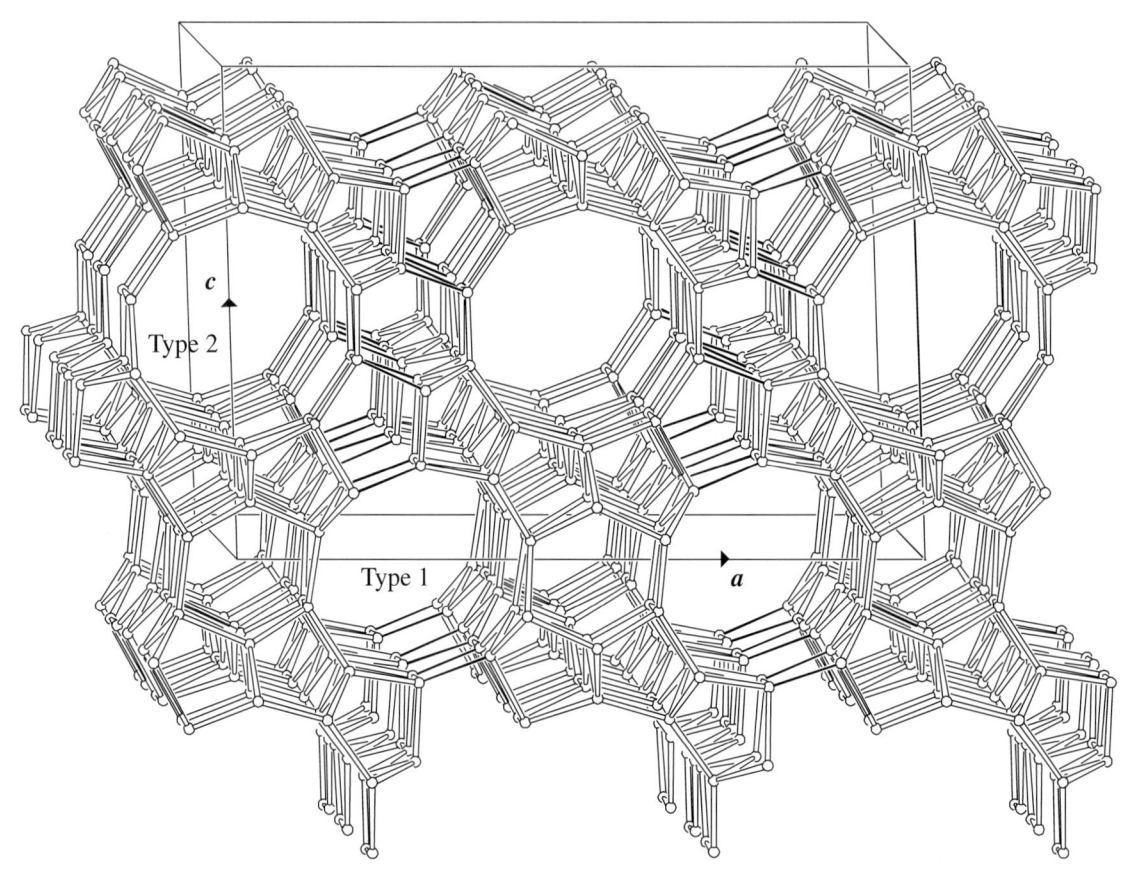

Fig. 2.   Connection mode (and unit cell content) viewed along $b$.

## 3. Channels and/or cages

Two types of 10-ring channels are parallel to $b$. The channels are interconnecting along [-101]. Pairs of channels of Type 1 are interconnecting along $a$. Non-interconnecting sinusoidal 10-ring channels are perpendicular to $b$. The two types of channels and their interconnection are depicted in Figure 3. The interconnection between pairs of channels of Type 1 along $a$ can be seen in Figure 2.

Type 1: {1 [$5^{16}6^410^410^{2/2}$] [010] (10-ring)]}

Type 2: {1 [$4^85^{16}6^{12}8^212^210^{2/2}$] [010] (10-ring)]}

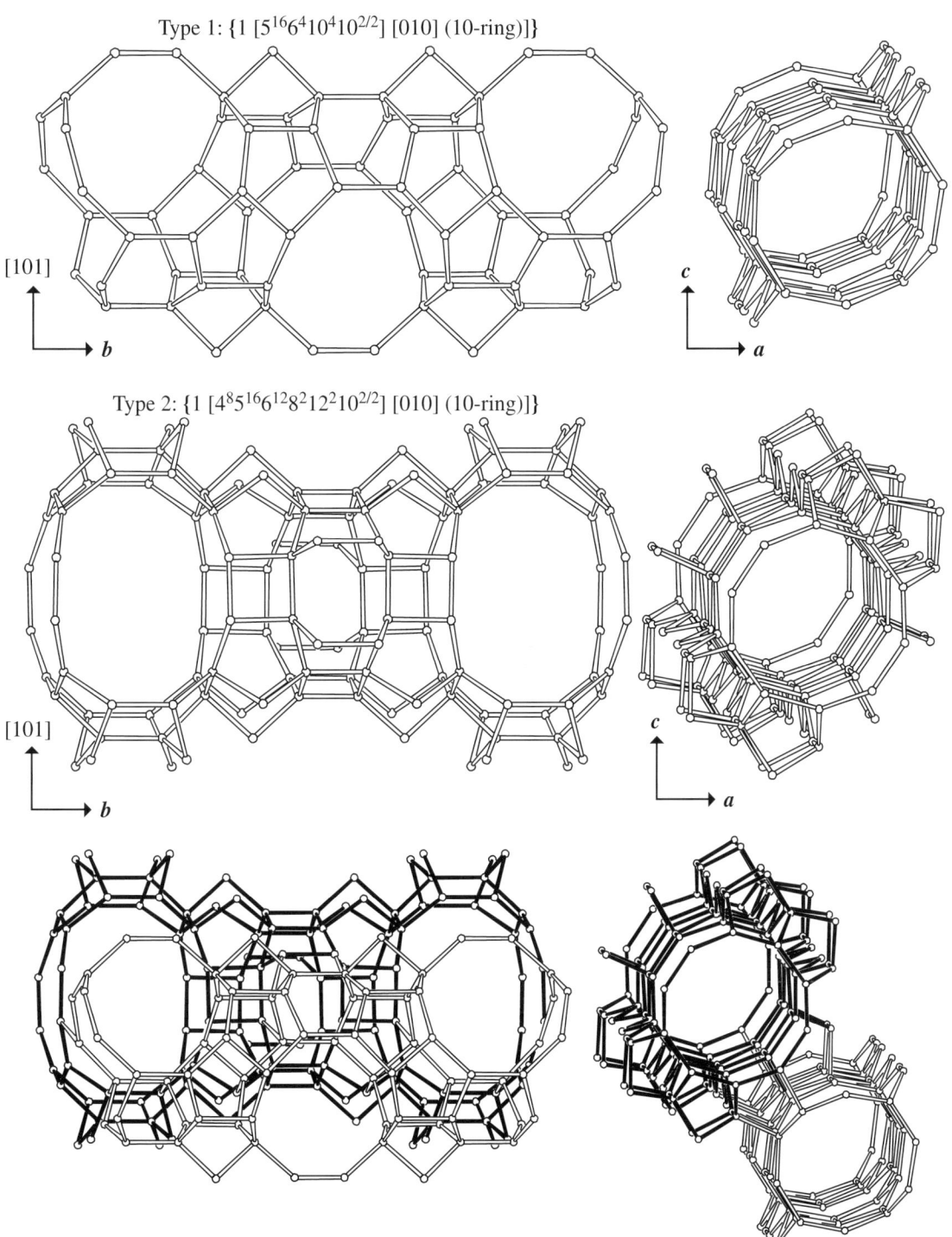

Fig. 3.    10-Ring channel of Type 1 (top) and of Type 2 (middle) viewed perpendicular to the channel axis (left) and along the channel axis (right). Bottom: interconnection of channels (Type 2 channel in bold).

# UEI

## Building Scheme

### 1. Periodic Building Unit

**UEI** can be built using the crankshaft chain (bold in Figure 1, left) running parallel to **b**. A one-dimensional PerBU is obtained when three crankshaft chains are connected into a tube with a 6-ring aperture. The tube wall consists of 4-, 6- and 8-rings. The repeat unit of the PerBU consists of a 3-fold (1,2,4)-connected double 6-ring of 12 T atoms (bold in Figure 1, right). (See also **AWO**.)

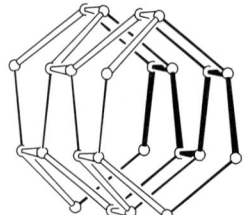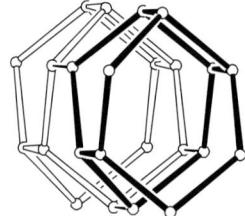

Fig. 1.   PerBU, constructed from crankshaft chains (left) and from 3-fold connected double 6-rings (right), viewed along **b**.

### 2. Connection mode

Neighboring PerBUs, related along **a** by a shift of $\frac{1}{2}(a+b)$ and along **c** by a shift of $\frac{1}{2}(c+b)$, are connected along **a** and **c** through 4-rings and double crankshaft chains, respectively (Figure 2).

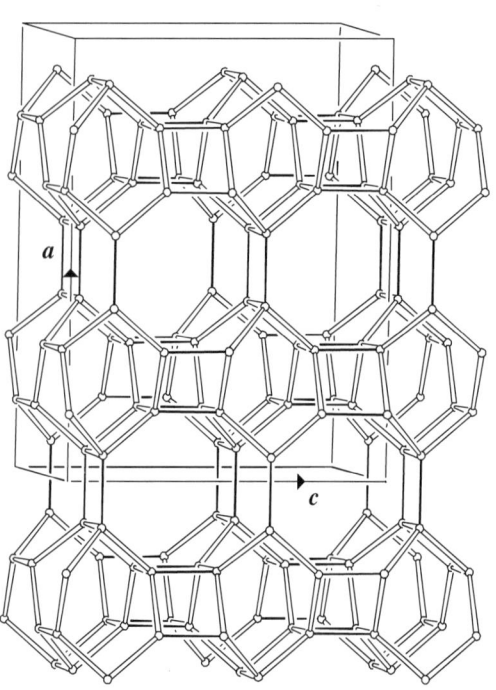

Fig. 2.   Connection mode (and unit cell content) viewed along **b**. For clarity, only $1\frac{1}{2}$ repeat unit along **b** of each PerBU is drawn.

## 3. Channels and/or cages

Interconnecting 8-ring channels are parallel to *b*. The channel is depicted in Figure 3.

{1 [4$^{12}$6$^{2}$8$^{4}$8$^{2/2}$] [010] (8-ring)}

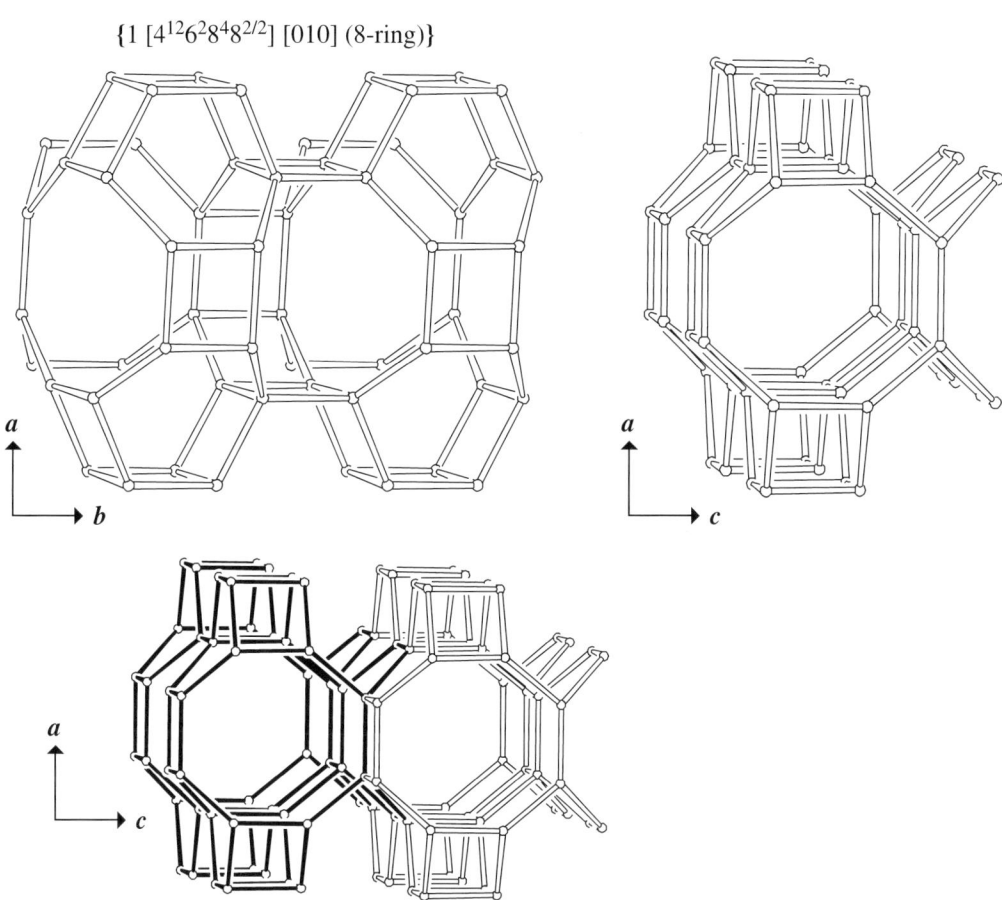

Fig. 3.   Top: channel viewed perpendicular to the channel axis (left) and along the channel axis (right). Bottom: channels (one channel in bold) are interconnecting through shared 8-rings perpendicular to *c* (bottom).

# UFI                    Building Scheme

## 1. Periodic Building Unit

Tetragonal **UFI** can be built using units of 32 T atoms. The T32-unit is composed of two, mirror related, "half-cages". A half-cage consists of four (fused) 6-rings (or a 12-ring and a 4-ring) and exhibits 4-fold symmetry (bold in Figure 1). T32-units, related along $a$ and $b$ by pure translations, are connected through 4-rings into the two-dimensional PerBU. *rho* Cavities are formed.

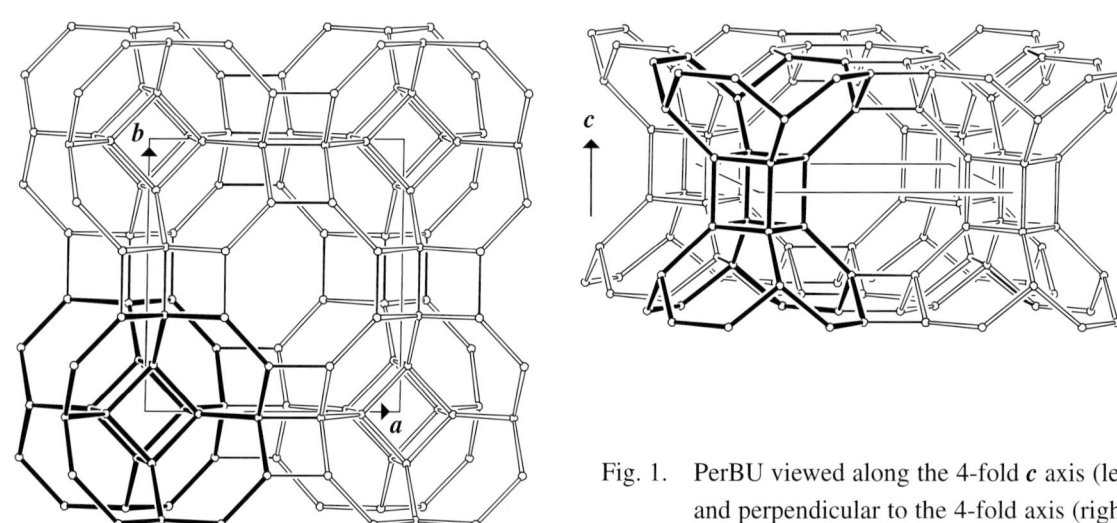

Fig. 1.    PerBU viewed along the 4-fold $c$ axis (left) and perpendicular to the 4-fold axis (right).

## 2. Connection mode

Neighboring PerBUs are related along $c$ by a shift of $\frac{1}{2}(a+b+c)$. 8-Rings of *rho* cavities, parallel to the $ab$ plane, are connected to the 12-rings of half-cages in neighboring PerBUs (Figure 2).

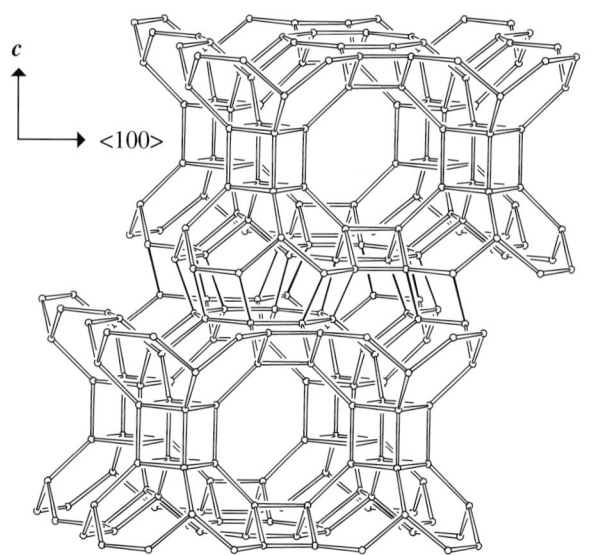

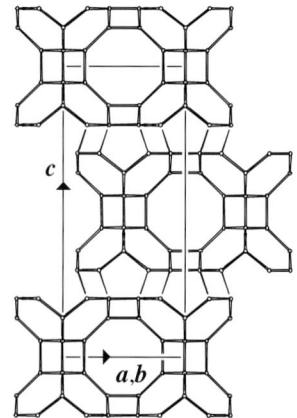

Fig. 2.    Connection mode (left) and unit cell content (right) viewed along <010>.

## 3. Channels and/or cages

Two non-interconnecting systems of intersecting 8-ring channels are parallel to <100>. The channel intersection, equal to the *rho* cavity, is depicted in Figure 3. Two "side-pockets" close the 8-ring windows (of the *rho* cavities) perpendicular to *c*. The *rho* cavity is also present in **-CLO**, **KFI**, **LTA**, **LTN**, **PAU**, **RHO** and **TSC**. The linkage of the cavities is also illustrated in Figure 3.

$\{3\ [4^{12}6^{8}8^{6}] <100> (8\text{-ring}) /0\ [4^{5}5^{4}6^{4}8^{1}]\}$

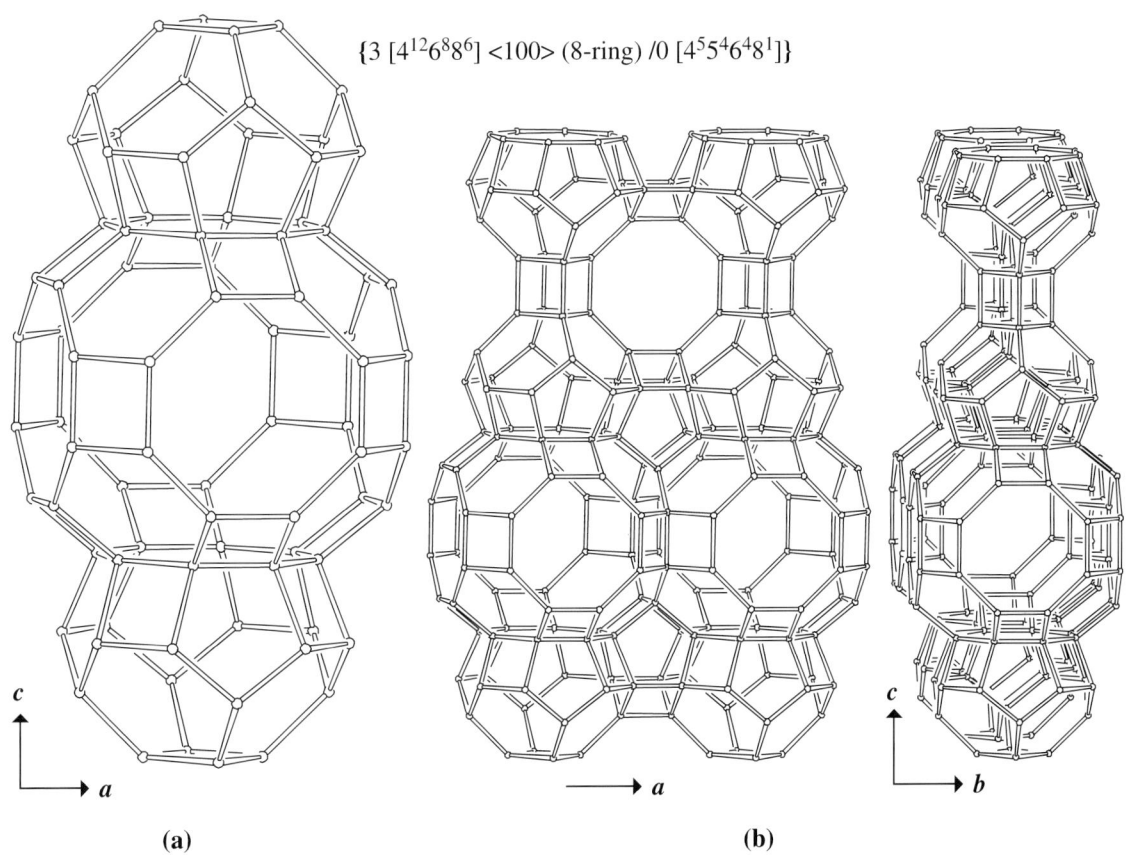

**(a)** **(b)**

Fig. 3. (a) Channel intersection and "side-pockets" viewed along *b* (or *a*); (b) Connection of cavities along *a* (or *b*) viewed along *b* (or *a*) (left) and along the 8-ring channel axis parallel to *a* (or *b*) (right).

# UOZ                    Building Scheme

## 1. Periodic Building Unit

Tetragonal **UOZ** can be built using the T20-unit consisting of two double 4-rings with two "dangling" T atoms (bold in Figure 1). Neighboring T20-units, related along **b** by a pure translation and along **c** by a screw rotation of (approximately) 180° about **c**, are connected along **b** and **c** through distorted (fused) 6-rings as shown in the drawing of the two-dimensional PerBU in Figure 1. (Compare with **AST** and **ASV**.)

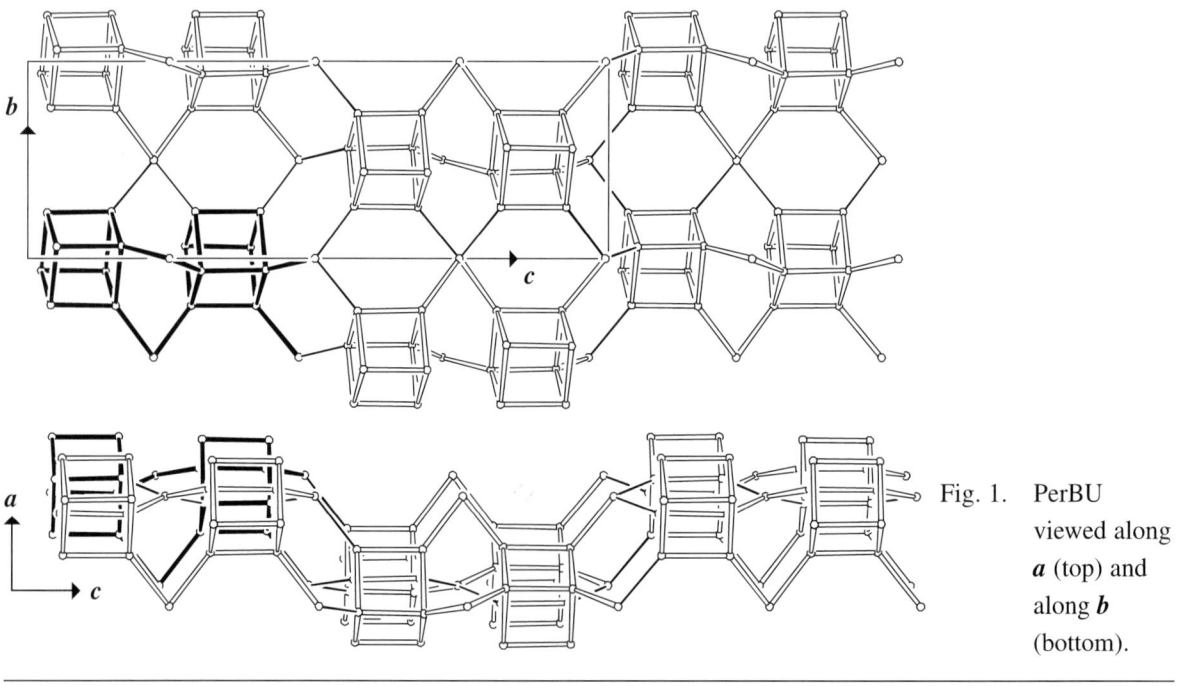

Fig. 1.  PerBU viewed along **a** (top) and along **b** (bottom).

## 2. Connection mode

Neighboring PerBUs, related along **a** by a pure translation, are connected along **a** through (fused) 6-rings. The connection of the units within the **ac** and **bc** layer is identical. (Compare Figures 1 and 2.)

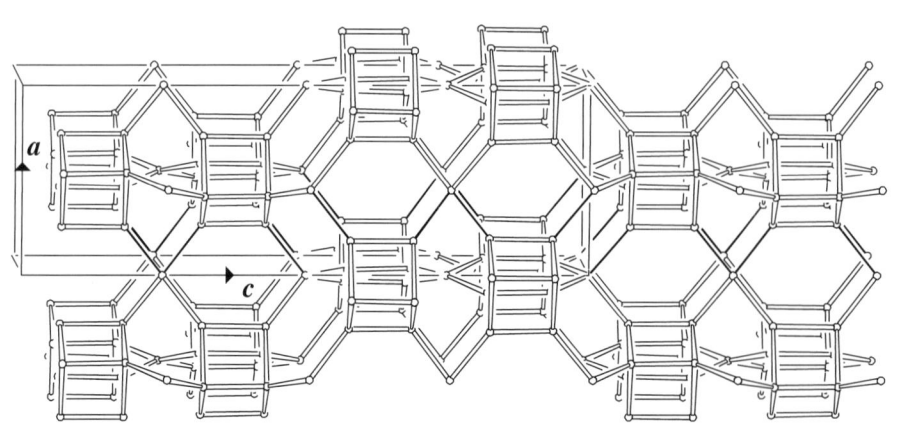

Fig. 2.  Connection mode (and unit cell content) viewed along **b**.

## 3. Channels and/or cages

The cage in **UOZ**, composed of (fused) 4- and 6-rings only, is depicted in Figure 3.

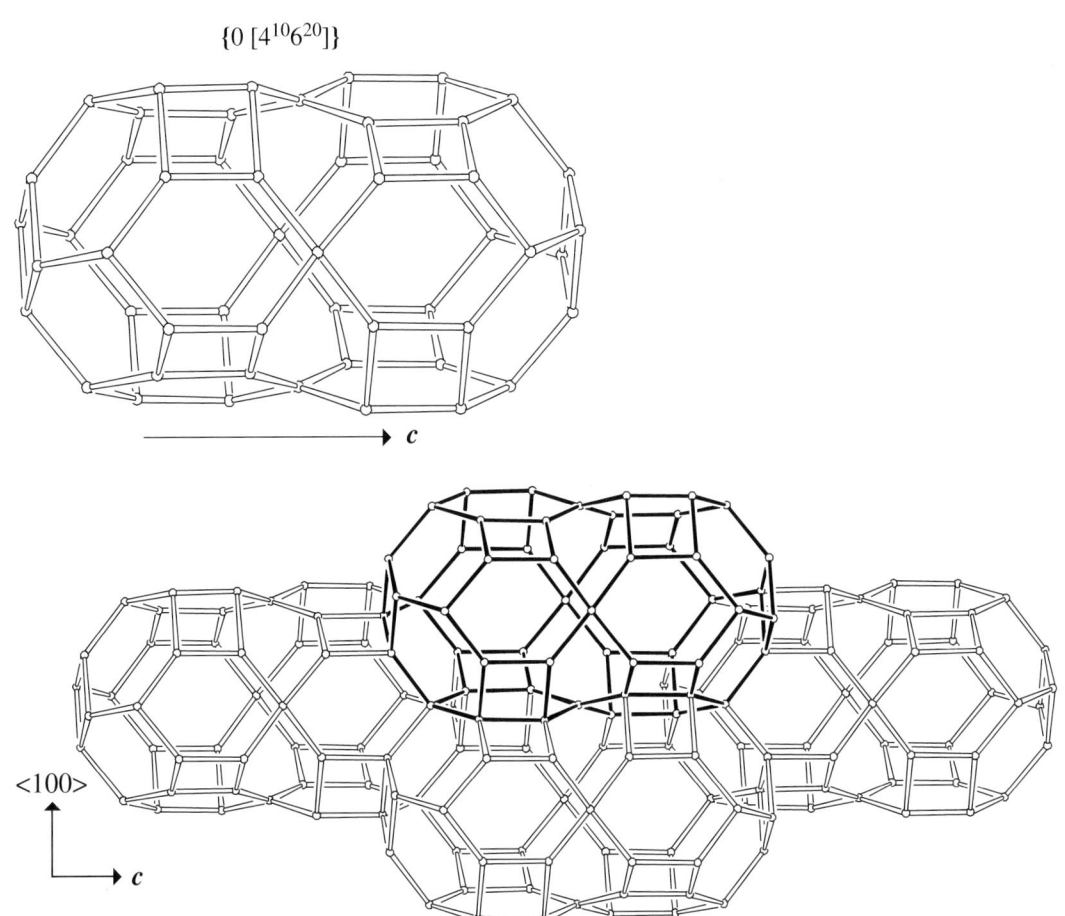

Fig. 3.    Cage (top) and fusion of cages (bottom; one cage in bold) viewed along <010>.

# USI

# Building Scheme

## 1. Periodic Building Unit

**USI** can be built using units of 10 T atoms. The T10-unit consists of doubly (1,3)-connected double 4-rings decorated with T2-dimers (bold in Figure 1). T10-units, related along **c** by a pure translation and along **b** by a screw rotation of 180° about **b**, are connected into the undulating **bc** layer. The two-dimensional PerBU is depicted in Figure 1.

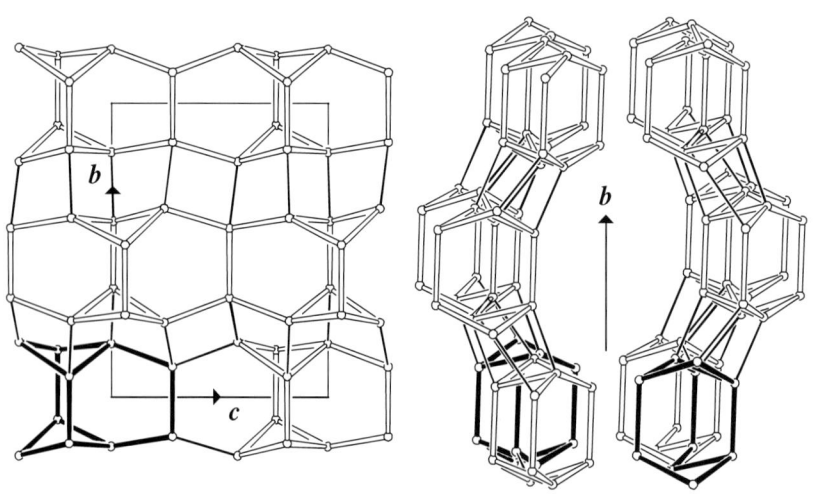

Fig. 1. PerBU viewed along **a** (left) and along **c** (right). The PerBUs at the right are related by a rotation of 180° about **b**.

## 2. Connection mode

Neighboring PerBUs, related along **a** by a rotation of 180° about **b**, are connected along **a** through 4-rings as depicted in Figure 2. 10- and 12-Ring channels (parallel to **b** and **c**) are formed.

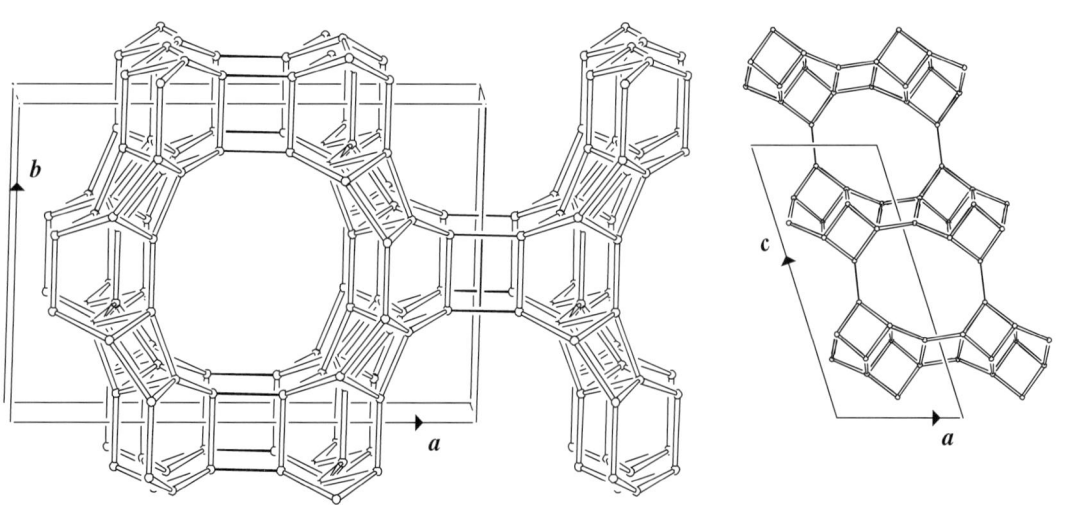

Fig. 2. Connection mode (and unit cell content) viewed along **c** (left) and unit cell content projected along **b** (right).

### 3. Channels and/or cages

Intersecting 12- and 10-ring channels are parallel to *c* and *b*, respectively. One repeat-unit of the 12-ring channel parallel to *c* and the linkage of these channels into a 10-ring channel parallel to *b* are illustrated in Figure 3.

$\{1\ [4^8 6^8 10^2 12^{2/2}]\ [001]\ (12\text{-ring})\}$

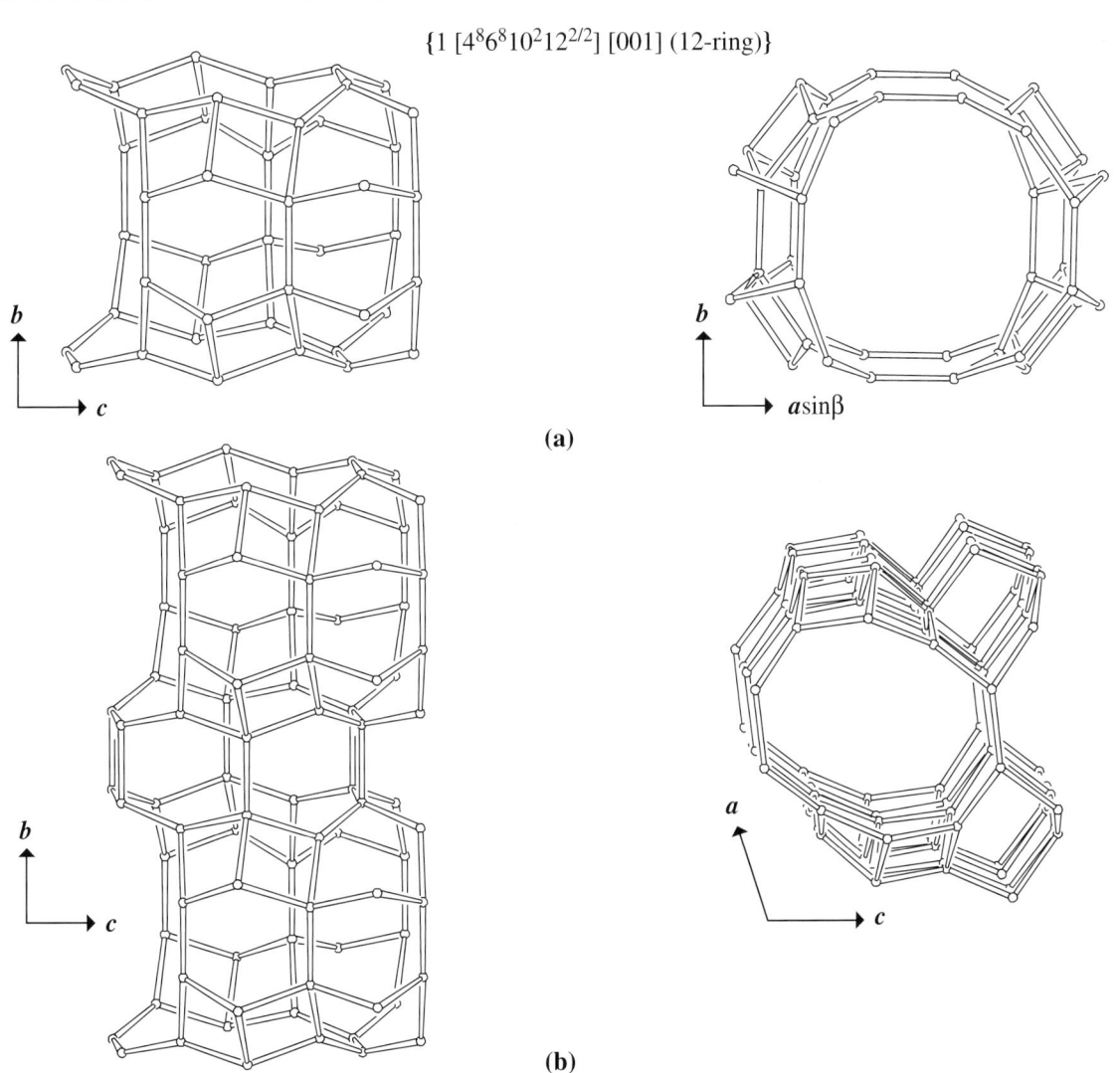

**(a)**

**(b)**

Fig. 3.  (a) 12-ring channel viewed perpendicular to the channel axis (left) and along the channel axis (right). (b) 12-ring channels are connected into 10-ring channels parallel to *b*. View along *a* (left) and along the 10-ring channel axis parallel to *b* (right).

# UTL

# Building Scheme

## 1. Periodic Building Unit

**UTL** can be built using a building unit consisting of 38 T atoms (bold in Figure 1). The T38-unit is composed of two subunits, related by a 2-fold axis parallel to **b**. The sub-units are connected through two 4-rings. The subunit itself consists of two 1-5-1 units, connected through a (finite) "saw-like" chain of 5 T atoms. Fused 5- and 6-rings are formed. A one-dimensional PerBU is obtained when T38-units, related along **c** by a pure translation, are connected as shown in Figure 1.

$$\{1 \ [4^2 5^{12} 6^4 12^2 14^{2/2}] \ [001] \ (14\text{-ring})\}$$

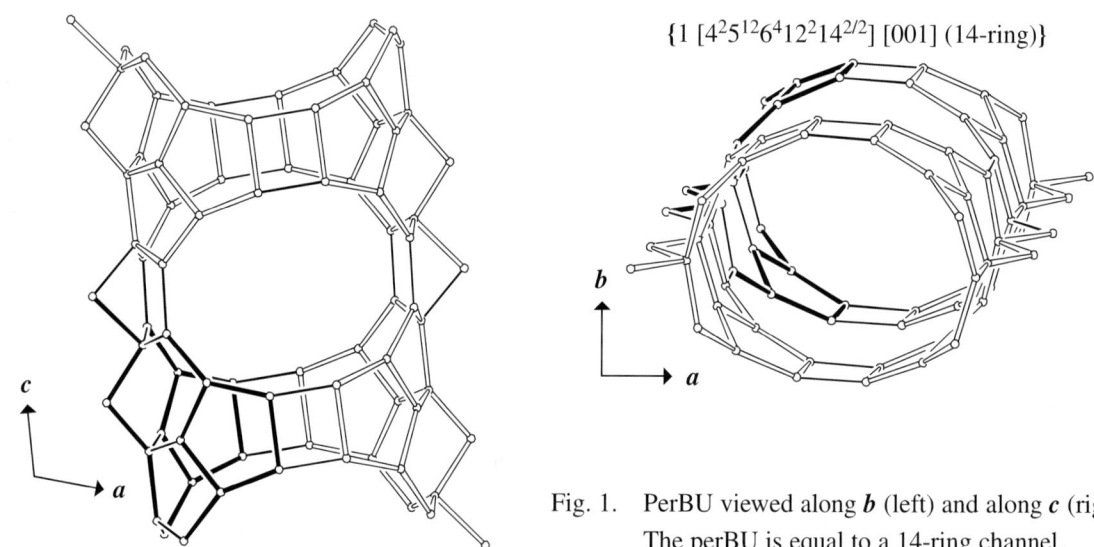

Fig. 1. PerBU viewed along **b** (left) and along **c** (right). The perBU is equal to a 14-ring channel.

## 2. Connection mode

Neighboring PerBUs, related along **b** by a pure translation and along **a** by a shift of $\frac{1}{2}(\boldsymbol{a}+\boldsymbol{b})$, are connected as shown in Figure 2. Only one repeat unit of the PerBU along **c** is drawn for clarity.

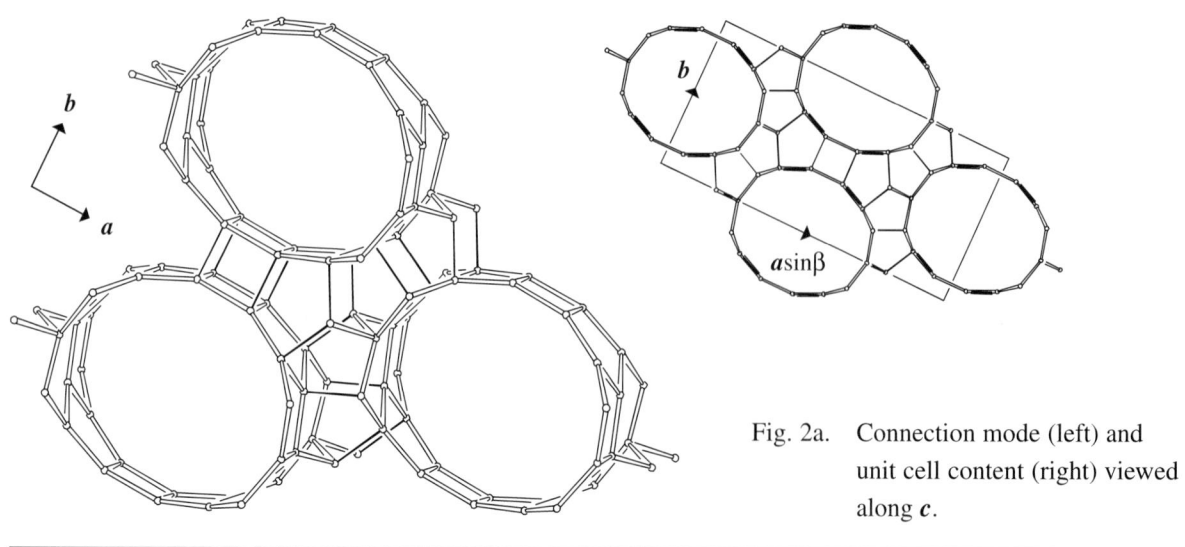

Fig. 2a. Connection mode (left) and unit cell content (right) viewed along **c**.

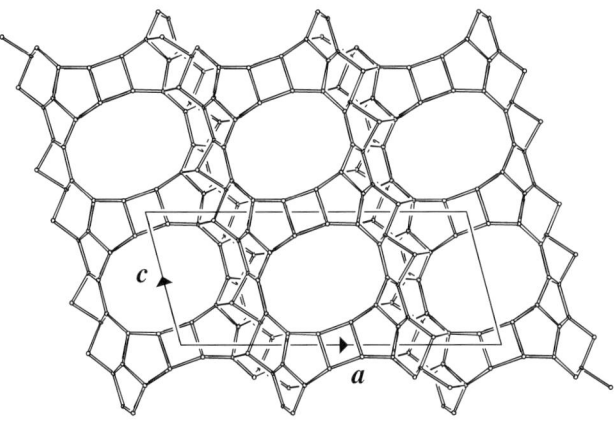

Fig. 2b. Unit cell content projected along $b$.

## 3. Channels and/or cages

Intersecting 14- and 12-ring channels are parallel to $c$ and $b$, respectively. The intersection is equal to the $[5^4 12^2 14^2]$-cavity. The 12-ring channel is depicted in Figure 3 (14-ring channel, see Figure 1).

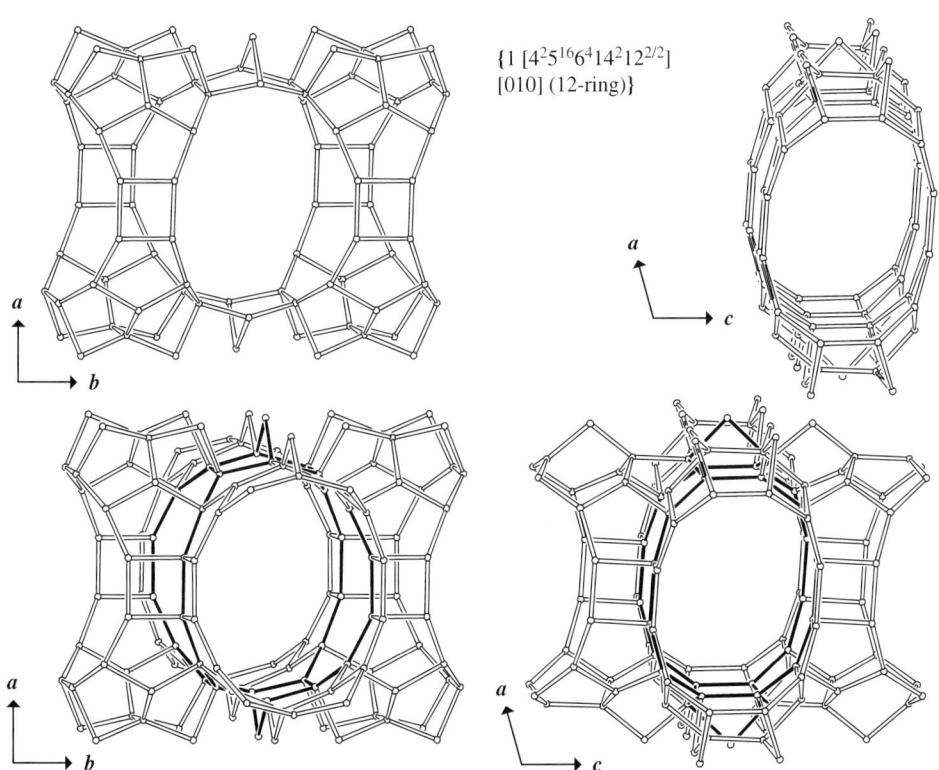

$\{1 \, [4^2 5^{16} 6^4 14^2 12^{2/2}] \, [010] \, (12\text{-ring})\}$

Fig. 3. Top: 12-ring channel viewed perpendicular to the channel axis (left) and along the channel axis (right). Bottom: linkage of channels viewed along the 14-ring channel axis parallel to $c$ (left) and along the 12-ring channel axis parallel to $b$ (right). The $[5^4 12^2 14^2]$-cavity is in bold.

# VET             **Building Scheme**

## 1. Periodic Building Unit

Tetragonal **VET** can be built using the zigzag chain (one in bold in Figure 1) running parallel to *c*. The one-dimensional PerBU consists of eight zigzag chains connected into a channel with a 12-ring aperture and an additional single T atom (drawn in Figure 2). The repeat unit consists of 16 T atoms: a 12-ring and 4 T atoms (bold in Figure 1). (See also **OSI**.)

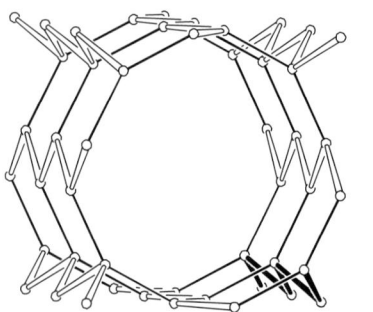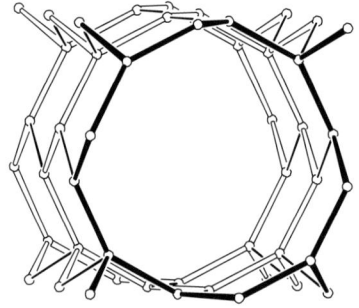

Fig. 1.   PerBU, constructed from eight zigzag chains (left) and from T16-units (right), viewed along *c*.

## 2. Connection mode

Neighboring PerBUs, related along *a* and *b* by pure translations, are connected along *a* and *b* through 5-rings. The connection mode exhibits a 4-fold inversion axis through the central single T atoms (bold bonded in Figure 2).

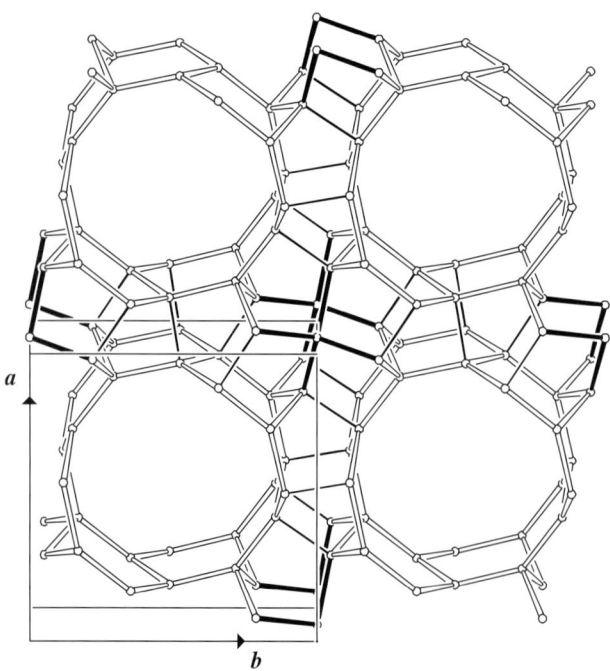

Fig. 2.   Connection mode (and unit cell content) viewed along *c*. Only two repeat units of the PerBUs are drawn for clarity.

### 3. Channels and/or cages

Non-interconnecting 12-ring channels are parallel to *c*. The channel wall consists of fused 6-rings as shown in Figure 3. The channel is topologically equivalent to the 12-ring channel in **OSI**.

$\{1\ [6^8 12^{2/2}]\ [001]\ (12\text{-ring})\}$

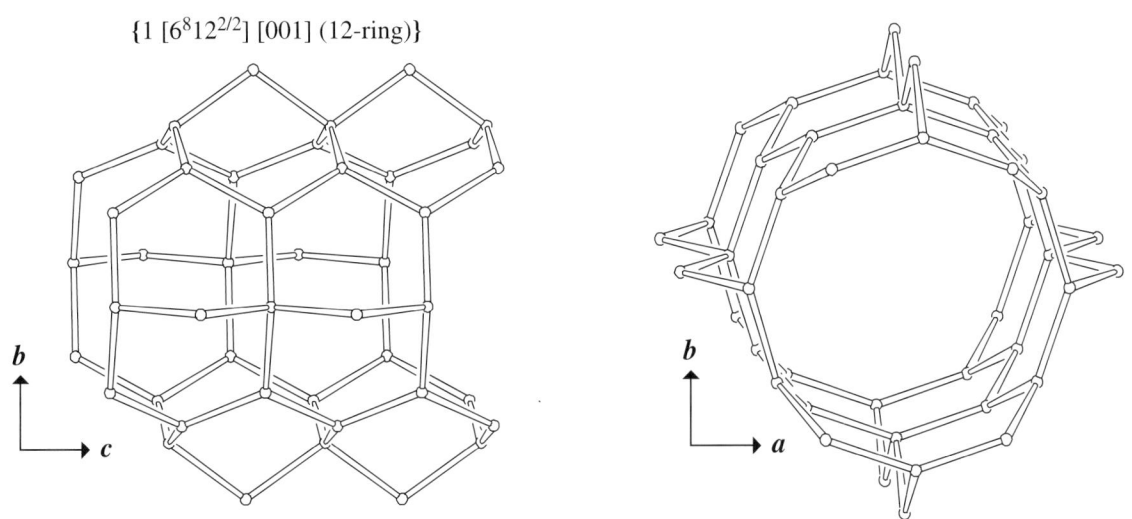

Fig. 3.    Channel viewed perpendicular to the channel axis (left) and along the channel axis (right).

# VFI                    Building Scheme

## 1. Periodic Building Unit

Hexagonal **VFI** can be built using the crankshaft chain (bold in Figure 1, left) running parallel to *c*. A one-dimensional PerBU is obtained when nine crankshaft chains are connected into a channel with an 18-ring aperture. The repeat unit consists of a 6-ring band of 36 T atoms.

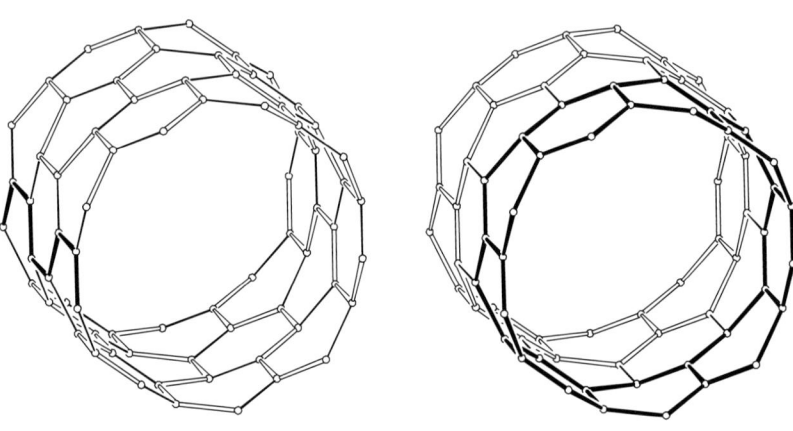

Fig. 1.   PerBU, constructed from nine crankshaft chains (left) and from 6-rings bands (right), viewed along *c*.

## 2. Connection mode

Neighboring PerBUs, related along *a* and *b* by pure translations, are connected along *a* and *b* (and $(a + b)$) through triple crankshaft chains (Figure 2).

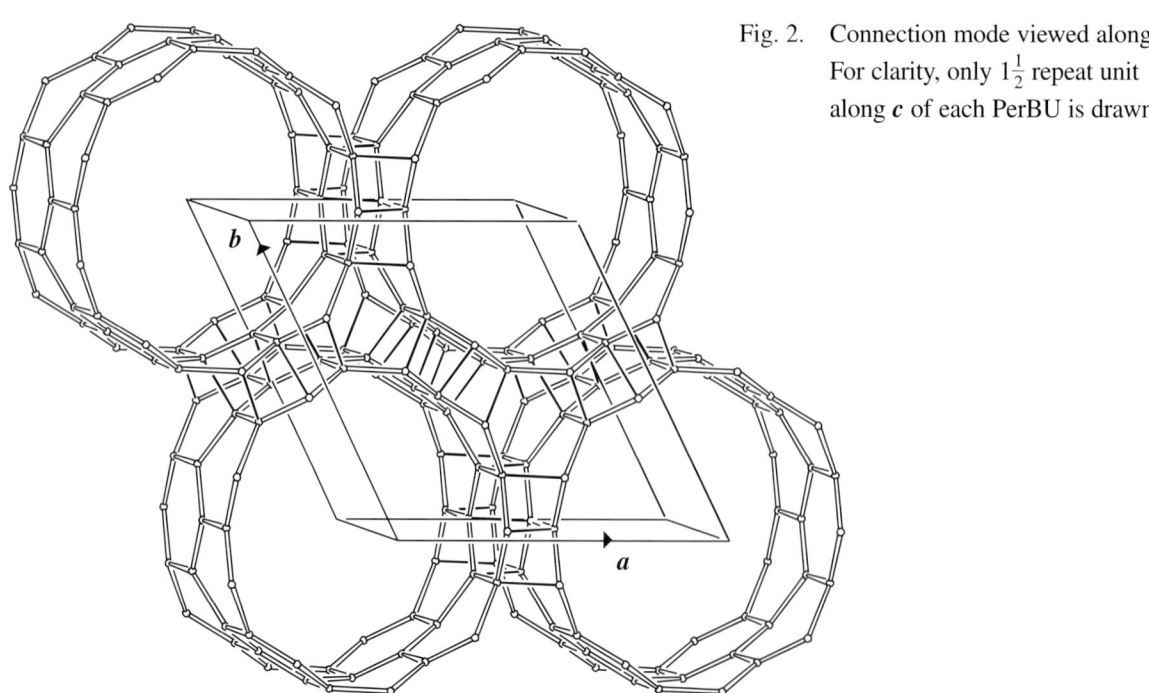

Fig. 2.   Connection mode viewed along *c*. For clarity, only $1\frac{1}{2}$ repeat unit along *c* of each PerBU is drawn.

### 3. Channels and/or cages

Non-interconnecting one-dimensional channels with 18-ring apertures are parallel to *c*. One channel is depicted in Figure 3. The channel wall consists of fused 6-rings.

{1 [6$^{18}$18$^{2/2}$][001] (18-ring)}

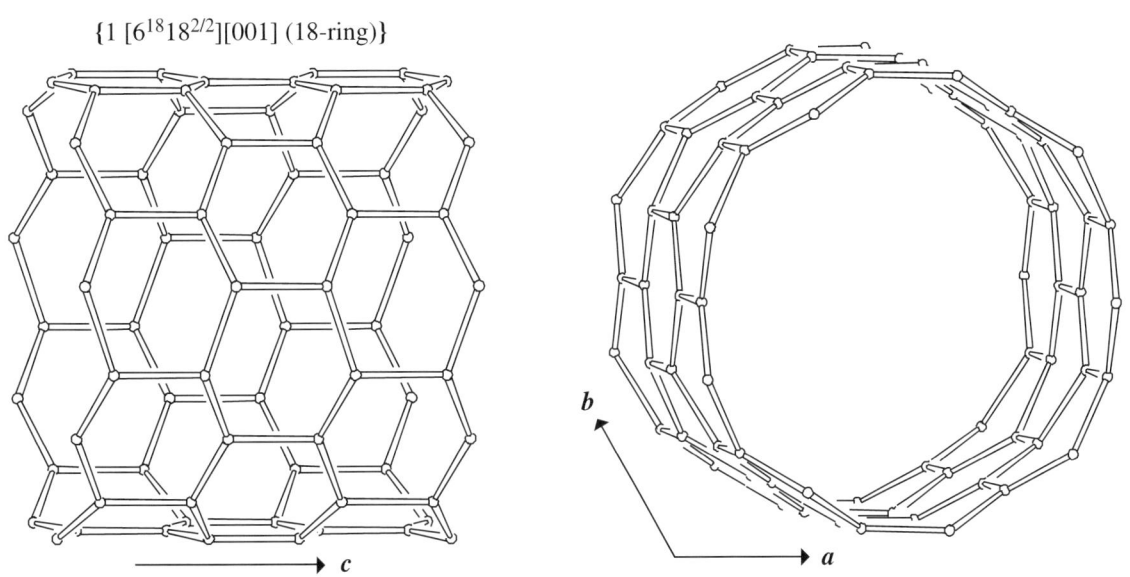

Fig. 3.   Channel viewed perpendicular to the channel axis (left) and along the channel axis (right).

# VNI                    Building Scheme

## 1. Periodic Building Unit

Tetragonal **VNI** can be built using two two-dimensional PerBUs. PerBU1 consists of chains of units of 11 T atoms parallel to [-110] (Figure 1, top). Each T11-unit is composed of two 4-rings and one 3-ring (in bold). T11-units in the chain are related by a screw rotation of 180° about the chain axis. Neighboring chains, related along *a* (or *b*) by a pure translation, are connected along [110] through single T–T bonds into PerBU1 depicted in Figure 1(middle). PerBU2, parallel to PerBU1, consists of 4-rings connected through single T–T bonds into the *ab* layer (Figure 1, bottom).

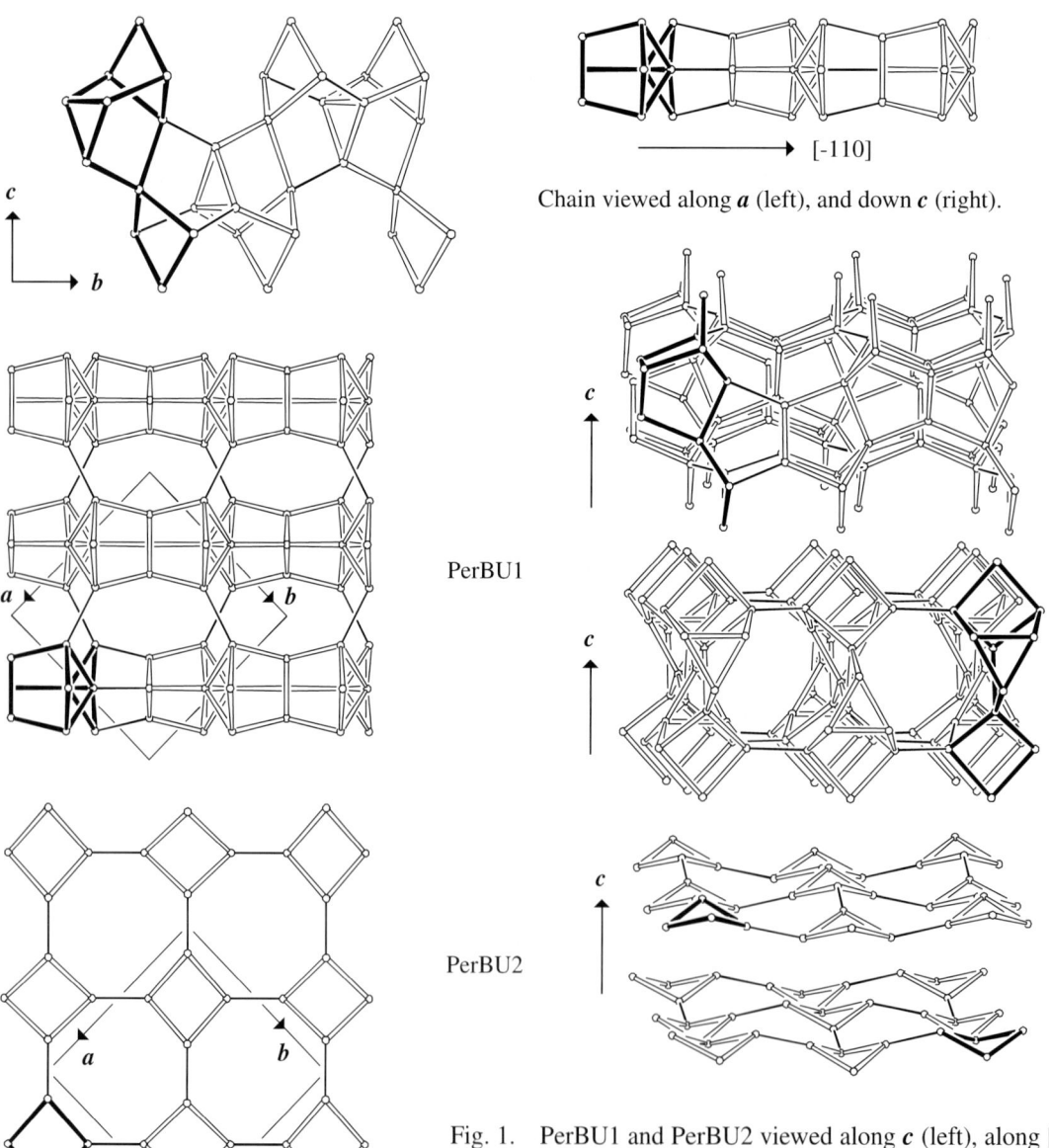

Chain viewed along *a* (left), and down *c* (right).

PerBU1

PerBU2

Fig. 1.  PerBU1 and PerBU2 viewed along *c* (left), along [110] (top right) and along [-110] (bottom right). The PerBUs depicted at the right, are related by a rotation of 90° about *c*.

# Building Scheme **VNI**

## 2. Connection mode

PerBU1 and PerBU2 alternate along $c$. Equal PerBUs are related by a rotation of 90° about $c$ and a shift of $\frac{1}{2}c$ (i.e. a $4_2$ axis). The PerBUs are connected along $c$ through 3-rings as shown in Figure 2.

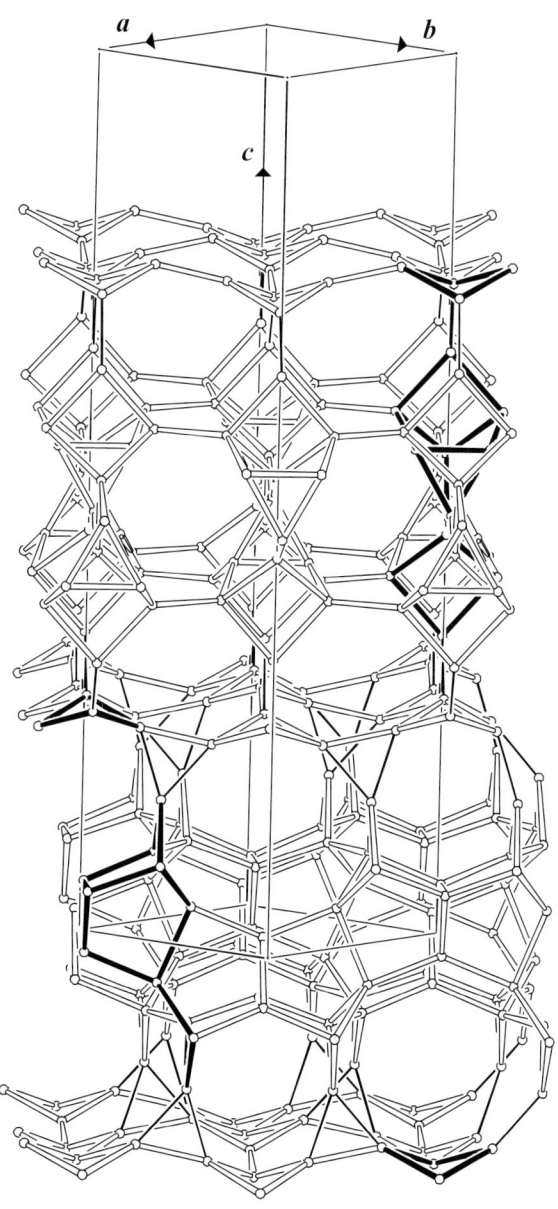

Fig. 2.   Connection mode (and unit cell content) viewed along <110>. (In order to illustrate a maximal number of contacts between the PerBUs, T11-units at the very back in the upper part of the drawing are skipped and those toward the reader are added; compare with Figure 1.)

# VNI

## Cage/Channel

### 3. Channels and/or cages

There are three types of 8-ring channels parallel to <110> as illustrated in Figure 3. The first two types intersect; the third type interconnects the other two along $c$.

Type 1: {1 [$4^4 5^2 8^6 8^{2/2}$] <110> (8-ring)}

Type 2: {1 [$3^4 8^4 10^2 8^{2/2}$] <110> (8-ring)}

Type 3: {1 [$5^8 8^4 8^{2/2}$] <110> (8-ring)}

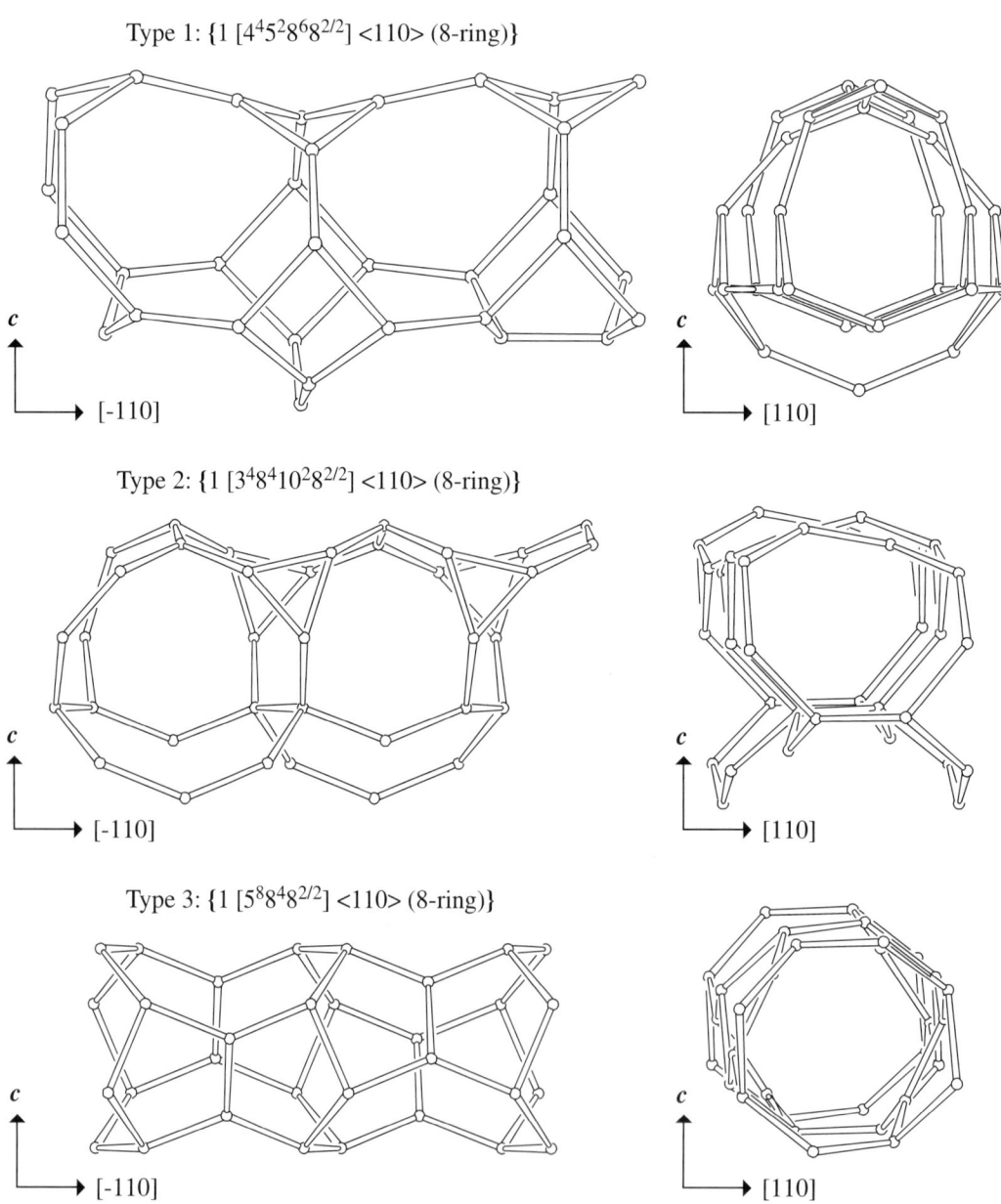

Fig. 3a.    8-ring channels viewed perpendicular to the channel axis (left) and along the channel axis (right).

# Cage/Channel **VNI**

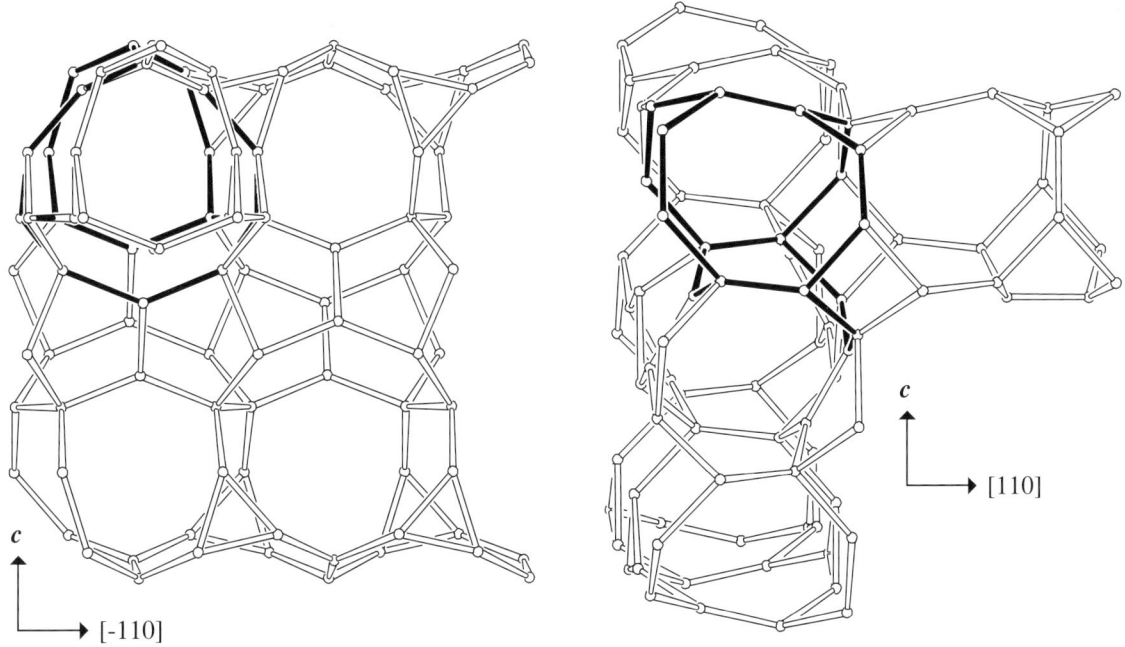

Fig. 3b.    Connection of channels viewed along [110] (left) and along [-110] (right). Type 3 channels interconnect Type 1 and Type 2 channels. The intersection between Type 1 and Type 2 channels, a $[8^4 10^1]$ cavity, is in bold.

# VSV                    Building Scheme

## 1. Periodic Building Unit

Tetragonal **VSV** can be built using units of 9T atoms: two 4-rings connected through a single T atom (bold in Figure 1). The two-dimensional PerBU, composed of T9-units related along $a$ and $b$ by pure translations, is equal to the layer depicted in Figure 1. (See also **LOV** and **RSN**.)

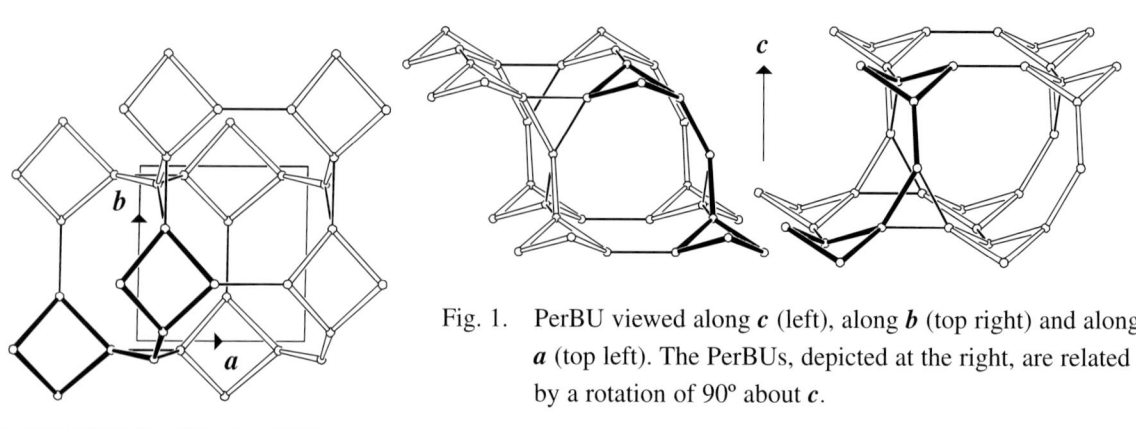

Fig. 1.    PerBU viewed along $c$ (left), along $b$ (top right) and along $a$ (top left). The PerBUs, depicted at the right, are related by a rotation of 90° about $c$.

## 2. Connection mode

Neighboring PerBUs are related along $c$ by a rotation of 90° about $c$ and a shift of $\frac{1}{4}c$ (i.e. a $4_1$-axis) followed by a lateral shift of (simultaneously) $\frac{1}{2}a$ and $\frac{1}{2}b$. The connectivity code, in fractions of $(a, b)$, is denoted as $(\frac{1}{2}, 0)$ or $(0, \frac{1}{2})$, respectively (Figure 2).

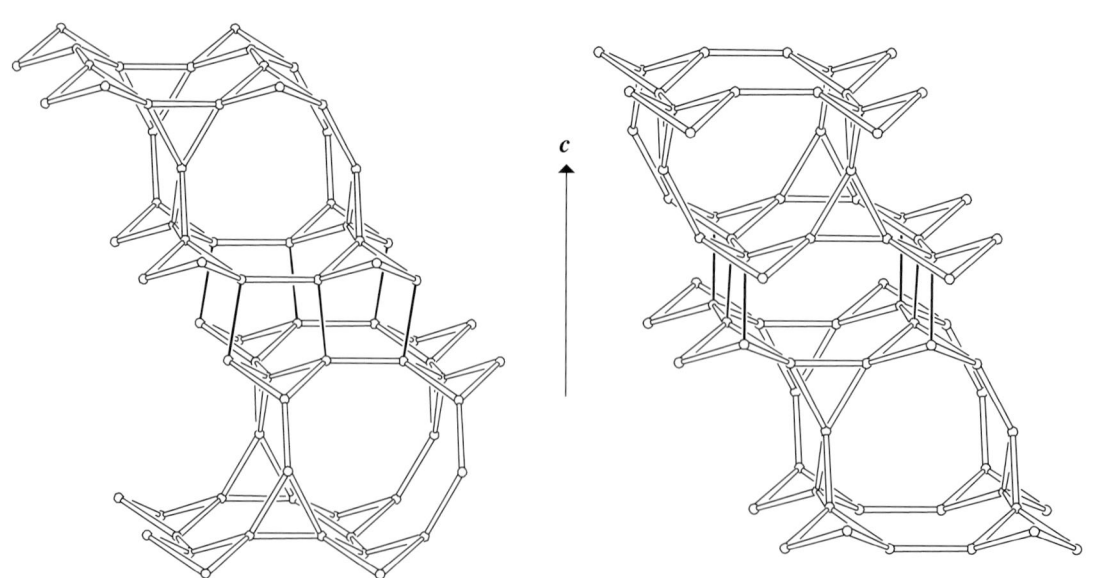

Fig. 2a.    Connection mode $(0, \frac{1}{2})$ viewed along $a$ (left) and along $b$ (right). The PerBUs are connected through 5-rings.

# Building Scheme

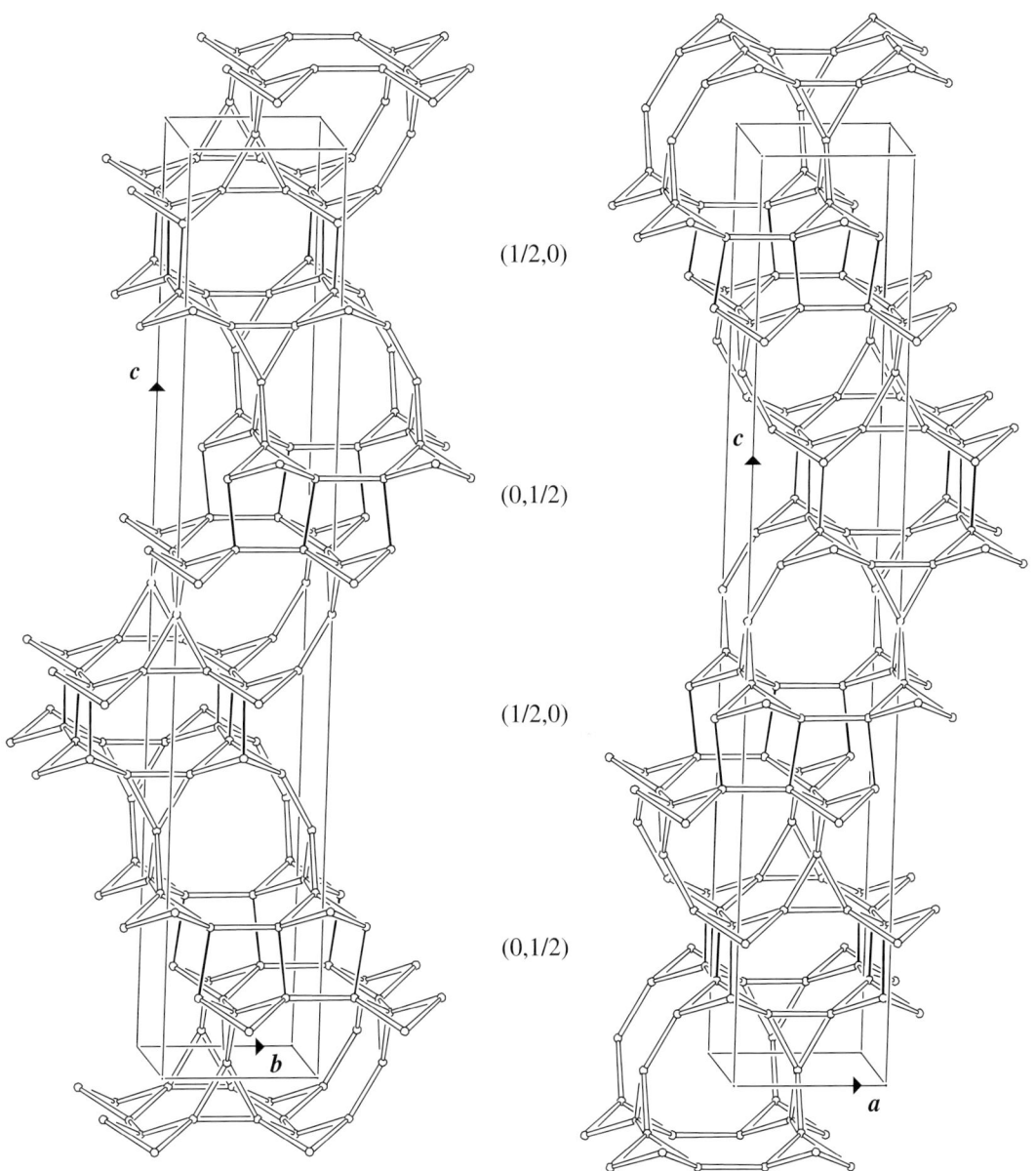

(1/2,0)

(0,1/2)

(1/2,0)

(0,1/2)

Fig. 2b.   Unit cell content viewed along **c** (left) and along **a** (right) The lateral shifts between neighboring
PerBUs, related by a rotation of 90° about **c**, are given in the drawings in fractions of (**a**, **b**).
Successive PerBUs are related by a $4_1$ axis.

# VSV

## Cage/Channel

### 3. Channels and/or cages

Intersecting 9-ring channels are parallel to <100>. The 9-ring channels are topologically equivalent to those in **NAB**, **LOV** and **RSN**. The channels are interconnected along $c$ through 8-ring channels parallel to <100>. The 8-ring channels are topologically equivalent to those in **EON**, **MAZ**, **MON**, **MOR** and **RSN**. The channels and their interconnection are depicted in Figure 3.

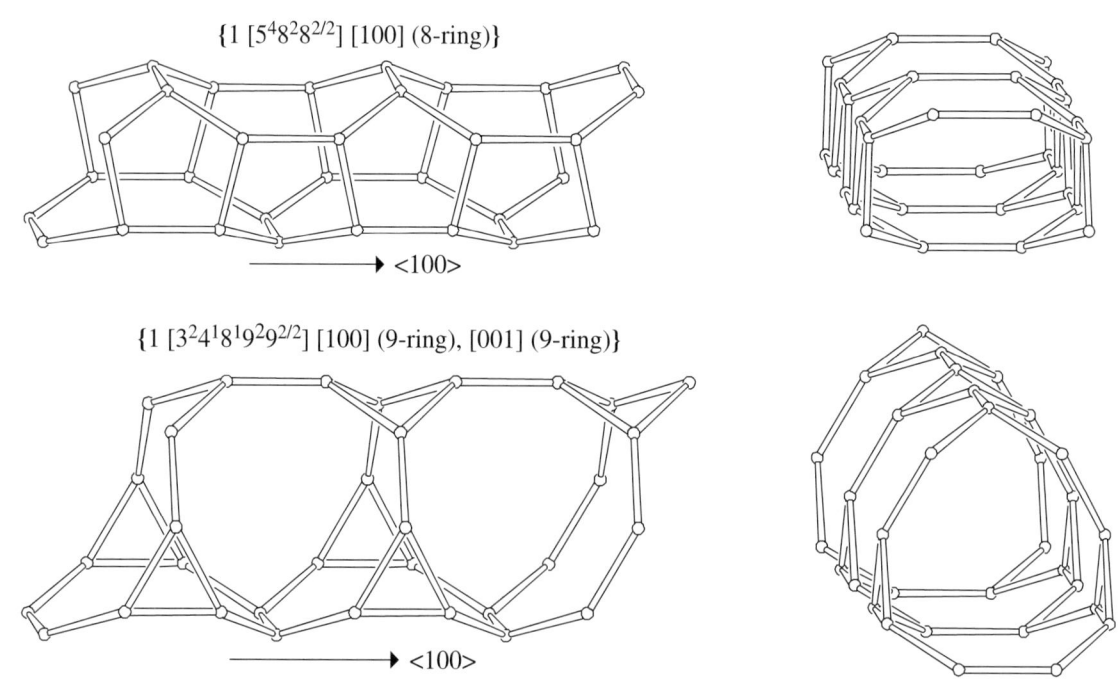

{1 [$5^4 8^2 8^{2/2}$] [100] (8-ring)}

⟶ <100>

{1 [$3^2 4^1 8^1 9^2 9^{2/2}$] [100] (9-ring), [001] (9-ring)}

⟶ <100>

Fig. 3a.  8-ring channel (top) and 9-ring channel (bottom) viewed perpendicular to the channel axis (left) and along the channel axis (right).

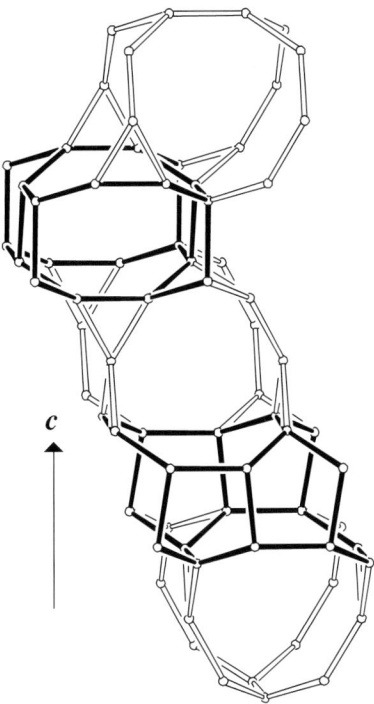

Fig. 3b.   Interconnection of channels along $c$ (8-ring channels in bold). Only one repeat unit in the channels is drawn for clarity. Intersecting 9-ring channels have $[9^4]$-cavities in common (see **NAB**).

# WEI                     Building Scheme

## 1. Periodic Building Unit

**WEI** can be built using units of 10 T atoms. The T10-unit consists of two 3-rings and a tetramer connected in such a way that a 4-ring is formed. The T10-units (one in bold), related along **b** and **c** by pure translations, are linked into the two-dimensional PerBU depicted in Figure 1. 4-Rings and 8-rings are formed.

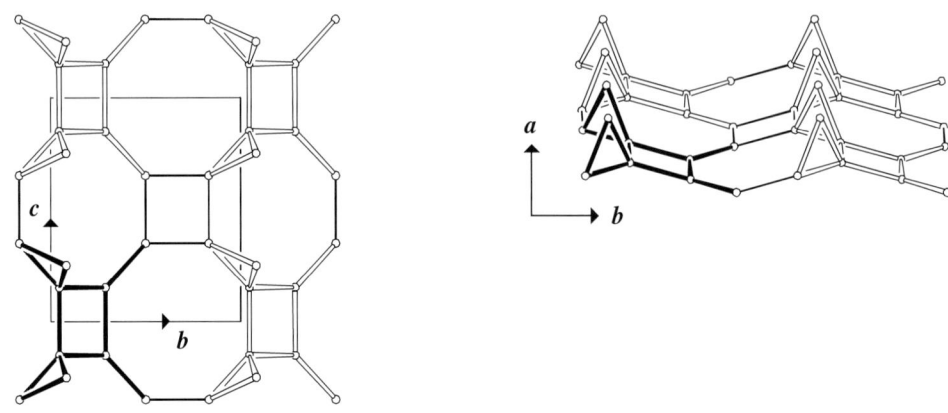

Fig. 1.   PerBU viewed along **a** (left) and along **c** (right).

## 2. Connection mode

Neighboring PerBUs, related along **a** by a shift of $\frac{1}{2}(a+b)$, are connected along **a** through 3-rings forming spiro-5-rings as shown in Figure 2 (one spiro-5-ring is in bold). Figure 2 illustrates that **WEI** can as well be constructed using spiro-5-rings.

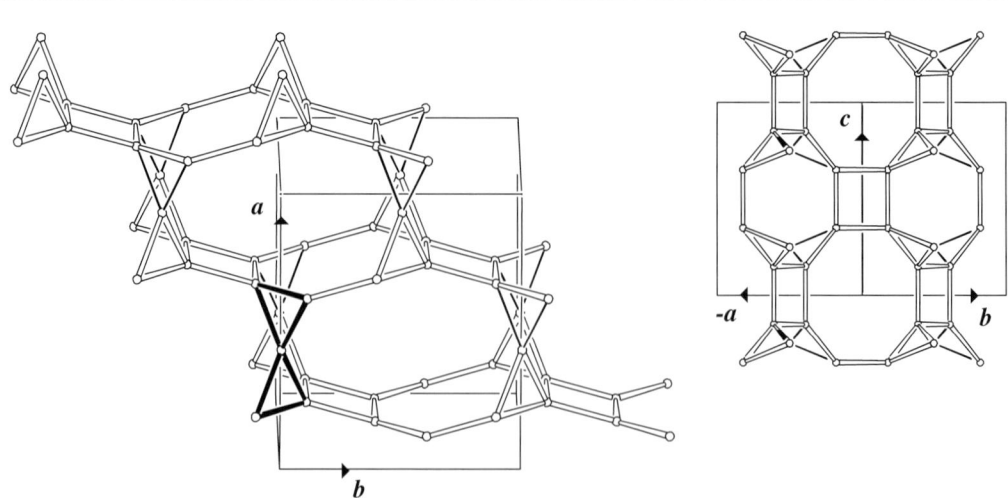

Fig. 2.   Connection mode (and unit cell content) viewed along **c** (left). For clarity, T10-units translated along **c**, are not shown. One spiro-5-ring in bold. Right: cell content projected along [−110].

### 3. Channels and/or cages

Interconnecting 10-ring channels are parallel to *c*. The 10-ring channels are interconnected through common 8-rings. Equal 8-ring channels parallel to [–110] and [110] are formed. 10-Ring channel and the linkage of 10-ring channels into an 8-ring channel parallel to [–110] is shown in Figure 3.

$\{1 \ [3^8 4^4 6^4 8^4 10^{2/2}] \ [001] \ (10\text{-ring})\}$

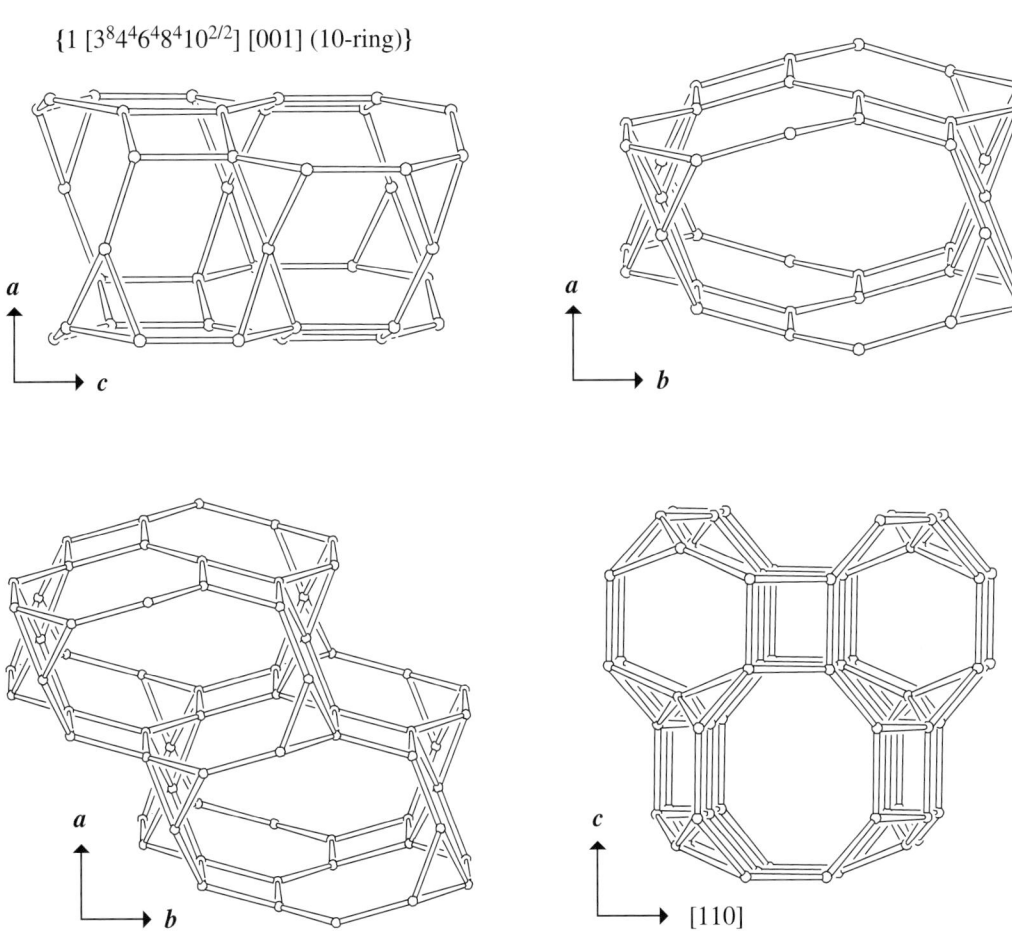

Fig. 3. Top: 10-ring channel viewed perpendicular to the channel axis (left) and along the channel axis (right). Bottom: connection of the channels into 8-ring channels along [–110] viewed along *c* (left) and along [–110] (right) (8-ring channels along [110] are identical).

# -WEN                Building Scheme

## 1. Periodic Building Unit

The interrupted hexagonal **-WEN** framework can be built using the saw chain (bold in Figure 1) running parallel to *c*. Six saw chains are connected into an one-dimensional PerBU consisting of a column of *can* cages connected through double 6-rings (Figure 1). Two additional T atoms, bearing the terminal oxygen atoms, are attached to each *can* cage.

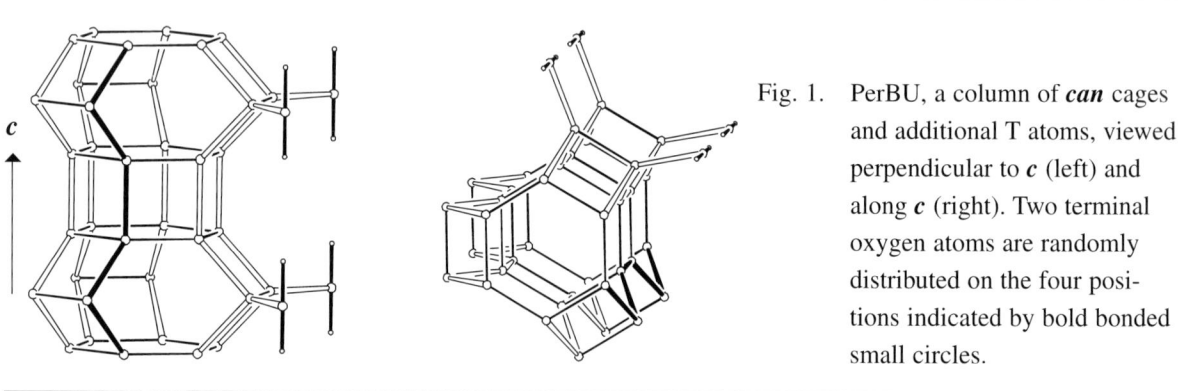

Fig. 1.    PerBU, a column of *can* cages and additional T atoms, viewed perpendicular to *c* (left) and along *c* (right). Two terminal oxygen atoms are randomly distributed on the four positions indicated by bold bonded small circles.

## 2. Connection mode

Neighboring PerBUs, related along *a* and *b* by pure translations, are connected along *a* and *b* (and (*a* + *b*) through single T–T connections using the additional T atoms as is depicted in Figure 2.

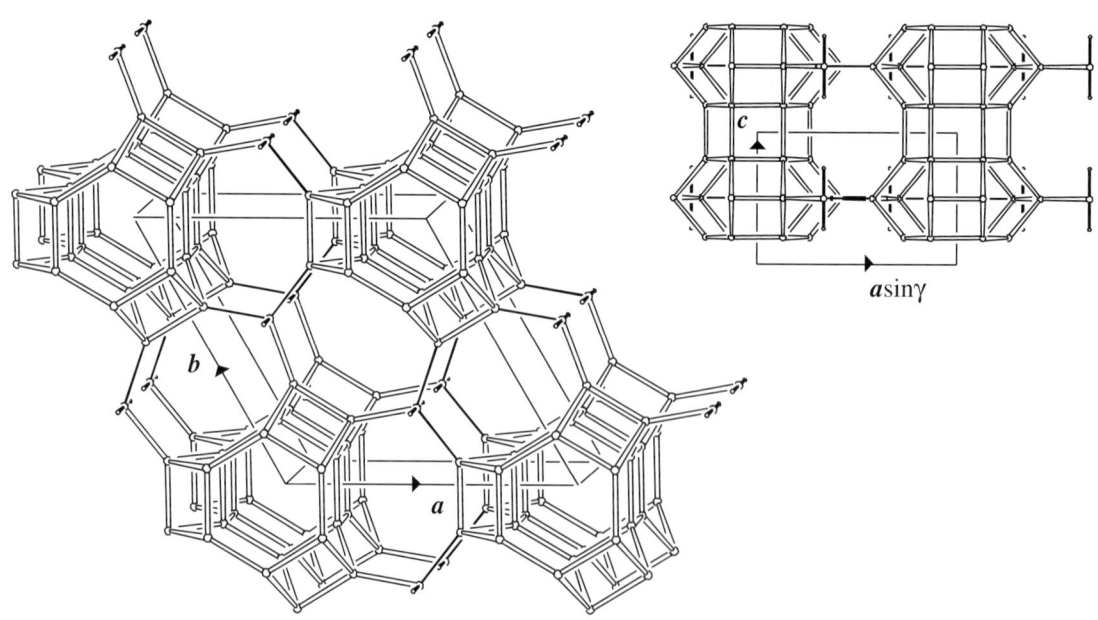

Fig. 2.    Connection mode (and unit cell content) viewed along *c* (left) and unit cell content projected along <010> (right).

### 3. Channels and/or cages

Interconnecting 8-ring channels are parallel to *c*. The 8-ring channels are connected into sinusoidal 10-ring channels parallel to <100>. Terminal oxygen atoms obstruct the free entrance to this 10-ring channel. The 8-ring channel and the linkage of these channels into sinusoidal 10-ring channels are illustrated in Figure 3.

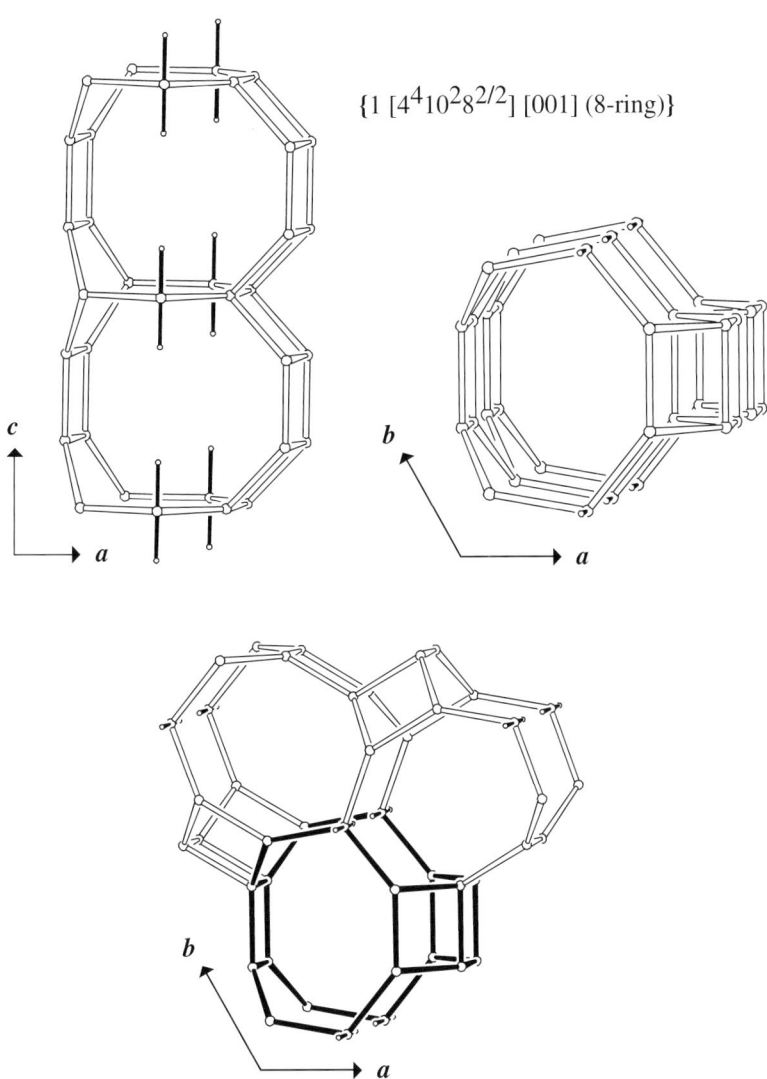

$\{1\ [4^4 10^2 8^{2/2}]\ [001]\ (\text{8-ring})\}$

Fig. 3.    Top: 8-ring channel viewed perpendicular to the channel axis (left) and along the channel axis (right). Bottom: linkage of 8-ring channels into sinusoidal 10-ring channels along <100> viewed along *c* (bottom). One 8-ring channel in bold.

# YUG　　　　　Building Scheme

## 1. Periodic Building Unit

The PerBU equals the layer of units of 8 T atoms depicted in Figure 1. These T8-units (one in bold), consisting of two singly connected 4-rings and related along **a** and **c** by pure translations, are connected through 5-rings that have a (twisted) zigzag chain in common.

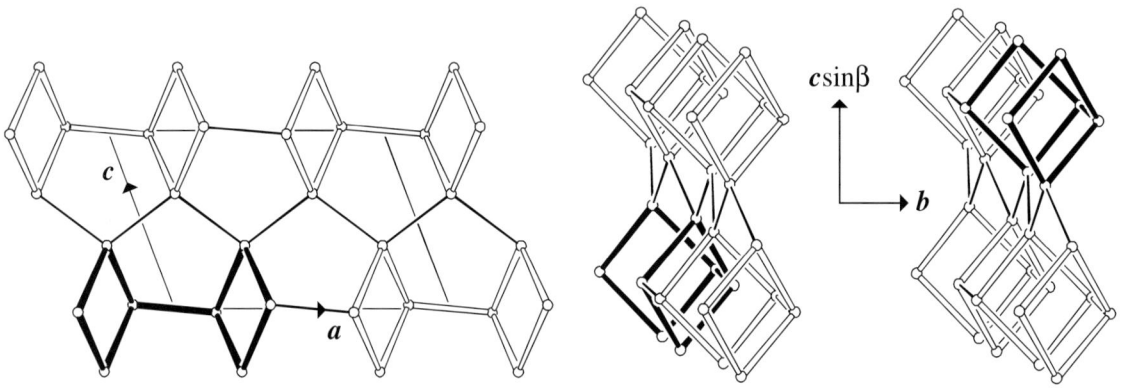

Fig. 1.　PerBU viewed along **b** (left) and along **a** (middle and right). The two PerBUs shown at the right differ by a rotation of 180° about **b**.

## 2. Connection mode

Neighboring PerBUs, related along **b** by a screw rotation of 180° about **b**, are connected along **b** through 4-rings as shown in Figure 2.

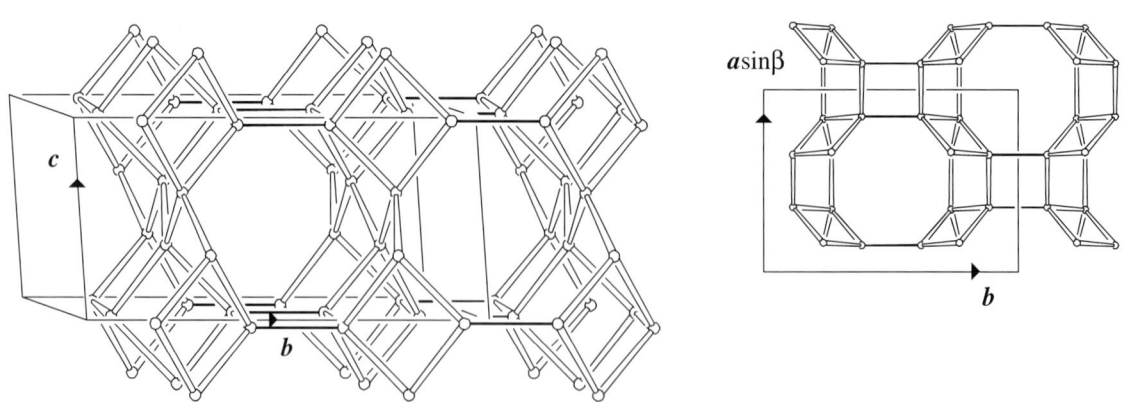

Fig. 2.　Connection mode (and unit cell content) viewed along **a**. Only $1\frac{1}{2}$ of the repeat unit along **a** of each PerBU is drawn for clarity. Right: unit cell content projected along **c**.

# Cage/Channel **YUG**

## 3. Channels and/or cages

Intersecting 8-ring channels are parallel to *a* and *c* as depicted in Figure 3.

{1 [$4^4 5^4 8^2 8^{2/2}$] [001] (8-ring)}

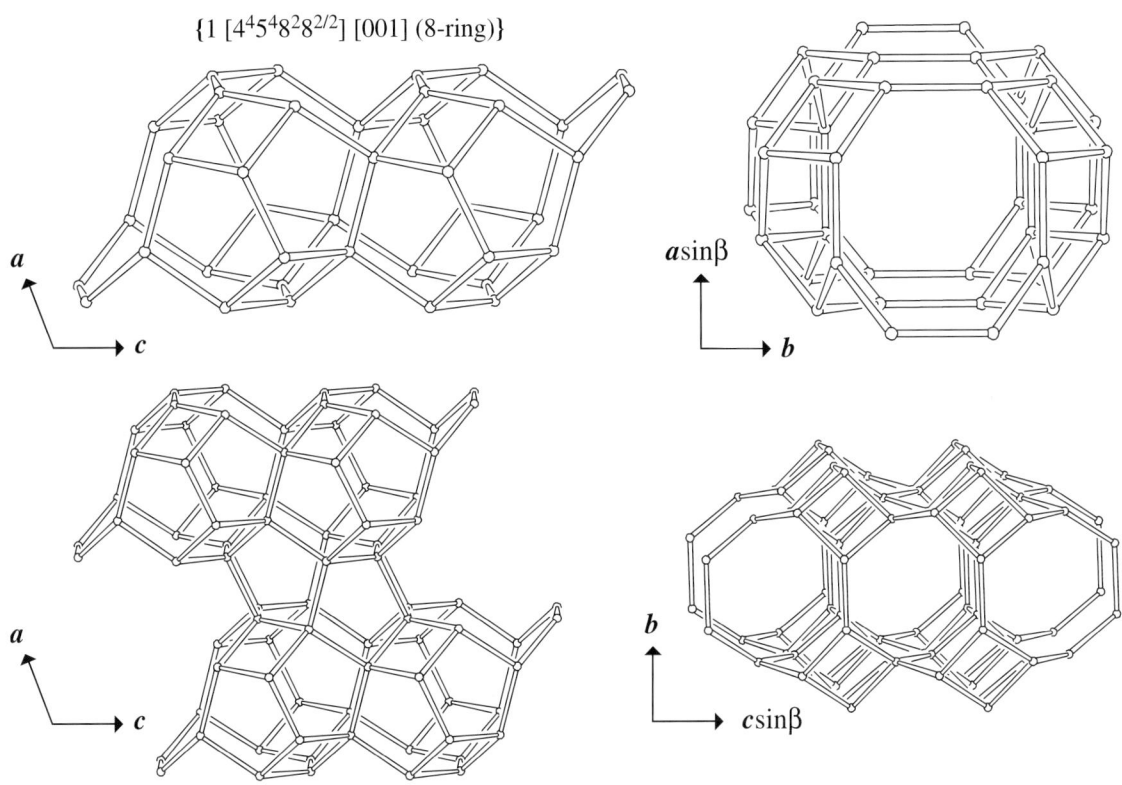

Fig. 3.  Top: 8-ring channel parallel to *c* viewed perpendicular to the channel axis (left) and along the channel axis (right). Bottom: channels parallel to *c* are connected into 8-ring channels parallel to *a*. View along *b* (left) and along the 8-ring channel axis parallel to *a* (right).

# ZON

## Building Scheme

### 1. Periodic Building Unit

**ZON** can be built using double 4-rings with one disconnected edge (a $[4^46^1]$-cage; bold in Figure 1). These T8-units, related along **a** by a pure translation, are connected into chains parallel to **a**. The two-dimensional PerBU is obtained when neighboring chains, related along **b** by a screw rotation of $180°$ about **b**, are linked into the flat **ab** layer as depicted in Figure 1. (See also **AFR** and **SFO**.)

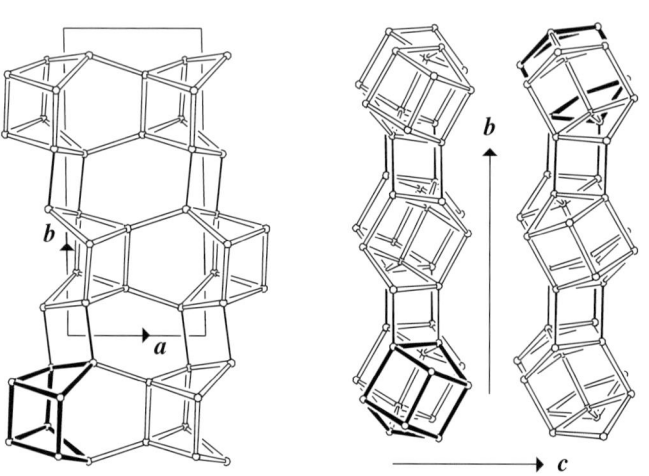

Fig. 1. PerBU viewed along **c** (left) and down **a** (middle and right). The PerBUs at the right differ by a rotation of $180°$ about **c**.

### 2. Connection mode

Neighboring PerBUs, related along **c** by a screw rotation of $180°$ about **c** and a shift of $\frac{1}{2}$ **b**, are connected along **c** through single T–T bonds. Intersecting 8-ring channels parallel to **a** and **b** are formed as shown in Figure 2.

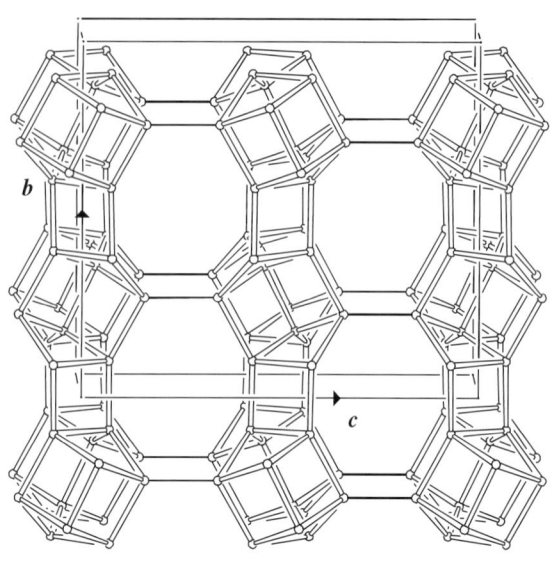

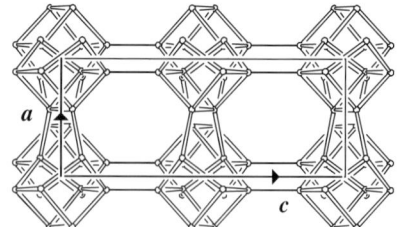

Fig. 2. Connection mode (and unit cell content) viewed along **a** (left) and unit cell content projected along **b** (right).

## 3. Channels and/or cages

Intersecting 8-ring channels are parallel to *a* and *b* as shown in Figure 3.

{1 [$4^6 6^4 8^2 8^{2/2}$] [100] (8-ring)}

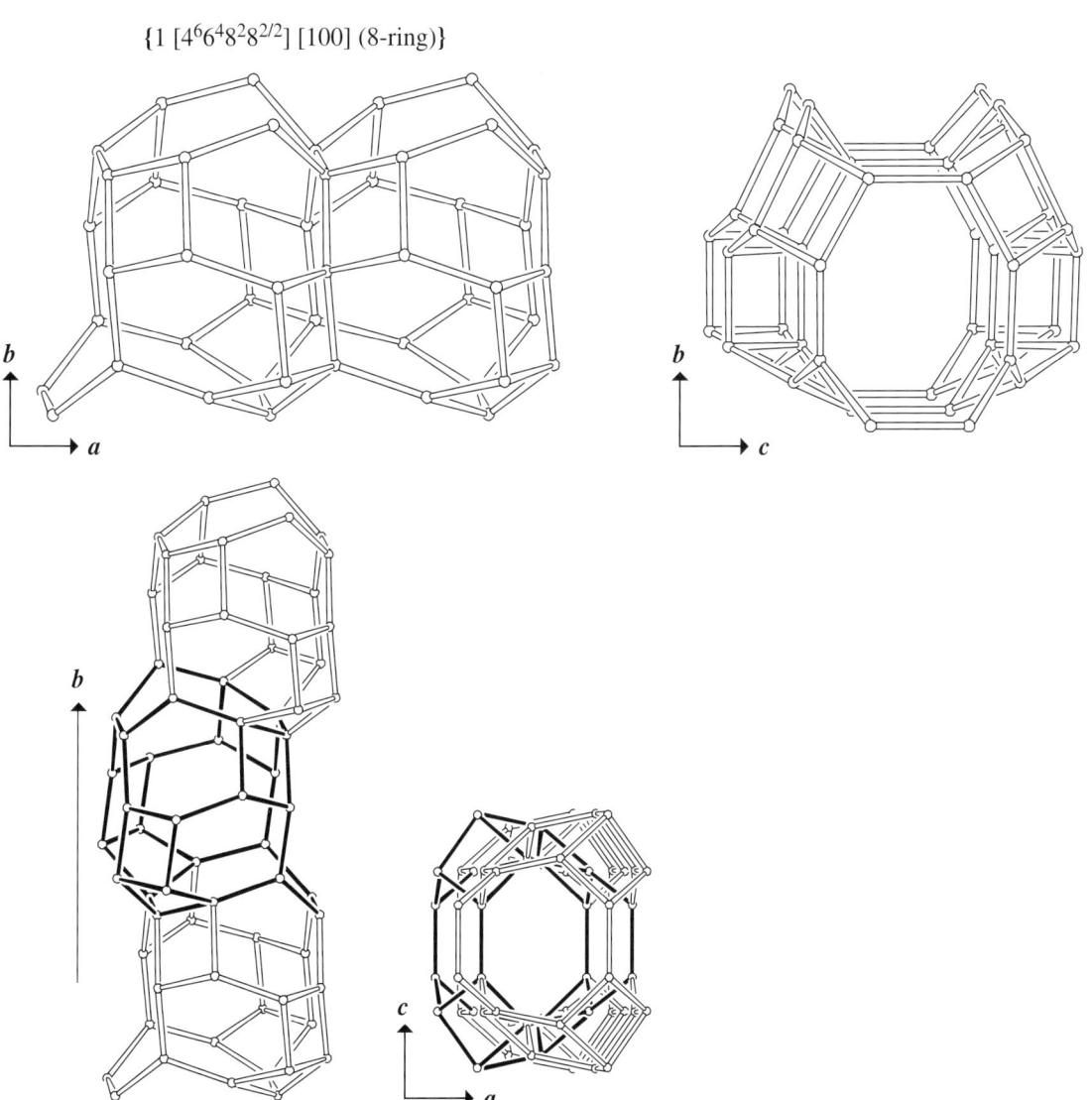

Fig. 3. Top: 8-ring channel parallel to *a* viewed perpendicular to the channel axis (left) and along the channel axis (right). Bottom: channels parallel to *a* are connected through common 8-rings into 8-ring channels parallel to *b*. Only one repeat unit in each channel along *a* is drawn. One channel in bold. View along *c* (left) and along the 8-ring channel axis parallel to *b* (right).

# APPENDICES

# APPENDIX 1

## Survey of Cages and of the Codes of Framework Types (FTCs) in which they Appear

The polyhedrons are arranged in increasing number of T atoms ($n$) defining the smallest faces of the polyhedron and in increasing number of faces ($\Sigma m_i$).

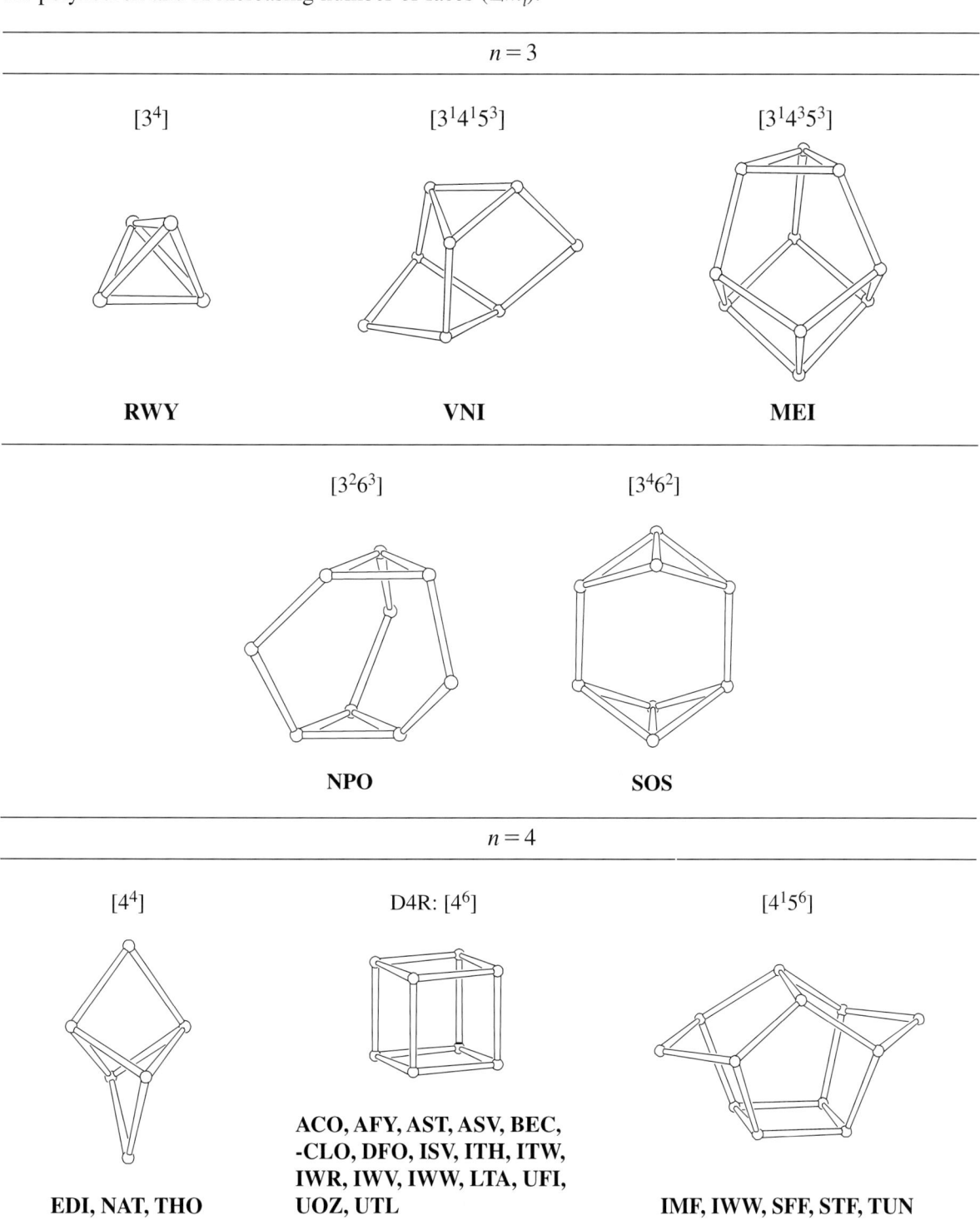

$n = 3$

$[3^4]$          $[3^14^15^3]$          $[3^14^35^3]$

**RWY**          **VNI**          **MEI**

$[3^26^3]$          $[3^46^2]$

**NPO**          **SOS**

$n = 4$

$[4^4]$          D4R: $[4^6]$          $[4^15^6]$

**EDI, NAT, THO**

**ACO, AFY, AST, ASV, BEC, -CLO, DFO, ISV, ITH, ITW, IWR, IWV, IWW, LTA, UFI, UOZ, UTL**

**IMF, IWW, SFF, STF, TUN**

# APPENDIX 1

(Cages and FTCs; continued)

| | |
|---|---|
| | *n* = 4 (Continued) |

$[4^1 5^8]$

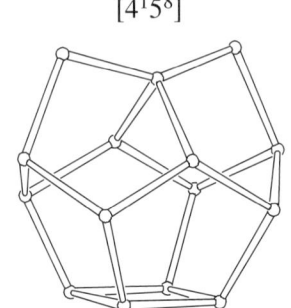

**EUO, IHW, IWV,
NES, NON, UTL**

$[4^1 5^2 6^4]$

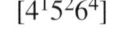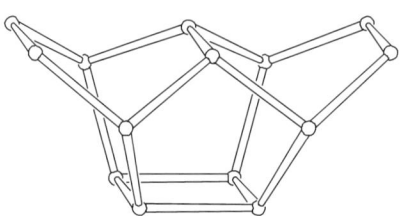

**CON, DON, ITH, IWR, IWW,
MEL, MFI, MWW, SFG**
see also: $[4^3 5^6 6^3]$-b

$[4^2 5^4]$

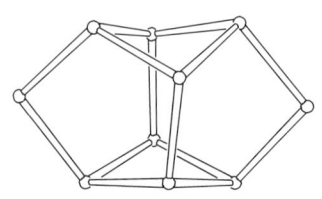

**BOG, BRE, CON,
HEU, IWR, IWW,
RRO, STI, TER**

$[4^2 5^2 6^4]$-a

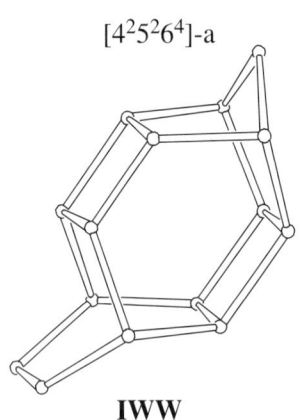

**IWW**

$[4^2 5^2 6^4]$-b

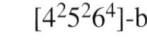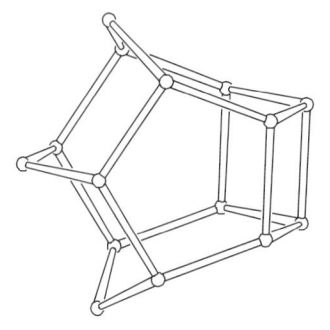

**CFI**

$[4^2 5^4 6^2]$-a

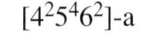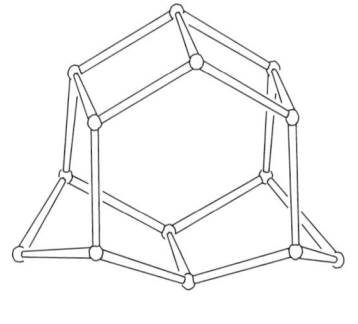

**SFG**

$[4^2 5^4 6^2]$-b

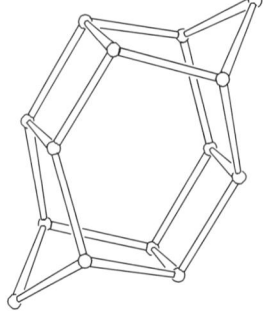

**\*BEA, BEC, GON, ISV,
MSE, MTW, SFH, SFN**

$[4^2 6^2]$

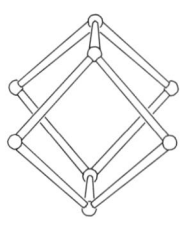

**APD, DFT, OWE**
see also:
$[4^2 5^4]$ and $[4^2 6^4]$-b

$[4^2 6^4]$-a

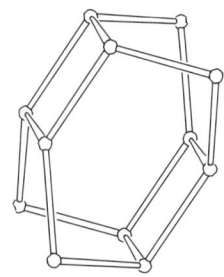

**ASV, BCT, -RON, SAO, UOZ**
see also:
$[4^2 5^2 6^4]$-a, $[4^2 5^4 6^2]$-b, $[4^2 6^6]$

# APPENDIX 1

(Continued)

| $n = 4$ (Continued) |

$[4^2 6^4]$-b

**AEL, AET, AFI,
AFO, AHT, ATV,
BOG, CGF, DFO,
LAU, TER, USI, VFI**

$[4^2 6^4]$-c

**ATS, IMF, OSI**

$[4^2 6^6]$

**ATO, CON, DFO, EZT, IFR, ITH,
IWR, LAU, MSO, OSI, TUN**

$[4^3 5^4]$

**\*BEA, MSE, STT**

$[4^3 5^6]$

**SGT**

$[4^3 5^2 6^2]$

**CON, IFR**

$[4^3 5^2 6^3]$

**DON**

$[4^3 5^4 6^1]$

**STT**

$[4^3 5^6 6^1]$

**DDR**

# APPENDIX 1

## (Continued)

| $n = 4$ (Continued) |
| --- |

$[4^3 5^6 6^3]$-a

**DOH**

$[4^3 5^6 6^3]$-b

**MWW**

$[4^3 6^1]$

**AFN, CGF, USI**
see also:
$[3^1 4^3 5^3]$, $[4^6 6^3]$

$[4^3 6^2]$

**GOO, PON**

$[4^4 5^4]$

**ITE, RTH, UFI**

$[4^4 5^4 6^2]$

**RTE, RUT**

$[4^4 5^4 6^3]$

**ESV**

$[4^4 6^1]$

**DFO, OWE, SBE, SBS, SBT, STI**
see also: $[4^4 6^2]$

$[4^4 6^2]$

**AFR, SAO, SFO, ZON**

# APPENDIX 1

## (Continued)

### $n = 4$ (Continued)

$[4^55^2]$

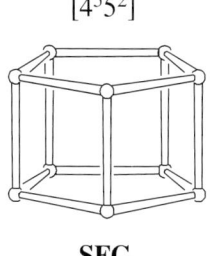

**SFG**

D6R: $[4^66^2]$

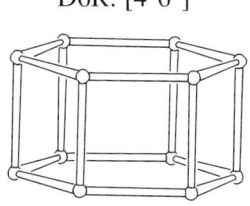

**AEI, AFT, AFX, CHA, EAB, EMT, ERI, FAU,
GME, KFI, LEV, LTL, LTN, MOZ, MSO, MWW,
OFF, SAS, SAT, SAV, SBS, SBT, SZR, TSC, -WEN**

$[4^66^3]$

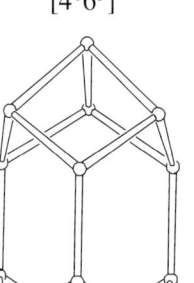

**AFS, BPH, EZT**

$[4^66^4]$

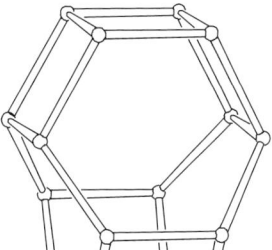

**AWW, SAO**

*can* cage: $[4^66^5]$

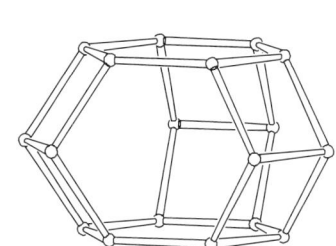

**AFG, CAN, ERI, FAR, FRA,
GIU, LIO, LOS, LTL, LTN,
MAR, MOZ, OFF, SAT, SBS,
SBT, TOL, -WEN**

*sod* cage: $[4^66^8]$

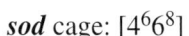
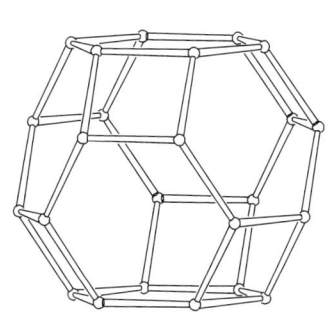

**EMT, FAR, FAU, FRA,
GIU, LTA, LTN, MAR,
SOD, TSC**

*los* cage: $[4^66^{11}]$

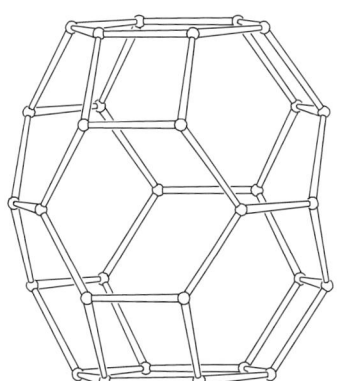

**FRA, LIO, LOS, TOL**

*ast* cage: $[4^66^{12}]$

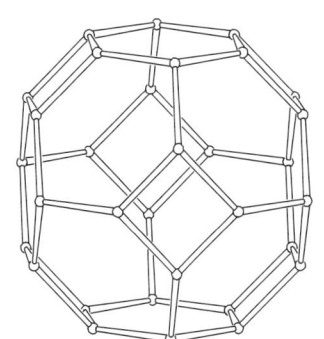

**AST**

# APPENDIX 1

## (Continued)

---

### $n = 4$ (Continued)

---

*lio* cage: $[4^6 6^{17}]$　　　　$[4^6 6^{20}]$　　　　*giu* cage: $[4^6 6^{23}]$

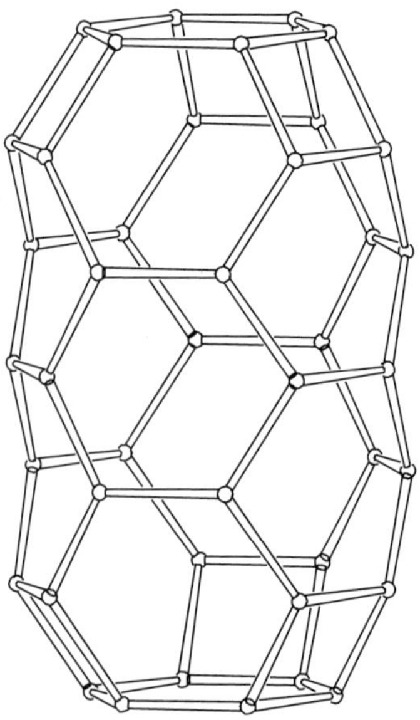

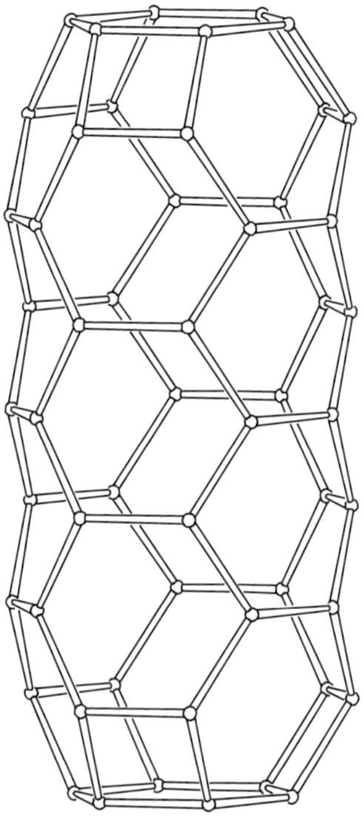

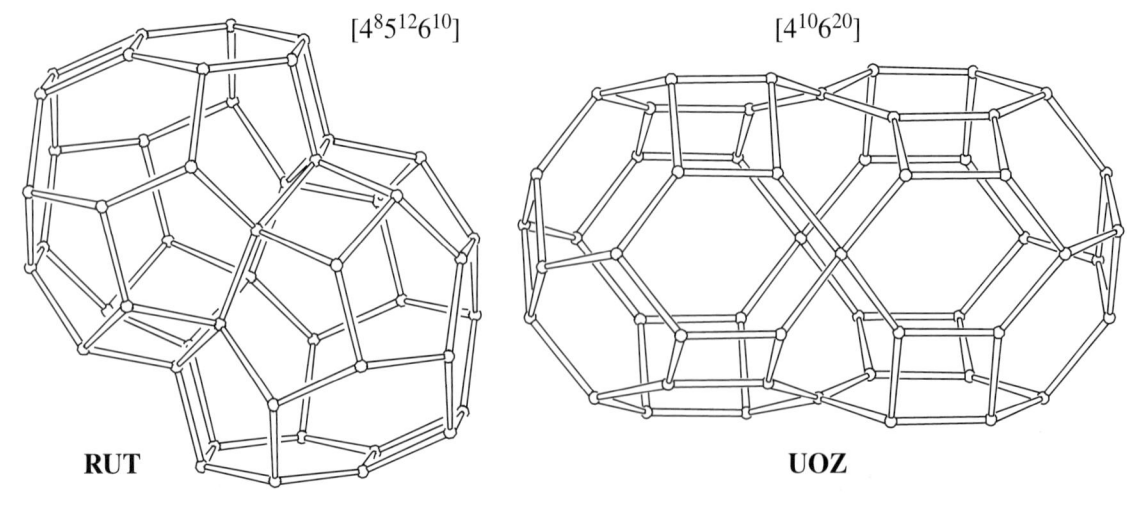

**AFG, FAR, LIO, MAR, TOL**　　　　**MSO**　　　　**GIU**

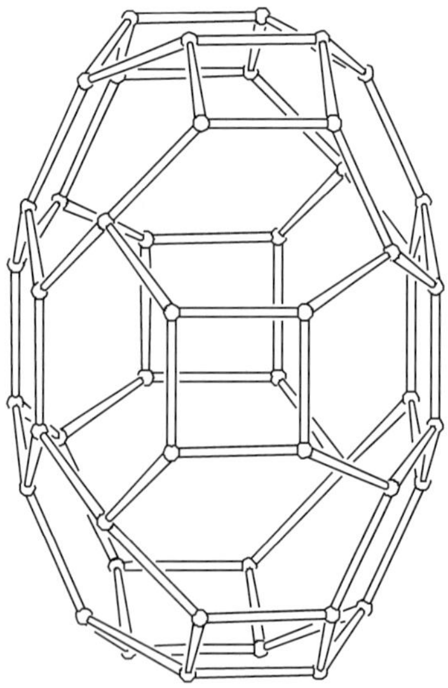

---

$[4^8 5^{12} 6^{10}]$　　　　$[4^{10} 6^{20}]$

**RUT**　　　　**UOZ**

---

# APPENDIX 1

(Continued)

---

$n = 5$

---

$[5^4]$

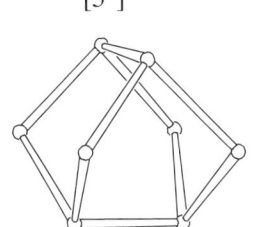

**\*BEA, BEC, CDO, DAC, EON, EPI, FER, IMF, ISV, IWW, MEL, MFI, MFS, MOR, MSE, RWR, TUN, UTL**

$[5^8]$

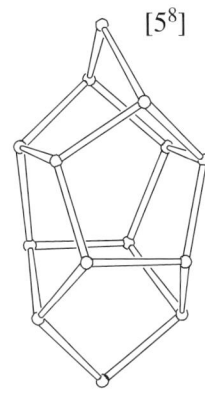

**MEL, MFI, MTF**

$[5^{12}]$

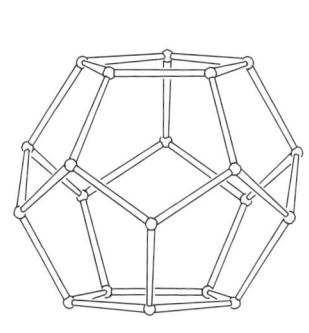

**DDR, DOH, MEP, MTN**

---

$[5^2 6^1]$

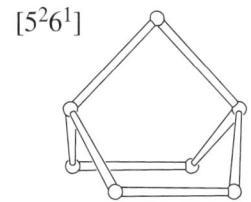

**BIK, CAS, CDO, CFI, DAC, EON, EPI, EUO, FER, IHW, IWV, MFS, MOR, MTT, NES, NON, NSI, SFE, SSY, SZR, TON, UTL, VET**
see also: $[4^1 5^6]$, $[4^2 5^4]$, $[4^2 5^2 6^4]$-a, $[4^2 5^4 6^2]$-a/b, $[4^3 5^4]$, $[5^8]$, $[5^{16} 6^4]$

$[5^2 6^2]$

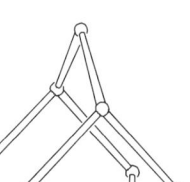

**BIK, MFS, MTT, SFE, SSY, TON**
see also: $[5^4 6^2]$

$[5^2 6^5]$

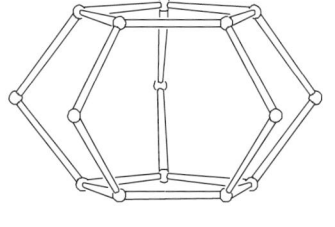

**SFG**

---

$[5^4 6^2]$

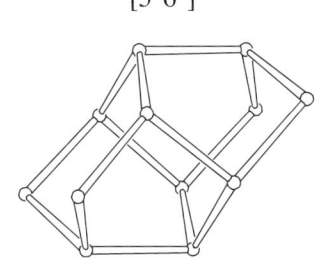

**BOG, CAS, CFI, EUO, GON, IHW, IWV, MFI, MTW, NES, NSI, SFF, SFH, SFN, STF, STT, TER, TUN, UTL, VET**

$[5^4 6^4]$-a

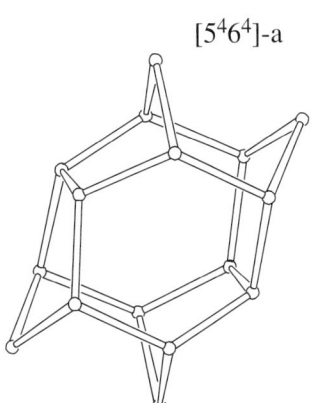

**TUN**

$[5^4 6^4]$-b

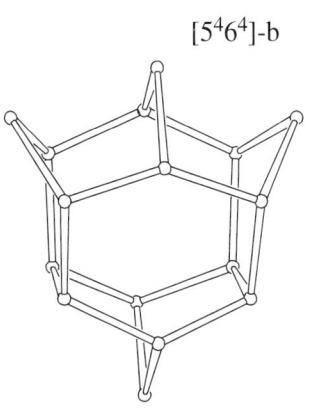

**IMF**

# APPENDIX 1

(Continued)

---

*n* = 5 (Continued)

---

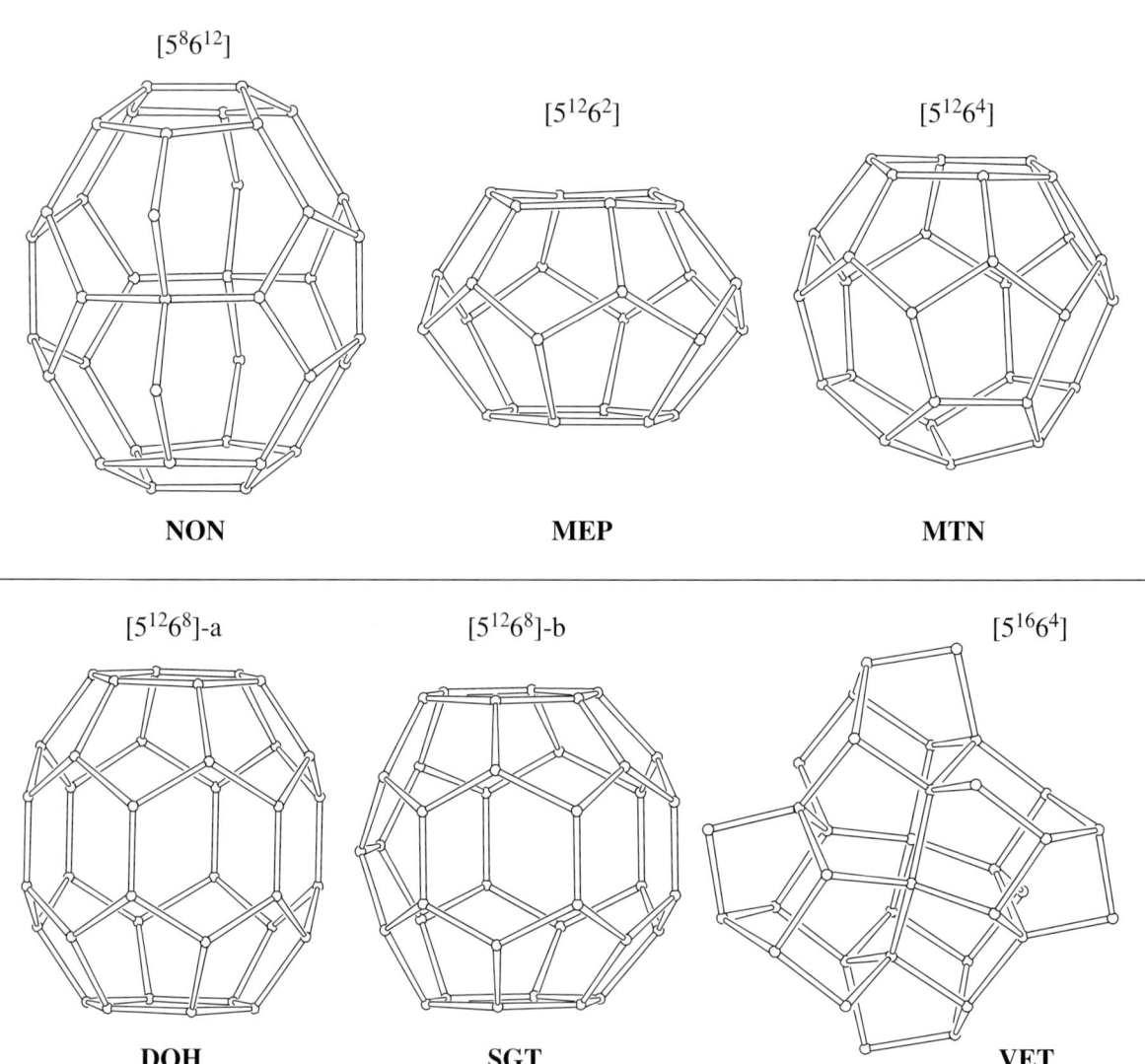

[$5^8 6^{12}$]

[$5^{12} 6^2$]

[$5^{12} 6^4$]

**NON**

**MEP**

**MTN**

[$5^{12} 6^8$]-a

[$5^{12} 6^8$]-b

[$5^{16} 6^4$]

**DOH**

**SGT**

**VET**

# APPENDIX 1

(Final page)

| $n = 6$ |
|---|

| $[6^3]$ | $[6^4]$ | $[6^5]$ |

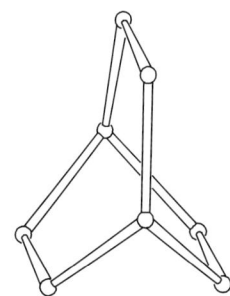

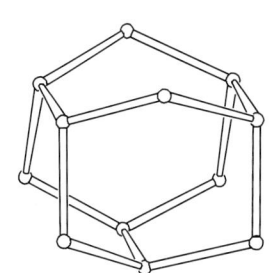

**AEN, APD, ASV, ATS,
-CHI, IMF, RWR**
see also:
$[4^15^26^4]$, $[4^26^4]$-b, $[4^26^6]$,
$[4^35^26^2]$, $[4^35^66^3]$-b,
$[4^46^2]$, $[6^8]$-a

**JBW, MTT, MTW,
SFE, SFN, SSY, TON**

**SFH**
see also:
$[4^66^3]$, $[6^8]$-a

$[6^8]$-a

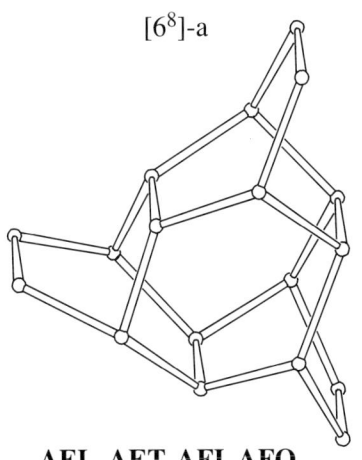

$[6^8]$-b

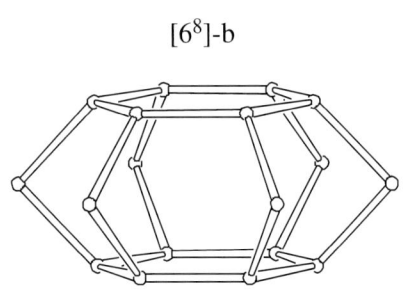

**AEL, AET, AFI, AFO,
ATV, DON, VFI**

**MSO, SZR**

# APPENDIX 2

## Pore Descriptors of Channels and Cavities that Appear in More than One Framework Type

| FTC | Channel Descriptor | Channel Intersection/Cavity Descriptor |
|---|---|---|
| **ABW/JBW** | $\{[4^46^48^{2/2}]\ (8\text{-ring})\}$ | $\{1\ [4^46^48^2]\}$ |
| **ACO/DFT/MER** | $\{[4^48^48^{2/2}]\ (8\text{-ring})\}$ | $\{2\ [4^28^4]\ (8\text{-ring})\}$ |
| **AEI: see also AFT, CHA** | | $\{3\ [4^{12}6^28^6]\ (8\text{-ring})\}$ |
| **AEL/AFO/AHT** | $\{[6^{10}10^{2/2}]\ (10\text{-ring})\}$ | $\{1\ [6^510^2]\}$ |
| **AET/DON** | $\{[6^{14}14^{2/2}]\ (14\text{-ring})\}$ | $\{1\ [6^714^2]\}$ |
| **AFO: see AEL** | $\{[6^{10}10^{2/2}]\ (10\text{-ring})\}$ | $\{1\ [6^510^2]\}$ |
| **AFR: see also SFO** | | $\{2\ [4^{10}6^48^212^2]\ (8\text{-ring}),\ (12\text{-ring})\}$ |
| **AFS/AFY, see also BPH** | | $\{3\ [4^{18}8^612^2]\ (8\text{-ring}),\ (12\text{-ring})\}$ |
| **AFT/AFX** | | $\{2\ [4^{15}6^28^9]\ (8\text{-ring})\};$ *aft* cavity |
| **AFT: see also CHA, GME** | | *chab* and *gmel* cavities |
| **AFX: see also AFT, GME** | | *aft* and *gmel* cavities |
| **AHT: see AEL** | $\{[6^{10}10^{2/2}]\ (10\text{-ring})\}$ | $\{1\ [6^510^2]\}$ |
| **ATN/BCT/GON** | $\{[6^48^{2/2}]\ (8\text{-ring})\}$ | $\{1\ [6^28^2]\}$ |
| **ATN: see also SBE** | $\{[4^86^48^{2/2}]\ (8\text{-ring})\}$ | $\{1\ [4^86^48^2]\};$ *atn* cavity |
| **ATO/CAN/NPO** | $\{[6^612^{2/2}]\ (12\text{-ring})\}$ | $\{1\ [6^412^2]\}$ |
| **ATT/GIS/SIV** | $\{[4^{12}8^48^{2/2}]\ (8\text{-ring})\}$ | $\{2\ [4^68^4]\ (8\text{-ring})\}$ |
| **AWW: see also -CLO** | $\{[4^86^88^{2/2}]\ (8\text{-ring})\}$ | $\{1\ [4^86^88^2]\};$ *aww* cavity |
| **BCT: see ATN** | $\{[6^48^{2/2}]\ (8\text{-ring})\}$ | $\{1\ [6^28^2]\}$ |
| **\*BEA/CON** | $\{[4^{25}8^6 2^212^212^{2/2}]\ (12\text{-ring})\}$ | $\{2\ [5^412^4]\ (12\text{-ring})\}$ |
| **BEC/IWR** | $\{[4^45^46^412^212^{2/2}]\ (12\text{-ring})\}$ | $\{2\ [4^212^4]\ (12\text{-ring})\}$ |
| **BIK/NSI** | $\{[6^68^{2/2}]\ (8\text{-ring})\}$ | $\{1\ [6^68^2]\}$ |
| **BPH: see also AFS, AFY** | | $\{3\ [4^{18}8^612^2]\ (8\text{-ring}),\ (12\text{-ring})\}$ |
| **CAN: see ATO** | $\{[6^612^{2/2}]\ (12\text{-ring})\}$ | $\{1\ [6^412^2]\}$ |
| **CHA/AFT, see also AEI** | | $\{3\ [4^{12}6^28^6]\ (8\text{-ring})\};$ *chab* cavity |
| **-CLO: see also AWW, RHO** | $\{[4^{28}6^{24}8^48^{2/2}]\ (8\text{-ring})\}$ | (channel of *aww* and *rho* cavities) |
| **CON: see \*BEA** | $\{[4^{25}8^612^212^{2/2}]\ (12\text{-ring})\}$ | $\{2\ [5^412^4]\ (12\text{-ring})\}$ |
| **CON/IWR** | $\{[4^66^412^210^{2/2}]\ (10\text{-ring})\}$ | $\{2\ [4^410^212^2]\ (10\text{-ring}),\ (12\text{-ring})\}$ |
| **DAC/FER** | $\{[5^46^28^210^{2/2}]\ (10\text{-ring})\}$ | $\{2\ [6^28^210^2]\ (8\text{-ring}),\ (10\text{-ring})\}$ |
| **DFT: see ACO** | $\{[4^48^48^{2/2}]\ (8\text{-ring})\}$ | $\{2\ [4^28^4]\ (8\text{-ring})\}$ |
| **DFT/LOV/LTL/MOZ/RSN** | $\{[4^26^28^28^{2/2}]\ (8\text{-ring})\}$ | $\{2\ [4^28^4]\ (8\text{-ring})\}$ |
| **DON: see AET** | $\{[6^{14}14^{2/2}]\ (14\text{-ring})\}$ | $\{1\ [6^714^2]\}$ |
| **EAB: see also GME** | | *gmel* cavity |
| **EDI: see also THO** | $\{[4^28^48^{2/2}]\ (8\text{-ring})\}$ | $\{3\ [8^6]\ (8\text{-ring})\}$ |

# APPENDIX 2

## (Continued)

| FTC | Channel Descriptor | Channel Intersection/Cavity Descriptor |
|---|---|---|
| **EON/MAZ/MON/MOR/ RSN/VSV** | $\{[5^48^28^{2/2}]\ (8\text{-ring})\}$ | $\{2\ [5^48^4]\ (8\text{-ring})\}$ |
| **EON: see also GME** | | *gmel* cavity |
| **FER: see DAC** | $\{[5^46^28^210^{2/2}]\ (10\text{-ring})\}$ | $\{2\ [6^28^210^2]\ (8\text{-ring}),\ (10\text{-ring})\}$ |
| **GIS: see ATT** | $\{[4^{12}8^48^{2/2}]\ (8\text{-ring})\}$ | $\{2\ [4^68^4]\ (8\text{-ring})\}$ |
| **GME/AFT/AFX/EAB/ EON/MAZ/OFF** | | $\{2\ [4^96^28^3]\ (8\text{-ring})\};$ *gmel* cavity |
| **GON: see ATN** | $\{[6^48^{2/2}]\ (8\text{-ring})\}$ | $\{1\ [6^28^2]\}$ |
| **GON/MTW** | $\{[6^812^{2/2}]\ (12\text{-ring})\}$ | $\{1\ [6^812^2]\}$ |
| **ITE: see also RTH** | | $\{2\ [4^65^86^48^4]\ (8\text{-ring})\}$ |
| **IWR: see BEC** | $\{[4^45^46^412^212^{2/2}]\ (12\text{-ring})\}$ | $\{2\ [4^212^4]\ (12\text{-ring})\}$ |
| **IWR: see CON** | $\{[4^66^412^210^{2/2}]\ (10\text{-ring})\}$ | $\{2[4^410^212^2](10\text{-ring}),(12\text{-ring})\}$ |
| **JBW: see ABW** | $\{[4^46^48^{2/2}]\ (8\text{-ring})\}$ | $\{1\ [4^46^48^2]\}$ |
| **KFI: see also MER, RHO** | $\{[4^{24}6^88^88^{2/2}]\ (8\text{-ring})\}$ | (channel of *mer* and *rho* cavities) |
| **LOV/NAB/RSN/VSV** | $\{[3^24^18^19^29^{2/2}]\ (9\text{-ring})\}$ | $\{2\ [9^4]\ (9\text{-ring})\}$ |
| **LOV: see DFT** | $\{[4^26^28^28^{2/2}]\ (8\text{-ring})\}$ | $\{2\ [4^28^4]\ (8\text{-ring})\}$ |
| **LTA: see also RHO** | $\{[4^{12}6^88^48^{2/2}]\ (8\text{-ring})\}$ | (channel of *rho* cavities) |
| **LTL/MOZ** | $\{[4^{18}8^612^{2/2}]\ (12\text{-ring})\}$ | $\{3\ [4^{18}8^612^2]\};$ *ltl* cavity |
| **LTL: see DFT** | $\{[4^26^28^28^{2/2}]\ (8\text{-ring})\}$ | $\{2\ [4^28^4]\ (8\text{-ring})\}$ |
| **MAZ: see also GME** | | *gmel* cavity |
| **MAZ: see EON** | $\{[5^48^28^{2/2}]\ (8\text{-ring})\}$ | $\{2\ [5^48^4]\ (8\text{-ring})\}$ |
| **MER/KFI/MOZ/PAU** | $\{[4^{20}8^48^{2/2}]\ (8\text{-ring})\}$ | $\{3\ [4^{12}8^6]\ (8\text{-ring})\};$ *mer* cavity |
| **MER: see ACO** | $\{[4^48^48^{2/2}]\ (8\text{-ring})\}$ | $\{2\ [4^28^4]\ (8\text{-ring})\}$ |
| **MON: see EON** | $\{[5^48^28^{2/2}]\ (8\text{-ring})\}$ | $\{2\ [5^48^4]\ (8\text{-ring})\}$ |
| **MOR: see EON** | $\{[5^48^28^{2/2}]\ (8\text{-ring})\}$ | $\{2\ [5^48^4]\ (8\text{-ring})\}$ |
| **MOZ/OFF** | $\{[4^36^38^312^{2/2}]\ (12\text{-ring})\}$ | $\{3\ [4^38^312^2]\ (8\text{-ring}),\ (12\text{-ring})\}$ |
| **MOZ: see DFT** | $\{[4^26^28^28^{2/2}]\ (8\text{-ring})\}$ | $\{2\ [4^28^4]\ (8\text{-ring})\}$ |
| **MOZ: see MER** | $\{[4^{12}8^48^{2/2}]\ (8\text{-ring})\}$ | (channel of *mer* cavities) |
| **MOZ: see LTL** | $\{[4^{18}8^612^{2/2}]\ (12\text{-ring})\}$ | (channel of *ltl* cavities) |
| **MTT: see also TON** | $\{[6^810^{2/2}]\ (10\text{-ring})\}$ | $\{1\ [6^810^2]\}$ |
| **MTW: see GON** | $\{[6^812^{2/2}]\ (12\text{-ring})\}$ | $\{1\ [6^812^2]\}$ |
| **NAB: see LOV** | $\{[3^24^18^19^29^{2/2}]\ (9\text{-ring})\}$ | $\{2\ [9^4]\ (9\text{-ring})\}$ |
| **NPO: see ATO** | $\{[6^612^{2/2}]\ (12\text{-ring})\}$ | $\{1\ [6^412^2]\}$ |

# APPENDIX 2

(Final page)

| FTC | Channel Descriptor | Channel Intersection/Cavity Descriptor |
|---|---|---|
| **NSI: see BIK** | $\{[6^68^{2/2}]$ (8-ring)$\}$ | $\{1\ [6^68^2]\}$ |
| **OFF: see also GME** | | *gmel* cavity |
| **OFF: see MOZ** | $\{[4^36^38^312^{2/2}]$ (12-ring)$\}$ | $\{3\ [4^38^312^2]\}$ |
| **OSI/VET** | $\{[6^812^{2/2}]$ (12-ring)$\}$ | $\{1\ [6^812^2]\}$ |
| **PAU: see also RHO, MER** | $\{[4^{60}6^88^{12}8^{2/2}]$ (8-ring)$\}$ | (channel of **mer** and **rho** cavities) |
| **PHI/SIV** | $\{[4^{12}8^48^{2/2}]$ (8-ring)$\}$ | $\{2\ [4^78^5]$ (8-ring)$\}$ |
| **RHO/-CLO/KFI/LTA/ LTN/PAU/TSC/UFI** | $\{[4^{20}6^88^48^{2/2}]$ (8-ring)$\}$ | $\{3\ [4^{12}6^88^6]$ (8-ring)$\}$; **rho** cavity |
| **RSN: see DFT** | $\{[4^26^28^28^{2/2}]$ (8-ring)$\}$ | $\{2\ [4^28^4]$ (8-ring)$\}$ |
| **RSN: see EON** | $\{[5^48^28^{2/2}]$ (8-ring)$\}$ | $\{2\ [5^48^4]$ (8-ring)$\}$ |
| **RSN: see LOV** | $\{[3^24^18^19^29^{2/2}]$ (9-ring)$\}$ | $\{2\ [9^4]$ (9-ring)$\}$ |
| **RTH: see also ITE** | | $\{2\ [4^65^86^48^4]$ (8-ring)$\}$ |
| **SBE: see also ATN** | $\{1\ [4^{28}6^88^812^48^{2/2}]$ (8-ring)$\}$ | (channel of **atn** (and **sbe**) cavities) |
| **SFE: see also SSY** | $\{[4^46^812^{2/2}]$ (12-ring)$\}$ | $\{1\ [4^46^812^2]\}$ |
| **SFF: see also STF** | $\{[4^45^86^610^{2/2}]$ (10-ring)$\}$ | $\{1\ [4^45^86^610^2]\}$ |
| **SFH: see also SFN** | $\{[4^46^814^{2/2}]$ (14-ring)$\}$ | $\{1\ [4^46^614^2]\}$ |
| **SFN: see also SFH** | $\{[4^46^814^{2/2}]$ (14-ring)$\}$ | $\{1\ [4^46^614^2]\}$ |
| **SFO: see also AFR** | | $\{2\ [4^{10}6^48^212^2]$ (8-ring), (12-ring)$\}$ |
| **SIV: see ATT** | $\{[4^{12}8^48^{2/2}]$ (8-ring)$\}$ | $\{2\ [4^68^4]$ (8-ring)$\}$ |
| **SIV: see PHI** | $\{[4^{12}8^48^{2/2}]$ (8-ring)$\}$ | $\{2\ [4^78^5]$ (8-ring)$\}$ |
| **SSY: see also SFE** | $\{[4^46^812^{2/2}]$ (12-ring)$\}$ | $\{1\ [4^46^812^2]\}$ |
| **STF: see also SFF** | $\{[4^45^86^610^{2/2}]$ (10-ring)$\}$ | $\{1\ [4^45^86^610^2]\}$ |
| **THO: see also EDI** | $\{[4^28^48^{2/2}]$ (8-ring)$\}$ | $\{3\ [8^6]$ (8-ring)$\}$ |
| **TON: see also MTT** | $\{[6^810^{2/2}]$ (10-ring)$\}$ | $\{1\ [6^810^2]\}$ |
| **TSC: see also RHO** | $\{[4^{36}6^{16}8^{20}8^{2/2}]$ (8-ring)$\}$ | (channel of **rho** (and **tsc**) cavities) |
| **UFI: see also RHO** | $\{[4^{24}6^{16}8^88^{2/2}]$ (8-ring)$\}$ | (channel composed of **rho** cavities) |
| **VET: see OSI** | $\{[6^812^{2/2}]$ (12-ring)$\}$ | $\{1\ [6^812^2]\}$ |
| **VSV: see EON** | $\{[5^48^28^{2/2}]$ (8-ring)$\}$ | $\{2\ [5^48^4]$ (8-ring)$\}$ |
| **VSV: see LOV** | $\{[3^24^18^19^29^{2/2}]$ (9-ring)$\}$ | $\{2\ [9^4]$ (9-ring)$\}$ |

**APPENDIX 3 starts on next page**

# APPENDIX 3

## Channel Intersections and Cavities that Appear in More than One Framework Type

The intersections and cavities are arranged in the order as they are tabulated in Appendix 2.

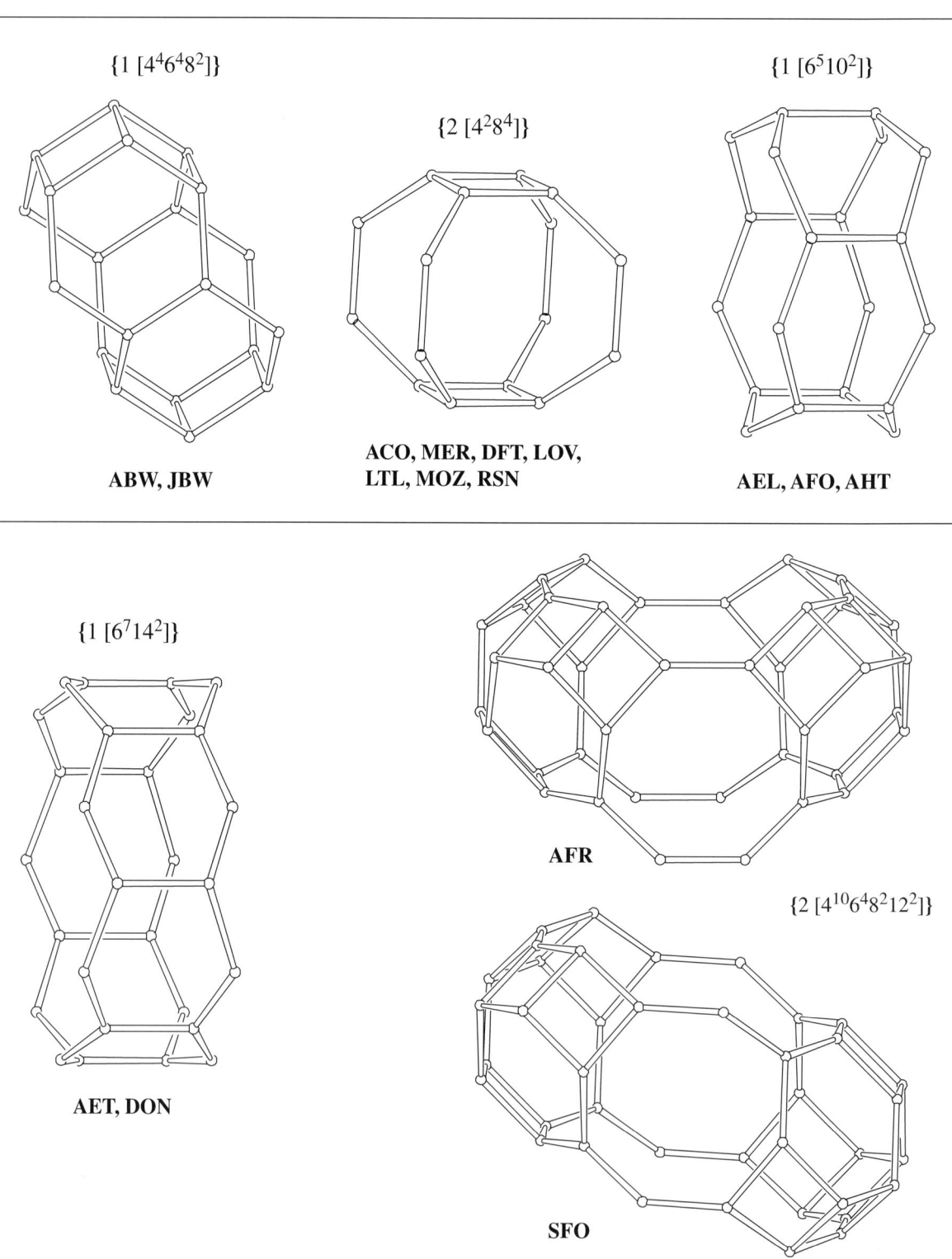

$\{1\ [4^4 6^4 8^2]\}$

$\{2\ [4^2 8^4]\}$

$\{1\ [6^5 10^2]\}$

**ABW, JBW**

**ACO, MER, DFT, LOV,
LTL, MOZ, RSN**

**AEL, AFO, AHT**

$\{1\ [6^7 14^2]\}$

**AFR**

$\{2\ [4^{10} 6^4 8^2 12^2]\}$

**AET, DON**

**SFO**

# APPENDIX 3

## (Continued)

$\{3\ [4^{18}8^612^2]\}$

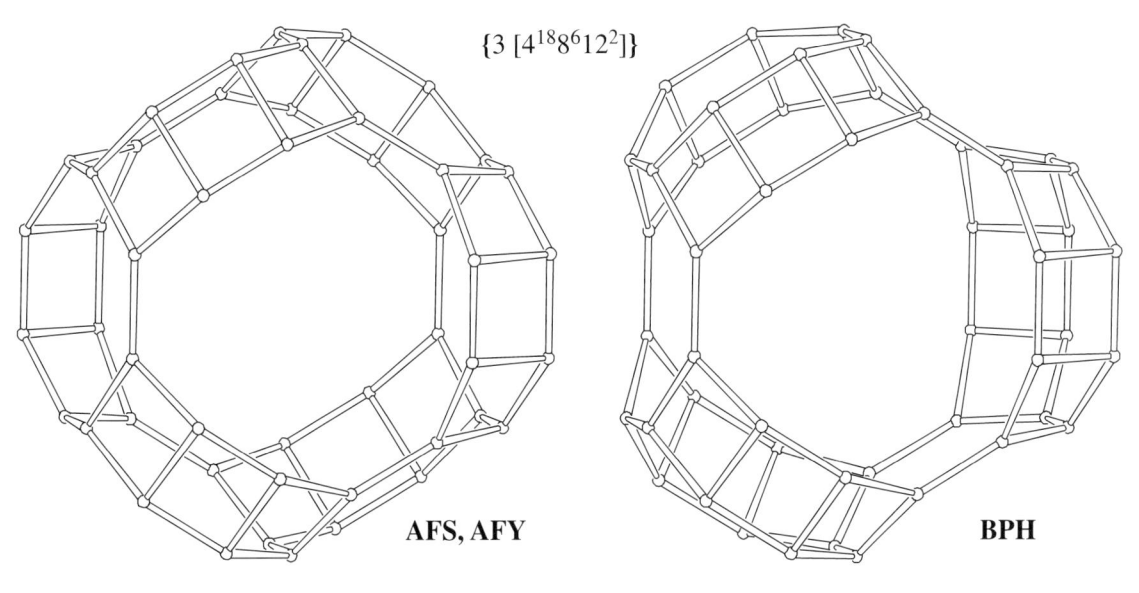

**AFS, AFY**

**BPH**

*aft* cavity: $\{2\ [4^{15}6^28^9]\}$

$\{1\ [6^28^2]\}$

$\{1\ [6^412^2]\}$

*atn* cavity: $\{1\ [4^86^48^2]\}$

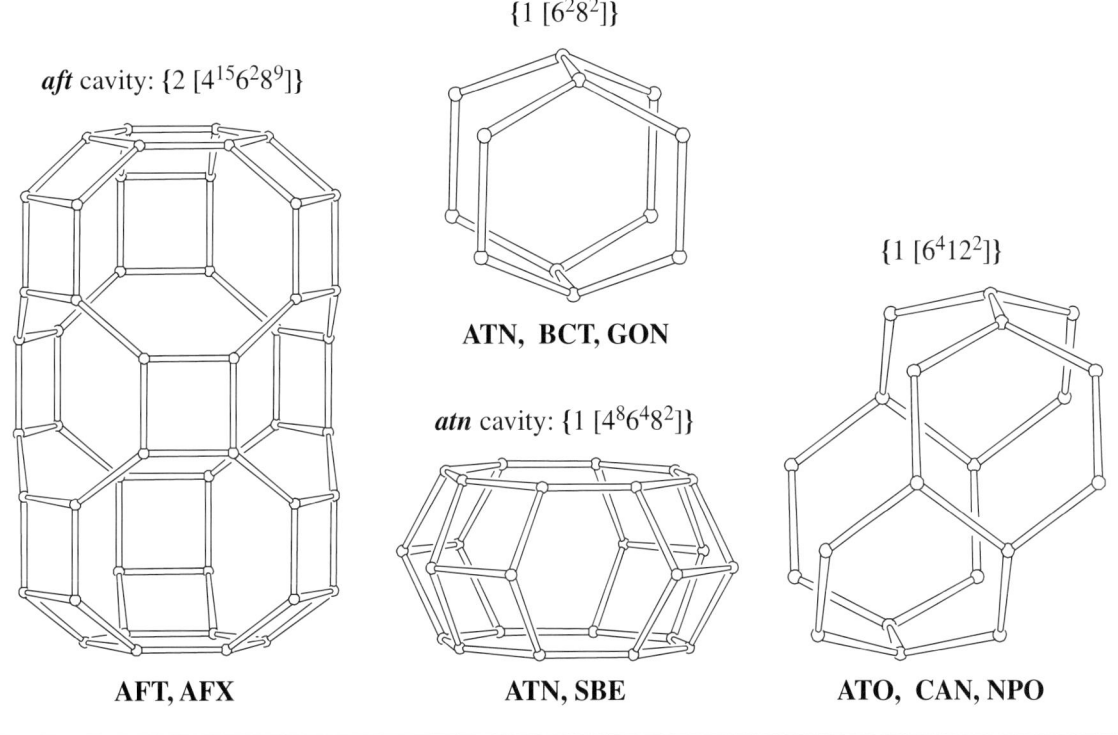

**AFT, AFX**

**ATN, BCT, GON**

**ATN, SBE**

**ATO, CAN, NPO**

# APPENDIX 3

## (Continued)

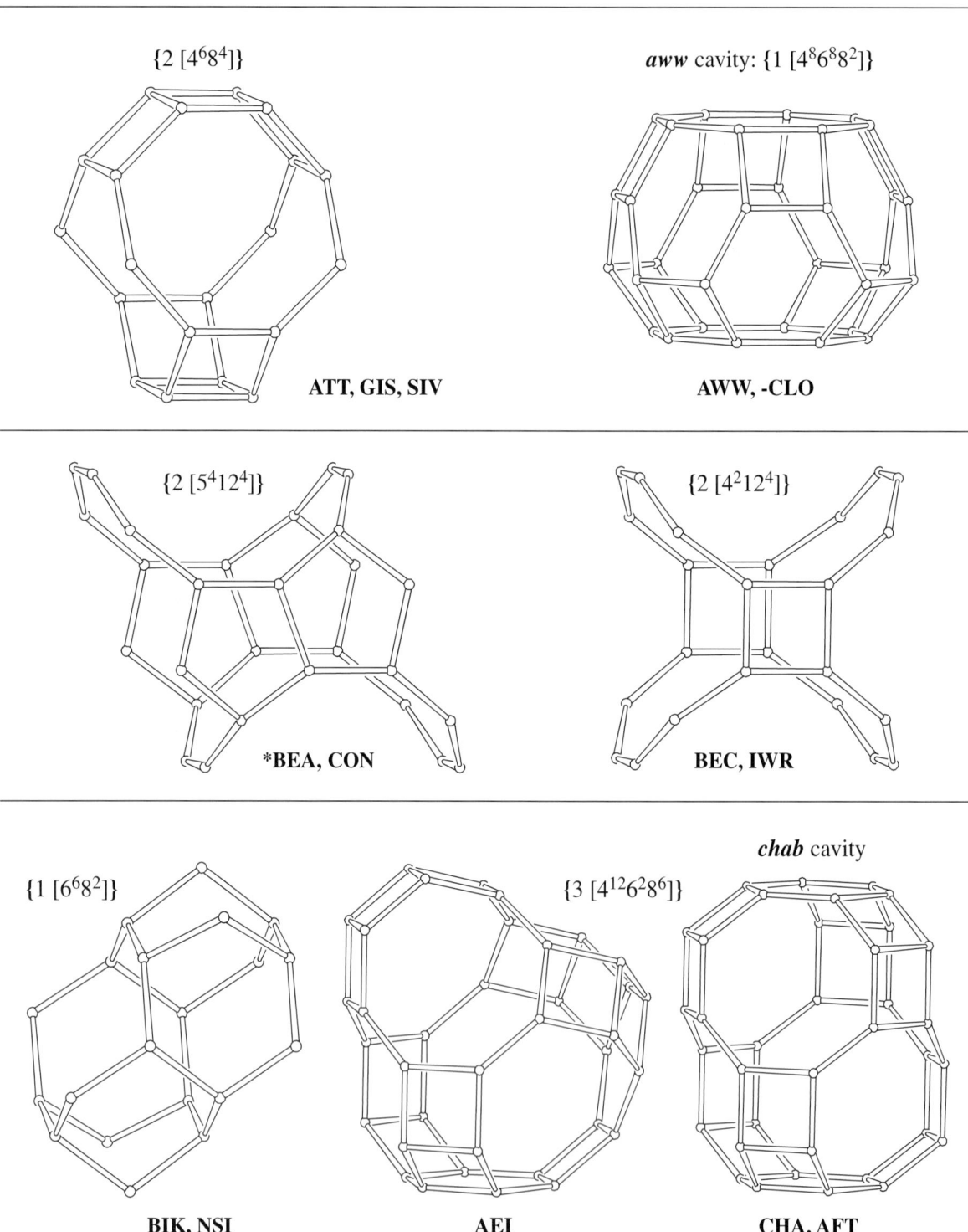

$\{2\ [4^6 8^4]\}$

**ATT, GIS, SIV**

*aww* cavity: $\{1\ [4^8 6^8 8^2]\}$

**AWW, -CLO**

$\{2\ [5^4 12^4]\}$

***BEA, CON**

$\{2\ [4^2 12^4]\}$

**BEC, IWR**

$\{1\ [6^6 8^2]\}$

**BIK, NSI**

$\{3\ [4^{12} 6^2 8^6]\}$

**AEI**

*chab* cavity

**CHA, AFT**

# APPENDIX 3

## (Continued)

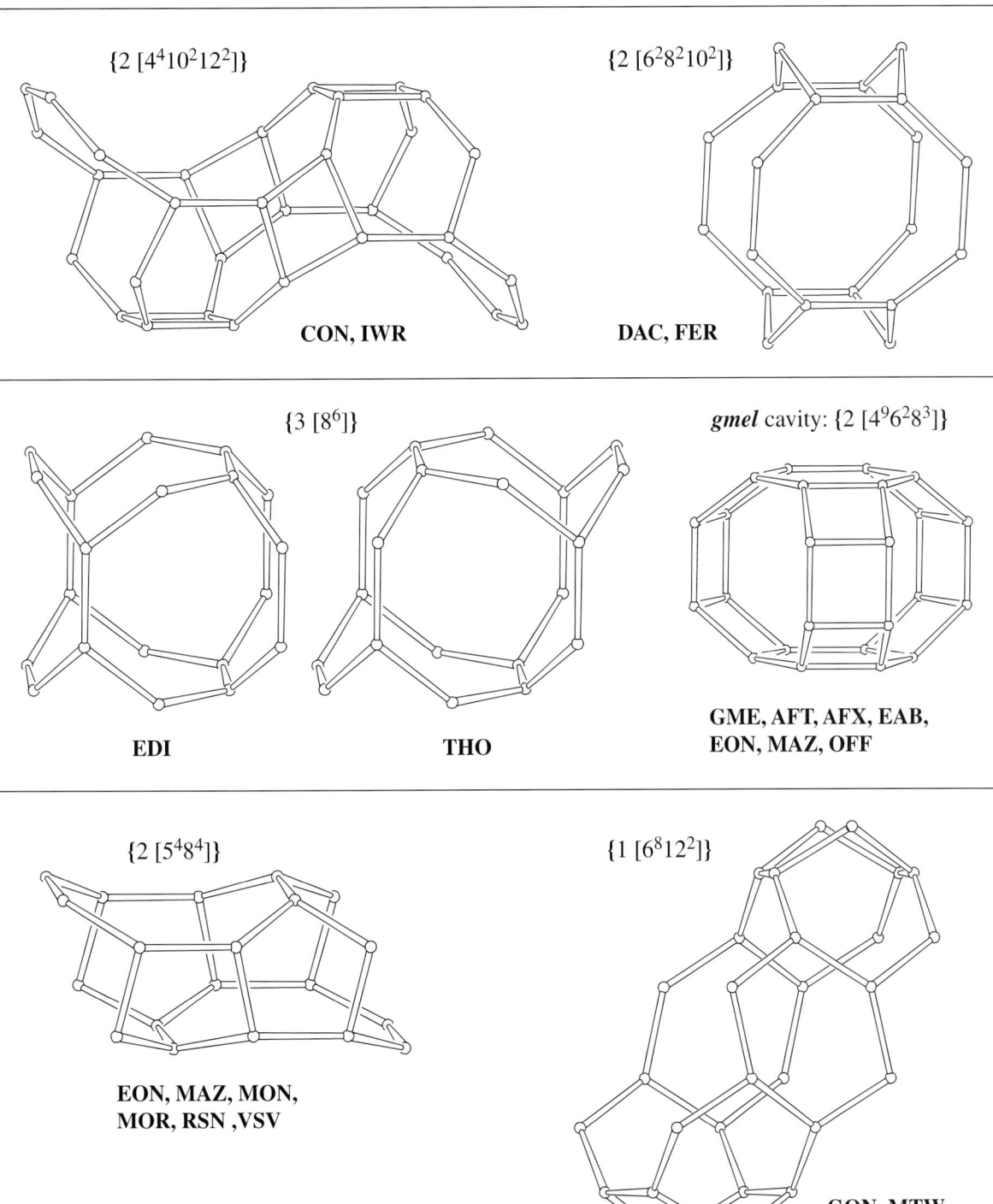

$\{2\,[4^410^212^2]\}$

CON, IWR

$\{2\,[6^28^210^2]\}$

DAC, FER

$\{3\,[8^6]\}$

EDI

THO

*gmel* cavity: $\{2\,[4^96^28^3]\}$

GME, AFT, AFX, EAB,
EON, MAZ, OFF

$\{2\,[5^48^4]\}$

EON, MAZ, MON,
MOR, RSN ,VSV

$\{1\,[6^812^2]\}$

GON, MTW

# APPENDIX 3

## (Continued)

$\{2\ [4^6 5^8 6^4 8^4]\}$

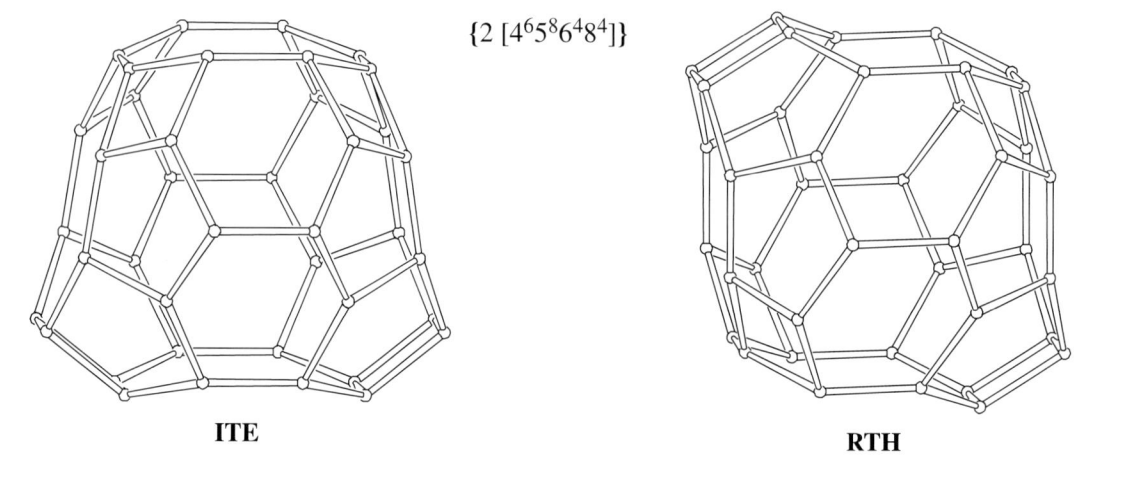

**ITE**

**RTH**

$\{2\ [9^4]\}$

*ltl* cavity: $\{3\ [4^{18} 8^6 12^2]\}$

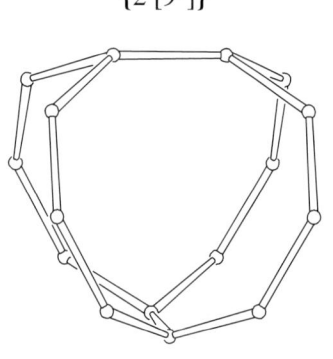

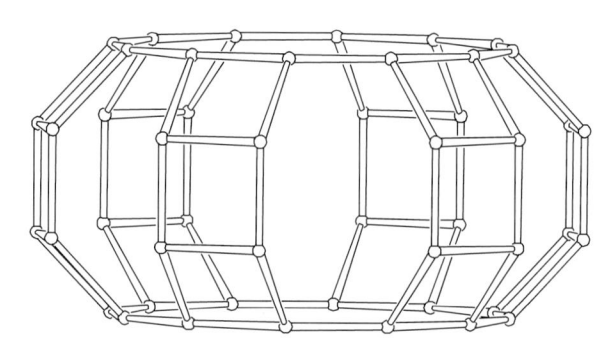

**LOV, NAB, RSN, VSV**

**LTL, MOZ**

*mer* cavity: $\{3\ [4^{12} 8^6]\}$

$\{3\ [4^3 8^3 12^2]\}$

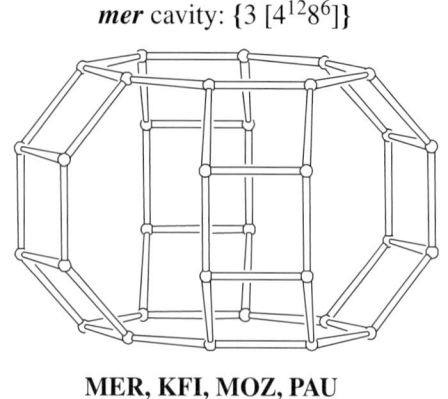

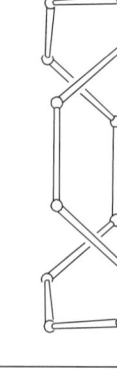

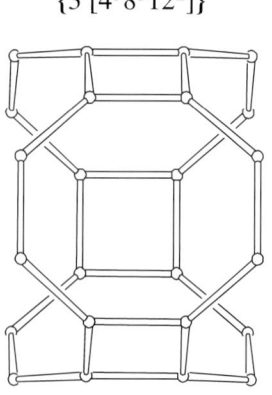

**MER, KFI, MOZ, PAU**

**MOZ, OFF**

# APPENDIX 3

## (Continued)

$\{1 \ [6^8 10^2]\}$

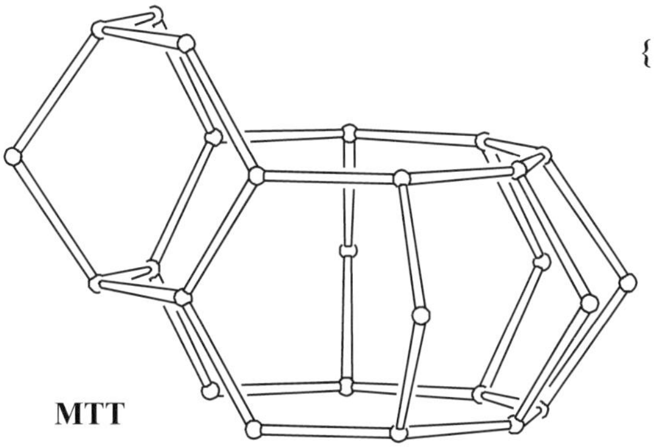

**MTT**

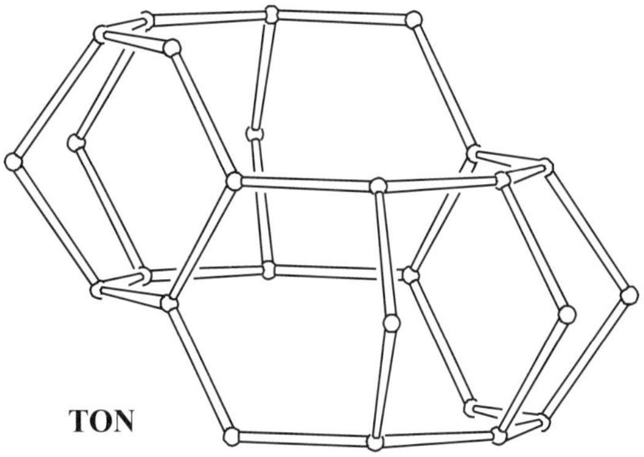

**TON**

$\{1 \ [6^8 12^2]\}$

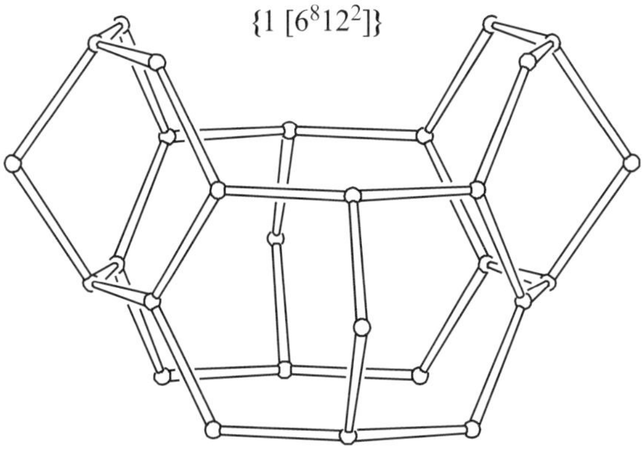

$\{2 \ [4^7 8^5]\}$

**OSI, VET**

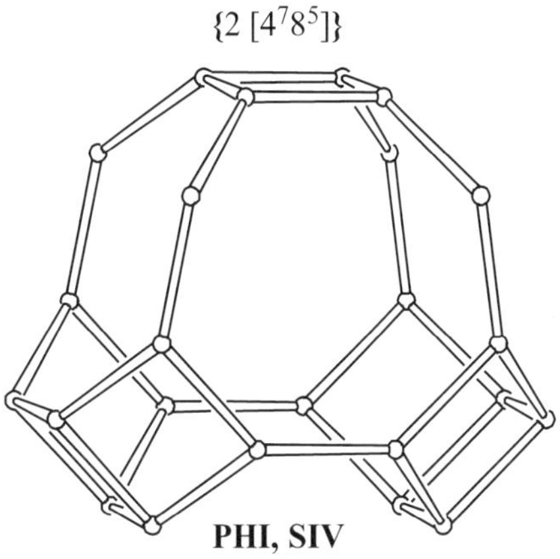

**PHI, SIV**

*rho* cavity: $\{3 \ [4^{12} 6^8 8^6]\}$

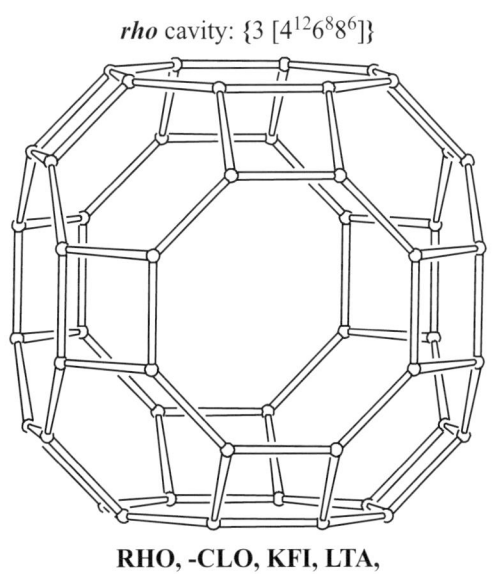

**RHO, -CLO, KFI, LTA,
LTN, PAU, TSC, UFI**

## APPENDIX 3

(Final page)

{1 [$4^4 6^8 12^2$]}

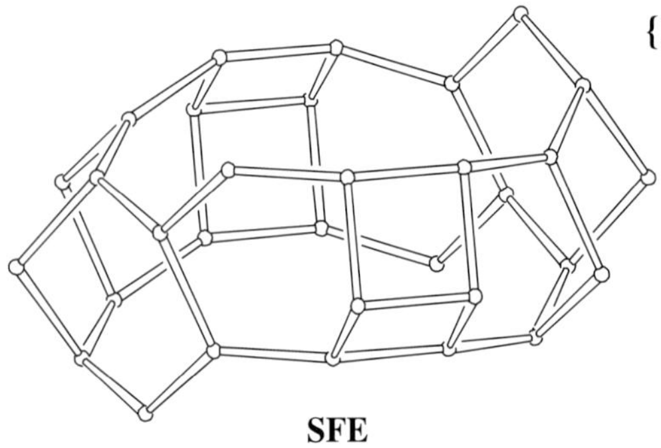

**SFE**

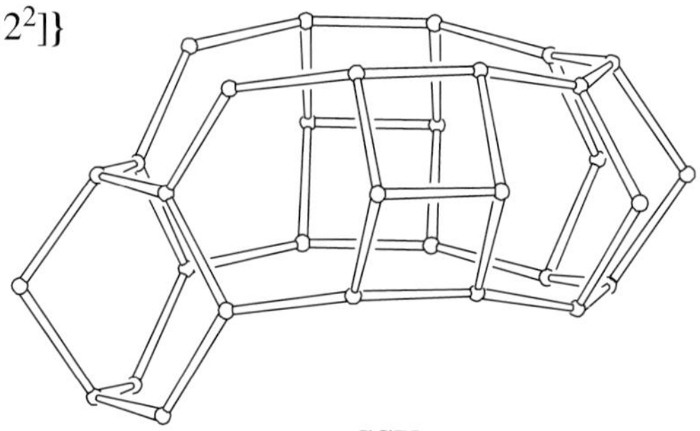

**SSY**

{1 [$4^4 5^8 6^6 10^2$]}

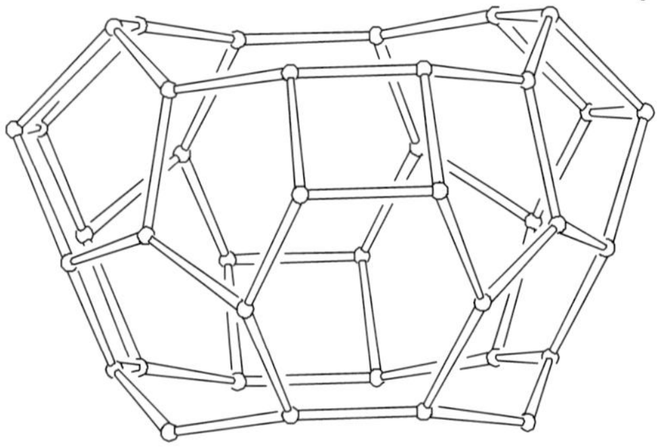

**SFF**

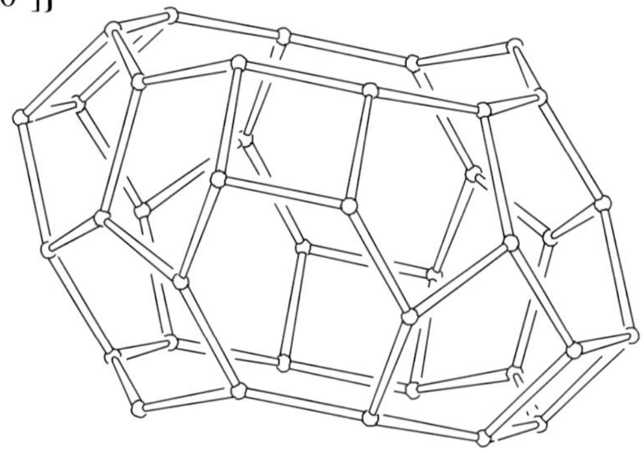

**STF**

{1 [$4^4 6^6 14^2$]}

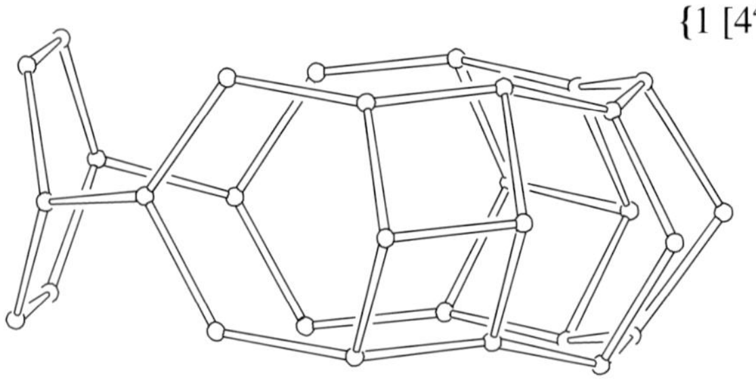

**SFH**

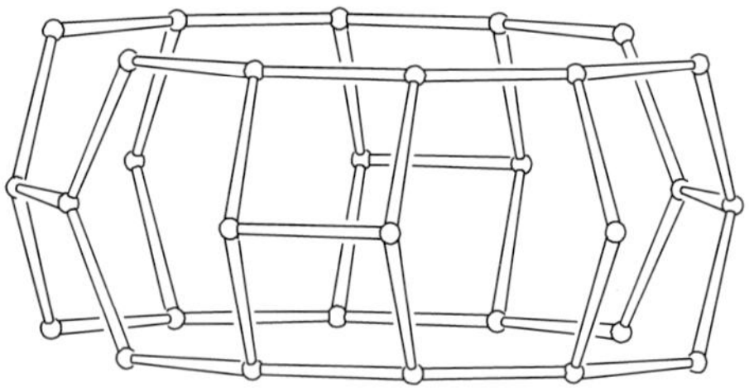

**SFN**